EDUCATION IN A COMPETITIVE AND GLOBALIZING WORLD

ENHANCING STEM MOTIVATION THROUGH CITIZEN SCIENCE PROGRAMS

EDUCATION IN A COMPETITIVE AND GLOBALIZING WORLD

Additional books and e-books in this series can be found on Nova's website under the Series tab.

EDUCATION IN A COMPETITIVE AND GLOBALIZING WORLD

ENHANCING STEM MOTIVATION THROUGH CITIZEN SCIENCE PROGRAMS

SUZANNE E. HILLER
AND
ANASTASIA KITSANTAS
EDITORS

nova science publishers
New York

Library of Congress Cataloging-in-Publication Data

ISBN: 978-1-53616-038-3

Published by Nova Science Publishers, Inc. † New York

CONTENTS

PREFACE

Just as with adults, school-age students appreciate when their ideas, work, and time has a purposeful use. Students benefit from meaningful tasks, which build competency, motivation, and enthusiasm for learning. In this regard, learners are like thermometers, quick to sense the difference between authentic activities as opposed to mundane tasks. With the coupling of naturalistic/environmental topics and technological advances, educators have a multifaceted arsenal to provide students with fulfilling and motivating activities, one of which is *citizen science*.

At first, working with students in the outdoors would seem simple. Yet, educational/ scientific goals, logistical requirements, and educational benefits are extensive, particularly in citizen science programs. Picture students in a citizen science program actively invested in the restoration and preservation of their surrounding ecological environment. They are taught how to monitor streams near their school grounds and devise methods to reduce waste. If their school does not have a stream on school grounds, schoolyards are used instead. Students look for refuse items (e.g. cigarette butts, plastic containers) that may disrupt or harm water quality, aquatic life, or recreational activities. Students are taught the saliency in finding the source of refuse deposits, referred to as litter-hot spots, in the streams or schoolyards as debris can flow overland or to nearby streams through storm drains.

The program example is referred to as the *Citizen Science Floatable Monitoring Program*. It is facilitated by the Stormwater Planning Division as a form of educational outreach, recently introducing the program to 170 fourth graders at an elementary school in the Mid-Atlantic region. Lasting the duration of several months, ecologists visited the school on several occasions to provide additional training and data collection support. The program evolved to allow teachers to facilitate data collection outings. At the end of the school year, a final presentation was given by the students to report their overall findings to the ecologists. This illustration typifies how citizen science programs, a collaboration between scientists and students, can encourage engagement, analytical skill acquisition, creative thinking exploration, and environmental awareness. It is through these types of citizen science programs that students become stewards of their environment gaining valuable knowledge as scientific researchers and becoming motivated to pursue careers in the STEM areas.

So, in summary, what is citizen science? Citizen science is an initiative in which volunteers work within the framework of professional research studies to collect data. More recently, hobbyists are being included in some or all aspects of research to contribute their talents more fully within the scientific realm of discovery. Through technology, and the recognition that children and adolescents are capable of creating, analyzing, and producing beyond the confines of a traditional curricula, this type of experience is available for learners in schools worldwide, and the elements for STEM motivation are compelling.

Citizen science programs encourage the public to help professional scientists collect data for research studies through training protocol. Public involvement is most helpful for scientists studying large-scale topics, such as animal migration patterns, plastic bead pollutants as a threat to environmental factors, and coastline changes due to storm events as some examples. Student involvement in these programs also has the potential to enhance student motivation in STEM. The overarching goal of this book is to (a) merge information on best practices for fostering student motivation in STEM through citizen science programs and (b) assist partnerships between motivational theorists, informal natural science educators and school personnel to integrate citizen science programs as part of formal school training and informal learning settings.

This book serves as a guide for educators, researchers, and stakeholders in learning how to apply various motivational theories to create effective citizen science training programs as well as assess learning and motivation outcomes (Bracey; Cleary, Slemp, & Waire; Shim & Lee). The chapters outline effective components of citizen science programs, which augment the formal classroom structure and describe exemplary programs (Cardamone & Cobb Kung; DiBenedetto & Schunk; Hiller & Reybold; Hiller, Whitehead, & Kitsantas; Huffling, Scott, Johnson, Gantt, Collins, & Weeks; Koomen, Hedenstrom, Moran, & Oberhauser; Lobel, Cianciola, & Erichsen; Rushton & Parker; Wallace & Bodzin). The work also emphasizes ways to evaluate the impact of citizen science programs on student development (Downey & Skidmore; Rushton & Parker). Few works provide ways to evaluate programs in the outdoors, which directly correlate with student achievement and motivation.

We extend our sincere appreciation to all our authors for their valuable contributions. We would also like to recognize the assistance we received in preparing this work from Jennifer Mischel and Aubrey Whitehead. We hope that readers will find this volume helpful in creating citizen science programs that promote and support student motivation and achievement in the STEM areas. This work would be particularly appealing to scientists, naturalists, teachers, administrators, and parents interested in providing students with authentic learning experiences in the outdoors as well as technology-based platforms.

Suzanne E. Hiller & Anastasia Kitsantas

SECTION I: OVERVIEW OF CITIZEN SCIENCE

In: Enhancing STEM Motivation through Citizen Science Programs ISBN: 978-1-53616-038-3
Editors: Suzanne E. Hiller and Anastasia Kitsantas

Chapter 1

ENHANCING STEM MOTIVATION THROUGH CITIZEN SCIENCE PROGRAMS: A FRAMEWORK FOR EDUCATORS AND RESEARCHERS

Suzanne E. Hiller*, Aubrey Whitehead and Anastasia Kitsantas
College of Education and Human Development, George Mason University, VA, US

ABSTRACT

Citizen science is an emerging educational platform with the potential to enhance student STEM motivation and achievement. In a citizen science capacity, volunteers work with scientists to participate in professional research studies. The current chapter provides an overview of citizen science programs and outlines the ways that these initiatives inherently foster motivation and self-regulation in STEM. Specifically, the goal of this chapter is to offer an alternative lens for citizen science, which encompasses dimensions related to student motivation and achievement. This chapter also provides an overview of the chapters within this book sequenced within the framework of psychological dimensions.

Keywords: career motivation, citizen science, interest, outcome expectations, self-efficacy, self-regulation, science achievement, STEM

In an era of unprecedented innovations, student learning and development is spontaneous and impressionable based on cultural advances, technological changes, and transnational interactions. In response, practitioners worldwide are adjusting to projections of the skills sets and cognitive abilities most supportive of students' personal and professional development. Specifically, upcoming vocational challenges will require that individuals grapple with extensive content, have strong critical thinking skills to facilitate problem solving, and draw connections across domains in a complex, global environment (Serageldin, 2013). Most recently, societal needs are spurring the inclusion of community members well-versed in

* Corresponding Author's E-mail: shiller@gmu.edu

social/political/scientific developments that reinforce policy decisions to support the public *writ large* (Bonney et al., 2014; Dick, 2017).

In efforts to solidify student preparedness, educators are adopting a variety of teaching techniques that foster critical thinking and problem solving, such as technology-enhanced models, inquiry-based learning, STEM (science, technology, engineering, and mathematics) initiatives, and self-regulatory practices (Belluigi & Cundill, 2017; DiBenedetto & Zimmerman, 2013; Haidar & Al Naqabi, 2008; Hiller, 2017; Hiller & Kitsantas, 2014; Loyens, Rikers, & Schmidt, 2008; Seraphin, Philippoff, Kaupp, & Vallin, 2012; Weigund, 2014). Technological discoveries continuously induce dynamic enrichment opportunities and greater interaction between students and community members.

The natural world, coupled with technological capabilities, imparts expansive learning options. These resources serve as viable and advantageous platforms for increasing student development. One such instructional learning medium is *citizen science* where members of the general public engage in data collection through training programs to support the research of professional scientists (Crall et al., 2011). The range of citizen science ventures available for youth education is wide-ranging and diversified with projects investigating historical documents, hummingbird populations, biodiversity, bee species assemblages, weather disturbances/partial lunar eclipses, and phenology that may occur on-site or via the web (Bendix, 2017; Givot, O'Connell, Hadley, & Betts, 2015; Grasser, Schunko, & Vogl, 2016; Le Féon et al., 2016; Portas, Barnard, Scott, & Harrison, 2016; Wallace & Bodzin, 2017).

Documentation of the role of hobbyists in citizen science programs is attributable to at least the 1800s (Dawson, Lintott, & Shuttleworth, 2015). More likely, non-scientists have engaged in scientific discoveries over the course of thousands of years, such as individuals who made observations of the stars. Possibly, the first citizen scientists recorded in modern history were colonialists in North America who studied the weather (CosmoQuest, 2018).

Within the 21st century, the scope and reciprocal benefits for contributor-scientist has expanded through changing technologies and societal interactions. Over the last two decades and across the globe, citizen science programs serve as a bridge between science professionals and the general public. From *Crowd Water* (n.d., a Swiss-directed program collecting and studying hydrological data) to *TurtleSAT* (2019, an Australian-led initiative tracking freshwater turtle populations) to the National Oceanic and Atmospheric Administration's *National Weather Service Cooperative Observer Program* (n.d., a meteorological program managed by the U.S. Department of Commerce), citizen science programs are relevant activities for students, teachers, administrators, stakeholders, and the general public (Jones, Childers, Stevens, & Whitley, 2012).

Citizen science, in its multifaceted applications, often spearheads global initiatives with the largest professional societies formed in the United States, Australia, and Europe (Ellwood, Crimmins, & Miller-Rushing, 2017). Using crowdsourcing as one strategy for the inclusion of hobbyists, citizen science programs have collected data from participants worldwide and resulted in considerable scientific findings (Wilson et al., 2017). As an illustration, more than 20% of adults in the European Union reported that they participate in volunteer or charitable activities (European Foundation for the Improvement of Living and Working Conditions, 2011). As an example of citizen science programs, which include school-age students, *Galaxy Zoo* (2007) boasted one of the most prolific and culturally-diverse citizen science programs, with over 200,000 volunteers from more than 100 countries. The project has contributed to the classification of over 100 million galaxies (Raddick et al., 2010). Currently, *Galaxy Zoo 2*

(Zooinverse, n.d.) extends the goals of the initial program by engaging volunteers in more fine-grained classifications of galaxies.

Specialized software and digital tools (e.g., sensors and probes), incorporated within citizen science, create greater interaction between students and professionals with novel programs. As a model of an interdisciplinary curriculum, eighth grade students in Copenhagen participated in a citizen science program geared toward training students to examine structural issues within declining urban areas. In conjunction with researchers, teachers, and Danish architects, citizen science students engaged in "Cities at Play," wherein individuals tackled community issues as urban planners using the environment-building video game, *Minecraft* (Mojang, 2019). A series of content measures indicated that adolescents transitioned in their content understanding by focusing on structural changes (e.g., the inclusion of street lights, limited community activities, and the restoration of dilapidated structures) rather than feelings related to safety and crime potential (Magnussen & Elming, 2015). These and similar results suggest citizen science programs may become more diversified in their uses for professional studies and to educate the public both formally and informally.

More than ever, an educationally-oriented trend in citizen science activities is the involvement of the world's youth in the production of authentic learning. The impact of citizen science programs on motivation and achievement has been assessed in varied regions of the world, including Australia, France, Qatar, South Africa, the Czech Republic, the United Kingdom, and the United States as archetypes (Belluigi & Cundill, 2015; Creel, Nite, & Almarri, 2017; Grasser et al., 2016; Le Féon et al., 2016; Roy, Baxter, Saunders, & Pocock, 2016; Zárybnicka, Sklenicka, & Tryjanowski, 2017). Within educational realms, citizen science programs are gaining recognition as useful learning channels (McKinley et al., 2017) given their documented impact on student motivation, science achievement, and STEM career motivation (Hiller & Kitsantas, 2014; Hiller, Whitehead, & Kitsantas, 2019; Raddick et al., 2010; Wiener & Matsumoto, 2014).

To depict the benefits of adolescent citizen science programs, in this chapter we provide an overview of citizen science while drawing comparisons between motivational factors from an educational psychology perspective. The writing in this chapter flows from the following sequence: First, we provide a broad overview of citizen science programs; then we describe the importance of protocol training as an integral component in maintaining scientific validity as well as teaching students; we define self-motivational beliefs as essential factors in examining citizen science motivation and achievement research; and finally, we provide a citizen science framework grounded in motivation and self-regulatory dimensions of learning. This perspective corresponds to the general layout of the current book. For the purposes of this work, we consider *adolescence* to include ages 10-25 due to the economic/social changes which have elongated this time frame within an individual's life cycle (Steinberg, 2015).

Citizen Science Programs

Citizen science programs typically consist of opportunities for volunteers from the public who, using data collection techniques and the scientific method, aid scientists in research across a myriad of areas (Raddick et al., 2013). Compelling citizen science programs allow a wider and broader network between volunteer data collectors and the scientific community. For professional scientists, this collaboration allows the dissemination of information on scientific findings (Losey, Perlman, & Hoebeke, 2007).

Cooper, Dickinson, Phillips, and Bonney (2007) define citizen science programs as "a method of integrating public outreach and scientific data collection locally, regionally, and across large geographical scales" (p. 1). The perception of citizen scientists in current society is evolving due to the changing needs of professional scientists. With changing technologies and increased volumes of data and volunteers, scientists not only need support collecting data but also in making inferences (CosmoQuest, 2018). Today, volunteer citizen scientists may participate in a range of activities including the determination of research questions, input in the project design, conducting experimental studies, collecting and analyzing data, interpreting results, creating technological advances and applications, identifying new discoveries, and problem solving (Crowdsourcing and Citizen Science Act, 2016).

Creating this collegial relationship between scientist and citizen diverges from scientific studies reserved solely for professionals, by involving the public to maximize resources (e.g., data collection hours), increasing scientific awareness, and educating the non-science community (Gardiner et al., 2012). In turn, increased public involvement in scientific activities allows for more informed management decisions (Bonney et al., 2014; Surasinghe & Courter, 2012; Vann-Sander, Clifton, & Harvey, 2016). Citizen science programs benefit those committed to scientific research as well as preserving certain environmental protections within their municipalities (Bonney et al., 2009). In a mutually reciprocal capacity, citizen science has sought to make science topics more meaningful, approachable, and tangible for members of the general community (Mueller, Tippins, & Bryan 2011). Non-scientific individuals function as valuable stakeholders in science research rather than singularly as consumers of scientific information (Schmeller, 2008). From the standpoint of heightening student motivation and achievement, citizen science opportunities offer an impressionable manifestation of the daily work of the scientist and in the learning cycle of adolescents (Eberbach & Crowley, 2009; Hiller & Kitsantas, 2015).

In citizen science programs, interested members of the public typically participate in data collection (*contributory*); however, some studies have extended hobbyist involvement to developing new questions based on exploratory findings (*collaborative*) or engaged volunteers in all parts of the research process through *co-created* research (Bonney et al., 2009; Pandya & Dibner, 2018). By including community members in research activities, the knowledge and contributions of local stakeholders provides valuable insights for research findings. For example, within the nexus of local gardeners, contaminated soil, and potential health risks, a team of local gardeners and university researchers affiliated with the University of Arizona (Departments of Soil, Water, and Environmental Science; Hydrology and Water Resources; and Pharmacology and Toxicology), engaged in the multiple aspects of citizen science research to pinpoint potential human hazards related to gardening and soil pollution. In this capacity, local residents became active stakeholders and informed decision-makers in identifying and solving complicated environmental issues (Ramirez-Andreatta, Brusseau, Antiola, Maier, & Gandolfi, 2015).

As the field of citizen science evolves, additional citizen science studies include volunteer participation in multifaceted aspects of collaborative work such as research design (Aristeidou, Scanlon, & Sharples, 2017). For adolescent students, citizen science programs further support learning by including individuals in data analysis, interpretation, and/or the research process. In recent years, some programs encourage student citizen scientists through co-created projects to personally engage in the scientific process from research design to the production of journal publications in a comprehensive way (Grasser et al., 2016; The Institute for Research in Schools, 2018).

Regardless of the extent to which citizen scientists engage in research design, data collection, data interpretation, and dissemination of findings, the cornerstone of citizen science programs is for volunteers to receive protocol training (Snäll et al., 2011). To encourage learners initially, students may benefit from record keeping in a journal, assistance through the classification processes, and an awareness that their work is meaningful in the scope of the project goals (The Cornell Lab of Ornithology, 2019). Further training may include lectures, small-scale activities, and modeling of experts (Hiller & Kitsantas, 2014, 2015; Wallace & Bodzin, 2017). To engage in research activities, guidance for students includes specific protocol requirements to maintain validity in data collection, such as a lady beetle study conducted in school yards across the United States and the United Kingdom in which students gathered viable data for scientific study (Gardiner et al., 2012).

Through protocol training, skilled researchers aim to substantiate the credibility of professional research studies. A critical concern in this regard is that volunteers receive training to collect data in a uniform manner. In this endeavor, field experts are essential models and mentors to provide consistency in data collection, which yields credible data analysis and interpretation for professional research studies. This component within citizen science programs is particularly relevant for the inclusion of adolescent students. Without consistent training, scientists may refrain from including children in scientific studies out of concern that data interpretations will be perceived as invalid (Hiller & Reybold, 2011a).

With a marked distinction from science instruction within formal science experiences, data collection required for citizen science necessitates that students adjust to unpredictable events (Eberbach & Crowley, 2009). For example, in a horseshoe crab study where students were measuring the distance between the eyes of horseshoe crabs, one team of students had to adjust when they found a one-eyed horseshoe crab (Hiller, 2012). With oversight, and age-appropriate, standardized data collection methods, professional researchers are including students in professional research studies. In the United Kingdom, researchers assessing the impact of abiotic factors, such as weather and soil moisture on macroinvertebrates, relied on a collaborative citizen science program with 410 schools sampling for 6 months over a two-year period. Student citizen scientists collected over 1,800 macroinvertebrate specimens in their school yard and counted and classified the organisms into 12 taxonomic groupings, such as Annelida (earthworm phylum) and lepidoptera (butterfly order) larvae. Findings from the study mirrored previous research on variations in the abundance of invertebrates based on soil status conducted by adults (Martay & Pearce-Higgins, 2018) suggesting proper data collection by the students.

While student participation in citizen science research provides scientists with large-scale variations and information which drives professional publications, from the perspective of educating youth, citizen science programs are immensely beneficial. From a motivational perspective, these types of activities are particularly potent from three standpoints of (a) modeling of field experts, (b) the development of scientific observation skills, and (c) the enhancement of interest and identity formation.

Scientific observation skills move beyond simply seeing to the use of all senses. These pivotal skills are instrumental in data collection and invariably influence data interpretation through the craftsmanship of the observer (Cartwright, 1989; Eberbach & Crowley, 2009;

Young, 2016). In fact, in one study of scientific observation skills for students from kindergarten through university levels, researchers concluded that young students utilize varying competencies of scientific observation skills and benefit from explicit training as early as preschool (Kolhauf, Rutke, & Neuhas, 2011).

Most importantly, the role of field experts in protocol training magnifies when considering the influence of citizen science programs on student development, particularly in terms of enhancing identities in STEM fields. Access to individuals who engage in scientific work enables the learner to observe and emulate required skills sets (Hiller & Kitsantas, 2015). Through repeated exposure, students have the opportunity to develop expertise and form identity perceptions as members of the scientific community (Eberbach & Crowley, 2009).

Citizen Science, Professional Research, and Impacts on Citizen Scientists

As citizen science programs have become more prevalent, researchers aim to highlight the impact of these programs in a variety of ways. One perspective is to view the research in terms of a continuum between the (a) impact on professional research and (b) impact on citizen scientists as shown in Figure 1. This consideration is important in terms of the citizen science program goals. If the primary goal is to educate individuals, some components of the citizen science program may need to be altered, thereby affecting overall scientific findings. For example, data collection techniques may be refined to correspond with students' developmental needs. Regardless of the focus of attention, training protocol of hobbyists and data collection accuracy is central to maintaining consistency in the collection of observations so that interpretations are credible (Aceves-Bueno et al., 2017; Crall et al., 2011; Cronje, Rohlinger, Crall, & Newman, 2011).

Figure 1. Citizen science research continuum.

A citizen science program focused on dog cognition is one example of an initiative dedicated to advancing professional knowledge. Through a website called *Dognition* (2019), 500 dog owners were recruited to conduct five cognitive tests on their pets. Pet owner-dog participants received a training video and written directions to assist as a form of protocol training. Next, participants followed a sequence of steps online to lead their dogs through cognitive tasks. Once steps were completed, the owners recorded their observations using a coding system. Participant bias and validity threats were accounted for prior to data interpretation, including a comparison with laboratory findings to assess multiple factors in dog cognition. The findings from the study illuminated that citizen science was a viable approach for studying dog cognition and psychology in tandem with laboratory studies (Stewart et al., 2015). In this work, emphasis rested on multiple factors, which affected dog behavior (e.g., human gestures and olfactory senses) and memory as well as the validity of findings in support of citizen science research. Distinct from educational studies, there was no analysis on the impact of the citizen science program on human volunteers.

A second study reflects citizen science programs geared to professional research aims as well as the impact on citizen scientists. Based in the Czech Republic, *The BirdsOnline Project*, used optical sensors (detecting motion, temperature, and light), microphones, and a video camera within songbird nest boxes (urban areas) and owl nest boxes (forest areas) as the backdrop for a citizen science program. Open to both the general public and primary and secondary students, researchers conjectured that engagement in this citizen science program increased student content knowledge gains, influenced identity formation, and developed conservation attitudes (Zárybnicka, Sklenicka, & Tryjanowski, 2017).

In the *BirdsOnline Project*, there was additional focus on professional research implications. Contributions from volunteers garnered new information about bird behavior, including nesting requirements, courtship and reproductive activity, parental-hatchling interactions, and diet needs. By incorporating the impact of citizen science work on both the volunteers and the scientific community, this work is reflective of citizen science programs centered on research outcomes as well as the impact on volunteers.

At the opposite end of the continuum depicted in Figure 1, researchers targeted the implications of a citizen science programs on volunteers. Situated within the context of migratory butterfly trends, Jeanpierre, Oberhauser and Freeman (2005) studied the impact of a monarch butterfly citizen science program known as the *Monarch Larva Monitoring Project* on a collaborative training program with professional researchers, teachers, and secondary students based in Minnesota and Texas. Professional researchers served as facilitators by offering training and support to the teacher-student teams. Data analysis and interpretation showed main effects for both teachers and students in content knowledge gains. Furthermore, individuals were more confident in their capabilities with tasks associated with professional research methods while studying monarch butterflies (Jeanpierre et al., 2005).

Citizen science program designers focus on the needs and the goals of the study; for some initiatives, the emphasis is on professional research while other study goals may center on learning outcomes. In the latter case, recent evidence suggests that there is potential to encourage student achievement and STEM career motivation through citizen science programs.

Citizen Science and Self-Motivational Beliefs

There are a variety of self-motivational beliefs integral to science research (including citizen science) focused on science achievement and career motivation. Students who have strong self-motivational beliefs are able to persevere in academic pursuits regardless of potential setbacks and barriers and tend to outperform their peers in academic pursuits, particularly when provided with adult support, autonomy in the learning environment, and challenging academic experiences (Bandura, 2005; Colby & Wooley, 2015).

To illustrate how the preceding self-motivational beliefs are contextualized in real-world citizen science programs, we refer to a group of students attending a canoeing trip to assess water quality of a neighboring river. These students are juniors and seniors who take Environmental Science/Biology and have volunteered to learn more about their local community, and whose experiences, are likely to be shaped by self-efficacy, interest, outcome expectations, and goal orientations.

A most influential construct in the overall development of a learner is *self-efficacy*. As described by Bandura (1997), self-efficacy is situated within the social/environmental learning structure and is definitively dependent on task requirements. Self-efficacy relates to an individual's self-perceptions and distinct tasks (Cleary, 2018). A student may perceive they are a proficient writer, yet consider themselves to be weak at completing word problems in math class; in this case, self-efficacy fluctuates based on the academic skill. Four sources of self-efficacy, which influence overall self-perceptions in this regard, include mastery experiences (opportunities to perfect skills/knowledge), vicarious experiences (verbal and nonverbal intimation from adult coaches/teachers/mentors), social persuasion (interactions with peers), and physiological states, such as positive affect and anxiety (Bandura, 1997). These sources are indicative of adolescent student achievement in science with mastery experiences as the most dominant source (Britner & Pajares, 2006). In addition, social persuasion is a predictor of achievement and motivation for adolescents exposed to citizen science (Hiller & Kitsantas, 2014).

In the canoeing excursion, those students who have canoed before, engaged in hikes, or participated in science work in/outside of the classroom are more likely to be self-efficacious in conducting field work (i.e., mastery experiences), particularly if they have worked with experts before (i.e., vicarious experiences). What is more, the verbal exchanges with classmates prior to and during the outing are likely to influence student's self-perceptions (i.e., social persuasion experiences). Some students may have anxiety about the trip; perhaps they think they will have trouble navigating the boat or will not be able to conduct the tests correctly (i.e., physiological state). The level of support, positive social experiences, and overall success in conducting science work is likely to influence a student's self-efficacy for conducting science tasks in the outdoors.

Affect inherently relates to self-efficacy and cognition. Those individuals who have positive affect are more likely to persist in tasks, have positive self-efficacy, and make cognitive strides. By examining the 2006 PISA, Ainley and Ainley (2006) found that 15-year-olds who expressed a positive affect (joy) for science, had positive associations with science as a discipline and expressed sustained interest for the topic. When speaking with naturalists about the overall aims of outdoor programs, these experts consistently described positive affect in the outdoors as a key factor in student success (Hiller & Reybold, 2011b).

During the canoe trip, students experience a range of sensory stimuli, such as the smell of plants, the taste of pawpaw fruit found along the banks, an exchange between hawks over a food source, or the triangular formations of water currents that meet shallow waters. These types of observations have the potential to spur strong affective reactions influencing moods and emotions. When affective states are positive, learners are likely to affiliate the experiences to subject matter and make stronger cognitive gains as well as become motivated to engage in other outdoor, science related experiences.

A third construct, *interest*, is most prescient in motivational studies as sustained interest in a topic is indicative of heightened motivation. Repeated involvement in activities with positive outcomes manifests in sustained interest and mastery experiences (Hidi & Renninger, 2006). In fact, the use of scientific observation skills, as required by citizen science programs, corresponds with student interest levels from early childhood through university levels (Kolhauf, Rutke, & Neuhaus, 2011).

To create long term interest, Bergin (2016) outlined a progression of steps. In generating interest for a topic of study (a) the initial step is an introduction to the subject (i.e., exposure), (b) a cognitively captivating follow-up activity to encourage sustained attention, (c) some form of engagement activity that spurs action, and (d) an opportunity for sustained, entrenched interest in the topic/project. Progression through this process, while it may or may not be linear, increases through the involvement of social group influencers (Bergin, 2016).

This progression is a useful tool when considering how to promote interest in outdoor and/or citizen science activities. Within citizen science, professional guides energize the volunteers by first exposing students to the environment while promoting fun and curiosity (The Cornell Ornithology Lab, 2019). One example, would be (a) leading adolescent students on a canoe trip down a river. As the introduction, (b) students benefit from being in the outdoors, learning to navigate the waters, and noticing through the five senses. Students could stop on the shoreline and (c) receive instruction on how to use a turbidity tube to estimate the water quality of a river as an example of a cognitively engaging introduction. Students could then apply the information by collecting samples, analyzing the water in the turbidity tube, and recording measurements. For sustained interest, (d) students could next learn about how micro/macro invertebrates are indicators of water quality through discussion, modeling, and collecting samples from the river. As another opportunity for sustained engagement, a variety of indicators could be used as part of a citizen science data base or an Environmental Science/Biology course project. Students could be encouraged to collect data, analyze and interpret data, and disseminate findings, depending on the overall aims of the course initiative or citizen science program.

In this example, each of these intermittent and continuous interventions help build a sustainable degree of STEM interest (Bergin, 2016). This process also creates a self-generating succession where the newly-established interest in the citizen science program feeds greater levels of engagement and attention, as well as further motivation to increase exposure in related areas. Notably, students who develop high interest levels are likely to also be self-efficacious for this type of science work. From a citizen science perspective rooted in motivational studies of adolescence, self-efficacy and interest have the potential to covary as predictors of student achievement and plans for career goal attainment (Hiller, 2012; Hiller & Kitsantas, 2014).

Another self-motivational belief, often paired with self-efficacy and interest, is *outcome expectations*. This term describes an individual's judgment about the potential to be successful when taking steps to reach a goal. The impact of a variety of STEM oriented interventions has been an overriding consideration for studies related to STEM career motivation in conjunction with self-efficacy, interest, and career goals for adolescent students (Navarro, Flores, & Worthington, 2007). In the canoeing experience, prior to volunteering for the trip, students likely weighed in on the benefits of attending such a trip as well as the likelihood of their success at participating in the tasks (i.e., canoeing, classifying organisms). Their outcome expectations were inherently related to their levels of interest and self-efficacy in engaging in the outdoor program.

Addressing *goal orientations* is another way to assess students' motivational beliefs. When tackling new tasks, students engage in a mastery approach (where the aim is to solidify in depth knowledge) or performance approach (where the aim is to impress others or earn a high grade/GPA). Learners with a mastery approach tend to demonstrate higher levels of achievement than their peers. Assessing the type of goal orientation exhibited by a student provides useful information on how to capitalize on motivational tendencies through instructional strategies. For example, during the canoeing excursion, some students may be more concerned with how they appear to their peers and teachers (performance approach), whereas other students center on learning the material fully as an example of mastery approach experiences. Based on research on goal orientations, those students with a mastery approach are more likely to make stronger academic gains over the long term as compared to their peers (Britner & Pajares, 2006; Lee, Hayes, Seitz, DiStefano, & O'Connor, 2016).

This section included definitions of key self-motivational beliefs with corresponding examples. The following section provides an overview of the emerging research on citizen science with the inclusion of these self-motivational beliefs to gauge student progress.

Adolescent Citizen Science Research and Motivation

As one trend in citizen science studies, researchers have addressed motivational factors, science achievement, and STEM career motivation. Self-efficacy and interest have been dominant factors in these types of studies. In citizen science programs which incorporated field expert guidance, students had higher self-efficacy and achievement (Jeanpierre et al., 2005). In a study of fifth graders engaged in a plant phenology citizen science program, Sutton (2009) found that exposure to the intervention positively affected science literacy gains, scientific attitudes, and interest in science activities.

In a more recent study on the impact of citizen science on students within formal education programs in Austria (Keleman-Finan, Schuech, & Winter, 2018), students completed measures related to interest in science and the environment, self-efficacy, motivation, nature of science understanding, scientific skills of inquiry, and stewardship for the environment. Pre- and post-measures garnered data on these constructs following studies on birds, insects, and hedgehogs with 309 participants ages 8-18. In this study, and particularly for primary school students, interest and self-efficacy corresponded at the highest level of impact following the interventions. These findings substantiate the claim that citizen science biodiversity programs are an effective means for increasing student knowledge acquisition, self-motivational beliefs, and attitudinal changes.

Wallace and Bodzin (2017) studied identity formation, interest, and STEM career motivation of ninth graders engaged in a phenological study (focus on the life cycles of plants in relationship to environmental factors), which integrated technology. Measures for the intervention group were considered in contrast to a comparison group, which worked in traditional classroom settings. The outcome of the study supported the hypotheses that citizen science activities had strong main effects for achievement, interest, and STEM career motivation.

With social cognitive career theory as an underlying framework, Hiller and Kitsantas (2014) examined the ways in which a horseshoe crab citizen science program influenced middle-school students' science achievement and STEM career motivation, while considering an array of self-motivational beliefs, including self-efficacy, interest, outcome expectations, and career choice goals. Using a quasi-experimental design, the comparison group received the state-sponsored curriculum in a public-school environment. The treatment group received a series of expert-led lectures, hands-on laboratory activities, on-site visits, and modeled a field biologist to collect data on the habits of horseshoe crabs. Before starting the intervention, the treatment group completed pre- and post-tests on prior knowledge and self-efficacy. Strong correlations between social persuasion and sources of science self-efficacy (namely, mastery experiences, vicarious learning, and anxiety) evidenced that peers positively influenced participant self-efficacy, while decreasing anxiety. Further path analyses suggested variables such as science self-efficacy, interest, and outcome expectations influenced STEM choice goals and academic performance.

A newer trend in citizen science/STEM career motivation literature is the impact of these types of programs on underrepresented groups in science-oriented occupations. For example, English language learners participating in a citizen science program based on plant phenology demonstrated increased science literacy, attitude formations, and interest in science as a field of study (Sutton, 2009). Cardamone and Lobel (2016) contend that citizen science programs are relevant methods of instruction to promote interest in STEM careers for individuals with varied backgrounds.

Further, Bombaugh (2000) posited that citizen science programs had the potential to motivate learners into engineering programs, particularly for African Americans in urban areas. Based on this interpretation, Hiller, Whitehead, and Kitsantas (2019) extended previous research to compare the impact of citizen science programs on minority students as compared to non-minority peers in an extension of the original horseshoe crab study. In this mixed method exploratory study, findings illuminated that the citizen science intervention influenced students in terms of self-efficacy and cognitive gains; there was a greater positive main effect on achievement and self-efficacy for scientific observation skills for the African American students. In addition, African American males, who were experiencing anxiety prior to the intervention, had no statistically significant differences in anxiety compared to peers following the citizen science program. Qualitative findings revealed that for minority and non-minority students, the program affected students in different ways and varying intensities. While both groups found the experience to be fun and helpful in determining career aspirations, the minority students detailed their use of scientific observation skills to a greater extent than their peers. In addition, the minority students, many of whom were planning to enter medical professions, viewed the program as one step in a progression to reach their goals, while the non-minority groups were not as specific in terms of how the program would help them in career aspirations. Overall, this study revealed that citizen science programs may benefit students of varying demographics in divergent ways, and assessing these differences may be beneficial in fostering STEM career motivation for underrepresented groups of students.

A unique dilemma in citizen science research is the use of these types of programs within formal instruction, the role of "student volunteers," and the assessment of interest. For the standard definition of citizen scientist includes the term "volunteer." Educators who incorporate citizen science programs within the school day are sometimes requiring students to participate in these programs, which delineates from the traditional volunteer status. From a research perspective, if citizen scientists are volunteers, they are likely to be interested prior to the start of the program. Therefore, gauging interest through self-measures may produce skewed results. For researchers conducting citizen science studies, particularly through school programs, the juxtaposition of student involvement and levels of interest based on volunteerism require consideration for program/study goals. This issue is similarly relevant for STEM programs, such as afterschool or summer camps, as students who are often underrepresented in STEM fields do not have access to this type of instruction outside of a traditional school program. In this case, assessing interest levels in research studies may not be applicable to groups of students who need the most support (Vallett, Lamb, & Annetta, 2017).

Considering the impact of citizen science programs on self-motivational beliefs within a given situation is very helpful when working with adolescent learners. For research and educational practices, framing citizen science programs as a function of psychological dimensions provides a broad and alternative means for addressing key issues within the novel citizen science learning structure. The following section outlines motivational and self-regulatory dimensions which are applicable to citizen science contexts.

Motivational and Self-Regulatory Dimensions of Citizen Science

As a participant in citizen science activities, individuals engaged in effective data collection must plan strategies for applying mechanisms to collect data, and subsequently monitor their accuracy within a pre-specified set of conditions. These behaviors are reflective of self-regulation in which an individual identifies a pursuit (e.g., task, skill, content), plans the method of involvement, and then reflects on their capabilities (Zimmerman, 1989). As identified across 40 years of research, self-regulatory practices are integral to achievement and motivation (Schunk, 2001) in many subject areas, such as mathematics, music, writing, and sports (Cleary & Chen, 2009; McPherson, Nielsen, & Renwick, 2013; Zimmerman & Kitsantas, 1997, 1999).

From an expansive, broader conceptual framework based on the writings of Zimmerman (1994), programs which support self-regulation include four underlying dimensions of learning: *Method, Motive, Environmental/Social Aspects*, and *Performance Outcomes* as shown in Figure 2. Students who function in a program structure which includes these dimensions have the potential to flourish by taking a proactive stance in the way they learn both in formal and informal learning settings.

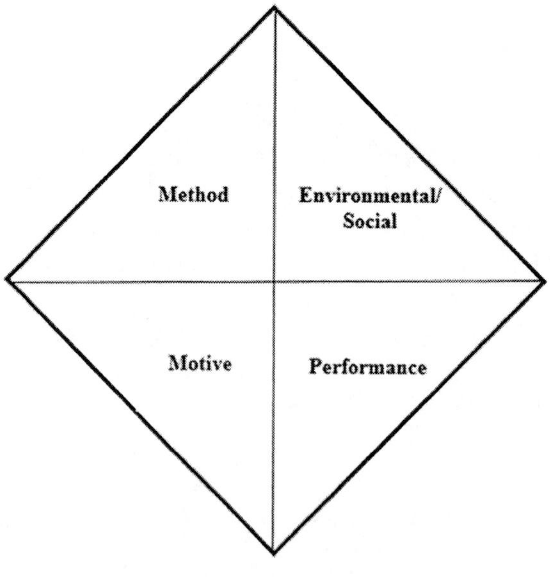

Figure 2. Dimensions of citizen science for adolescents; adapted from Zimmerman's model (1994).

For a student in control of their learning, the *Method* dimension addresses how the student tackles a prespecified goal. An example of a self-regulatory strategy reflective of the *Method* dimension would be task analysis, wherein the student assesses the types of activities needed to reach an outcome. The necessary criteria of citizen science contexts of protocol training and the use of scientific observation skills, to some extent, inherently dictate student choice of method selection. For instance, if a student participated in a wildflower citizen science program, they would need to consider the program goals and protocol for data collection. Once these criteria were in place, the student volunteer would decide factors such as how to identify wildflower species. Students might refer to a field guide or training handouts. If the focus was on a description of leaf characteristics, students could choose to make an initial identification and then confer with other volunteers in the group.

A crucial element which steers the *Method* dimension and is a mainstay of the *Motive* dimension is the establishment of self-motivational beliefs in planning/determining the method of progression. Key self-motivational beliefs, which influence student performance in citizen science programs, include self-efficacy, interest, and outcome expectations (Zimmerman, 2013). An individual who believes they have strong capabilities in a particular task would have high self-efficacy (Bandura, 1994). Of the three constructs, self-efficacy strongly correlates with science achievement and motivation (Britner & Pajares, 2006). In several citizen science studies, self-efficacy had strong main effects for science achievement and STEM career motivation. In addition, self-efficacy for scientific observational skills was instrumental in high achievement and motivation within citizen science programs (Hiller & Kitsantas, 2014, 2016; Hiller et al., 2019). Similarly, self-efficacy and interest are consistent influential factors in citizen science research studies of adolescents (Jeanpierre et al., 2005; Sutton, 2009; Wallace & Bodzin, 2017.)

Outcome expectations is a judgment an individual makes about the possibility of their success in a given task (Bandura, 2005). In education research, the triad of self-efficacy, interest, and outcome expectations is a key feature of achievement and career motivation studies (Lent & Brown, 2006; Lent et al., 2016; Navarro, Flores, & Worthington, 2007; Rogers & Creed, 2011). These mirrored effects are evident in citizen science research. From a social cognitive perspective, strong self-motivational beliefs are essential to self-regulatory processes. Those individuals with positive self-efficacy, interest, and outcome expectations are more likely to persevere when confronted with setbacks, thereby focusing on readjusting their methods for additional attempts at a task (Cleary & Labuhn, 2013). By contrast, a student with low self-efficacy in reading may be uncomfortable using a field guide to identify specimens. Instead, during training they may guess or refer to memory from class instruction. After several failed attempts during practice sessions, the learner may begin to lose interest in the activity as they are sure they will not be able to identify any other wildflowers (outcome expectations). The student is sure they are not good at science work and cannot be successful. Instead of recognizing that the strategies they are using are not effective, the student level of resiliency diminishes, and interest is lost.

Method, Motive, and *Environmental/Social Aspect* dimensions are inseparable within citizen science activities. As discussed earlier, the role of the mentor in citizen science programs influences student performance to a great extent. Student method selection follows the training through a field expert's guidance. Volunteers aim to emulate their guides as they approximate expert levels of scientific observation skills (Eberbach & Crowley, 2009). For this reason, citizen science programs rooted in field work are social endeavors. A student engaged in a wildflower citizen science project emulates the naturalist by making observations about petal size, formation, and color, while differentiating between more detailed distinctions. Students may engage in other forms of environmental/social interactions reflective of self-regulation, including help seeking. If a student finds an anomaly or has difficulty referring to a field guide, the onsite expert stands ready for support. This type of social exchange fosters self-efficacy, interest, and outcome expectations, creating the potential for mastery experiences, sustained interested, and new perceptions about identity (Eberbach & Crowley, 2009; Estrada et al., 2016; Hiller & Kitsantas, 2014; Wallace & Bodzin, 2017).

With new advances in technology, social support is not exclusive to outdoor work. Online citizen science programs offer help resources or provide a comment box so that volunteers may express concerns about discrepancies in their observations. For both outdoor and technology platforms, mentor availability offers a helpful form of scaffolding. Most recently, online citizen science initiatives have included natural language generation (NLG) technology as a support for classification systems. Using online classification keys, volunteers identify species based on photos and receive immediate online feedback based on their work (van der Wal, Sharma, Mellish, Robinson, & Siddharthan, 2015).

The fourth dimension, *Performance Outcomes*, addresses how the learner reflects on their performance (Zimmerman, 1994). In this case, student citizen scientists monitor their progress as they collect information and make adjustments in strategies when encountering difficulties. Students consider how well they identified wildflowers in terms of the project criteria throughout the process of data collection. This self-regulatory process, referred to as *calibration,* mirrors other data accuracy concerns (Bol, Hacker, Walck, & Nunnery, 2012). Here, self-regulated students tend to monitor their progress accurately in line with predetermined standards. However, lower achieving students tend to over predict their abilities (Chen & Rossi, 2013). For students engaged in self-monitoring, accurate calibration through feedback and practice leads to more productive self-reflective strategies (Labuhn, Zimmerman, & Haselhorn, 2010). By first considering how well they identified wildflowers and then comparing their findings with field experts or classmates, students have the opportunity to improve calibration skills. Again, in this example, the intersection of the domains of *Method*, *Motive*, *Environmental/Social Aspects*, and *Performance Outcomes* are fundamentals of well-designed citizen science programs for adolescents.

Based on Zimmerman's (1994) conceptual framework of dimensions of learning, Table 1 summarizes this dynamic within a citizen science context. Zimmerman attributed key scientific questions to correspond with each psychological dimension. Self-regulatory strategies and citizen science examples extend his work on the dimensions of learning to the evolving instructional medium of citizen science.

Scientific observation skills form the basis of this summary as a keystone element of citizen science programs. The statement, "How are students using scientific observation skills?" illustrates the *Method* dimension. The sample self-regulatory processes of goal setting and task analysis are mirrored in the description of Megan, a student, who understands the overall project aims and has received protocol training in wildlife data collection. She sets a data collection goal and decides to use informational guides to identify wildflowers in the form of goal setting and task analysis.

Table 1. An Extension of Zimmerman's Dimensions of Learning in Citizen Science (Zimmerman, 1994, p. 8)

Scientific question	Psychological dimension	Sample self-regulatory processes	Student citizen science example
How are students using scientific observation skills?	Method	1. Goal setting (short and long term) 2. Task analysis	1. After learning about the citizen science program goals and protocol, Megan decides to identify at least five wildflowers within a given time frame. 2. She uses a field guide with classification keys and then double checks information sheets from class.
Why are students using scientific observation skills?	Motive	1. Self-efficacy 2. Interest 3. Outcome expectations	1. After identifying wildflowers in class, Megan is very confident she can identify wildflowers in nature based on characteristics. 2. She wants to study more about wildflowers. 3. She feels she will be very successful identifying wildflowers through the use of resources.
Where are students using scientific observation skills and what resources are available for social support?	Environmental/ social aspect	1. Help seeking 2. Environmental structuring 3. Time management	1. Megan is having difficulty identifying a wildflower and asks a naturalist to show her how to identify the specimen. 2. She finds a quiet place to continue working. 3. In order to be more efficient, she decides to use a bookmark in some of the guides.
What are the ways students are reflecting on the scientific observation skills?	Performance outcomes	1. Self-monitoring 2. Self-reflection 3. Calibration	1. Megan checks her identifications for the first flowers to see if she is correct. 2. She decides the technique of checking two sources is effective. 3. After she identifies a few wildflowers she thinks are correct, she asks a field expert to check her work.

As a leading question for the *Motive* dimension, "Why are students using scientific observation skills?" this technique is reflective of the motivational factors which influence student learning, including self-efficacy, interest, and outcome expectations as self-regulatory factors. In this example, Megan has had a lot of training and is self-efficacious about wildflower identification. Her interest in the topic is growing, and she believes she will likely be successful in collecting data through the field guides.

Yet even with this background and training, Megan encounters some challenges. "Where are students using scientific observational skills, and what resources are available for support?" serves as a representation of the *Environmental/Social Aspect* dimension. Megan finds limitations in her identification strategies so she finds help from an individual with expertise. Once she returns to her field work, Megan locates a quiet place to work (environmental structuring) and uses time management to reach her goals by shortening her lunch.

As a student assessment of *Performance Outcomes*, the guiding scientific question is "What are the ways students are reflecting on the scientific observation skills?" Sample self-regulatory processes in this arena include self-monitoring, self-reflection, and calibration. Megan self-monitors her work, especially in the first few attempts. She reflects on her accuracy in identifying wildflowers. Then she asks a field expert to double check her accuracy as a form of calibration. As a highly self-efficacious student, Megan will likely readjust her method of work if her calibration is inaccurate.

Zimmerman's psychological dimensions present an alternative way of viewing citizen science programs. As an extension of this notion, the following book presents theoretical discussions, research studies, and guidelines for citizen science evaluations as representations of each dimension.

Chapter Highlights

The chapters within this book represent an intersection between two research fields: citizen science and educational psychology. To create a dialogue between the two realms, the chapters in this book are organized in five sections. The works include an array of theoretical writing, research studies, curricula overviews, and evaluation frameworks. Regardless of the structure of the writing, self-motivational constructs are embedded within all the works and are essential in linking the two domains of research.

Section I, *Overview*, includes Hiller, Whitehead, and Kitsantas (Chapter 1) and Bracey (Chapter 2), which outline the components of citizen science programs while relating self-motivational beliefs and authentic learning through citizen science programs as a way to reshape student identities as "scientist."

Section II, *Motives for Learning in Citizen Science Programs*, provides guidance on how to incorporate motivation within citizen science research studies based on the writing of Cleary, Slemp, and Waire (Chapter 3), while Shim and Lee (Chapter 4) describe ways of encouraging healthy social exchanges during citizen science programs from the perspective of four differing theoretical perspectives. In addition, Rushton and Parker (Chapter 5) offer specific guidelines on how to implement co-created citizen science programs for adolescents while recognizing motivational factors.

Section III, *Creating Effective Collaborations among Educators and Citizen Science Programs*, considers the interactions and perspectives of collaborators in citizen science programs as a social context. In Chapter 6, Hiller and Reybold present a qualitative research study on naturalists' perspectives of effective practices in outdoor initiatives and citizen science programs. DiBenedetto and Schunk (Chapter 7) outline citizen science programs for youth, which foster motivation in collaboration with teachers. Huffling, Scott, Johnson, Gantt, Collins, and Weeks (Chapter 8) report on a participatory research study focused on teacher perspectives on the integration of citizen science within formal education. This section concludes with a focus on collegiate instruction by Cardamone and Cobb Kung (Chapter 9), describing how to integrate an online citizen science program within a collegiate introductory Astronomy course from the perspective of faculty members.

Section IV, *The Impact of Citizen Science Programs on STEM Careers,* centers on student performance in citizen science. Koomen, Hedenstrom, Moran, and Oberhauser (Chapter 10) present a study centered on the *Monarch Larvae Monitoring Project* in terms of student interest and identity formation. Wallace and Bodzin follow suit (Chapter 11) by relaying a qualitative study on the impact of a mobile learning phenology citizen science program on student interest levels and identity formation. Lobel, Cianciola, and Erichsen (Chapter 12) conclude this section by studying the impact of a water quality program in the Charles River Watershed on student self-efficacy and interest.

Finally, Section V, *Assessment and Evaluation of Citizen Science Programs*, outlines ways to evaluate citizen science programs. Downey and Skidmore (Chapter 13) outline essential components in evaluating citizen science programs. Rushton and Parker (Chapter 14) conclude with a specific example of how a co-created citizen science program for adolescent students, known as the Institute for Research in Schools (IRIS), was evaluated for program effectiveness.

In this progression, the following chapters assimilate the two fields of citizen science and educational psychology. With the above structure of the book, our goal is to offer stakeholders, educators, and researchers a framework for citizen science programs and research studies to harness student STEM achievement and career motivation.

REFERENCES

Aceves-Bueno, E., Adeleye, A. S., Feraud, M., Huang, Y., Tao, M., Yang, Y., & Anderson, S. E. (2017). *Bulletin of the Ecological Society of America, 98*(4), 278-290. doi:10.1002/bes2.1336

Ainley, M., & Ainley, J. (2006). Student engagement with science in early adolescence: The contribution of enjoyment to students' continuing interest in learning about science. *Contemporary Educational Psychology, 36*, 4-12. doi:10.1016/j.cedpsych.2010.08.001

Aristeidou, M., Scanlon, E., & Sharples, M. (2017). Profiles of engagement in online communities of citizen science participation. *Computers in Human Behavior, 74*, 246-256. doi:10.1016/j.chb.2017.04.044

Bandura, A. (1997). *The exercise of control.* New York, NY: W. H. Freeman and Company.

Bandura, A. (2005). Adolescent development from an agent perspective. In T. Urdan & F. Pajares' (Eds.), *Self-efficacy beliefs of adolescents* (pp. 2-43). Greenwich, CT: Information Age Publishing.

Belluigi, D. Z., & Cundill, G. (2017). Establishing enabling conditions to develop critical thinking skills: A case of innovative curriculum design in Environmental Science. *Environmental Education Research*, *23*(7), 950-971. doi:10.1080/13504622.2015.1072802

Bendix, H. (2017). Citizen science: Help a historian. *ScienceNordic*. Retrieved from http://sciencenordic.com/citizen-science-help-historian

Bergin, D. A. (2016). *Social influences on interest. Educational Psychology,* 51(1), 7-22. doi:10.1080/00461520.2015.1133306

Bol, L., Hacker, D. J., Walck, C. C., & Nunnery, J. A. (2012). The effects of individual or group guidelines on the calibration accuracy and achievement of high school biology students. *Contemporary Educational Psychology, 37*, 280-287. doi:10.1016/j.cedpsych.2012.02.004

Bombaugh, R. (2000). From citizen scientist to engineers. *Journal of Professional Issues in Engineering Education and Practice, 126*(2), 64-68. doi:10.1061/(ASCE)1052-3928(2000)126:2(64)

Bonney, R., Cooper, C. B., Dickinson, J., Kelling, S., Phillips, T., Rosenberg, K. V., & Shirk, J. (2009). Citizen science: a developing tool for expanding science knowledge and scientific literacy. *BioScience*, *59*(11), 977-984. doi:10.1525/bio.2009.59.11.9

Bonney, R., Shirk, J. L., Phillips, T. B., Wiggins, A., Ballard, H. L., Miller-Rushing, A. J., & Parrish, J. K. (2014). Next steps for citizen science. *Science,* 343(6178), 1436-1437. doi:10.1126/science.1251554

Britner, S. L., & Pajares, F. (2006). Sources of science self-efficacy beliefs of middle school students. *Journal of Research in Science Teaching*, *43*(5), 485-499. doi:10.1002/tea.20131

Cardamone, C., & Lobel, L. (2016). Using citizen scient to engage introductory students: From streams to the solar system. *Journal of Biology & Microbiology Education, 17*(1), 117-119. doi:10.1128/jmbe.v17i1.1082

Cartwright, N. (1989). *Nature's capacities and their measurements*. Oxford, United Kingdom: Clarendon Press.

Chen, P. P., & Rossi, P. D. (2013). Utilizing calibration accuracy information with adolescents to improve academic learning and performance. In H. Bembenutty, T. J. Cleary, & A. Kitsantas (Eds.), *Applications of self-regulated learning across diverse disciplines. A tribute to Barry J. Zimmerman* (pp. 263-297). Charlotte, NC: Information Age Publishing, Inc.

Cleary, T. J. (2018). *The self-regulated learning guide: Teaching students to think in the language of strategies*. New York, NY: Routledge.

Cleary, T. J., & Chen, P. P. (2009). Self-regulation, motivation, and math achievement in middle school: Variations across grade level and math context. *Journal of School Psychology, 5*, 291-314. doi:10.1016/j.jsp.2009.04.002

Cleary, T. J., & Labuhn, A, S. (2013). Application of cyclical self-regulation interventions in science-based contexts. In H. Bembenutty, T. J. Cleary, & A. Kitsantas (Eds.), *Self-regulated learning applied across diverse disciplines. A Tribute to Barry J. Zimmerman* (pp. 89-124). Charlotte, NC: Information Age Publishing Age Publishing, Inc.

Colby, P. S., & Woolley, M. E. (2015). Testing a model of environmental risk and protective factors to predict middle and high school students' academic success. *Children and Schools, 37*(1), 135-143. doi:10.1093/cs/cdv014

Cooper, C. B., Dickinson, J., Phillips, T., & Bonney, R. (2007). Citizen science as a tool for conservation in residential ecosystems. *Ecology and Society*, *12*(2). doi:10.5751/ES-02197-120211

CosmoQuest (2018). *History of citizen science.* Retrieved from https://cosmoquest.org/x/about-cosmoquest/history-of-citizen-science/

Crall, A. W., Newman, G. J., Stohlgren, T. J., Holfelder, K. A., Graham, J., & Waller, D. M. (2011). Assessing citizen science data quality: An invasive species case study. *Conservation Letters, 4*, 433-442. doi:10.1111/j.1755-263X.2011.00196

Creel, B., Nite, S., & Alamarri, J. E. (2017). *Inspiring interest in STEM education among Qatar's youth.* Paper presentation at the annual conference of the American Society for Engineering Education. Columbus, Ohio. Paper # 20731.

Cronje, R., Rohlinger, S., Crall, A., & Newman, G. (2011). Does participation in citizen science improve scientific literacy? A study to compare assessment methods. *Applied Environmental Education and Communication, 10*, 135-145. doi:10.1080/1533015X.2011.603611

Crowdsourcing and Citizen Science Act, H. R. 6414 (2016). Retrieved from https://www.govtrack.us/congress/bills/114/hr6414

CrowdWater (n.d.). Retrieved from https://www.minecraft.net/en-us/

Dawson, G., Lintott, C., & Shuttleworth, S. (2015). Constructing scientific communities in the nineteenth and twenty-first centuries. *Journal of Victorian Culture, 20*(2), 246-254. doi:10.1080/13555502.2015.1022053

DiBenedetto, M. K., & Zimmerman, B. J. (2013). Construct and predictive validity of microanalytic measures of students' self-regulation of science learning. *Learning and Individual Differences, 26*, 30-41. doi:10.1016/j.lindif.2013.04.004

Dick, D. M. (2017). Rethinking the way we do research: The benefits of community-engaged, citizen science approaches and nontraditional collaborators. *Alcoholism: Clinical and Experimental Research*, *41*(11), 1849-1856. doi:10.1111/acer.13492

Dognition (2019). Retrieved from https://www.dognition.com

Eberbach, C., & Crowley, K. (2009). From everyday to scientific observation: How children learn to observe the biologist's world. *Review of Educational Research*, *79*(1), 39-68. doi:10.3102/0034654308325899

Ellwood, E. R., Crimmins, T. M., & Miller-Rushing, A. J. (2017). Citizen science and conservation: Recommendations for a rapidly moving field. *Biological Conservation*, *208*, 1-4. doi:10.1016/j.biocon.2016.10.014

Estrada, M., Burnett, M., Campbell, A. G., Campbell, P. B., Denetclaw, W. F., Gutiérrez, C. G., . . . Zavala, M. E. (2016). *CBE Life Sciences Education, 15*(3), es5. doi:10.1187/cbe.16-01-0038

European Foundation for the Improvement of Living and Working Conditions (2011). *Second European quality of life survey: Participation in volunteering and unpaid work.* Luxembourg: Publications Office of the European Union.

Galaxy Zoo (2007). Retrieved from http://zoo1.galaxyzoo.org/Default.aspx

Gardiner, M. M., Allee, LL., Brown, P. M. J., Losey, J. E., Roy, H. E., & Smyth, R. R. (2012). Lessons from lady beetles: Accuracy of monitoring data from US and UK citizen-science programs. *Frontiers in Ecology and the Environment, 10*(9), 471-476. doi:10.1890/110185

Givot, R., O'Connell, K., Hadley, A. S., & Betts, M. (2015). Hummingbird citizen science. *Science Teacher, 82*(8), 25-3. doi:10.2505/4/tst15_082_08_25

Grasser, S. Schunko, C., & Vogl, C. R. (2016). Children as ethnobotanists: methods and local impact of participatory research project with children on wild plant gathering in the Grosses Walsertal Biosphere Reserve, Australia. *Journal of Ethnobiology and Ethnomedicine, 12*(46), 1-16. doi:10.1186/s13002-016-0119-6

Haidar, A. H., & Al Naqabi, A. K. (2008). Emirati high school students' understanding of stoichiometry and influence of metacognition on their understanding. *Research in Science and Technological Education, 26*(2), 215-237. doi:10.1080/02635140802 037393

Hidi, S., & Renninger, A. K. (2006). The four-phase model of interest development. *Educational Psychologist, 41*(2), 111-127. doi:10.1207/s15326985ep4102_4

Hiller, S. E. (2012). *The impact of a citizen science program on student achievement and motivation: A social cognitive career perspective.* (Dissertation thesis). Retrieved from Proquest Dissertations and Theses Global. (UMI Number 3547088)

Hiller, S. E. (2017). *Mastering science with metacognitive and self-regulatory strategies: A teacher-researcher dialogue of practical applications for adolescent students. Education in a competitive and globalizing world* (Vol. 2). Hauppauge, NY: Nova Science Publishers, Inc.

Hiller, S. E., & Kitsantas, A. (2014). The effect of a horseshoe crab citizen science program on middle school student science performance and STEM career motivation. *School Science and Mathematics, 114*(6), 302-311. doi:10.1111/ssm.12081

Hiller, S. E., & Kitsantas, A. (2015). Fostering student metacognition and motivation in STEM through citizen science programs. In A. Peña-Ayala's (Ed.), *Metacognition: Fundaments, applications, and trends* (pp. 193-221). Cham, Switzerland: Springer.

Hiller, S. E., & Kitsantas, A. (2016). The validation of the citizen science self-efficacy scale (CSSES). *The International Journal of Environmental and Science Education, 11*(5), 543-558. doi:10.12973/ijese.2016.405a

Hiller, S. E., Whitehead, A., & Kitsantas, A. (2019, April). *The role of citizen science on African American students' STEM achievement and motivation*. Paper presentation at the annual conference of the American Educational Research Association. Toronto, Canada.

Hiller, S. E., & Reybold, L. E. (2011a, October). *Field expert's perceptions of scientific observations in the natural world.* Paper presentation at the annual research symposium of the North American Association for Environmental Education, Raleigh, NC.

Hiller, S. E., & Reybold, L. E. (2011b, October). *The role of scientific observations on cognition, affect, and competence.* Paper presented at the annual conference of the North American Association of Environmental Education, Raleigh, NC.

Jeanpierre, B., Oberhauser, K., & Freeman, C. (2005). Characteristics of professional development that effect change in secondary science teachers' classroom practices. *Journal of Research in Science Teaching 42*(6), 668-690. doi:10.1002/tea.20069

Jones, G., Childers, G., Stevens, V., & Whitley, B. (2012). Investigating science in the community: Providing meaningful contexts for students to engage in the processes of science. *The Science Teacher, 79*(9), 36.

Keleman-Finan, J., Schuech, M., & Winter, S. (2018). Contributions from citizen science to science education: An examination of a biodiversity citizen science project with schools in Central Europe. *Journal of Science Education, 40*(17), 2078-2098. doi:10.1080 /09500693.2018.1520405

Kolhauf, L., Rutke, U., & Neuhaus, B. (2011). Influence of previous knowledge, language skills and domain-specific interest on observation competency. *Journal of Science Education and Technology, 20*, 667-678. doi:10.1007/s10956-011-9322-3

Labuhn, A. S., Zimmerman, B. J., & Haselhorn, M. (2010). Enhancing students' self-regulation and mathematics performance: The influence of feedback and self-evaluative standards. *Metacognition Learning,* 5, 173-194. doi:10.1007/s11409-010-9056-2

Lee, C. S., Hayes, K. N., Seitz, J., DiStefano, R., & O'Connor, D. (2016). Understanding motivational structures that differentially predict engagement and achievement in middle school science. *International Journal of Science Education, 38*(2), 192-215. doi:10.1080/09500693.2015.1136452

Le Féon, V., Guilbaud, L., Coiffait-Gombault, C., Dufrêne, E., Kolodziejczyk, E., Kuhlman, M., . . . Vaissière, B. E. (2016). An expert-assisted citizen science program involving agricultural high schools provides national patterns on bee species assemblages. *Journal of Insect Conservation, 20*, 905-918. doi:10.1007/s10841-016-992-1

Lent, R. W., Miller, M. J., Smith, P. E., Watford, B. A., Lim, R. H., & Hui, K. (2016). Social cognitive predictors of academic performance in engineering: Applicability across gender and race/ethnicity. *Journal of Vocational Behavior, 94*, 79-88. doi:10.1016/j.jvb. 2016.02.012

Lent, R. W., & Brown, S. D. (2006). On conceptualizing and assessing social cognitive constructs in career research: A measurement guide. *Journal of Career Assessment, 14*(1), 12-35. doi:10.1177/1069072705281364

Losey, J. E., Perlman, J. E., & Hoebeke, E. R. (2007) *Journal of Insect Conservation, 11*(415). doi:10.1007/s10841-007-9077-6

Loyens, S. M. M., Rikers, R. M. J. P., & Schmidt, H. G. (2008). Relationships between students' conceptions of constructivist learning and their regulation and processing strategies. *Instructional Science, 36*, 445-462. doi:10.1007/s11251-008-9065-6

Magnussen, R., & Elming, A. (2015). *Cities at play: Children's redesign of deprived neighborhoods in Minecraft.* In R. Munkvold & L. Kolås (Eds.), Proceedings of the ninth European Conference on Games Based Learning (pp. 331-337). Reading, United Kingdom: Academic Conferences and Publishing International. Academic Bookshop Proceedings Series.

Martay, B., & Pearce-Higgins, J. W. (2018). Using data from schools to model variation in soil invertebrates across the UK. The importance of weather, climate, season, and habitat. *Pedobiologia-Journal of Soil Ecology, 67*(1-9). doi:10.1016/j.pedobi.2018.01.002

McKinley, D. C., Miller-Rushing, A. J., Ballard, H. L., Bonney, R., Brown, H., Cook-Patton, S. C., . . . Soukup, M. A. (2017). Citizen science can improve conservation science, natural resource management, and environmental protection. *Biological Conservation, 208*, 15-28. doi:10.1016/j.biocon.2016.05.015

McPherson, G. E., Nielsen, S. G., & Renwick, J. M. (2013). Self-regulation interventions and the development of music expertise. In H. Bembenutty, T. Cleary, & A. Kitsantas (Eds.), *Applications of self-regulated learning across diverse disciplines: A tribute to Barry J. Zimmerman* (pp. 355-382). Charlotte, NC: Information Age Publishing, Inc.

Mojang (2019). *Minecraft.* Retrieved from https://www.minecraft.net/en-us/

Mueller, M. P., Tippins, D., & Bryan, L. A. (2011). The future of citizen science. *Democracy and Education, 20*(1). Retrieved from https://democracyeducationjournal.org/ home/vol20/iss1/2

National Oceanic and Atmospheric Administration, Department of Commerce (n.d.) *Cooperative observer program.* Retrieved from https://www.weather.gov/coop/

Navarro, R. L., Flores, L. Y., & Worthington, R. L. (2007). Mexican American middle school students' goal intentions in mathematics and science: A test of social cognitive career theory. *Journal of Counseling Psychology, 54*(3), 320-335. doi:10.1037/0022-0167.54.3.320

Pandya, R., & Dibner, K. A. (2018). *Learning through citizen science. Enhancing opportunities by design. A consensus study of the National Academies of Science, Engineering, and Medicine.* National Academic Press: Washington, DC. Retrieved from https://www.nap.edu/catalog/25183/learning-through-citizen-science-enhancing-opportunities-by-design?utm_source=NASEM+News+and+Publications&utm_campaign=1ea596a1e6-NAP_mail_new_2018-11-06&utm_medium=email&utm_term=0_96101de015-1ea596a1e6-104793173&goal=0_96101de015-1ea596a1e6-104793173&mc_cid=1ea596a1e6&mc_eid=9a43192017

Portas, A. M., Barnard, L., Scott, C., & Harrison, R. G. (2016). The National Eclipse Weather Experiment: Use and evaluation of a citizen science tool for schools outreach. *Philosophical Transactions of the Royal Society: Mathematical, Physical & Engineering Sciences, 374*(2077), 1-13. doi:10.1098/rsta.2015.0223

Raddick, M. J., Bracey, G., Gay, P. L., Lintott, C. J., Cardamone, C., Murray, P., . . . Vandenberg, J. (2013). Galaxy Zoo: Motivations of citizen scientists. *Astronomy Education Review, 12*(1). doi:10.3847/AER2011021

Raddick, M. J., Bracey, G., Gay, P. L., Lintott, C. J., Murray, P., Schawinski, K., . . . Vandenberg, J. (2010). Galaxy Zoo: Exploring the motivations of citizen science volunteers. *Astronomy Education Review, 9*(1). doi:10.3847/AER2009036

Ramirez-Andreatta, M. D., Brusseau, M. L., Antiola, J., Maier, R. M., & Gandolfi, A. J. (2015). Building a co-created citizen science program with gardeners neighboring a superfund site: The Gardenroots case study. *International Public Health Journal, 7*(1), 1-18. Retrieved from https://www.ncbi.nlm.nih.gov/pmc/articles/PMC4420190/pdf/nihms633688.pdf

Rogers, M. E., & Creed, P. A. (2011). A longitudinal examination of adolescent career planning and exploration using a social cognitive career theory framework. *Journal of Adolescence, 34*, 163-172. doi:10.1016/j.adolescence.2009.12.010

Roy, H. E., Baxter, E., Saunders, A., & Pocock, M. J. (2016). Focal plant observations as a standard method for pollinator monitoring: Opportunities and limitations for mass participation citizen science. *PloS One, 11*(3): e0150794. doi:10.1371/journal.pone.0150794

Schunk, D. H. (2001). Social cognitive theory and self-regulated learning." In B. J. Zimmerman & D. H. Schunk's (Eds.), *Self-regulated learning and academic achievement: Theoretical perspectives* (pp. 129-151). New York, NY: Routledge.

Schmeller, D. K. (2008). European species and habitat monitoring: where are we now? *Biodiversity and Conservation, 17*(14), 3321-3326. doi:10.1007/s10531-008-9514-1

Serageldin, I. (2013). Tomorrow's universities and the seven pillars of knowledge revolution. *Cadmus, 2*(1), 8-25.

Seraphin, K. D., Philippoff, J., Kaupp, L., & Vallin, L. M. (2012). Metacognition as means to increase the effectiveness of inquiry-based science education. *Science Education International, 23*(4), 366-382.

Snäll, T., Kindvall, O., Nilsson, J., & Pärt, T. (2011). Evaluating citizen-based presence data for bird monitoring. *Biological Conservation, 144*, 804-810. doi:10.1016/j.biocon. 2010.11.010

Steinberg, L. (2015). *Age of opportunity: Lessons from the new science of adolescence.* Boston, MA: Houghton Mifflin Harcourt Publishing Company.

Stewart, L., MacLean, E. L., Ivy, D., Woods, V., Cohen, E., Rodriguez, K., . . . Hare, B. (2015). Citizen science as a new tool in dog cognition research. *PLOS ONE, 10*(9): e0135176. doi:10.1371/journal.pone.0135176

Surasinghe, T., & Courter, J. (2012). Using eBird to Integrate Citizen Science into an Undergraduate Ecology Field Laboratory. *Bioscene: Journal of College Biology Teaching, 38*(2), 16-20.

Sutton, S. (2009*). Increasing science literacy among English language learners through plant phenological monitoring: A citizen science program at Sequoia and Kings Canyon National Parks*. (Master's thesis). Retrieved from Proquest Dissertation and Theses Global. (UMI No. 1473523)

The Cornell Lab of Ornithology (2019). *BirdSleuth K-12.* Retrieved from http://www.birdsleuth.org/tips-for-cit-sci-engagement/

The Institute for Research in Schools (2018). *The MoEDAL experiment.* Retrieved at http://www.researchinschools.org/MoEDAL/home.html

TurtleSAT (2019). Map turtles in your waterways. Retrieved from https://www.turtlesat.org.au/turtlesat/default.aspx

Vallett, D. B., Lamb, R., & Annetta, L. (2017). After-school and informal STEM projects: The effect of participant self-selection. *Journal of Science Education and Technology, 27*, 248-255. doi:10.1007/s10956-017-9721-1

van der Wal, R., Sharma, N., Mellish, C., Robinson, A., & Siddharthan, A. (2015). The role van der Wal, R., Sharma, N., Mellish, C., Robinson, A., & Siddharthan, A. (2015). The role of automated feedback in training and retaining biological recorders for citizen science. *Conservation Biology, 30*(3), 550-561. doi:10.1111/cobi.1270

Vann-Sander, S., Clifton, J., & Harvey, E. (2016). Can citizen science work? Perceptions of the role and utility of citizen science in a marine policy and management context. *Marine Policy, 72*, 82-93. doi:10.1016/j.marpol.2016.06.026

Wallace, D., & Bodzin, A. (2017). Developing scientific citizenship identity using mobile learning and authentic practice. *Electronic Journal of Science Education, 21*(6), 46-71.

Weigund, M. (2014). The digital woodlouse-scaffolding in science-related scratch projects. *Informatics in Education, 13*(2), 293-305. doi:10.15388/nfedu.2014.09

Wiener, C. S., & Matsumoto, K. (2014). Ecosystem pen pals: Using place-based marine science and culture to connect students. *Journal of Geoscience Education, 62*(1), 41-48. doi:10.5408/12-401.1

Wilson, A., Bacher, K., Breckheimer, I., Lundquist, J., Rochefort, R., Theobald, E., . . . HilleRisLambers, J. (2017). *Monitoring phenology using traditional science, citizen science, and crowdsourcing approaches. National Park Services.* Retrieved from https://www.nps.gov/articles/parkscience33-1_17-26_wilson_et_al_3850.htm

Young, M. T. (2016). Technology and technique: The role of skill in the practice of scientific observation. *Perspectives on Science, 24*(4), 396-415. doi:10.1162/POSC_a_00213

Zárybnicka, M., Sklenicka, P., & Tryjanowski, P. (2017). A Webcast of bird nesting as a state-of-the-art citizen science. *PLoS Biol, 15*(1), 1-9. doi:10.1371/journal.pbio.2001132

Zimmerman, B. J. (1989). A social cognitive view of self-regulated academic learning. *Journal of Educational Psychology, 81*(3), 329-339. doi:10.1037/0022-0663.81.3.329

Zimmerman, B. J. (1994). Dimensions of academic self-regulation: A conceptual framework for education. In D. H. Schunk, & B. J. Zimmerman (Eds.), *Self-regulation of learning and performance* (pp. 3-21). Hillsdale, NJ: Lawrence Erlbaum Associates, Publishers.

Zimmerman, B. J. (2013). From cognitive modeling to self-regulation: A social cognitive career path. *Educational Psychology, 48*(3), 135-147. doi:10.1080/00461520 2013.794676.

Zimmerman, B. J., & Kitsantas, A. (1997). Developmental phases in self-regulation: Shifting from process to outcome goals. *Journal of Educational Psychology, 89*, 29-36. doi:10.1037//0022-0663.89.1.29

Zimmerman, B. J., & Kitsantas. A. (1999). Acquiring writing revision skill. Shifting from process to outcome self-regulatory skills. *Journal of Educational Psychology, 91*(2), 241-250. doi:10.1037//0022-0663.91.2.241

Zooinverse (n.d.). *Galaxy zoo.* Retrieved from http://zoo1.galaxyzoo.org/Default.aspx

In: Enhancing STEM Motivation through Citizen Science Programs ISBN: 978-1-53616-038-3
Editors: Suzanne E. Hiller and Anastasia Kitsantas

Chapter 2

KEEPING IT REAL: MAXIMIZING THE AUTHENTICITY OF CITIZEN SCIENCE EXPERIENCES TO STRENGTHEN YOUTH STEM INTEREST, IDENTITY, AND MOTIVATION

Georgia Bracey*
Southern Illinois University Edwardsville, IL, US

ABSTRACT

The potential for citizen science to increase youth interest, identity, and motivation related to STEM has inspired both formal and informal educators to bring these types of programs to their students. While finding a relevant citizen science program is becoming easier, implementation of the program to maximize its science authenticity, and thereby its benefit to students, can still pose challenges. This chapter explores what is meant by "authenticity" in the context of citizen science and how it connects to the concepts of interest, identity, and motivation. Five elements of citizen science projects that contribute to their authenticity are described and then illustrated through an exemplar project, *Vital Signs*. Finally, implications for education and some directions for future research are offered.

Keywords: authentic science, citizen science, identity, informal science learning, interest, motivation, science community, STEM, youth

Citizen science, broadly defined as the public participation in scientific research (PPSR), holds great potential to engage students of all ages in a variety of scientific activity that may positively impact their interest, identity, and motivation related to science, technology, engineering, and mathematics (STEM). Both formal and informal educators have seen this potential and are working to bring programs like *eBird*, *MonarchWatch*, *Globe at Night*, and

* Corresponding Author's E-mail: gbracey@siue.edu

Supernova Hunters into classrooms, museums, and nature centers. To aid this work, repository websites like CitSci.org and Cornell's Citizen Science Central (The Cornell Lab of Ornithology, 2018) keep tabs on existing programs, offer guidance and support to those looking to create their own, and often provide or point to supplemental resources for educators. With hundreds of programs to choose from, or the option of creating one, there is a good chance that most educators can find a citizen science activity in which to engage their students.

However, finding (or creating) a citizen science program is only the first step. The type of project, its associated tasks, and how it is implemented can significantly impact the experience for students. For example, a co-created project, where participants work alongside scientists in deciding what to study and how, may feel very different to students compared to a contributory project, where scientists have already chosen the research questions and designed the study. There may be more opportunities for meaningful interactions between student and scientist, and the students may better understand and relate to the scientific questions being examined. The science task itself—from data collection in the field to data analysis online to co-authoring a paper—may also make a difference in a student's citizen science experience. Projects vary greatly, and much of this is not under the control of the teacher, scout leader, museum docent, or park ranger. Fortunately, for educators who decide to involve their students in citizen science, there is much they can do to maximize the authenticity of any citizen science program. This chapter will discuss several strategies for providing students with an authentic science experience through citizen science, take a close look at how one citizen science program (*Vital Signs*) uses these strategies, and finally outline some challenges encountered by classroom educators that are also applicable to more informal environments.

Overview of Chapter

The purpose of this chapter is to provide guidance to informal educators looking to use citizen science as an opportunity for youth to participate in "real" or "authentic" science. As discussed in more detail in the following section, the opportunity to engage students in "authentic" scientific work is a chief reason that many educators implement citizen science programs into existing curricula (Bracey, 2018). Many teachers feel that the chance to make a difference in the world and to generate data and analyses that are actually used by others (compared to traditional "cookbook" lab work and most classroom science activities) is highly motivating, interesting, and relevant to students, and a growing number of research studies support this.

This chapter has five objectives. First, I discuss the idea of "authenticity" and provide some context from the education research literature, describing how authenticity connects to and strengthens youth interest, identity, and motivation in STEM (definitions, theoretical frame). Second, I describe five aspects of citizen science, which are critical to ensuring an authentic science experience for youth participants. These aspects are not specific to any particular citizen project but are implementation strategies that can be applied (to some degree) to any project in any educational setting. Third, I provide a model of how these aspects might look in action through the description of an exemplary program, *Vital Signs* (Gulf of Maine Research Institute, n.d.). I also share results from a recent study (Bracey, 2018), highlighting excerpts from interviews with classroom teachers who have used this program with their students. Fourth, I give some practical suggestions from educators who have brought citizen science programs into their classrooms, sharing their experiences of overcoming challenges and maximizing student benefits. Fifth, I examine and recommend areas of future research. The phenomenon of citizen science, as it appears in educational settings, is only beginning to be studied with many areas waiting to be explored.

In discussing citizen science as an authentic science experience, and providing guidance to educators, both formal and informal, I will use a specific definition of citizen science. There are many definitions currently in use by practitioners and researchers (Bonney et al., 2009; Eitzel et al., 2017), and the term itself is still not used or even liked by some groups, with many preferring "participatory science" or "community science". To keep things consistent in this chapter, I will stay with the most common and probably most familiar term, "citizen science," and use the Oxford English Dictionary (2018) definition: "scientific work undertaken by members of the general public, often in collaboration with or under the direction of professional scientists and scientific institutions." For me, the most important phrase in this definition is "scientific work". Participants must be contributing in a meaningful, intelligent, useful way, paralleling, to the greatest extent possible, the work of professional scientists. Participants must be engaged in work that professionals would recognize as *scientific*. Here is where the key to maximizing authenticity lies.

CITIZEN SCIENCE AS AN "AUTHENTIC" SCIENCE EXPERIENCE

Teachers often speak of wanting their students to do "real science," to be involved in "authentic science activities". Many teachers also note that although classroom science activities might teach important skills and concepts, they aren't real science; they aren't authentic. But what do these terms "real" and "authentic" mean? What do they mean for educators, and how does that compare to what they mean to education researchers? Why is authenticity so important? I begin the discussion by first looking at the idea of authenticity in the context of citizen science and then examining its connection to STEM interest, identity, and motivation.

Authenticity and Citizen Science

Science can be considered to be a community of practice, with its members (professional scientists) sharing histories, experiences, knowledge, and accepted practices that are very different from other groups (Lave & Wenger, 1991; Wenger, 2000). As a community, science and its activities ". . . are framed by its culture. Their meaning and purpose are socially constructed through negotiations among present and past members" (Brown, Collins, & Duguid, 1989, p. 34). Some of these cultural, and often dispositional, elements fall under the umbrella of the Nature of Science (NOS), and there is large body of work devoted to understanding the teaching and learning of NOS (e.g., Lederman & Lederman, 2014). Some of these cultural elements, such as observing the natural world in a systematic way or reasoning based on evidence, form the foundation of science education standards along with the more traditional acquisition of science content knowledge (Next Generation Science Standards, 2013; National Science Education Standards, 1996). Using these standards as a guide, educators have designed and implemented science activities that aim to be "inquiry-based" and involve science practices as well as elements of NOS. However, for many educators, even these types of activities often lack something; they are somehow not real science. Bound by the walls, the time constraints, and the culture of a typical classroom or museum experience, students are engaged in science activities but not in authentic science.

Even with activities that are inquiry-based, students do not generally experience or, more importantly, enter into the community or culture of science. For an activity or experience to be authentic, it must involve the "ordinary practices of the culture" (Brown, Collins, & Daguid, 1989, p. 34). The activities must be recognizable and meaningful to members of the culture, in this case, professional scientists. Practicing science tasks in disconnected bits and pieces, without a sense of contribution to the larger endeavor, or working through a set of pre-defined steps to achieve a pre-determined answer as part of a "cookbook" lab are not things that professional scientists typically do. As noted by Brown, Collins, and Daguid, most educational activities ". . . are simply not the activities of practitioners and would not make sense or be endorsed by the cultures to which they are attributed" (1989, p. 34).

Fortunately, one need not become a professional scientist in order to do authentic science, and this is where citizen science can, in theory, play a significant role in both formal and informal science education. Thoughtfully implemented, citizen science can act as a bridge between the everyday world and the world of professional science, allowing students to experience and enter into the culture of science gradually with guidance from educators and professional scientists. In that way, citizen science can create for students an opportunity for "legitimate peripheral participation" in the scientific community, where newcomers are able to enter the community by engaging in its practices to a greater extent, similar to an apprenticeship (Lave & Wenger, 1991). Ideally, as "science apprentices," students who participate in citizen science programs will experience the practices, the knowledge, and the culture of the science community in a way that professional scientists would recognize as their own.

As peripheral community members, students would not only be doing recognizable (i.e., authentic) science tasks but also interacting socially with each other and scientists within the context of the research study. Students would discuss how what they are doing contributes to the larger study, and develop an understanding of the larger issues that the study may impact. On a smaller scale, students would get feedback on their performance and also give advice to others. Problems encountered during the course of the study would be shared and solved together. As time goes on, students would move from peripheral involvement in the science community towards full involvement, as their skills, knowledge, and interests develop. This connection to and entry into the community of science is a key aspect of citizen science for many teachers. As one teacher put it:

> And that's the whole point of citizen science, right? The whole community is getting together to help, and there are probably professionals in the community who are willing to help the students during their data collection. There's a lot of work to get it done, but it's so worth it to have the kids out there and collecting the data and being a part of the larger science community. (Bracey, 2018, p. 81)

Developing Interest

Generating interest in STEM, and particularly in science, has long been a goal of science educators and is becoming an area of focus for many science education researchers. Interest, defined as "the psychological state of engaging or the predisposition to reengage with particular classes of objects, events, or ideas [i.e., content] over time" by Hidi and Renninger (2010, p. 111), has been linked to student motivation, achievement, and career choices in STEM. For example, Maltese and Thai (2011) found that high school students who chose to concentrate in STEM made that choice because of a growing interest in math and science. Interest has also been found to positively influence learning, especially regarding student attention (Hidi, Renninger, & Krapp, 2004).

Although it is generally acknowledged that interest plays a significant role in a student's educational experience, many teachers may feel that it is something they cannot influence; students either have it, or they don't (Lipstein & Renninger, 2006). Hidi and Renninger (2010), however, have proposed a model of interest as an emerging and developing phenomenon, and one that can be impacted by educators and educational practices. Their model consists of four phases of interest development: triggered situational interest, maintained situational interest, emerging individual interest, and well-developed individual interest. As their names indicate, in the first two phases, it is a person's environment or situation that generates interest. Something outside themselves catches their attention or draws them into an activity. The last two phases are internal but can still be impacted by outside influences such as others' opinions and support (or lack thereof) from others. In this way, interaction with teachers can affect student interest in science and STEM.

Strengthening Science Identity

The concept of "identity" has been studied extensively in the social sciences. Identity can be defined as "the 'kind of person' (i.e., 'science person') one is seeking to be and enact in the here and now" (Gee, 1999, p. 13). By this definition, identity can be a goal that one is seeking to achieve rather than what one is or how one acts currently. For example, many teachers speak of wanting to help their students to become scientists or at least to be that kind of person in the future (Bracey, 2018). There is the sense that a person can develop a particular identity through experience and education, even if it's not how they see themselves now.

Gee (2000) outlined four ways that one could achieve an identity: it could be part of one's nature, it could come from one's social or professional position, it could come from others recognizing one's achievements, or it could come from experiences one has with certain groups. Banks et al. (2005) gives a similar description of the concept, but adds the elements of flexibility and individual choice; a person may take on different identities in different situations, with different groups, and in different places. Finally, Carlone and Johnson (2007) posited that in order to be a certain kind of person, one needed to have a certain level of competence, to be able to demonstrate it, and to be acknowledged for it by others. They created a model of science identity to use in their study of the science experiences of women of color. This model has three dimensions: performance, recognition, and competence. To develop a science identity, one must perform scientific practices, be recognized as a science person, and be competent in knowledge of science content. They also noted that a particular identity—in this case science identity—can interact with and be affected by other identities (e.g., racial, ethnic, and gender), making the development of a strong science identity particularly challenging for underrepresented groups.

This identity disconnect can potentially affect students' interest in science. Several teachers felt that participation in citizen science sparked or strengthened their students' interest in science, citing an increase in asking questions in class or participation in the project at home. A strengthening of science identity may be the mechanism by which this increased interest occurs. A student who doesn't identify with science or as a scientist would have little reason to be interested, as described by Stromholt and Bell (2017):

> When learners view what happens in formal science education as being systematically disconnected from their everyday lives, it can result in issues of identity conflict where learners find themselves disinterested in or unable to access science-linked identities. In other words, some learners do not see science as a desirable or accessible resource that contributes to who they are or who they want to be in the future. People try to engage in activities they see as part of who they want to be and avoid activities they perceive to be misaligned with who they see themselves as. (Stromholt & Bell, 2017, p. 4)

Students who initially do not see how science (or the outcomes of scientific studies) affects them or relates to them as an individual may begin to see some connection as they participate in citizen science. This connection may come through discussion and other types of social interaction with others, through learning and developing expertise in science content, or through seeing the results of their completion of scientific practices. As students become more connected, they may begin to see science as a part of them, something relevant to their lives, something that they can be a part of and in which they could find interest.

Connecting Interest and Identity to STEM Motivation and Achievement

Interest and identity may both play a significant role in student motivation and achievement in STEM. Using an expectancy-value lens, Eccles (2009) described how one's identity can be considered a set of self-perceptions about one's characteristics, abilities, values, and goals. These perceptions would comprise the extent to which someone expects that they will succeed at a particular task and the value that they attach to completing that task. These two constructs—expectancy and value—work together to influence an individual's choice of action. Seen through this lens, science identity development is the forming of expectations and values related to the self and science, and thus an important motivator for students participating in citizen science.

In Eccles' model, interest-enjoyment is one of the constructs that contributes to the overall value that an individual places on a task. According to Eccles (2009, p. 82), "interest value" relates to the "anticipated enjoyment of engaging in the activity or behavior itself." A student, for example, would be interested in participation in citizen science if he/she thought it would be enjoyable. The activity would then have some value to the student; this value, along with the student's expectations about participating in the activity, would be a factor in motivating the student to choose to engage in the activity.

The interplay of interest, identity, and motivation is a critical aspect of a citizen science experience. Initial interest sparked at the introduction of the activity can develop into long-term individual interest through meaningful tasks and interaction; these are elements of authenticity that connect participants to the science community and strengthen their science identity. With a strong science identity, interest is likely to be sustained and students will view citizen science as enjoyable and valuable, motivating students to continue their participation and move closer to full participation in the science community.

Five Critical Elements of Citizen Science That Increase Authenticity

Given that authenticity is a desirable element of a student's citizen science experience, it is necessary to understand how maximum authenticity might be achieved. In this section, I describe five elements that are critical to implementing citizen science in the most authentic way possible. This list comes from a series of interviews with middle school science teachers who were working to give their students an authentic science experience through citizen science (Bracey, 2018). These elements comprise the issues that the teachers considered to be most important. Sometimes, the citizen science projects were lacking one or more of these elements, and the teachers were seeking out ways to fill the gap. Other times, the teachers were highlighting an element that was going particularly well in a project. For each element, I will give a brief description and then let the teachers share their ideas in their own words.

Element One: An Explicit and Relevant Scientific Question

At the core of each citizen science project is a scientific question, and often, a problem that is trying to be solved. The question drives the data collection and analyses, and therefore, the citizen science tasks. Yet many times, the participants are unaware of the significance of their contribution. They may sense that what they are doing is important and that it is helping a larger scientific effort, but there may be very little understanding or appreciation of the driving question and the scientific context. Participants may enjoy their involvement in the project but still be wondering "Why does what we're doing matter?"

Educators need to ensure that the scientific question is explicit, understandable, and relevant to the participants. This can be achieved through guided discussion before beginning the project, with reinforcement during and after participants engage. Teachers, docents, and even experienced participants can lead short conversations that elicit the project's driving question and how citizen science is helping to find an answer. These conversations can also be part of the interactions participants may have with a project scientist, whether face-to-face or online. Both educators and scientists can help participants to understand what the question is, why the question is important, and how it relates to their lives. This gives an opportunity to deepen their learning of science content and how science works.

In addition to understanding the scientific question, participants need to know how their particular task makes a meaningful contribution towards answering that question. Often, it is difficult to see the connection between the task of the citizen scientist and the generation of findings from the study. Making these connections explicit for participants will help them to understand why they are doing what they do and why it must be done in a certain way. Again, this is an opportunity for deeper learning of science content and process, but it also enhances authenticity as participants will be doing tasks and using methods that professional scientists recognize.

Element Two: Meaningful Feedback to Participants

Educators need to give meaningful feedback to participants, letting them know how well they are doing in the citizen science task and offering guidance when needed. This formative feedback helps participants to build competency and develop insight into their activity, similar to what happens during apprenticeships. In this way, participants, as newcomers to the science community, can begin to make their entry.

Most citizen science participants desire feedback on their activity; they want to know if they have correctly identified a bird, accurately marked a crater on the Moon, or placed their data in the right spot on a report sheet. This desire stems, in part, from a general sense of wanting to perform a task well, or as well as possible, but also from motivation to make a contribution to science and scientific research. Often, participants worry that if they perform a task incorrectly or inaccurately, the larger project will suffer. Some projects that involve large amounts of participant data [e.g., *GalaxyZoo* (Zooinverse, n.d.), *Moon Mappers* (Cosmoquest, n.d.)] have analyzed results from their volunteers, finding that in general, volunteers are at least as accurate as professionals for many tasks (Droege, 2007). Statistical methods are also used by scientists to produce an accurate value from the input of a large number of participants with a wide range of skill levels and expertise (Swanson, Kosmala, Lintott, & Packer, 2016). However, even when these ideas are communicated to project volunteers, they still express a real need to know if their own particular data are accurate and correct.

In addition to information about accuracy and correctness, participants need to be given feedback about *how* and *why* what they did was right or wrong, good or bad, accurate or way off the mark. As with any learner who is trying something new, citizen scientists will need time and formative feedback to build their skills and expertise within the context of the project. Even science hobbyists, who may have plenty of knowledge in a particular scientific domain, will need to learn specific data collection protocols or become familiar with online analysis tools that are required for successful participation in a particular citizen science program. Effective learning of these types of authentic science skills requires meaningful information given to volunteers by professionals that not only evaluates but supports and guides, leading to knowledge gain, improved task performance, and increased confidence in participating in and contributing to science.

Feedback may come in the form of the modeling of science skills by the educator, especially and ideally if the educator participates in the citizen science project along with the students. Most projects involve some type of observation, measurement, and recording. Educators can use their own participation experiences to not only model these skills (among others) but also to reflect on how their own performance of the skills helps the science to progress. Sharing reflections will help students to see nuances that are often missed or glossed over, such as type and level of detail needed, why specific units are used and what might happen if they are not, things to pay attention to and things to ignore, and why consistency (reliability) and checking initial results with others (validity) matter. The educator may want to keep a scientific notebook along with their students, share his or her own pictures or sketches, and take a turn at writing a blog post or newsletter article about how science skills are developing throughout the duration of the project. This type of reflective modeling will help students to be more evaluative and mindful about making their own observations.

Element Three: Interaction of Participants and Scientists as Collaborators

In many citizen science projects, there is limited opportunity for participants and scientists to interact. Most projects tend to be *contributory* in nature (Bonney et al., 2009; Shirk et al., 2012), with professional scientists initiating and managing the project while volunteer participants collect and analyze data at some distance. This structure can make meaningful, collaborative interactions between scientists and volunteers challenging to achieve. It may be difficult to find time for participants to ask questions, make suggestions, and discuss outcomes of their work with project scientists. Participants may receive periodic reports on the progress of the research and thanks for their help, but in order to have an authentic experience, participants and scientists need to interact more like collaborators.

Collaborative interaction involves scientists treating participants more like fellow professionals, elevating them to a higher level and a more complex role than typical volunteers. Participants may not have a level of expertise comparable to the professional scientists (although some may), but they can still discuss the progress of the study, the analyses, and the results using scientific language. There can be a respectful and intelligent exchange of ideas rather than a didactic presentation of information. Scientists can ask the youth what they are seeing in the data, what do they think it means, and what patterns do they notice. Have they seen anything new or unusual? Are things going as they expected or differently? What challenges or problems have arisen and what would they advise? Participants spend a lot of time collecting data and looking at it; scientists can take advantage of that.

Many teachers noted that their students seemed to "rise to the occasion" and take on a more confident demeanor when working with scientists on these projects. Even just knowing that scientists are "out there," expecting them to do their part accurately, is energizing and motivating for the students:

> Some of it is a confidence builder for kids, especially when you're taking part in projects where they know . . . the scientists that are running them are trusting everyone to be as accurate and reliable as possible. So, it's kind of giving them that extra responsibility. You're 12 or 13, but you're still being held accountable to these high scientific standards. I think a lot of kids kind of take that, and it kind of gives them a bit of a boost. (Bracey, 2018, p. 76)

Element Four: Interaction among Participants as Collaborators

Science is a social activity. Professional scientists talk to each other about all aspects of their research, and citizen science participants should do the same. Similar to how scientists and participants need to talk as collaborators, participants should be encouraged to talk to each other about the data they've collected, unusual or unexpected results, patterns that may be developing, and how their activities are making progress towards answering the study's question. Depending on the group, this may need to be a more structured and guided conversation, but it can also be quick and casual as the opportunity arises.

Opportunities for this interaction can be provided through online venues as well as face-to-face encounters. Blogs, virtual galleries, discussion boards, and social networking can all be used to give a platform for interaction among participants. With educator and/or scientist guidance and oversight, these online opportunities can take place outside of the classroom or museum as well as within, allowing students to continue their participation at home.

Element Five: Opportunity to Present Results to the Public

Scientific findings must be shared, adding to the general understanding of the world around us. Whether it be through a written article in a peer-reviewed journal or a presentation at a conference, professional scientists publish their findings so that others may build upon them in future work. Citizen scientists are not often able to be directly involved in this aspect of the science process; however, many projects try to keep their participants informed of publications that come out of their work, and some try to include or acknowledge citizen scientists as authors of the work.

Often, educators must find creative ways for publication to happen. Blogs, virtual galleries, discussion boards, and social networking are all possible ways for youth to get their results out to the public. This is a critical part of an authentic science experience, as this teacher explains:

> I think it can be frustrating for the students who understand that the logical ending to a research project is that you write a paper on it and someone reads about your research, whether it's your teacher reading it or just the scientific community at large reading it, they understand that that is the end . . . But that's why . . . I think one of the ways to alleviate that is I always make my kids write a report at the end of whatever little project we were doing so regardless of whatever data we contributed to the project at large, any data that they contributed they have to write a report on. It was small, and of course their trends were not noticeable, but it got them understanding and feeling that sort of closure for that particular project, even if they didn't get a full closure. (Bracey, 2018, p. 87)

Just as professional scientists work to get their research results published, students want others to know what they have accomplished, what they have discovered, and that their work can make a difference.

Whether sharing of results happens online through blogs and virtual discussion boards, or face-to-face at science fairs and conferences, or through a traditional peer-reviewed journal article, finding opportunities for publication is a critical element of science as practiced by the scientific community. Ideally, project scientists and educators will work together to find the best route to publication for the young citizen scientists.

AN EXEMPLARY PROGRAM - *VITAL SIGNS*

Overview of Program

Vital Signs (Gulf of Maine Research Institute, n.d.) is an environmental field-based citizen science program run by the Gulf of Maine Research Institute (GMRI) that focuses on invasive species. The program brings community members, educators, students, and scientists together to examine how invasive species are impacting a variety of local habitats (i.e. freshwater, coastal, and upland). Citizen scientists of all ages collect data on both invasive and native species. The educational component of *Vital Signs* is designed for middle school students and teachers.

Highlights Illustrating the Five Critical Elements that Increase Authenticity

Element one: Explicit and relevant scientific question. Vital Signs offers participants many "missions" from which to choose. Each mission page begins with a clearly stated research question, plus information about the scientist (or teacher, in some cases) and why he/she is conducting the study (Figure 1). Then, at the bottom of each mission page is a "Why This Mission Matters" paragraph explaining how the study may help solve problems and address issues that are relevant to the participants and the world they live in.

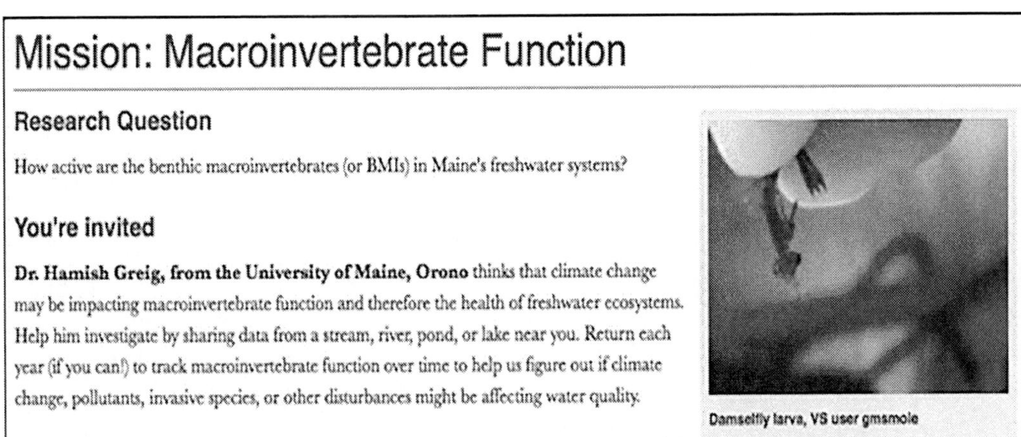

Figure 1. Gulf of Maine Research Institute (n.d.) Mission: Macroinvertebrate function. Retrieved from http://vitalsignsme.org/mission-macroinvertebrate-function.

Element two: Meaningful feedback to participants. Participants in Vital Signs can receive feedback on their activities in several ways. As examples of species are found and pictures submitted to the program's website, scientists will check and confirm correct identifications (Figure 2).

Figure 2. Participant identifications and scientist feedback. Gulf of Maine Research Institute (n.d.). Explore data.
Retrieved from http://vitalsignsme.org/explore/search

In many cases, it is the project scientist who provides feedback more directly through email or a comment on the website, validating their results or encouraging them to try again, as described by this teacher:

> Once the kids post their observations to the Internet and whether they found what they're looking for or not, if they think they found it, then a scientist from the Gulf of Maine Research Institute or one of their people who's been working with this for a long period of time, will go on and comment about whether they think this is an accurate description, or if students should look at something else. (Bracey, 2018, p. 76)

Not only do participants find out whether or not their contribution was accurate and/or correct, but they often receive suggestions on how to proceed.

Element three: Interaction of participants and scientists as collaborators. Meaningful feedback from scientists to participants is one aspect of collaborative interaction in Vital Signs, but there is another option. Participants may initiate their own study to investigate a question that they have generated by creating a Vital Signs Field Mission (Figure 3). This co-created type of citizen science lets participants assume a larger role, leading the research effort as a professional scientist might do (Bonney et al., 2009). Scientists and staff from the Gulf of Maine Research Institute provide support and guidance, as in the regular missions, but now the participants have greater involvement and greater responsibility on par with the professionals. This higher level of collaboration comes with an increase in the authenticity of the experience, since participants are working more like professional scientists and progressing towards greater participation in the scientific community.

Element four: Interaction among participants as collaborators. Student participants are able to connect with each other via posting and commenting on the Vital Signs community blog, conducting peer review for student-written articles in the research journal Findings from the Field, and through the many online opportunities to share and comment on other participants' field notes, scientific notebooks, sketches, and photos. Through all these venues, participants share their own project experiences, discuss findings and results using scientific language, and ask and answer questions about the research.

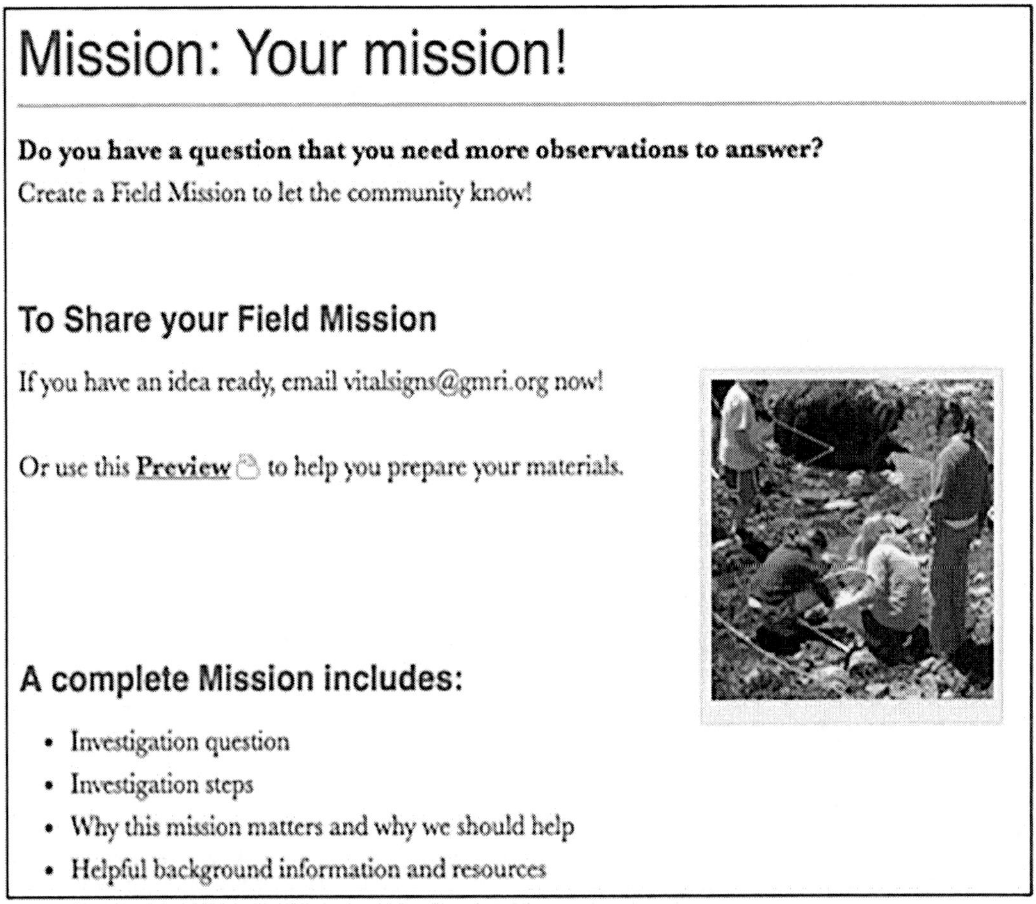

Figure 3. Creating your own project. Gulf of Maine Research Institute (n.d.). Mission: Your mission! Retrieved from http://vitalsignsme.org/mission-your-mission.

Element five: Opportunity to present results to the public. Citizen scientists involved in Vital Signs have several ways to share their results with the wider community. The website hosts a blog, where students and educators can post stories of their experiences and discuss their findings. Other participants can comment on the posts, giving feedback and generating discussion.

Also, on the website is a "best of VS" page where the public can go to see highlights of recent findings, photos, and sketches. The program also shares findings through social media, a peer-reviewed journal for middle school students, and online interactive data maps.

Beyond these, there are various creative options for youth to share their projects and results. Through the *Vital Signs* website, students can create a public service announcement, create posters and flyers, develop a set of species identification cards, draw a comic strip, program an educational game using *Scratch*, and author a how-to guide. These educational science products let students practice and develop critical communication skills that are vital to professional scientists, who are always looking for ways to effectively present their science to the public.

Challenges and Advice from Formal Educators

It is perhaps no surprise that, when asked about the challenges involved with implementing citizen science projects in the classroom, teachers often speak about issues with time, materials, and student management. Teachers face these types of problems every day, and these issues are often exacerbated whenever a new and somewhat complex activity is brought into the classroom. Informal educators face the same challenges, typically having limited time and resources with which to engage their audience and having to quickly adapt to new projects and activities. However, there are some challenges that are particular to citizen science activities and to maximizing their authenticity. Although these challenges come from formal educators, much of this information is also applicable to those working in informal environments.

Working with "Negative" Experiences

A common challenge is concerned with how to address a situation where participants do not find what they are looking for, or they obtain negative or null results. Not finding any evidence for something was a particular challenge for educators in two ways. First, students became frustrated or disappointed when they felt they should be finding a particular species or seeing some type of result and did not. This often happened in environmental projects when students were trying to see an event or find certain animals or plants, as described by this teacher involved in the *MonarchWatch* program:

> I think sometimes when a particular student couldn't find certain things in their area, that they would be disappointed. So, in the case of the monarchs, the kids were coming from all over the city but yet only a certain number of kids got to see the monarchs because they travel in such a narrow path when they go up and down Kansas City. They're the ones who are disappointed when they didn't see it, when they didn't get a chance. (Bracey, 2018, p. 85)

Students may feel that they are missing out when they do not experience all aspects of a project, and it can be challenging for a teacher to help them overcome this disappointment and understand that not observing something is often just as important, scientifically, as observing it. Secondly, as in the case of looking for invasive species or environmental contaminants when *not* finding something can be a very good thing, it was still hard for students to be excited about, in their view, coming up empty. One *Vital Signs* teacher said, "Explaining to the students that not finding anything when you're looking for an invasive is a really good thing. That's kind of hard for them to get their heads around sometimes."

Making Connections to the Larger Scientific Endeavor

Teachers overwhelmingly described authentic data collection and analysis as two positive parts of citizen science involvement. When students engaged in the tasks of citizen science, they were engaged in authentic scientific activities that contributed to a larger scientific endeavor. However, the reasons for these activities, the connections to the bigger picture, were not always apparent to the students. Although students were participating, they didn't always know how their involvement contributed to the study. This teacher describes her students' disappointing experience with a mapping project:

> And with that one, they didn't see a result, they were just like, "Oh . . . that's it? Ok . . ." and they didn't realize that even counting craters was making a huge difference. So, I think if I would've been able to communicate that better to them, they would've been in a better place. (Bracey, 2018, p. 84)

For her students, the activity of mapping the craters on the Moon had no connection to anything meaningful. Students were expecting some type of response or result from their actions and received nothing. Without guidance or feedback, the citizen science project was simply a repetitive task without any meaning.

For example, if the students' participation consists of simply "checking in" to a project to add their observations and then "checking out" again, there will be no opportunity to see how their one-piece fits into, and supports, the research study. Participation should not be sporadic or isolated from any context. Citizen science should not be a lesson supplement or an add-on; it should be an intentional part of the curriculum with a storyline that builds student interest and helps them to connect to the larger scientific study. Furthermore, whenever possible, students need to experience the *entire* research process, especially the culminating activity of sharing the results with a wider audience. When students communicate their results to others, they will feel that what they have accomplished matters and that they have truly made a contribution.

Building and Sustaining Collaborations

Science and citizen science are social and collaborative endeavors. When groups come together and work together (i.e., educators, administrators, scientists, families, students) more science happens, more learning happens, and authenticity is increased as these various communities have numerous opportunities to interact. Getting these collaborations going and keeping them going requires energy, flexibility, and adaptability. Educators had the most success with this when they sought out and, in some cases, created multiple points of involvement for these groups in the citizen science project. More involvement led to greater awareness and understanding of the project and its impacts.

For example, parents and families can provide assistance with activities, donate supplies, and even be a resource to help students take their projects beyond the classroom, the museum, and the science center. One educator had a student whose parent was able to work with the students to write a bill that would support their project. She said,

> One of the parents of the students is a lobbyist, and he saw his son was pretty into the [citizen science] topic. So, he came in and worked with one of the students as part of a green team, and they ended up partnering with the assemblywoman in our district. They're working on writing a bill to provide a tax incentive for restaurants in the district to donate their oyster shells to the *Billion Oyster Project* so that they can build more reefs. (Bracey, 2018, p. 78)

Not only do the students learn about how science works, but also, thanks to this parent, they get a glimpse of another way that they can impact their environment and work for change. Sometimes the youth participants can help to develop relationships with community members and build support for a program. For example, during a beach restoration project, one teacher had a problem with the students' signs being vandalized. The students put an editorial in the local paper, letting the community know how bad the vandalization made them feel and how important the beach was to the community. This problem became an opportunity for the community to become aware of the importance of the project and the value of their beaches. As community members began to help keep an eye on the beaches, the vandalization eventually stopped (Bracey, 2018). By reaching out to the local community, the students gained local support for their work. The community became project collaborators, taking part in the work and seeing its value.

EDUCATIONAL IMPLICATIONS AND FUTURE DIRECTIONS FOR RESEARCH

Maximizing authenticity in youth citizen science experiences through the five elements discussed in this chapter will require changes in how science is typically taught and how learning environments are usually structured, with important implications for science education in both formal and informal settings. In this section I discuss three areas of impact: (1) roles of the educator, (2) administrative support, and (3) types of educator and administrator training. Finally, I describe some directions for future research in this area.

Educator Roles as Co-Investigator and Guide

For decades, as the focus on inquiry-based teaching has grown, educators have been encouraged to be less of the "sage on the stage" and more of the "guide on the side". In science education particularly, the focus has shifted from knowing/transmitting a set of facts to engaging students in a systematic set of practices and increasing their understanding of how evidence can support reasoning and conclusions. However, many science educators are still not comfortable with this less-structured approach. It is more satisfying to know and be able to provide a definite answer to students. Although new science education standards, as well as a growing community of educators and administrators have embraced the emphasis of practices as well as content, change is slow and many educators are reluctant to move from a more familiar and traditional way of teaching.

Authentic science, however, is about exploring the unknown. To do citizen science with authenticity is to follow a path that has an unknown end, and educators must become comfortable walking that path along with their students. Educators must become *co-investigators*, exploring and learning at the same level as the youth they are teaching. They must be comfortable with and even excited by *not* knowing the answers; this is a significant challenge for many educators.

Yet, the educator is more than just another participant in the citizen science project. The educator must ensure that the project is implemented effectively and authentically, meeting curricular objectives as well as the scientific goals of the project. As such, educators act as guides, stepping outside the role of participant in order to help students have an authentic science experience. Educators must organize opportunities for students to interact with peers and scientists, to connect what they are doing to the larger work, and to share their findings with a larger community. Educators need to participate in citizen science but also understand what the project is doing at a higher level. As a co-investigator and guide, an educator is a critical part of providing a successful and authentic citizen science experience for youth.

Administrative Support

If educators are to transition successfully into the roles of co-investigator and guide as they provide youth with these authentic science experiences, the support of administrators will be key to their success. Implementing citizen science with maximum authenticity requires flexible scheduling and long-term stability in participation, two things that may only happen with assistance at the administrative level.

Depending on the project, citizen scientists may need the flexibility to collect and analyze data at particular times of the year, (i. e., when certain species of plants and animals are present) and be observed, even when certain weather conditions prevail. Administrators can help with arranging schedules so that the time and the staff needed are available. If students need to go out into the field, administrators can help find transportation and chaperones. If educators and students need to spend time on a project beyond a class period or a school year, administrators would be able to give those permissions. Administrators can more easily work across departments, offices, and institutions, providing a level of coordination that is often impossible for educators.

In addition to logistical assistance, administrators can play a key role in supporting and facilitating the change in role of the educator. Students and the general public are not always comfortable with an educator who has stepped out of the traditional teacher role, especially in a formal educational environment. Principals, superintendents, and museum directors can all help to educate the wider community about the importance and the necessity of this different role. Through newsletters, blog posts, meeting presentations, and other types of community interaction, administrators can show their support for educators who are using citizen science and lead a shift in how the public views science education.

Training for Educators and Administrators

These shifts in roles and ideas about science education will be difficult to achieve; educators and administrators often operate in environments where traditions are strong and lasting change is elusive. However, making changes to the training of pre- and in-service teachers and, to some extent, administrators may help this change happen over time. Undergraduate and graduate education programs should place more focus on the nature of science (NOS) and the introduction of citizen science, ideally having educators participate in a citizen science project as part of the course. Similar courses or workshops could be offered to administrators.

Educators and administrators may want to participate in research opportunities often offered in connection with universities and museums. These programs offer a more formal, structured research experience compared to most citizen science projects, and participants would get a more comprehensive look at the process and community of science. Having this type of formal research experience would enable educators to be expert guides for their students and administrators to be supporters and advocates for their educators.

Future Research Directions

As the field of citizen science continues to grow, education researchers will have many opportunities to study how the experience affects the participants, the educators, and even the scientists. The construct of authenticity and how it manifests itself in different types of citizen science projects needs to be researched as very little work has been done in this area. The study of authenticity in citizen science can take place within the context of youth identity and interest development, as discussed in this chapter, as well as within other contexts such as relevance and agency.

Conclusion

Citizen science has great potential to engage youth in meaningful science activity. Thoughtfully implemented, these activities can spark and sustain youth interest in STEM, help to develop STEM identity, and motivate students to deepen their involvement in STEM both in and out of school. The level of authenticity of the citizen science experience plays a key role in achieving these positive outcomes for youth participants, and this chapter has outlined five elements critical to maximizing authenticity. Realistically, it may not be possible to incorporate all five perfectly into every citizen science program; citizen science, like professional science, is a complex, rich, and very human activity. However, these elements provide important goals and guidelines for educators, scientists, and program designers.

References

Banks, J. A., Banks, C. A., Cortés, C. E., Hahn, C. L., Merryfield, M. M., Moodley, K. A., . . . Parker, W. C. (2005). *Democracy and diversity: Principles and concepts for educating citizens in a global age.* Panel discussion Global Citizenship Education, University of Washington, Seattle, WA. Retrieved from https://www.gcedclearinghouse.org/resources/

democracy- and- diversity- principles- and- concepts- educating- citizens- global-age?language=en

Bonney, R., Ballard, H., Jordan, R., McCallie, E., Phillips, T., Shirk, J., & Wilderman, C. C. (2009). *Public participation in scientific research: Defining the field and assessing its potential for informal science education. A CAISE inquiry group report.* Washington D.C.: Center for the Advancement of Informal Science Education (CAISE). Retrieved from http://www.birds.cornell.edu/citscitoolkit/publications/CAISE-PPSR-report-2009.pdf

Bracey, G. (2018). *Teaching with citizen science: An exploratory study of teachers' motivations & perceptions.* (Doctoral dissertation). Retrieved from Proquest Dissertations and Theses Global. (UMI No. 10791145)

Brown, J. S., Collins, A., & Duguid, P. (1989). Situated cognition and the culture of learning. *Educational Researcher, 18*(1), 32-42. doi:10.3102/0013189X018001032

Carlone, H. B., & Johnson, A. (2007). Understanding the science experiences of successful women of color: Science identity as an analytic lens. *Journal of research in science teaching, 44*(8), 1187-1218. doi:10.1002/tea.20237.

Cosmoquest (n.d.). *Moon mappers.* Retrieved from https://cosmoquest.org/x/science/moon/

Droege, S. (2007). Just because you paid them doesn't mean their data are better. In McEver, C., Bonney, R., Dickinson, J., Kelling, S., Rosenberg, K., & Shirk, J., (Eds.), *Citizen science toolkit conference* (pp. 13-26). Ithaca, NY: Cornell Laboratory of Ornithology.

Eccles, J. (2009). Who am I and what am I going to do with my life? Personal and collective identities as motivators of action. *Educational Psychologist, 44*(2), 78-89. doi:10.1080/00461520902832368

Eitzel, M., Cappadonna, J., Santos-Lang, C., Duerr, R., West, S. E., Virapongse, A., . . . Metcalfe, A. (2017). Citizen science terminology matters: Exploring key terms. *Citizen Science: Theory and Practice, 2*(1), 1-20. doi:10.5334/cstp.96

Gee, J. P. (1999). *An introduction to discourse analysis: Theory and method.* New York, NY: Routledge.

Gee (2000). Identity as an analytic lens for research in education. *Review of Research in Education, 25*(1), 99-125. doi:10.2307/1167322

Gulf of Maine Research Institute (n.d.). Vital signs. Retrieved from vitalsignsme.org

Hidi, S., & Renninger, K. A. (2006). The four-phase model of interest development. *Educational Psychologist, 41*(2), 111-127. doi:10.1207/s15326985ep4102_4

Hidi, S., Renninger, K. A., & Krapp, A. (2004). Interest, a motivational variable that combines affective and cognitive functioning. In D. Y. Dai & R. J. Sternberg (Eds.), *Motivation, emotion, and cognition: Integrative perspectives on intellectual functioning and development* (pp. 89-115). Mahwah, NJ: Lawrence Erblaum Associates.

Lave, J., & Wenger, E. (1991). *Situated learning: Legitimate peripheral participation.* Cambridge, United Kingdom: Cambridge University Press.

Lederman, N. & Lederman, J. (2014). Research on teaching and learning of nature of science. In N. G. Lederman & S. K. Abell (Eds) *Handbook of research on science education* (pp. 600-620). Abingdon, United Kingdom: Routledge.

Lipstein, R. L., & Renninger, K. A. (2007). Interest for writing: How teachers can make a difference. *English Journal*, 79-85. doi:10.2307/30047170.

Maltese, A. V., & Tai, R. H. (2011). Pipeline persistence: Examining the association of educational experiences with earned degrees in STEM among US students. *Science Education, 95*(5), 877-907. doi:10.1002/sce.20441.

National Science Education Standards. (1996). *National science education standards.* Washington, DC: The National Academies Press. Retrieved from https://www.nap.edu/catalog/4962/national-science-education- standards.

NGSS Lead States. (2013). *Next generation science standards: For states, by states.* Washington, DC: National Academies Press. Retrieved from https://www.nap.edu/catalog/18290/next-generation-science-standards-for-states-by-states.

Oxford English Dictionary (OED). (2018). *Citizen Science.* Retrieved from: http://www.oed.com/view/Entry/33513?redirectedFrom=citizen+science#eid316619123

Shirk, J., Ballard, H., Wilderman, C., Phillips, T., Wiggins, A., Jordan, R., . . . Bonney, R. (2012). Public participation in scientific research: a framework for deliberate design. *Ecology and Society*, *17*(2): 29. doi:10.5751/ES-04705-170229.

Stromholt, S., & Bell, P. (2017). Designing for expansive science learning and identification across settings. *Cultural Studies of Science Education*, 1-33. doi:10.1007/s11422-017-9813-5

Swanson, A., Kosmala, M., Lintott, C., & Packer, C. (2016). A generalized approach for producing, quantifying, and validating citizen science data from wildlife images. *Conservation Biology*, *30*(3), 520-531. doi:10.1111/cobi.12695

The Cornell Lab of Ornithology (2018). *Citizen science central.* Retrieved from birds.cornell.edu/citscitoolkit/projects/find

Wenger, E. (2000). Communities of practice and social learning systems. *Organization*, *7*(2), 225-246. doi: 10.1177/135050840072002

Zooinverse (n.d.). *Galaxy zoo.* Retrieved from https://www.zooniverse.org/

SECTION II: MOTIVES FOR LEARNING IN CITIZEN SCIENCE PROGRAMS

In: Enhancing STEM Motivation through Citizen Science Programs ISBN: 978-1-53616-038-3
Editors: Suzanne E. Hiller and Anastasia Kitsantas

Chapter 3

MOTIVATION PROCESSES IN CITIZEN SCIENCE PROGRAMS: CURRENT STATUS AND FUTURE DIRECTIONS

Timothy J. Cleary*, Jacqueline Slemp and Jamal Waire
Rutgers University, NJ, US

ABSTRACT

Despite the growing interest among citizen science (CS) researchers regarding the key motivational processes underlying volunteer participation, there appears to be many gaps in this literature. In this chapter, the authors provide an overview of current models of CS engagement and participation and discuss the benefits of integrating these models with established motivational theories and constructs emphasized in the motivation and educational psychology literature. In addition to providing explicit descriptions of common motivational processes (i.e., subjective task values, self-efficacy, goal orientation, and intrinsic/extrinsic motivation), the authors use case scenarios to illustrate the natural link between these traditional motivational constructs and CS research. The authors also provide guidelines and recommendations for optimizing the design and evaluation of CS projects, specifically in terms of volunteer motivation and self-regulatory processes.

Keywords: citizen science, engagement, motivation, self-regulation, volunteer participation

Citizen science (CS) programs emphasize collaboration between professional scientists and citizen volunteers in science-based activities addressing biodiversity, the environment, and other important societal issues (Buytaert et al., 2014; Cappa, Laut, Nov, Giustiniano, & Porfiri, 2016; Carballo-Cardenas & Tobi, 2016; Hiller & Kitsantas, 2016). Although public engagement in science projects has a long history, CS projects have become increasingly popular in recent years because of the benefits gained by researchers, the field of science in

* Corresponding Author's E-mail: timothy.cleary@gsapp.rutgers.edu

general, and society as a whole (Cappa et al., 2016; Kobori et al., 2015). For example, assembling a large team of volunteers can greatly expand the quantity and geographic reach of data collection activities for a given project while also reducing costs. This non-traditional research paradigm also enables researchers to generate knowledge about emerging, real-world problems, such as the loss of biodiversity, climate change, or pollution, and to use such knowledge to develop potential solutions (Kobori et al., 2015).

Of particular interest to this chapter are the perceived benefits of CS participation to prospective volunteers, such as enhanced science literacy skills, self-awareness about environmental challenges, and feelings of satisfaction and pride in making an important contribution to the world in which they live (Cappa et al., 2016). In recent years, CS researchers have begun to delve deeper into the specific factors underlying CS volunteer motivation (Cappa et al., 2016; Carballo-Cardenas & Tobi, 2016; Dickinson, Crain, Yalowitz, & Cherry, 2013; Hiller & Kitsantas, 2014). Despite these recent efforts, motivation is still considered a fairly understudied topic in the CS literature, with Nov, Azary, and Anderson (2014) noting, "Within the contexts of online citizen science, research on motivations for participation is still in an early phase and empirical evidence is scarce" (p. 2). This paucity of research is somewhat surprising given that educational psychologists and motivational researchers in various fields have examined human motivation for decades.

Described as a "process whereby goal-directed activities are instigated and sustained" (Schunk, Meece, & Pintrich, 2014, p. 5), motivation reflects an internal process that results in particular actions, such as choice, effort, and persistence. Importantly, this definition distinguishes individuals' attempts to *initiate* or mobilize their effort to begin participating in an activity from their desire to *sustain* those efforts over time. This distinction parallels a recent focus in CS research on the specific factors or motives underlying short-term and long-term motivation among volunteer citizen scientists (Rotman et al., 2014). It is also important to underscore that motivation is closely intertwined with, yet distinct from, the concept of self-regulated learning (SRL), a process through which individuals seek to purposefully and strategically control their behaviors, thoughts, and environments to optimize goal attainment (Cleary, 2018; Zimmerman, 2000). Targeting both motivation and SRL should be of value to CS researchers given that each construct addresses key questions relevant to volunteer participation: "Why do citizens decide to volunteer?" (i.e., motivation) and "How skilled are volunteers at strategically and purposefully directing their own behaviors during assigned activities?" (i.e., SRL).

The primary objectives of this chapter are to address volunteer motivation and to illustrate the importance of integrating current models of CS engagement and participation with established motivational theories and constructs emphasized in educational contexts. We begin by providing a general overview of emerging motivation frameworks within the CS literature. We then summarize several core motivational processes emphasized in the educational psychology and motivation literatures and briefly highlight whether such processes have been targeted by CS researchers. Finally, we provide recommendations for optimizing the design and evaluation of CS projects, specifically in terms of volunteer motivation and SRL processes.

Models of Participation Engagement and Motivation in Citizen Science

CS researchers have identified several factors that influence the motivation of citizens to volunteer. Commonly cited motivators include the opportunity to express important values, to learn about and improve the world in which one lives, to further career options, and to enhance self-perceptions (Carballo-Cardenas & Tobi, 2016; Leao & Izadpahani, 2016; Nov et al., 2011). Other factors affecting volunteer motivation include the opportunity for enhanced social interaction and logistical or practical constraints, such as time availability or access to resources. To integrate and organize these motivation factors, CS researchers have proposed a variety of frameworks (Carballo-Cardenas & Tobi, 2016; Leao & Izadpahnai, 2016; Nov et al., 2014; Rotman et al., 2014). In this section, we provide a brief overview of a sampling of these emerging frameworks, underscoring the core motivational categories, factors, or processes emphasized within each.

Carballo-Cardenas and Tobi (2016) developed a model of participation in conservation CS that includes *personal motivations* (e.g., personal values reflected in projects, socializing with others, self-enhancement) as well as contextual factors classified as either *enabling* (e.g., time availability, access to equipment) or *constraining* (e.g., limited time, lack of communication with lead researcher). They argue that the specific motivations of volunteers to participate may vary across contexts and time; thus, it is critical to examine how personal values and meanings intersect with both positive and negative situational factors (Carballo-Cardenas & Tobi, 2016).

Leao and Izadpahnai (2016) developed a framework to delineate the reasons underlying participation or non-participation among volunteers for a mobile sensing project. This framework was similar to Carballo-Cardenas and Tobi's model in that it focuses on both personal (subjective) and contextual (means) factors; however, they use different terminology and assume that these two broad classes of factors (subjective and means) influence behavioral intentions, which in turn, influence the decision to participate. In this model, subjective factors include both *beliefs* (e.g., how can the project improve the quality of their lives?) and *concerns* (e.g., to what extent do volunteers value the environmental focus of the project?). Conversely, means factors pertain to logistical or contextual factors, such as having knowledge of and access to *technology* and the *time* needed to perform key roles on the project. Collectively, this set of factors can influence one's intentions and actual engagement or level of participation.

Nov et al.'s (2014) model of CS participation is unique in that it draws from two existing models: Klandermans' (1997) model of volunteer participation in social movements and Deci and Ryan's (2002) theory of self-determination that emphasizes intrinsic motivation. Using these two theoretical paradigms, Nov et al. (2014) emphasized four key motivational factors that influence both the quality and quantity of participation of CS volunteers, specifically in online projects: *collective motives* (i.e., perceived value of the collective goals of the project), *norm-oriented motives* (i.e., favorable reactions from others), *intrinsic rewards* (i.e., inherent enjoyment or interest in the project) and *extrinsic motives* (i.e., desired outcomes). The authors argued that these four broad motivational factors are key antecedents to participation in online CS projects.

Other CS researchers have conducted research with the potential to inform existing conceptual models of volunteer participation. As an example, in a recent case study analysis targeting citizen volunteers across three countries, Rotman et al. (2014) examined factors underlying initial and sustained motivation to engage in CS projects. A total of 44 interviews of volunteers across the three countries were conducted. Although this study did not examine a specific theoretical model or framework, their findings are relevant to expanding and refining existing frameworks. More specifically, they identified and categorized different motivations across the two broad dimensions of volunteer motivation: *initial engagement* and *sustained engagement*. Although some motivators appear to fit within each of the two broad categories, there were differences that emerged. Key initial participation motives included personal interest, self-promotion, self-efficacy, and social responsibility, whereas long-term participation appeared to be most heavily influenced by social factors or relationships cultivated within or external to the project. In developing models of CS engagement, the authors argued that one must not only consider the positive motivators for short-term and long-term participation but also the demotivating factors and cultural contexts (Rotman et al., 2014).

Each of the existing models of CS engagement is important in its own right because it organizes and integrates multiple motivators within a single framework that can stimulate and guide research. From our perspective, however, there is value in examining the extent to which the CS literature targets constructs emphasized in the motivation literature (e.g., subjective task values, self-efficacy, goal orientation, and intrinsic/extrinsic motivation; Schunk et al., 2014; Wentzel & Miele, 2016). Using a fairly similar nomenclature or set of motivational labels across fields and contexts can enhance the quality of communication among researchers and practitioners and may help to bridge gaps between different literatures.

Before exploring this issue, we would like to make two qualifying points. First, our intention for introducing specific constructs or theories in the broader motivation literature is not meant to critique or minimize the importance of emerging models of volunteer participation in the CS literature. Rather, we wish to underscore key motivational constructs that have been thoroughly researched over the past couple of decades and that we believe are relevant to CS research. Another qualifying point involves our recognition that several of the motives identified in the CS literature overlap to some degree with constructs targeted in the motivation literature; however, they often differ in terms of the specific labels or descriptors. By underscoring and discussing several well-established motivation constructs along with the broader theoretical paradigms or traditions from which they originate, we hope that CS researchers can broaden the theoretical foundation from which they conduct their research.

Overview of Motivation Processes

In the past decade, several edited and authored volumes have illustrated the importance of student motivation and engagement in educational contexts and settings (Christenson, Reschly, & Wylie, 2012; Elliot & Dweck, 2005; Schunk et al., 2014; Wentzel & Miele, 2016). Within the broader motivation literature, several important constructs have been shown to influence student behavior, ways of thinking, and achievement, such as competency or self-efficacy perceptions, subjective task values, intrinsic/extrinsic motivation, goal orientation, causal attributions, self-theories or conceptions of ability, and need for autonomy and relatedness. Typically, these constructs have been studied in relation to achievement motivation and striving for competence. However, we believe that many of them can also help to address why individuals ultimately decide to become CS volunteers.

Table 1. Definitions of Motivational Constructs

Construct	Definition	Applied example
Subjective task value	Value or importance of an activity. Includes attainment value, intrinsic/interest value, utility value, and cost	
Attainment value	Perceived importance of an activity based on personal identity or sense of self	Doug perceives the overall mission/goals of the project to be aligned with his core values and sense of identity.
Intrinsic/interest value	Perceived level of enjoyment or interest for a given activity	Doug finds the data collection protocols to be fun and interesting.
Utility value	Perceived usefulness or importance of activity	Doug's participation opens doors to further his career and teaches him skills necessary for graduate school.
Cost	Perceived effort or negative experience resulting from participation	Doug will need to spend much time and money to support travel to participate in the program.
Self-efficacy	Task and context-specific perceptions of one's capability to perform behaviors at a specific level of performance	Doug believes he is good at working with others but doubts his abilities to operate the necessary data-recording devices.
Goal orientation	Pattern of beliefs, attitudes, and emotions underlying why and how individuals approach and engage in an activity; includes mastery and performance goals	
Mastery goal orientation	Beliefs that ability is malleable, challenging activities are fun, and that self-improvement, growth, and developing knowledge and insight are important	Doug wants to deepen his knowledge of conservation activities and appreciates a good challenge. He has confidence that he will learn requisite skills over time.
Performance goal orientation	Beliefs that outperforming others and receiving praise/recognition from others is important and that challenging activities are aversive and anxiety-provoking	John believes he is not talented in science, and he is worried that others will perceive him as "dumb." He does not like taking on difficult challenges.
Intrinsic/extrinsic motivation	Types of motivation that reflect a distinction between being motivated for the sake of the activity or to obtain a desired outcome	
Intrinsic motivation	The desire to engage in an activity for the sake of participating	Doug participates due to his love and interest in science and conservation activities.
Extrinsic motivation	The desire to engage in an activity as a means to an end	Doug participates due to the desire for a reward (external regulation) or to enhance his standing among colleagues (introjected regulation).

Given that it is beyond the scope of this chapter to review all motivational constructs relevant to CS initiatives, we elected to conduct a narrower review emphasizing the following four constructs: subjective task values, self-efficacy, goal orientation, and intrinsic/extrinsic motivation (Table 1).

In addition to providing definitions and brief descriptions of these four motivational constructs, we further convey their meaning through a case scenario of Doug, a 26-year-old science teacher at a local high school. Doug was recently identified by school administrators as an outstanding teacher and had previously expressed an interest in becoming the district's science curriculum supervisor. Recently, Doug's best friend told him about an opportunity to participate in a conservation CS project examining the habitats of endangered fish. This project was of interest to Doug, and he was considering whether or not to participate. Doug has a wife and two children and spends much of his leisure time coaching his childrens' sports teams. We purposefully use this vignette to clarify the meaning of each motivational construct and to help illustrate how such constructs can be applied to real-life CS projects and research activities.

Subjective Task Values

Eccles and colleagues developed a general expectancy-value model of achievement choices. From this perspective, the decisions individuals make in educational, vocational, and other contexts are most heavily influenced by the individuals' expectations for success (similar to self-efficacy but more future-oriented) and the value or importance of an activity to those individuals. Given that we discuss expectations (i.e., self-efficacy) in a subsequent section, we focus our attention in this section on subjective task values. Subjective task values is an umbrella term that includes four general types: attainment value, intrinsic value, utility value, and cost. *Attainment value* is defined as the importance of an activity based on personal identity or sense of self (Wigfield, Hoa, & Klauda, 2008). For example, Doug would exhibit a greater chance of volunteering in the conservation CS project if he perceives the overall mission and goals of the project to be closely aligned with his core values and sense of identity, or if participation helped to fulfill one or more of his needs (Eccles, 2005).

Intrinsic or *interest value* is typically defined in terms of the level of enjoyment or interest that individuals experience when performing an activity (Eccles, 2005; Hidi, 2006). Thus, if Doug believes that the specific data collection protocols are fun and interesting, he may be more likely to initiate and sustain his desire to participate.

Utility value represents a third subjective task value and is defined as the usefulness or importance of an activity to the person. For Doug, the conservation activity could have utility value if it provides opportunities to further his career (i.e., to become the curriculum supervisor in his school district) or to develop scientific knowledge or skills that can assist him in getting accepted into a graduate degree program (Eccles, 2005).

Finally, subjective task values also include the *costs* associated with participating in a given activity. Similar to constructs used in current models of volunteer participation in CS (Carballo-Cardenas & Tobi, 2016; Rotman et al., 2014), a cost is defined in terms of what the individual has to relinquish or give up, or as an experience perceived in a negative way by that person, such as required effort and mental energy or fear of negative evaluations by others. For Doug, the key costs for participation include the time commitment and the expenditure of money to support his travel to the data collection sites.

Collectively, the four task values address the question of, "What about this activity makes me want to engage?" or more simply, "Why should I participate in this citizen science project?" Research shows that although subjective task values are related to achievement, they also greatly influence decision-making and choice behaviors (Eccles, 2005; Schunk et al., 2014). Further, considering and assessing *all four* of these values is important given that any one of them can lead individuals to decline a request to participate or sustain that interest. For example, even though Doug believes that engaging in conservation activities is a core aspect of who he is as a person (attainment value), participating in the current conservation project will enhance his knowledge and skills as a science teacher (utility value), and collecting data in the natural world is enjoyable (interest), he may ultimately decide to decline to participate because he would not have time due to working full-time and coaching his childrens' sports teams (time cost) and because he would not get reimbursed for the travel associated with the project (financial cost).

Self-Efficacy

Self-efficacy beliefs pertain to one's perceptions of his/her capability to perform specific behaviors or to attain a certain level of performance (Bandura, 1997). It represents the things that individuals perceive that they can capably perform in the present moment, rather than what they think they can do in general or what they intend to do in the future (Zimmerman, 2000). In short, self-efficacy addresses the question, "Can I perform the required activities successfully?" Although these types of self-perceptions are emphasized in various theoretical frameworks, they play a central role in Bandura's social-cognitive paradigm (Bandura, 1986, 1997). Decades of research have shown the influential role of self-efficacy perceptions on decision-making and choice behaviors in students (Bandura, 1997; Hiller & Kitsantas, 2014; Lent, Brown, & Hackett, 2000; Zimmerman & Cleary, 2006). When students believe that they possess the capability to perform key behaviors relative to a particular activity, they are more likely to choose to participate in that activity and to display the necessary effort and persistence to succeed (Zimmerman & Cleary, 2006).

Self-efficacy perceptions are also task- and context-specific. That is, individuals may possess adequate confidence to perform certain behaviors or skills for one activity but weak efficacy beliefs for other skills or in other situations (Cleary, 2009; Pajares & Urdan, 2006). Consider Doug and his decision to participate in the conservation project. For this project, individuals need to efficiently use sophisticated data-recording technology and are expected to carefully adhere to specific procedures regarding when and what to record during a three-hour time block. Given that individuals need to work in small groups during data collection, it is also important that they be able to interact effectively with others. From Doug's perspective, he works extremely well with others and is very knowledgeable about the scientific method in general. However, he has some self-doubts about his ability to operate the data-recording devices appropriately and often struggles when asked to sustain his attention over long periods of time. In short, it is important for CS researchers to not only understand the various types of activities that they ask volunteers to perform but to also be cognizant of how volunteers may differ in their perceptions of capability to perform these activities (Cleary, 2009).

Goal Orientation

Goal orientation is conceptualized as a multi-faceted concept involving a pattern of beliefs, attitudes, and emotions that lead individuals to approach activities in a certain way (Pintrich, 2003; Senko & Tropiano, 2016). This concept is a key aspect of achievement motivation theory and traditionally has included two broad types of goals (i.e., mastery and performance). Each goal type is represented by a unique pattern of reasons for engaging in an activity, types of standards used to judge performance (e.g., normative, mastery), and overall attitudes about learning (Dweck & Leggett, 1988; Elliott & Dweck, 1988). Individuals with a *mastery goal orientation* believe that ability is malleable, challenging activities are fun, and that self-improvement, growth and developing knowledge and insight are of utmost importance. Conversely, a *performance goal orientation* involves a pattern of beliefs and emotions characterized by a focus on outperforming others, receiving praise and recognition from others, viewing challenges as aversive, and fearing mistakes or failure. Generally speaking, mastery goals tend to be viewed as more adaptive than performance goals, although there is evidence that performance goals can be effective in certain situations or contexts (Senko & Tropiano, 2016).

Suppose Doug asked his friend John, a social studies teacher at the same high school, to participate in the conservation project. Doug was excited to begin the project because he wants to deepen his knowledge of conservation activities, and he appreciates a good challenge, particularly when the challenge involves science-related tasks or information. Even though Doug realizes that he does not possess all of the requisite skills to perform the data collection tasks, he has confidence that he will naturally learn to do so over time. Unlike Doug's mastery goal approach, John believes he is not very talented in science. He worries that he will be working with people who are smarter and more knowledgeable than him and thinks he will look "dumb." It is also problematic that John does not envision himself as a "numbers guy" and does not like taking on difficult challenges. In short, Doug's more adaptive goal orientation stimulates his enthusiastic and positive approach to the conservation activity, whereas John's pattern of beliefs and attitudes could potentially dissuade him from volunteering.

Intrinsic and Extrinsic Motivation

Intrinsic and extrinsic motivation represent another spoke of the motivation umbrella, although they are conceptualized as types of motivation rather than sources of motivation (Eccles, 2005). Generally speaking, intrinsic motivation refers to the desire to engage in some activity simply for the sake of participating in that activity, whereas extrinsic motivation reflects more of a "means to an end" concept (Deci & Ryan, 2002; Schunk et al., 2014). Self-determination theory, which can be thought of as an integration of multiple mini-theories, has played a key role in enhancing our understanding of this intrinsic versus extrinsic distinction (Deci & Ryan, 2002). From this perspective, motivation operates along a continuum ranging from amotivation (i.e., lacking intention to act), to several forms of extrinsic motivation (i.e., external regulation, introjected regulation, identified regulation, and integrated regulation) to intrinsic motivation. This model provides an extensive framework from which to assess the nature and level of an individual's autonomous and self-determining behaviors as they pertain to volunteer participation. As an example, if Doug wanted to participate in the conservation project in order to obtain some type of reward, he would be considered externally regulated, the most extreme type of extrinsic motivation. But if his primary motivator for participation involved his desire to enhance his standing or perception among science colleagues at the high school, then this would be considered a more advanced type of extrinsic motivation called introjected regulation. Finally, if Doug's participation was derived largely from his love for and interest in science and conservation activities, then he would be considered intrinsically motivated. The major implication from this theoretical paradigm is that the extent to which an individual's sense of autonomy and self-determination can be supported will play an influential role in the development of intrinsic motivation and subsequent decisions to participate in a given project.

ADDRESSING MOTIVATION CONSTRUCTS IN CITIZEN SCIENCE RESEARCH

We conducted a systematic literature search to examine the extent to which CS research specifically references common motivational constructs emphasized in the motivational literature. It is important to underscore that this search did not use multiple methods (e.g., database search, reference lists from identified studies) to identify all possible motivation studies in CS, as is typically done when conducting comprehensive systematic reviews or meta-analyses. Rather, our goal was simply to develop a broad understanding of whether the CS literature addresses many of the key motivational constructs introduced in this chapter and the potential implications of findings from our analysis.

We conducted an online search using the following broad databases: EBSCOhost (e.g., Academic Search Premier, ERIC, Teacher Reference Center), ProQuest (e.g., Natural Science Collection, Agricultural & Environmental Science Database), Scopus, and Web of Science (e.g., Web of Science Core Collection, SciELO Citation Index). Although we included general search terms (e.g., motivation, engagement), we were primarily interested in identifying the number of studies that targeted specific motivational constructs (e.g., self-efficacy, task value). Asterisks were also used in order to ensure our search was not restricted to specific words. For example, we used "motiv*" to cover potential variations in the term motivation, such as motivate, motive, motivation, and motivations. The specific search term combinations included: "Citizen Science" AND (motiv* OR engage*), "Citizen Science" AND ("task value" OR "task interest"), "Citizen Science" AND "self-efficacy", "Citizen Science" AND ("goal orientation" OR "goal setting"), "Citizen Science" AND ("intrinsic motivation" OR "extrinsic motivation"), and "Citizen Science" AND ("growth mindset" OR "fixed mindset" OR "mindset" OR "growth motivation").

For the broad search terms (i.e., citizen science, engagement, motivation), we found 3,252 hits. Although we did not review this entire list of studies to eliminate duplicate studies across databases, it appears the general issue of motivation or engagement is an important one in the

CS literature. Of greater relevance for this paper, however, was the number of studies referencing particular motivational constructs, such as task value/interest, self-efficacy, goal setting/orientation, intrinsic/extrinsic motivation, and mindset. A total of 176 studies were identified across the different databases. The second and third authors for this chapter independently reviewed all study abstracts to determine acceptability for our review. Given our interest in determining the extent to which empirically-focused motivation CS studies have been conducted, the reviewers used the following eligibility criteria: (a) descriptive and/or inferential statistics were conducted (qualitative studies, case studies, or review articles were not included), (b) volunteer motivation was the target, and (c) formal measures of the motivation constructs were utilized. Using these criteria, only six studies were retained. Collectively, the six studies addressed self-efficacy (with two of these studies also including task interest), and intrinsic or extrinsic motivation. Although two of the studies addressing self-efficacy (Hiller & Kitsantas, 2014, 2016) also included measures of subjective task values (i.e., interest, outcome expectations), no additional CS studies that targeted subjective task values, goal setting, goal orientation, or mindset met eligibility criteria. In the following section, we briefly review the six studies and their findings, and discuss the two Hiller and Kitsantas (2014, 2016) studies within the context of self-efficacy.

Self-Efficacy in Citizen Science Programs

Self-efficacy has been evaluated in the field of CS as an outcome for volunteers who participate in these programs. Collectively, three of the applicable studies aimed to not only better understand self-efficacy as a result of CS programs but to also understand its' relationship to other variables, such as student achievement and science, technology, engineering, and mathematics career motivation (Hiller & Kitsantas, 2014, 2016; Lynch, Dauer, Babchuk, Heng-Moss, & Golick, 2018). Two of these three studies evaluated middle school students' self-efficacy for scientific observation skills in an outdoor learning context (a horseshoe crab CS program; Hiller & Kitsantas, 2014, 2016) but also included measures of task interest and outcome expectations (i.e., a type of task value). The third study addressed adult CS volunteers in an entomology CS program and used a mixed methods approach, including a pretest-posttest design and interviews (Lynch et al., 2018).

Hiller and Kitsantas (2014) conducted a pretest-posttest quasi-experimental study with 86 eighth-grade students targeting the relations between a horseshoe crab CS program and student achievement and career motivation. The duration of this CS intervention was one day, which included participants receiving content instruction and modeling from field experts and data collection protocol training from teachers. After modeling occurred, students participated in data collection which involved measuring specific parts of the horseshoe crabs and to identify their gender and relative age. The authors used the Citizen Science Self-Efficacy Scale (CSSES) to evaluate student beliefs about their ability to engage in scientific observation skills (Hiller, 2012). Relative to the students' perceptions of efficacy, two key results emerged. Students who participated in this CS project exhibited statistically significant higher levels of self-efficacy for scientific observation skills when compared to the control group ($p < .001$). Further, students' self-efficacy perceptions were found to exhibit strong relations to their overall interest in CS projects as well as their content knowledge.

Hiller and Kitsantas (2016) conducted a validation study of the CSSES. For this measure, students were asked to use a 5-point Likert scale to rate how well they can conduct specific activities relevant to the target project, such as locating horseshoe crabs or finding differences between horseshoe crabs. Factor analyses revealed a one-factor scale with a Cronbach's alpha reliability coefficient of .87. The authors also showed that self-efficacy perceptions were significantly correlated with task interest, outcomes expectations, and career goals.

In a separate study that focused on adult citizen scientists, Lynch et al. (2018) used a mixed-methods approach to evaluate how an entomology CS program affected participant science self-efficacy and self-efficacy for environmental action, among other variables. Their approach included a pretest-posttest quasi-experimental design and a qualitative analysis of data generated from semi-structured interviews. There were six CS programs that were included in this study. Though the exact details of the programs were variable, they all required the volunteers to monitor a target organism, collect entomological data, and make submissions to their leaders. The 28 volunteers were all over 19 years old and new to their respective programs. The authors noted discrepancies in changes in self-efficacy perceptions based on the measures deployed. As part of an interview protocol, the adults reported higher *perceived* science self-efficacy and self-efficacy for environmental action following participation in the CS projects. However, when self-report measures were used, no statistically significant change in self-efficacy was noted; a finding that could potentially be explained by the very small sample used in the study. In short, although researchers have begun to explore the role of self-efficacy in CS participation, additional research is needed to more fully understand its influence in general and across a variety of CS programs and contexts.

Intrinsic and Extrinsic Motivation in Citizen Science Programs

Three CS studies were identified that specifically addressed intrinsic or extrinsic motivation (Beza et al., 2017; Nov et al., 2014; Tiago, Gouveia, Capinha, Santos-Reis, & Pereira, 2017). Collectively, these studies support the notion that several factors often underscore or promote volunteer participation in CS projects. Two of these studies suggest that differences in the value placed on egoistic motivational factors (e.g., intrinsic, extrinsic) may be attributed to geographical location (Beza et al., 2017; Tiago et al., 2017). Tiago et al. (2017) also added that the assessment of motivational factors may yield different results depending on the degree of initial engagement in CS projects.

Beza et al. (2017) assessed farmers from three different countries (e.g., India, Ethiopia, Honduras) and their motivation to participate in a CS project with the purpose of observing differences within and across the countries. The researchers conducted semi-structured interviews with 426 farmers involved in the *Seeds for Needs* initiative. Both quantitative and qualitative analyses were conducted. The questionnaire utilized in this study was developed based on previous literature of motivational factors influencing volunteer participation in CS programs. The researchers identified seven unique motivational factors embedded within three broad types of motivation: egoistic (e.g., intrinsic, extrinsic), collectivistic, and altruistic. This study revealed an exhaustive list of differences among the motives that were most salient to the farmers based on several characteristics, including the country in which they were located and level of education. For the purposes of this chapter, the findings on the differences among the egoistic motives are of particular importance. For instance, the Ethiopian and Honduran farmers both ranked the egoistic intrinsic type of motivation (i.e., sharing information) higher than collectivistic, altruistic, or egoistic extrinsic types of motivation. On the other hand, the Indian farmers valued collectivistic motives more than both egoistic intrinsic and egoistic extrinsic motivations; these results suggest that CS researchers may need to consider various sociocultural variables to more fully understand the differential influence of motivational processes on the the participation of citizen scientists (Beza et al., 2017).

As previously discussed, Nov et al. (2014) conducted a longitudinal study to examine the association between intrinsic and extrinsic motivation with the quantity and quality of participant contribution. Data were collected from three CS projects, including *Stardust@home* ($n = 139$), the *Citizen Weather Observer Program* (CWOP; $n = 2390$), and the *Berkeley Open Infrastructure for Network Computing* (BOINC; $n = 1843$). The authors reported that intrinsic motivation was associated with greater *quantity* of contribution in the projects. There was variable support indicating that a specific type of extrinsic motivation (e.g., reputation) was associated with greater *quality* of contribution. The authors hypothesized that this latter relationship is dependent on how performance is measured and made publicly available. However, there was evidence to suggest a "crowding in" effect, as extrinsic motivations were found to potentially influence intrinsic motivation in a positive way. In other words, the reward or feedback system improved the participants' connection to the organization; thus, it was argued that the intervention improved intrinsic motivation.

Tiago et al. (2017) conducted a study that focused specifically on intrinsic motivation. They surveyed 149 participants involved in *BioDiversity4All*, a Portuguese CS project, and examined differences in intrinsic motives relative to gender, age, level of education, and level of participation. For this study, the researchers adapted the Intrinsic Motivation Inventory (IMI) to make it more relevant to CS and biodiversity activities. The IMI is an instrument designed to assess several constructs related to intrinsic motivation, including interest/enjoyment, perceived choice, perceived competence, pressure/tension, effort, value/usefulness, and relatedness. The authors did not find any statistical differences in the intrinsic motivation categories across gender, age, or level of education. Conversely, significant differences were found across levels of participation (i.e., never participate, participate occasionally, participate regularly, and participate a lot) for each of the IMI categories, with the exception of interest/enjoyment. Level of participation in the project was assessed through a self-report survey. The group of participants who had the highest level of participation in the project also reported the highest levels of intrinsic motivation in all categories except effort/importance and perceived choice. These findings further demonstrate the link between intrinsic motivation and level of participation/quantity of contribution.

In conclusion, although an emerging segment of the CS literature conveys that researchers are highly interested in the topic of volunteer engagement and motivation, our review of the CS literature suggests that there is a limited number of studies directly targeting and assessing the relations among common constructs emphasized in the motivation literature. As we noted previously, we acknowledge that our review process may have resulted in an underestimate of the motivation-related CS studies that do exist. However, what is clear is that the empirical CS literature is only beginning to scratch the surface regarding volunteer motivation and that much more work is needed within this realm. In the following section, we outline recommendations for designing programs that enhance citizen participation.

Tips for Designing Motivation-Enhanced Citizen Science Programs

Recommendation One – Using a Common Motivational Nomenclature

As previously emphasized, CS researchers recognize the importance and need to examine the motivation of volunteer participants. Further, many models of volunteer participation have been developed, with each touching upon key motives underlying such participation. In reviewing these models, it is evident that many of these motives closely correspond with established constructs emphasized in the motivation literature including subjective task values (e.g., interest, utility, attainment cost), self-efficacy, goal orientation, and intrinsic/extrinsic motivations. As one example, Carballo-Cardenas and Tobi's (2016) model highlights general categories, such as personal motivations, enabling factors, and constraining factors. Specific motives within these categories appear to correspond closely to subjective task values: attainment value (i.e., acting on important personal values), utility value (e.g., career building; self-enhancement), and cost (e.g., time constraints).

Along a similar vein, the Nov et al. (2014) model identifies motivators that appear to overlap with goal orientation (i.e., norm-oriented motives or expectations about the reactions of others), intrinsic/extrinsic motivation (i.e., intrinsic motivators, reputation) and subjective task values including task interest (i.e., intrinsic motives or the enjoyment of the activity) and attainment value (i.e., collective motives or the overall importance of the project to the person). Importantly, across most models, social interaction and relationships represent key motivators for many citizen scientists.

Despite this level of overlap between the CS and motivation literatures, the timing is ripe for considering ways to expand and further refine this level of integration. There are several advantages for using established motivational constructs in CS research. In addition to enhancing communication and cross-discipline collaborations among researchers, using a common motivational nomenclature would enable researchers to assess whether specific motivational processes operate in similar ways across traditional (e.g., K-12 contexts) and nontraditional learning contexts (e.g., citizen science; Rosenzweig & Wigfield, 2016). Examining the generalizability of research findings across contexts, situations, and/or cultures has been recognized as a contemporary issue of value to both researchers and practitioners (McInerney & King, 2017; Rotman et al., 2014). Further, because the specific constructs emphasized in the motivation literature (e.g., self-efficacy, intrinsic motivation, subjective task values) are typically embedded within broad, comprehensive theoretical models (Schunk & Zimmerman, 2008; Wentzel & Miele, 2016), CS researchers can greatly enrich the theoretical and conceptual foundation from which they conduct their research.

Recommendation Two – Designing Motivationally-Robust Citizen Science Programs

We encourage CS researchers to think about two overarching issues when developing the design, methodology, and infrastructure of their projects: (a) the need to identify multiple motivational constructs most relevant to the project, and (b) the need to systematically and directly target these motivational processes as a natural part of the project.

Identifying multiple motivational constructs. Motivation is an individualized process involving the intersection of multiple within-person and contextual factors. To put it simply, a group of citizens may volunteer to participate in the conservation project for different reasons. Whereas Doug may be primarily motivated because of the close alignment between his personal identity or core beliefs with the essence of conservation projects, another individual may be motivated to participate due to their need for affiliation or relationship. At a project-specific level, individuals may also show differences in their motivation to perform the various roles or tasks for the given project, such as Doug being more motivated to perform data collection tasks than data entry (Carballo-Cardenas & Tobi, 2016; Rotman et al., 2014). Thus, consistent with recent trends in the motivation literature focusing on multi-process motivation interventions (Cleary, Velardi, & Schnaidman, 2017; Linnenbrink-Garcia & Patall, 2016; Rosenzweig & Wigfield, 2016), we encourage CS researchers to identify and target multiple motivational processes that are most germane to their study. Rosenzweig and Wigfield (2016) underscored the importance of multi-construct motivation approaches noting, "We believe an advantage of these interventions compared to single-construct interventions is that targeting multiple motivational constructs at once might 'reach' more students in classrooms where different students face different motivational challenges" (p. 155). Although this conclusion was based on research in traditional classroom settings, we believe that a multiple-motivation process focus in CS environments is particularly valuable because of the complexity, diversity, and nature of such environments.

To assist CS researchers in the selection of target motivational processes, we encourage them to consider example guiding questions for motivational constructs presented in Table 2. For example, suppose the researchers leading the conservation project with Doug and other volunteers are fully aware of the complexity of their data collection technology and procedures. In reviewing the list of motivation processes and related questions in Table 2, they concluded that they needed to take extra steps to ensure that all volunteers: (a) feel efficacious to perform the different tasks (i.e., self-efficacy), (b) recognize that their abilities or skills will naturally improve and grow with practice (i.e., growth mindset), and (c) receive some type of reimbursement for travel expenses (i.e., reduced cost). Further, because the CS researchers also recognized that some participants prefer to work in small groups (i.e., high need for relationships) whereas others are motivated to work alone (i.e., low need for relationships), they sought to devise ways to clearly communicate the social dimensions of each potential activity.

Applying motivation principles to your project. To effectively apply motivation principles within a CS project, researchers need to possess adequate knowledge of specific motivational processes and should consider ways to concurrently target several of them. Relative to the CS conservation project with Doug, suppose the researchers identified three motivation processes as most desirable to target: self-efficacy, cost, and relatedness. Using self-efficacy as an example, Bandura (1997) underscored four sources of self-efficacy: masteryexperiences, verbal persuasion, vicarious experiences, and physiological responses, with mastery experiences or opportunity to demonstrate success being the most powerful (Bandura, 1997; Schunk et al., 2014). Using this knowledge, the CS researchers took several steps to bolster the confidence of their volunteers as they participated in the project.

Table 2. Guiding Questions for the Selection of Motivational Processes in CS Projects

Motivation construct	Questions that address conceptual meaning of motivational processes	Questions that prompt motivationally-enhanced practices
Self-efficacy	Do the volunteers perceive that they possess the capability to perform all tasks and activities required for the project?	Are there ways in which I can help volunteers experience success or help them believe that they will be successful on the various activities in the project?
Mastery goal orientation	Do the volunteers want to participate because they love learning and simply want to better their scientific skills?	What can I do to prompt volunteers to focus primarily on the quality of their learning and use of scientific skills?
Performance goal orientation	Are the volunteers primarily focused on making sure they appear competent or capable to others?	Are there ways in which I can minimize normative comparison between the volunteers?
Interest/affect	Do the volunteers perceive the assigned tasks or activities to be fun and enjoyable?	Are there ways in which I can increase the enjoyment or level of interest among volunteers in the project?
Attainment value	Do the volunteers believe that the project closely aligns with their identity or core personal characteristics and beliefs?	Are there ways to help volunteers recognize the link between their core identity and the essence/goals of the project?
Utility value	Do the volunteers realize that participating in the project will lead to important outcomes or can help them attain future goals?	What can I do to help volunteers to clearly recognize that their participation will have value or importance to them in some way?
Cost	Do the volunteers perceive many drawbacks or "negatives" to participating in the project?	Is there anything I can do to minimize or lessen the costs associated with the project?
Growth mindset	To what extent do volunteers believe that their "science intelligence" or capabilities can improve with practice?	Are there ways in which I can help volunteers recognize that their "science" or "data collection" abilities are changeable?
Need for relationships	To what extent do the volunteers prefer to engage in collaborative, social exchanges as part of the project?	Can I identify volunteer activities that involve social and independent components, and then allocate task assignments based on that information?
Extrinsic factors	To what extent are the volunteers excited to receive things like praise, money or other tangible items due to the participation?	What types of extrinsic supports can I realistically provide to increase volunteer participation?

In addition to providing modeling and training opportunities for the volunteers to practice their data collection skills, the CS researchers purposefully created multiple opportunities for their volunteers to experience success. They provided periodic feedback to volunteers emphasizing growth in their accuracy in using the data collection equipment. They also broke down the data collection activity into discrete skills so that volunteers could more easily recognize their strengths and weaknesses when using the technology (Cleary, 2018).

To tap into the volunteers' need for affiliation or relationship, the researchers identified three activities that were most amenable to social interaction and collaboration among volunteers and two activities that were more solitary or individualized in nature. Upon agreeing to participate in the project, the CS researchers gave the volunteers the opportunity to choose the type of activity based on their preferences for social interaction. Finally, because time constraints were identified by CS researchers as a major cost to participating in prior projects, the researchers devised ways to shorten the time blocks when observations would be required and to allow greater flexibility regarding when the individuals could volunteer.

Recommendation Three – Developing Specific Motivation Assessment Tools

In addition to optimizing the motivationally-enhancing features of CS programs, researchers should also strive to systematically assess volunteer motivation before, during, and after participation. To our knowledge, there are very few reliable and valid assessment tools to measure the vast array of motivation processes relevant to CS volunteers, although CS-related self-efficacy measures have been emphasized in recent years (Hiller & Kitsantas, 2014, 2016; Lynch et al., 2018). As noted previously, Hiller and Kitsantas (2016) developed and validated a self-efficacy scale in the context of outdoor learning with middle school students. The nature of the items used in this measure underscore a core principle in self-efficacy assessments; that is, self-efficacy items should reflect a range of specific behaviors or skills that individuals are expected to display during a given activity (Hiller & Kitsantas, 2016). It is important for researchers to continue addressing the need for developing CS-specific assessment tools that address the wide range of motivational processes underlying volunteer engagement across contexts.

Recommendation Four – Combining Motivational "Will" with Regulatory "Skill"

From our perspective, motivation is a necessary but not sufficient condition for enhancing the production and work of CS volunteers. Given that many CS activities require a high level of autonomy and independence on the part of volunteers, it is important that volunteers possess adequate self-regulated learning (SRL) skills. SRL is conceptualized as a cyclical, goal-directed process whereby individuals strive to exert strategic control over their learning and performance (Cleary, 2018; Panadero, 2017). It consists of a series of sub-processes, such as goal-setting, strategy use, motivational beliefs, self-monitoring, and self-reflection, that work in concert to help individuals control and manage their thoughts, behaviors, feelings, and learning contexts.

Although each of these SRL sub-processes are important to CS research and should be used in combination when possible to support CS-related activities and initiatives (Cleary, 2018), it is well beyond the scope of this chapter to address each in detail. From our perspective, however, we believe that CS researchers should pay particular attention to self-monitoring. Defined as a process of tracking one's performance, personal process (e.g., thoughts, affect), and behavior, as well as the contextual factors surrounding learning and performance (Zimmerman, 2000), self-monitoring is an effective way for volunteers to enhance their awareness of their behaviors and performance. That is, it helps people generate information about their personal strengths and weaknesses when performing an assigned CS activity. This information is of high functional value because it can be used to help volunteers make more effective decisions regarding how to adapt or improve their skills for that activity.

Because CS activities often involve people independently engaging in research-related activities without access to immediate feedback, researchers can help to create a type of "self-regulatory structure" in their projects via self-monitoring tools. For example, suppose Doug was asked to operate a data-recording device that involved a series of steps and procedures. To help Doug monitor the core aspects of the required activity, the researcher gave Doug a checklist with each of the key procedures listed and two additional columns for recording purposes. The first column prompted Doug to check off whether he completed a given step of the procedure, while the second column provided space for Doug to note any challenges he encountered during a given step. By utilizing the worksheet in this way, Doug was naturally prompted to focus on the core aspects of his assigned role while concurrently gathering useful formative assessment data that he and/or the researchers could use to identify and overcome barriers to participation.

CONCLUSION

In this chapter, we underscored the value of integrating current CS research with the broader motivation literature. Based on our review of the CS literature, it clear that participant motivation is an emerging topic of interest among researchers, but much more work is needed. To start, researchers can adopt a more parallel or common nomenclature regarding target motivational constructs. Such an initiative can enhance communication and collaboration between experts across disciplines, increase opportunities to conduct more theoretically-grounded research, and provide a foundation for examining the generalizability of motivation principles across fields.

Ultimately, we recommend that CS researchers consider using motivational theory and constructs to: (a) guide the development of CS program components and infrastructure, and (b) assess changes in volunteer motivation during and following the CS project. Thus, it is important for researchers to be mindful of how CS projects can be designed to optimize the thinking and perceptions of volunteers regarding their confidence to perform assigned roles, understanding of the importance and value of the project, the cost-benefit ratio of participation, and how the project aligns with personal identity or core beliefs. By developing assessment tools specific to CS projects, researchers would also be able to identify the specific factors causing high levels of attrition and to gain knowledge about specific areas of motivation that are of critical importance to various groups or types of volunteers. For example, if a CS researcher administered a baseline set of motivation measures to volunteers to assess their self-efficacy and subjective task values (i.e., perception of interest, utility, cost), the researcher may be able to determine specific areas or issues that need to be considered or addressed before the project begins.

It is also relevant for CS researchers to distinguish between the *will* and *skill* needed to perform at an effective level as a volunteer participant (Weinstein & Acee, 2013). Volunteers need to be motivated (i.e., will) in order to mobilize and sustain their effort over time, yet they also need to be able to manage, direct, and adapt their behaviors and thinking during their CS experiences (i.e., regulatory skill).

REFERENCES

Bandura, A. (1986). *Social foundations of thought and action: A social cognitive theory*. Englewood Cliffs, NJ: Prentice Hall.

Bandura, A. (1997). *Self-efficacy: The exercise of control*. New York, NY: W.H. Freeman and Company.

Beza, E., Steinke, J., van Etten, J., Reidsma, P., Fadda, C., Mittra, S. & Kooistra, L. (2017). What are the prospects for citizen science in agriculture? Evidence from three continents on motivation and mobile telephone use of resource-poor farmers. *PloS ONE*, *12*(5). doi:10.1371/journal.pone.0175700

Buytaert, W., Zulkafli, Z., Grainger, S., Acosta, L., Alemie, T. C., Bastiaensen, J. ... & Zhumanova, M. (2014). Citizen science in hydrology and water resources: Opportunities for knowledge generation, ecosystem service management, and sustainable development. *Fronteirs in Earth Science*, *2*(26). doi:10.3389/feart.2014.00026

Cappa, F., Laut, J., Nov, O., Giustiniano, L. & Porfiri, M. (2016). Activating social strategies: Face-to-face interaction in technology-mediated citizen science. *Journal of Environmental Management*, *182*, 374-384. doi:10.1016/j.jenvman.2016.07.092

Carballo-Cardenas, E. C. & Tobi, H. (2016). Citizen science regarding invasive lionfish in Dutch Caribbean MPAs: Drivers and barriers to participation. *Ocean & Coastal Management*, *133*, 114-127. doi:10.1016/j.ocecoaman.2016.09.014

Christenson, S. L., Reschly, A. L. & Wylie, C. A. (2012). *The handbook of research on student engagement.* New York, NY: Springer Science.

Cleary, T. J. (2009). Monitoring trends and accuracy of self-efficacy beliefs during interventions: Advantages and potential applications to school-based settings. *Psychology in the Schools*, *46*(2), 154-171. doi:10.1002/pits.20360

Cleary, T. J. (2018). *The self-regulated learning guide: Teaching students to think in the language of strategies.* New York, NY: Routledge.

Cleary, T., J., Velardi, B. & Schnaidman, B. (2017). Effects of the Self-Regulation Empowerment Program (SREP) on middle school students' strategic skills, self-efficacy, and mathematics achievement. *Journal of School Psychology*, *64*, 28-42. doi:10.1016/j.jsp.2017.04.00.

Deci, E. L. & Ryan, R. M. (2002). *Handbook of self-determination research.* Rochester, NY: The University of Rochester Press.

Dickinson, J. L., Crain, R., Yalowitz, S. & Cherry, T. M. (2013). How framing climate change influences citizen scientists' intentions to do something about it. *The Journal of Environmental Education*, *44*(3), 145-158. doi:10.1080/00958964.2012/742032

Dweck, C. S. & Leggett, E. L. (1988). A social-cognitive approach to motivation and personality. *Psychological Review*, *95*(2), 256-273. doi: 10.1037//0033-295x.95.2.256.

Eccles, J. S. (2005). Subjective task value and the Eccles et al. model of achievement-related choices. In A. J. Elliot & C. S. Dweck (Eds.), *Handbook of competence and motivation* (1st ed., pp. 105-121). New York, NY: The Guilford Press.

Elliot A. J. & Dweck, C. S. (2005). *Handbook of competence and motivation* (1st ed.). New York, NY: The Guilford Press.

Elliott, E. S. & Dweck, C. S. (1988). Goals: An approach to motivation and achievement. *Journal of Personality and Social Psychology*, *54*, 5-12. doi:10.1037//0022-3514.54.1.5

Hidi, S. (2006). Interest: A unique motivational variable. *Educational Research Review*, *1*(2), 69-82. doi:10.1016/j.edurev.2006.09.001

Hiller, S. E. (2012). *The impact of a citizen science program on student achievement and motivation: A social cognitive career perspective*. (Dissertation thesis). Retrieved from Proquest Dissertations & Theses Global. (UMI Number 3547088)

Hiller, S. E. & Kitsantas, A. (2014). The effect of a horseshoe crab citizen science program on student science performance and STEM career motivation. *School Science and Mathematics Journal*, *114*(6), 302-311. doi:10.1111/ssm.12081

Hiller S. E. & Kitsantas, A. (2016). The validation of the citizen science self-efficacy scale (CSSES). *International Journal of Environmental & Science Education*, *11*(5), 543-558. doi:10.12973/ijese.2016.405a

Klandermans, B. (1997). *The social psychology of protest*. Oxford, England: Blackwell Publishers.

Kobori, H., Dickinson, J. L., Washitani, I., Sakurai, R., Amano, T., Komatsu, N. & Miller-Rushing, A. J. (2015). Citizen science: A new approach to advance ecology, education, and conservation. *Ecological Research*, *31*(1), 1-19. doi:10.1007/s11284-015-1314-y

Leao, S. & Izadpahani, P. (2016). Factors motivating citizen engagement in mobile sensing: Insights from a survey of non-participants. *Journal of Urban Technology*, *23*(4), 85-103. doi:10.1080/10630732.2016.1175824

Lent, R. W., Brown, S. D. & Hackett, G. (2000). Contextual supports and barriers to carreer choice: A social cognitive analysis. *Journal of Counseling Psychology*, *47*(1), 36-49. doi:10.1037/0022-0167.47.1.36

Linnenbrink-Garcia, E. & Patall, E. (2016). Motivation. In L. Corno & E. Anderman (Eds.), *Handbook of educational psychology*, (pp. 91-103). New York, NY: Routledge.

Lynch, L. I., Dauer, J. M., Babchuk, W. A., Heng-Moss, T. & Golick, D. (2018). In their own words: The significance of participant perceptions in assessing entomology citizen science learning outcomes using a mixed methods approach. *Insects*, *9*(16). doi:10.3390/insects9010016

McInerney, D. M. & King, R. B. (2017). Culture and self-regulation in educational contexts. In D. H. Schunk & J. A. Greene (Eds.), *Handbook of self-regulation of learning and performance*, (2nd ed., pp. 485-502). New York, NY: Routledge.

Nov, O., Arazy, O. & Anderson, D. (2011). Technology-mediated citizen science participation: a motivational model. Proceedings of the *AAAI International Conference on Weblogs and Social Media* (ICWSM 2011). Barcelona, Spain, July 2011.

Nov, O., Azary, O. & Anderson, D. (2014). Scientists@Home: What drives the quantity and quality of online citizen science participation? *PloS ONE*, *9*(4). doi:10.1371/journal.pone.0090375

Pajares, F. & Urdan, T. C. (2006). *Self-efficacy beliefs of adolescents.* Greenwich, CT: Information Age Publishing, Inc.

Panadero, E. (2017). A review of self-regulated learning: Six models and four directions for research. *Frontiers in Psychology*, *8*, 422. doi:10.3389/fpsyg.2017.00422

Pintrich, P. R. (2003). A motivational science perspective on the role of student motivation in learning and teaching contexts. *Journal of Educational Psychology*, *95*(4), 667-686. doi:10.1037/0022-0663.95.4.667

Rosenzweig, E. Q. & Wigfield, A. (2016). STEM motivation interventions for adolescents: A promising start, but further to go. *Educational Psychologist*, *51*(2), 146-163. doi:10.1080/00461520.2016.1154792

Rotman, D., Hammock, J., Preece, J., Hansen, D., Boston, C., Bowser, A. & He, Y. (2014). Motivations affecting initial and long-term participation in citizen science projects in three countries. In *iConference 2014 Proceedings*, 110-124. doi:10.9776/14054

Schunk, D. H., Meece, J. L. & Pintrich, P. R. (2014). *Motivation in education: Theory, research, and applications.* Boston, MA: Pearson.

Schunk, D. H. & Zimmerman, B. J. (2008). *Motivation and self-regulated learning: Theory, research, and applications.* New York, NY: Lawrence Erlbaum Associates.

Senko, C. & Tropiano, K. L. (2016). Comparing three models of achievement goals: Goal orientations, goal standards, and goal complexes. *Journal of Educational Psychology*, *108*(8), 1178-1192.

Tiago, P., Gouveia, M. J., Capinha, C., Santos-Reis, M. & Pereira, H. M. (2017). The influence of motivational factors on the frequency of participation in citizen science activities. *Nature Conservation*, *18*, 61-78. doi: 10.3897/natureconservation.18.13429

Weinstein, C. E. & Acee, T. W. (2013). Helping college students become more strategic and self-regulated learners. In H. Bembenutty, T. J. Cleary, & A. Kitsantas (Eds.), *Applications of self-regulated learning across diverse disciplines: A tribute to Barry J. Zimmerman*, (pp. 197–236). Charlotte, NC: Information Age Publishing, Inc.

Wentzel, K. R. & Miele, D. B. (2016). *Handbook of motivation at school* (2nd ed.). New York, NY: Routledge.

Wigfield, A., Hoa, L. W. & Klauda, S. L. (2008). The role of achievement values in the regulation of achievement behaviors. In D. H. Schunk, & B. J. Zimmerman (Eds.), *Motivation and self-regulated learning: Theory, research, and applications*, (pp. 169-195). New York, NY: Erlbaum.

Zimmerman, B. J. (2000). Attaining self-regulation: A social cognitive perspective. In M. Boekart, P. R. Pintrich, and M. Zeidner (Eds.) *Handbook of self-regulation.* New York, NY: Academic Press.

Zimmerman, B. J. & Cleary, T. J. (2006). Adolescents' development of personal agency: The role of self-efficacy beliefs and self-regulatory skill. In F. Pajares & T. C. Urdan (Eds.), *Self-efficacy beliefs of adolescents.* Greenwich, CT: Information Age Publishing, Inc.

In: Enhancing STEM Motivation through Citizen Science Programs ISBN: 978-1-53616-038-3
Editors: Suzanne E. Hiller and Anastasia Kitsantas

Chapter 4

ENSURING HEALTHY SOCIAL DYNAMICS AND MOTIVATION IN YOUTH CITIZEN SCIENCE PROGRAMS

Serena Shim[1,*] and Hyunju Lee[2]
[1]Ball State University, IN, US
[2]Ewha Womans University, Korea

ABSTRACT

Citizen science programs have become increasingly popular over the last few years, and have been a popular model for various informal science education programs. Recent research evidence indicates that such programs are effective in enhancing students' science motivation and engagement. For citizen science programs to be maximally effective, students' social motivation and the social dimensions of their learning environment must be taken into account. Accordingly, the current chapter features four theories of motivation and adolescent development, specifically self-determination theory, achievement goal orientation theory, stereotype theory, and social capital theory, and provides practical instructional recommendations for educators who aim to implement citizens science programs in formal school settings. Based on these theories, we offer recommendations on instructional planning, assessment, social climate, and working and building partnerships with the community. We also emphasize how citizen science programs can provide a supportive learning environment for underprivileged youth.

Keywords: academic motivation, adolescent development, citizen science program, social motivation

With the radical development of science and technology, we are facing various unexpected risks and conflicts almost every day. For example, communities are grappling with the danger

[*] Corresponding Author's Email: sshim@bsu.edu

of nuclear power accidents (e.g., 2012 in Fukushima Japan), pollution and destruction of the environment, dangers associated with genetic engineering, and irreversible chemical contamination. Many countries around the world are looking for solutions to solve the problems stemming from advancements in science and technology. Some citizens are calling for those at risk from these problems to directly participate in the problem-solving and policy-making related to the issues. Accordingly, non-governmental organizations (NGO) and other organizations were founded to engage the general public and experts in carrying out scientific activities to resolve issues. Such collaborative research efforts are referred to as citizen science. Scientists and volunteers often collaborate through scientific data collection or other forms of participation in authentic research projects (Dickinson et al., 2012; Raddick et al., 2013). Most citizen science projects have been carried out with adult volunteers who are concerned about the issue at hand or are very enthusiastic about scientific research and want to continue developing their interests, knowledge, and skills such as *Budburst*, (Chicago Botanic Garden, n.d.) and *The Great Sunflower Project* (n.d.).

Over the past few decades, many science educators and citizen scientists have looked into the possibility of involving students in citizen science activities in a more formalized setting (e.g., classrooms, afterschool programs; Lorenz, 2016). Educating students to become responsible citizens and change agents who actively discuss science and technology issues and enact solutions for them is critical (Elmose & Roth, 2005; Roth, 2003; Roth & Lee, 2004). Citizen science can be effective in boosting students' scientific literacy, problem solving skills, and interest in scientific issues (Zeidler, Sadler, Simmons, & Howes, 2005). There is also initial empirical evidence supporting the effects of citizen science programs on student science motivation, achievement, and STEM career interests (Hiller & Kitsantas, 2014; Jeanpierre, Oberhauser, & Freeman, 2005; Wallace & Bodzin, 2017).

However, special attention must be taken if citizen science programs are to be conducted with youth in school settings (Shah & Martinez, 2016; Sutton, 2009). Although many educators promote introducing citizen science into a more formalized setting based on the prior research showcasing the educational benefits of citizen science projects, the results from studies about adult volunteer-based citizen science projects may not necessarily be generalizable to student populations. There are likely to be logistical and practical challenges to be addressed before citizen science programs can be successfully integrated into formalized school settings (Gray, Nicosia, & Jordan, 2012; Mueller, Tippins, & Bryan, 2012). To date, many important questions remain unanswered: How can we as educators motivate students' interest and participation in citizen science activities? What teaching strategies should be used for students who are not interested in science? How will we encourage their participation in learning by promoting interaction between a teacher and students or students and students? What classroom environment should we create? In this chapter, we aim to answer these key questions drawing from the theories of motivation and adolescent development.

To answer these guiding questions, we featured four major theories of motivation and adolescent development (i.e., self-determination theory, achievement goal orientation theory, stereotype threat theory, social capital theory) and provide recommendations and guidelines for best practices. Our ultimate goal is to offer guidelines to optimize educational benefits of citizen science programs for students. While doing so, we will place particular emphasis on the importance of healthy social relationships and interpersonal ambience within the learning environment. As we will elaborate later, social-relational dimensions of the learning environment greatly impact the academic outcomes of the students (Baumeister & Leary, 1995;

Reis, Collins, & Berscheid, 2000; Roseth, Johnson, & Johnson, 2008; Ryan & Patrick, 2001), and academic motivation often has social roots (Juvonen, 2006; Urdan & Maehr, 1995). All theories featured in this chapter emphasize social elements as one of the key dimensions for students' adjustment, motivation, and learning.

STUDENT INVOLVEMENT IN CITIZEN SCIENCE

Citizen science typically refers to research collaboration between scientists and volunteers, often through scientific data collection or other forms of participation. Many organizations and nonprofit organizations have made efforts to involve lay people and scientists concerned about societal issues closely connected to science (e.g., environmental issues) in research and scientific activities aimed at resolving the problems. For instance, SAFECAST (n.d.) is an international organization launched by voluntary citizens who recognized the danger of nuclear power plants after the explosion of Fukushima Daiichi reactor in 2011. The primary aim of the organization is to raise people as scientifically and technically competent citizens who are able to monitor their environment and manage their lives in healthier and safer ways. Toward this end, they continue monitoring and collecting data on environmental radiation and sharing the information with the public and experts.

Across Europe and North America, there have been efforts to implement educational programs offering students experience in citizen science activities. Although student participation is not as critical as the adult volunteers' role in the citizen science projects, these programs offer students invaluable opportunities to conduct scientific research and/or have debates with their peers, community members, or experts to resolve issues. PARRISE (Promoting Attainment of Responsible Research and Innovation in Science Education, 2014-2017) is a good example. This is a four-year project aiming to introduce the concept of "Responsible Research and Innovation (RRI)" in primary and secondary education. This project combines inquiry-based learning and citizenship education with socio-scientific issues in science education and has been implemented in European Union countries such as Sweden, Hungary, and Germany. It emphasizes that the various actors of society work together to ensure the research processes and results are well-suited to the values, needs, and expectations of the European Community. In the PARRISE project, students have the opportunity to participate in the research process as well as discuss the issues surrounding this process of inquiry (e.g., science- and technology-related innovations). The knowledge gained from these discussions allows them to grow into democratic citizens with the scientific literacy needed to effectively discuss the issues. Another example is STEPWISE (Science and Technology Education Promoting Wellbeing for Individuals, Societies, and Environments, n.d.) in Canada. The STEPWISE program emphasizes the need for students to cultivate activism, which enables them to critically recognize and solve problems posed by science and technology so they can grow into truly competent citizens. The fundamental assumption is that students should understand the nature of problems inherent in science and technology (e.g., imbalance, inequality, and capitalist viewpoint) through conducting their own research.

Citizen science is increasing its offerings to a wide range of content areas, beyond ecology and environmental sciences. There is a clear upward trend of building partnerships among citizen scientists, community organizations (e.g., 4D, Scouts), and schools to involve students in citizen science programs. Nonetheless, there has been a lack of effort to use motivation and adolescent development theories to guide the design and implementation of the citizen science programs. In the following sections, we introduce four theories of motivation and adolescent development (i.e., self-determination theory, achievement goal theory, stereotype threat theory, social capital theory) and provide key recommendations drawing from these theories. In the field of motivation, there is no shortage of theories (See Linnenbrink-Garcia & Patall, 2015 for the major theories in the field of motivation). As it will become clear in the following sections, we have selected these four theories because they consider social elements as a central piece in explaining how and why students are motivated to do well and become interested in science.

Although these guidelines are paired with a specific theory in this chapter, many of the guidelines are supported by more than one theory. We also described the essence of the theories, followed by the instructional strategies and recommendations drawn from each theory. Note that it was not our intention to offer detailed and comprehensive overviews of the theories, as they are beyond the scope of this chapter and also available elsewhere (Kindermann, 2015; Linnenbrink-Garcia & Patall, 2015). Although informal educators (e.g., youth workers, community educators, afterschool program coordinators) don't identify themselves as teachers and are referred to with various titles (e.g., social worker, program directors, coordinators), we referred to all individuals who assume the role of instructor as "teacher" for brevity. For the same reason, we will also refer to the various learning contexts (e.g., workshop, camp, afterschool program site) as "classroom" in this chapter. We used the terms "participants" and "students" interchangeably.

Self-Determination Theory

Overview of the Theory

Self-determination theory (Deci & Ryan, 2000), one of the predominant theories of motivation, is useful to understand the various kinds of motivation that students may bring to the table and how to help students to develop interest. In this theory, broadly, three levels of motivation are distinguished: amotivation, autonomous motivation, and controlled motivation. Amotivation represents a lack of motivation. Autonomous motivation indicates voluntary and willful participation. This occurs when students fully accept the value of the activity (autonomous extrinsic motivation) or truly enjoy the activity (intrinsic motivation). Controlled motivation indicates that the participation is due to external or internal rewards and punishments (the value of the activity is not fully internalized by the student, non-autonomous extrinsic motivation; Deci & Ryan, 2000).

Expectedly, amotivation (lack of motivation) poses the most serious challenge to the educators, as students who are not motivated are likely to withdraw from the activity or participate half-heartedly if withdrawal is impossible. They may engage in disruptive behavior out of boredom. On the contrary, the students with autonomous motivation will be easy to work with. The students with the autonomous form of motivation are likely to show adaptive academic outcomes, such as high effort, deep conceptual learning strategies, and achievement (e.g., Barron, 2006; Corpus, McClintic-Gilbert, & Hayenga, 2009; Csikszentmihalyi, Rathunde, & Whalen, 1993; Lepper, Corpus, & Iyengar, 2005; Lipstein & Renninger, 2006;

Otis, Grouzet, & Pelletier, 2005; Renninger & Hidi, 2002; Walker, Greene, & Mansell, 2006). Students with controlled motivation (e.g., students participate to earn reward or avoid failure, or out of sense of obligation or guilt) will participate and may show some level of engagement. However, given that they did not truly internalize the value of the task, their engagement might not be high quality. Students with controlled motivation tend to show negative outcomes such as the use of maladaptive learning strategies, negative attitudes, high anxiety, inability to cope with challenges, low academic achievement, and even school dropout (e.g., Lepper et al., 2005; Ryan & Connell, 1989; Vansteenkiste, Zhou, Lens, & Soenens, 2005; Walker et al., 2006). From this theory, the key question is how to move students' motivation from amotivation or controlled motivation to more autonomous motivation. Needless to say, student engagement and learning outcomes are maximized when they participate out of intellectual curiosity and genuine interest in the topic.

The type of motivation exhibited by citizen science participants may differ between adult volunteers and students. Adult volunteers who *choose* to participate in the projects are already interested in the project per se. When citizen science projects are incorporated into school settings (e.g., enrichment program to regular curriculum, homework assignment, or afterschool program), *all* students (regardless of their intention to participate in a specific project) are likely to participate due to the structure and nature of formal school settings. This means, unlike adult volunteers, not all students will bring the enthusiasm and initial interest for the project.

Then the key question is how teachers can help students develop a more desirable form of motivation. Citizen science projects tackle real-world problems, and thus have high potential to draw students' interest and engagement. Nonetheless, educators cannot assume that all students will bring autonomous motivation to the citizen science projects. Some students may show amotivation or undesirable controlled motivation. Due to their more advanced cognitive abilities, older children may have an advantage in adopting more autonomous regulation compared to younger children (Loevinger & Blasi, 1991). However, autonomous motivation is not a natural maturation process and does not always occur as children get older. Some students may like science not for its content and learning activities but for social reasons, such as a desire to please the teacher or parents (in the self-determination theory term, introjected regulation; see Deci & Ryan, 2000 for more information). For teachers, granting students autonomy and relinquishing control may not be easy. Science teachers who had been used to lecture-based instruction have to change how they teach their students and may feel afraid to hand over control to their students (Lee & Yang, 2019).

Practical Applications and Recommendations

Make projects personally meaningful. When students find the activity fun, it will lead to autonomous motivation. Science might be an extremely fun subject for some students but not for others. Considering the developmental trend of domain-specific interest and self-concept (Marsh, 1990a, 1990b, 1993; Marsh, Trautwein, Lüdtke, Köller, & Baumert, 2005), most children and early adolescents do not possess a fully developed domain-specific interest or self-concept in science. Thus, teachers must take actions to support students' developing science self-concepts and nurture their interest in science. This should be viewed as one of the key objectives when implementing citizen science programs. However, depending on the end-goal of the instructional activity, this might not always be possible. For example, it might be difficult to make continuous data collection fun as students need to follow predetermined procedures closely. The good news is that students can still be autonomously motivated if they see the meaning or purpose behind the activities. If teachers provide the rationale for doing something (even if it is a boring task), students feel respected and their sense of autonomy is enhanced, which in turn leads to intrinsic motivation (Niemiec & Ryan, 2009). Although it seems to be simple to do, prior research has found that teachers rarely provide reasons for their activities or lessons (Urdan, 2004).

Citizen science projects vary in the activities in which students are involved. Some citizen science projects are hypothesis-driven, and students collect the data to answer specific questions (e.g., *Lost Ladybug Project*, Cornell University, 2018). Other kinds of projects may focus on environmental monitoring (e.g., water quality monitoring; see Jollymore, Haines, Satterfield, & Johnson, 2017), and students collect the data to depict the status quo or the trend of change. In either case, teachers cannot assume that the ultimate goal or meaning of the project would be self-evident to students. Teachers can explicitly communicate to the students what the goals of projects are and how they impact them personally. Teachers can also ask students to come up with meaningful rationale instead of offering one (Hulleman, Godes, Hendricks, & Harackiewicz, 2010). Teachers can consider administering a short reflective writing exercise (e.g., having students generate and write about how the topic covered in the class is beneficial or useful to them personally), followed by group discussion (see Hulleman et al., 2010, for research evidence for this writing exercise) for this purpose.

Many citizen science projects have meaningful causes (e.g., conservation and environmental sustainability) which impact students themselves, their families, and communities. In these kinds of citizen science projects, students are likely to show autonomous motivation. Roth and Lee (2004) provide a great example. They conducted water projects for seventh grade students in a small city in Henderson Creek, Canada. In recent years, mechanical maintenance shops and biotechnology laboratories have been built on the shore in the region, which has resulted in coastal erosion and pollution around farmland. Such changes had worsened coastal erosion during the winter rainy season, and local residents had to move their personal belongings to avoid flooding. Also, during the dry season in summer, the water level of the river was lowered and the temperature of the river increased, so the residents had to pay extra tax for the irrigation system. Water shortage in the summer required a limit on water use by local residents. To solve the problem, a scientist began a water project at Henderson Creek with local residents and local environmental law and policy experts at a local university. As members of the community, the students understood the seriousness of the water problem, so they defined the purposes and goals of the project, used local resources and tools, and shared roles in conducting the project. Throughout the project, personal relevance and interdependence were emphasized. Such emphasis of personal relevance led students to take an initiative, beyond simply following teacher direction.

Research by Sperling and Bencze (2010) is another great example. They conducted a case study of seventh graders who were involved in self-determined investigations on the topic of waste management. They emphasized action-taking to promote personal, social, and/or environmental well-being. Personal relevance was emphasized through receiving positive response and feedback about their projects from their family or community members. This feedback loop reinforced their role as a leader and change-agent. Participating students reported that they have witnessed changes in family members' actions to reduce waste, which further enhanced students' own motivation and engagement in the project. In addition, many students reported an increased sense of self-efficacy.

Respect and empower students. To support sense of autonomy, teachers must trust and respect students. Allowing choice is an effective way to empower students and support their sense of autonomy. The provision of choice and ultimately students' experience of autonomy leads to enhanced motivation, achievement, and subsequent learning (Cordova & Lepper, 1996; Deci & Ryan, 1985; Iyengar & Lepper, 1999; Patall, Cooper, & Wynn, 2010). However, one may find issues with the number of choices that can be offered given that students' roles are limited to data collection in most citizen science projects. Furthermore, the data collection must be done reliably, and students must adhere to the data collection protocol to ensure data quality. In that case, teachers can devise and incorporate other pedagogical activities and experiences with some built-in choices. For example, teachers can allow students to choose the medium to present their data (e.g., paper, poster, PowerPoint), let them come up with the panels on their own for the upcoming discussion, or choose the community setting to share their findings, such as city hall, school event day, or museums (Kim & Lee, 2017a; Kim, Lee, & Kim, 2013). However, it is important to note that meaningless choices, choices that students find irrelevant, or choices with subtle pressures with them should not be offered (Assor, Kaplan, & Roth, 2002).

When students are given flexibility and choice in what and how they would do their research, they become autonomous learners, and their learning outcomes are enhanced. Sperling and Bencze (2010) supported the link between student autonomy and motivation. A student's quote demonstrated the internal sense of autonomy, afforded by the project, "We all got to show *our point of view on things*. If we had a problem with something, we got a chance to fix it and we got a chance to *make our own plan* how to fix it" (Sperling & Bencze, 2010, p. 263). There are many ways that teachers can support students' autonomy and incorporate student perspectives in designing and implementing the projects. Participation by choice is inherent in the adult volunteer-based citizen science projects. However, students do not always participate in school-based citizen science programs by their own choice, and thus teachers must make an effort to offer viable choice of activities that can appeal to a variety of students.

ACHIEVEMENT GOAL ORIENTATION THEORY

Overview of the Theory

Classrooms can vary in terms of quality of emotional and affective interaction among students within the classroom. Some classrooms may have hostile social climates, and other classrooms may have more supportive and caring social climates. What might explain the differences in the quality of collective peer relationships across classrooms? Achievement goal orientation theory provides a useful lens to explain such classroom variability. This theory focuses on the "why" and "underlying purpose" of engagement which is classified as either performance goal oriented or mastery goal oriented (Linnenbrink-Garcia & Patall, 2015; Schunk, Meece, & Pintrich, 2014).

Teachers can create an environment where either mastery goals (desire to learn as much as one can and get better) are valued or performance goals (desire to outperform others or match their success with less effort) are emphasized and encouraged. Such different goal focuses set in motion different patterns of beliefs and behaviors (Senko & Hulleman, 2013). Mastery versus performance goal focused classrooms convey different messages to the students about the meaning of success (mastery of content versus doing better than others), value of effort (effort is valuable in its own right versus success with minimum effort is preferred because it signals high ability), and view of errors and mistakes (natural part of learning versus indication of low ability; Ames, 1992; Anderman & Maehr, 1994; Dweck, 1986; Nicholls, 1984; Schunk et al., 2014; Urdan, 1997). When students focus on learning, mastery, and making progress, without being fettered by concerns about social comparison and others' judgement of their ability, they tend to choose challenging tasks, persist through setbacks, and experience positive affect during the learning process. On the contrary, under the performance goal framework, students tend to believe that errors and mistakes reflect a lack of innate ability, which cannot be altered by effort. Hence, they may choose tasks that are overly easy or difficult to avoid having people infer that they lack innate ability (Meece, Anderman, & Anderman, 2006). Performance approach goal students are less likely to enjoy the learning process and show resilience after experiencing failure. Prior research has consistently found the pedagogical benefits of mastery goal focused classrooms. Mastery goal-oriented students are likely to persist, use deep processing, seek challenge, and enjoy the learning process while performance goal-oriented students are likely to show a maladaptive approach to the task, withdraw effort, give up prematurely, avoid challenging tasks, and experience negative affect (Harackiewicz, Barron, Tauer, Carter, & Elliot, 2000; Meece & Miller, 2001; Midgley et al., 1998; Wolters, 2004).

Then, the critical question is how teachers can promote mastery goals and discourage performance goals among their students in their classrooms. A variety of instructional practices can directly and indirectly influence students' achievement goal adoption. Mastery goals can be promoted through meaningful learning rather than rote memorization, student autonomy rather than teacher control, collaboration among peers rather than competition, use of heterogeneous grouping rather than homogenous grouping, and adoption of a grading system rewarding personal progress and effort rather than social comparison among students. Use of competitive grading system (e.g., curving the grades), use of ability grouping, and public display of the students' relative ranking on the exam are likely to foster performance goals among students.

A sizable literature attests to the benefits of mastery goal focused classroom environments (Meece, Anderman, & Anderman, 2006). When a classroom is mastery focused, students tend to show adaptive patterns of motivation and engagement (e.g., high persistence, effort, challenge seeking, and adaptive learning strategies but low disruptive behavior and cheating). On the contrary, classroom emphasis of performance goals has been linked to maladaptive patterns of student outcomes and avoidant behaviors (e.g., high self-handicapping, cheating, and procrastination but low persistence; for reviews, see Anderman & Wolters, 2006; Meece et al., 2006). Furthermore, emphasis on performance goals in the classroom obstructs the formation of supportive and collaborative social climate (Patrick, Anderman, Ryan, Edelin, & Midgley, 2001; Patrick, Kaplan, & Ryan, 2011). Classroom performance goal focus predicts hostile peer climates (Shim, Kiefer, & Wang, 2013). It seems that classroom goal focus shapes students' working models about how they should relate to and interact with others in the classroom. Mastery goal orientation relates to healthy peer relationships (e.g., trust and sharing), and performance goals relates to mistrust, inconsideration, and tension between friends (Levy-Tossman, Kaplan, & Assor, 2007). Competitive learning environments tend to trigger anxiety, which could hinder students' performance (Aronson & Patnoe, 1997; Cohen,

1994). Such effects are consistent with the premise of social interdependence theory (Johnson & Johnson, 1989, 2005), which postulates that more healthy and desirable peer interactions prevail when cooperative goals are imposed in the learning context (Roseth et al., 2008, for a meta-analysis). These findings are also consistent with the broader literature on teachers' roles in guiding and developing the classroom social ecology and students' peer interactions (Farmer, Lines, & Hamm, 2011).

Mastery versus performance goal focus in the classroom affects the nature of interactions among students. When a classroom emphasizes normative ability and comparison, students are likely to be concerned about appearing incompetent in front of others rather than actual learning. Such concerns make students unwilling to ask questions or share what they think with others, and hinder free exchange of ideas and discussion, which is critical to knowledge construction (Butler, 1998; Karabenick, 2004; Nelson-Le Gall, 1985; Newman, 2000; Ryan & Pintrich, 1997; Ryan, La Guardia, Solky-Butzel, Chirkov, & Kim, 2005; Shim et al., 2013). Research has shown that maladaptive peer interaction around academic tasks can be prevalent and problematic, especially during early adolescence in which the concerns about self-worth and fitting into peer group peak (Brown, Bakken, Ameringer, & Mahon, 2008; Steinberg & Silverberg, 1986).

Practical Applications and Recommendations

Emphasize learning, not competition. Citizen science programs, with their focus on hands-on investigations and real-world problems, are likely to promote mastery goals among students. These kinds of projects will help teachers challenge students' concept of science learning as memorization and regurgitation and encourage them to embrace and welcome the uncertainties and difficulties of generating scientific knowledge.

In the traditional science classroom, especially in Asian countries, academic achievement in science is often heavily emphasized. In order to effectively implement citizen science, teachers need to create an atmosphere where expression of ideas and feelings, diverse perspectives, mutual respect, and reflection on the learning process are valued and promoted. In the context of informal citizen science programs, discussion activities or collaborative activities are often utilized as a wrap-up or enrichment activity. During debates or discussion activities, students sometimes feel a sense of rivalry and get carried away with the desire to win the debate. Whenever such a pattern is observed, teachers can gently remind students to focus on learning and exchange of ideas. It is important that students see these projects as important processes in solving the problems facing the community, rather than merely something to "win." In all stages of citizen science projects, teachers should emphasize what they have learned during the process and how newly learned insight or knowledge can contribute to the community in solving the real-world problems.

Frequent use of authentic and informal assessment for learning. Often the term assessment is equated with test. However, assessment is a much broader term. Assessment can serve various purposes, not just to gauge students' current levels of understanding but also to promote self-regulation and learning. For this to happen, various forms of assessment should be considered. For example, in addition to formal assessments (e.g., quiz for all students), frequent informal assessments (e.g., observation, questioning of some students to check understanding) can be conducted to better support students' learning. To promote mastery goals, teachers can offer frequent progress feedback to students and avoid making social-comparative references when evaluating students' performance (e.g., student A did better than student B). Students have gaps in their knowledge and skills, and such gaps can be filled through frequent feedback. For example, despite the provision of protocols and training, students still make errors resulting in unreliable data (Galloway, Tudor, & Haegen, 2006; Newman, Buesching, & MacDonald, 2003; Penrose & Call, 1995). When students in Grades 3-10 were trained to survey an Oregon white oak in the state of Washington, their measurements and tree counts turned out to be accurate. However, they overestimated the rare pines and larger oaks (Galloway et al., 2006). This is not surprising, as students are not professionals after all. Some citizen science projects have built-in feedback loops to offer critical feedback to the students. When involving students in citizen science projects, such assessment must be pre-planned.

The provided feedback should be more than a single score summing up the overall student performance. The assessment should provide useful information to fix errors or address misconceptions. For that purpose, detailed rubrics can be used to inform students about the gap in their knowledge or skills. That way, students can learn to evaluate their own performance and take necessary actions to remedy their knowledge or skill gap (e.g., seek help, deliberate practice).

One of the great strengths of citizen science projects is that teachers do not have to be the only source of feedback. Verbal interactions with the experts and external partners can provide great feedback for students' performance. Critical feedback must be framed as reflection of high standards and teachers' confidence in students' ability to meet them (Cohen & Steele, 2002; Cohen, Steele, & Ross, 1999; Yeager, Trzesniewski, Tirri, Nokelainen, & Dweck, 2011).

Assessment does not have to be limited to knowledge and skills. Behaviors, affect, attitudes, and motivation should be an integral part of assessment to promote students' long-term interest and achievement in science.

Involve students in a variety of meaningful activities. The best way to promote mastery goals is to engage students in a variety of meaningful tasks. Citizen science projects showcase the concept of authentic learning. Through participating in citizen science projects, students can connect what they are taught in school to real-world problems. The learning activities they participate in closely mirror the complexities of real life. Students not only acquire knowledge, but also participate in discourse and performance, solve problems, create products, and spread knowledge.

Bouillion and Gomez (2001) provide an excellent example. In this project, elementary school students worked to solve a problem facing their community (i.e., contaminated water on the Chicago River overflowing into their residential areas). The students explored the area in cooperation with the Chicago River officials, returned to the school, wrote down their observations, and discussed with their peers. By participating in these diverse activities, the students learned important scientific concepts (e.g., water quality, soil erosion), became more interested in science, and had the experience of working together to create useful living knowledge (i.e., a practical knowledge that connects lived experience on the ground with the wider body of national and international critical knowledge) in their community. A discussion activity can be easily incorporated into instructional activities, especially if the project involves the issue with broader societal consequence. Roth and Lee (2004) had students discuss the issues with the community members. An excerpt taken from the student interviews clearly shows the student's authentic interest and deep concern for the issue that they have been researching:

> I worked very hard on the map and proceedings. During this course I learned about fieldwork: I learned how to collect samples of the creek and take temperatures and speed. I also did some work with the community. It taught me about working with others and working in the community. I noticed that ever since our Henderson Creek article was published in the Peninsula News Review that the public has begun to notice the creek. (Roth & Lee, 2004, p. 284)

Taken together, involving students in various learning activities (beyond the data collection) can spark students' interest in the topic and boost their willingness to act on the issues facing their communities.

STEREOTYPE THREAT THEORY

Overview of the Theory

The number of jobs in science, technology, engineering, and math fields has continued to increase since 2000. However, females and racial minorities are under-represented in these jobs, relative to their presence in the overall US workforce (Funk & Parker, 2018). According to 2016 U. S. News/Raytheon STEM index (Neuhauser & Cook, 2016), the gap in the STEM field persists: Overall, women lag behind men in the number of STEM degrees granted, exam scores, and general interest in the STEM fields. On those three indicators, Black, Hispanic, and American Indian students lag behind White and Asian students.

Many of these gaps, especially gender differences, cannot be fully explained by differences in achievement and are likely to be related to motivational processes (Thoman, Smith, Brown, Chase, & Lee, 2013). Stereotype threat theory (Steele & Aronson, 1995) postulates that minority students' performance is compromised by the apprehension that their performance might confirm the negative stereotypes held against their group in the society (termed as stereotype threat). Such stereotypes may include the beliefs that women are not as talented in science or Blacks are intellectually inferior than Whites. Stereotypes cause students to underperform on tests due to stress and test anxiety (Bell, Spencer, Iserman, & Logel, 2003; Osborne, 2001). The stereotype threat effects have been shown to impact a variety of stereotyped groups. Blacks and Latinos underperform when they are in the minority compared to Whites, and low SES students' performance was compromised on tests when the tests were framed as a diagnostic test of global intellectual ability (Aronson, Quinn, & Spencer, 1998; Croizet, & Claire, 1998; Gonzales, Blanton, & Williams, 2002; Steele & Aronson, 2000). Asian womens' performance was impacted either positively or negatively depending on whether they were negatively stereotyped for being a woman or positivity stereotyped for their ethnicity (Shih, Pittinsky, & Ambady, 1999).

A substantial body of research suggests that socialization and discrimination contribute to these gender and racial differences in scientific domains. Influential adults (e.g., parents and teachers) who espouse stereotypes are likely to remind their children and students of such stereotypes, and hence, they may unintentionally induce stereotype threat. Teachers' implicit "science is for males" stereotypes are positively related with males' self-concept and intrinsic value but negatively associated with females' motivational beliefs (Thomas, 2017). If continuously reminded of the negative stereotype that their group has limited competence, and thus does not belong in mathematics or physical science, students are likely to become discouraged and dis-identify with these subjects (Good, Rattan, & Dweck, 2012; Steele, 1997).

However, such negative effects can be reduced by altering educational environments. Toward this end, science educators must be aware of their own gender and racial biases and keep them in check. For example, female teachers with high math anxiety can affect girls' math achievement by bolstering the stereotype of math as a male discipline (Beilock, Gunderson, Ramirez, & Levine, 2010). On the opposite end of the spectrum, teachers can be too wary of critiquing minority students and overcompensate for racial prejudice by grading leniently (i.e., positive feedback bias). Such overcompensation can end up undermining minority students' self-esteem and interest in schoolwork (Harber et al., 2012; Harber, Stafford, & Kennedy, 2010). Unconscious bias is prevalent, and good intentions to remedy the situation can backfire. Implicit biases of teachers can exert powerful influence on minority students. However, informed teachers can reduce the effect of negative stereotypes held in the broader society and make a positive impact on minority students' academic motivation.

The good news is that underprivileged students tend to do well in citizen science programs (Birmingham & Calabrese Barton, 2014; Kim & Lee, 2017b). Kim and Lee (2017b) reported that some students who were quiet and less engaged in traditional lecture-based science classrooms emerged as leaders in citizen science programs. Students are exposed to different instructional activities and real-world related instructional goals. Citizen science activities tend to spark interest among students who might have been otherwise passive and nudge them to engage.

Practical Applications and Recommendations

Promote gender and racial equality through participating in community-related projects. In addition to avoiding negative stereotypes, teachers must also avoid making blatant statements of positive stereotypes (e.g., Asians are good at math), as such statements tend to induce negative emotions among students (Czopp, Kay, & Cheryan, 2015; Shih, Ambady, Richeson, Fujita, & Gray, 2002). Adolescents tend to think that success with minimal effort signals high intelligence. Such beliefs tend to serve as a double-edged sword for the students: If you don't work hard, you may not do well. But if you do work hard, other people may think you are not so smart. Teachers should teach students that effortful accomplishment is important and putting in effort is important in its own right.

Whenever feasible, teachers should try to represent various gender and ethnic groups in the group of experts/scientists. It is important to increase the visibility and representation of people from minority groups in the field (Murphy, Steele, & Gross, 2007; Purdie-Vaughns, Steele, Davies, Ditlmann, & Crosby, 2008) and in positions of authority (Marx & Goff, 2005; Marx & Roman, 2002; McIntyre, Paulson, & Lord, 2003).

Birmingham and Calabrese Barton (2014) conducted a case study with six African American females in fifth, sixth, and seventh grade, who were from lower-income or minority communities. These students were disinterested in science. In this project, the students were asked to connect what they were learning in the classroom to the real-world community through conducting science experiments, collecting data research, analyzing and interpreting data, and presenting what they learned. When students participated in educational activities with meaningful goals outside of school and were asked to take action to improve their community (e.g., leading middle school students to participate in Green Carnival), their interest in science was greatly enhanced. These activities opened up new spaces for the disadvantaged students to have dialogues with the community members and experts on green energy and ultimately provided a safe space to feel a sense of belonging in the community. As a result, they explained that it was beneficial for learning to be connected to action and practice.

Kim and Lee (2017a) designed and implemented Community-Based Socioscientific Issues Programs (SSI-COMM) on three community issues (i.e., abandoned pets, air quality, recycling of materials) that students encounter in their everyday lives. SSI-COMM encouraged seventh graders to apply and practice what they learned about science in their everyday lives. It made a strong connection between schools and local communities. Thus, the participating students not only explored the issues in the classroom, but also visited local organizations and centers and met with experts and community residents to solve the community issues.

SSI-COMM were designed to urge middle school students to contribute to their local communities by raising their interest in issues that may arise in the communities, exploring and examining the issues at the local level, and cultivating citizenship. During the program, the teachers and invited lecturers who led SSI-COMM were pleasantly surprised to find that students who had been disengaged or underachieving in the regular classroom now showed leadership and learning potential in SSI-COMM. For example, when visiting a health center, Student A who had been typically disengaged in the science class was pleasantly engaged in meeting with local residents and actively interviewed them about the idea of fine dust. In addition, he helped other hesitant friends to conduct interviews and when he visited a senior center, he took initiative and confidently explained how to cope with fine dusts to the local elders. He voluntarily came to the teacher after class and expressed his sense of accomplishment through SSI-COMM. Since such learning is experiential rather than academically oriented and students are not competitive, even low-achieving students are able to voluntarily engage in learning. It increases students' confidence and self-esteem. Another teacher also began to observe that students participated in the study differently than they did in the regular science classroom and said, "I was so surprised!" The teacher confessed that he mistakenly thought that students were not actively engaged in science classes because they lacked understanding of science content. However, after observing such students actively raising their voices and participating in the SSI-COMM, he became aware of the potential of individual students.

Social Capital Theory

Overview of the Theory

Needless to say, social network is critical to students' academic and social adjustment. According to social capital theory, a supportive social network provides psychosocial and academic resources that enhance individual students' academic outcomes (Altermatt, & Pomerantz, 2003; Berndt & Keefe, 1995; Coleman, 1961; Coleman, 1988; Dika & Singh, 2002; Ryan, 2000). There is substantial evidence in the literature demonstrating that students with successful social relationships do better in school. The students with social capital have access to peers for academic or social support, and the classroom academic activities may be more available to them (Wentzel & Wigfield, 1998). Both individual friendships and social network connections are important for academic adjustment (Furrer & Skinner, 2003; Wentzel, Battle, Russell, & Looney, 2010).

When students are disconnected and alienated, they are prone to negative educational outcomes (Becker & Luthar, 2002; Berndt, Hawkins, & Jiao, 1999; Ladd, 1990; McDougall & Hymel, 1998; Parker & Asher, 1987; Wentzel, Barry, & Caldwell, 2004). Nonetheless, the importance of fostering healthy peer relationships can be easily overlooked, especially in informal science education programs. Some informal education programs are too short for the peer system to be stabilized, and additionally, the instructors in those informal education programs do not have the same level of control and authority as classroom teachers might have. Teachers often do not consider managing peer issues in the classroom as part of their responsibilities (Ryan, Kuusinen, & Bedoya-Skoog, 2015), and this might be particularly true for teachers in informal education programs. Having close friends within the classroom not only offers psychological and emotional safety nets but also offers instrumental support, which is conducive to learning. Having healthy friendships in the classroom is linked to positive perceptions of the overall classroom environment (Ladd, Kochenderfer, & Coleman, 1996), emotional security, and academic engagement (Connell & Wellborn, 1991; Wentzel et al., 2010).

In informal science learning programs, group collaboration is often adopted. Social relationships among students can impact student learning disproportionally. Students with quality friendships tend to have positive social interactions and exhibit prosocial behaviors (Azmitia & Montgomery, 1993; Barry & Wentzel, 2006). In contrast, research has shown that the more children are viewed as rejected, the more classmates exclude them from activities. If exclusion is impossible, then classmates tend to assign minimal roles or degrade contributions of these rejected students. Rejected children may get excluded or not be able to take a substantial role in the classroom activities. Accordingly, they have less opportunities to learn due to exclusion from peer supported learning processes, which impacts their eventual learning and achievement (Buhs & Ladd, 2001; Buhs, Ladd, & Herald, 2006). Peer rejection relates to negative school attitudes, school avoidance, and underachievement and tends to have long-term negative consequences in students' academic and social development (Ladd, 2006; Ladd, Herald-Brown, & Reiser, 2008).

A classroom's overall relational structure also impacts the students' adjustment and engagement. Students tend to thrive in classrooms where the classroom's relational *structure* is less hierarchical. In classrooms where peer relationships are hierarchically structured, only a few students are highly popular or socially connected. On the contrary, in an egalitarian classroom structure, many children are socially connected in friendship and "hang out" with many other children in the classroom. In an egalitarian classroom, students tend to be more engaged and show less aggressive behavior (Ahn, Garandeau, & Rodkin, 2010; Cappella, Kim, Neal, & Jackson, 2013). An interconnected and inclusive classroom environment creates a safe space in which students feel comfortable to explore new ideas and take intellectual risks (Duke, Borowsky, & Pettingell, 2011; Smith, Osgood, Caldwell, Hynes, & Perkins, 2013). However, it might be harder for teachers to accurately capture the peer group affiliations (e.g., who hangs out with whom) compared to individual students' peer status (e.g., rejected or isolated students; Ahn, Rodkin, & Gest, 2013).

The concept of social capital can be extended to the social resources outside of the classroom as well. Since citizen science programs involve community members and scientific organizations, these programs can be opportunities for students to expand their social capital outside of the classroom. Such projects can create opportunities for students to share social resources with various entities in the local communities or groups of experts. Scientists and local community members can be invited to speak or work with students. For example, in the "Students, Teachers, and Rangers and Research Scientists" (STARRS) program, all group members collaborate in an inquiry-based investigation of Mammoth Hot Springs, a unique hydrothermal feature in Yellowstone National Park (Houseal, Abd, & Destefano, 2014).

Practical Applications and Recommendations

Support all students, especially the ones who are not well accepted by peers. Teachers recognize students' peer statuses, and their assessments are quite accurate, as they constantly observe how students interact with each other in the classroom (Gest, Farmer, Cairns, & Xie, 2003; Leff, Kupersmidt, Patterson, & Power, 1999). Such peer dynamics can easily go unnoticed by the teacher in informal education settings. Teachers in informal science programs can make informal observations by noting the seating patterns (e.g., who sits with whom) and interaction patterns among students (e.g., who acts friendly/unfriendly to whom). Based on these observations, teachers can discuss this issue with the class (e.g., who is rejected and excluded in the classroom). When deemed necessary, a more systematic collection of data can be used to supplement the observation. For example, teachers can conduct a sociometry study in their classroom (Moreno, 1934), which can identify informal leaders, social rankings, and isolated students. Sobieski and Dell'Angelo (2016) offers an informative guide to use sociometry for classroom teachers (see also, Leung & Silberling, 2006).

Given the importance of peer relationships noted in the literature, the extra time given to understanding a classroom's relational patterns may be worthwhile for the eventual success of a citizen science program. When teachers have a pretty good understanding of the peer relationship dynamics, they can use this information for instructional purposes (e.g., grouping) and create contexts where all students, regardless of their peer status, can learn and enjoy learning. Teachers can be highly intentional when creating student work pairs so that positive peer interactions can be ensured without conflict (e.g., not pairing the victim and bully). Such intentional and considerate teachers are more successful in supporting all students' academic achievement and social adjustment (Farmer, Estell, Bishop, O'Neal, & Cairns, 2003; Hamm, Farmer, Dadisman, Gravelle, & Murray, 2011; Hamm et al., 2010).

Collaborate with the communities and build relationships. Many citizen science programs include opportunities to connect to and involve the community. Formation of relationships with the community can facilitate the sharing of social capital by providing community members with the opportunity to contribute their knowledge, skills, and resources. Birmingham and Calabrese Barton (2014) recommended building systematic connections with the local community. When students have continuous connections and lasting partnerships, their interest in investigating and resolving the problems in the local community is greatly enhanced.

Even the students who lack prior science knowledge or skills can participate in solving problems in different ways and learn from others throughout the process. Birmingham and Calabrese Barton (2014) reported that the students who have been marginalized in science learning were eager to share their learning experiences with others. This statement implies that these students could obtain and also share social capital through the formation of new relationships with members of the community. As in Roth and Lee's (2004) study, various individuals (farmers, civil servants, secondary school students and college students, parents, and experts in the area) can all work together and contribute to the knowledge building in different ways. An interview with a student who participated in Birmingham and Calabrese Barton's (2014) program/study showed how knowledge sharing could be expanded via social network and vice versa (social network can facilitate the knowledge sharing):

> I want to teach other kids about how bad the Earth could be or how bad it is and how you can make a change in the world. Someone might say, ok, it's just coal or it's just gas and we can get more, but now they can teach someone a lesson. (p. 296)

Collaboration with the community contributed to the science literacy of the students and all citizens involved in the project, effectively illustrating collective praxis (Bowen & Roth, 2007; Roth & Lee, 2004).

CONCLUSION

The purpose of this chapter was to provide guidelines to support informal educators who aim to implement citizen science programs in formal school settings. Many researchers have begun utilizing K-12 students for data collection (Galloway et al., 2006), which is great as citizen science projects have a high potential to yield a number of educational benefits: Students are exposed to a wide range of approaches for scientific inquiry, apply content knowledge in real-world scientific investigations, and participate in sharing results. They learn to reason scientifically to understand the world around them. When students examine the issues directly affecting their immediate communities (e.g., risk and health threats in a neighborhood), their scientific citizenship, sense of place, environmental science agency, and stewardship are greatly enhanced (Ballard, Dixon, & Harris, 2017; Barton, 2012; Bonney, Phillips, Ballard, & Enck, 2015; Irwin, 1995; Krasny, Russ, Tidball, & Elmqvist, 2014; Lee, 2018; Mueller et al., 2012). However, citizen science projects must be carefully designed and implemented with educational outcomes in mind. Citizen science programs in informal science education settings have a clear motivational advantage. However, only when teachers are guided by sound psychological and educational theories can such projects serve as a springboard for learning.

Offering a wide range of evidence-based learning and instructional activities, policies, and programmatic approaches to support students' positive outcomes in informal citizen science projects is important in fostering their motivation. Such efforts should be continued. Note that not all of the research that these recommendations are drawn from have been conducted in the context of informal citizen science programs. There is emerging empirical evidence that citizen science programs could possibly affect mastery skills, science literacy, content knowledge and interest, and self-efficacy (Cronje, Rohlinger, Crall, & Newman, 2011; Harrison, Underhill, & Barnard, 2008; Hiller, 2012; Sutton, 2009; Tomasek, 2006). However, more research attention and effort should be directed to empirically testing the effectiveness of the citizen science programs on a variety of students' learning and motivation outcomes using rigorous study design and rich data (e.g., multiple informants, objective outcomes data on student achievement and the impact on the community).

REFERENCES

Ahn, H.-J., Garandeau, C. F., & Rodkin, P. C. (2010). Effects of classroom embeddedness and density on the social status of aggressive and victimized children. *The Journal of Early Adolescence, 30*(1), 76-101. doi:10.1177/0272431609350922

Ahn, H.-J., Rodkin, P. C., & Gest, S. D. (2013). Teacher-student agreement on "bullies and kids they pick on" in elementary school classrooms: Gender and grade differences. *Theories into Practice, 52*(4), 257-263. doi:10.1080/00405841.2013.829728

Altermatt, E. R., & Pomerantz, E. M. (2003). The development of competence-related and motivational beliefs: An investigation of similarity and influence among friends. *Journal of Educational Psychology, 95*(1), 111-123. doi:10.1037/0022-0663.95.1.111

Ames, C. (1992). Classrooms: Goals, structures, and student motivation. *Journal of Educational Psychology, 84*(3), 261-271. doi:10.1037/0022-0663.84.3.261

Anderman, E. M., & Maehr, M. L. (1994). Motivation and schooling in the middle grades. *Review of Educational Research, 64*(2), 287-309. doi:10.2307/1170696

Anderman, E. M., & Wolters, C. A. (2006). Goals, values, and affect: Influences on student motivation. In P. A. Alexander & P. H. Winne (Eds.), *Handbook of educational psychology* (pp. 369-389). Mahwah, NJ: Lawrence Erlbaum Associates Publishers.

Aronson, E., & Patnoe, S. (1997). *The jigsaw classroom: Building cooperation in the classroom* (2nd ed.). New York: Addison Wesley Longman.

Aronson, J., Quinn, D. M., & Spencer, S. (1998). Stereotype threat and the academic underperformance of minorities and women. In J. K. Swim & C. Stangor (Eds.), *Prejudice: The target's perspective* (pp. 83-103). San Diego, CA: Academic Press.

Assor, A., Kaplan, H., & Roth, G. (2002). Choice is good, but relevance is excellent: Autonomy-enhancing and suppressing teacher behaviours predicting students' engagement in schoolwork. *British Journal of Educational Psychology, 72*(2), 261-278. doi:10.1348/000709902158883

Azmitia, M., & Montgomery, R. (1993). Friendship, transactive dialogues, and the development of scientific reasoning. *Social Development, 2*(3), 202-221. doi:10.1111/j.1467-9507.1993.tb00014.x

Ballard, H. L., Dixon, C. G., & Harris, E. M. (2017). Youth-focused citizen science: Examining the role of environmental science learning and agency for conservation. *Biological Conservation, 208*, 65-75. doi:10.1016/j.biocon.2016.05.024

Barron, B. (2006). Configurations of learning settings and networks: Implications of a learning ecology perspective. *Human Development, 49*(4), 229-231. doi:10.1159/000094370

Barry, C. M., & Wentzel, K. R. (2006). Friend influence on prosocial behavior: The role of motivational factors and friendship characteristics. *Developmental Psychology, 42*(1), 153-163. doi:10.1037/0012-1649.42.1.153

Barton, D. (2012). Participation, deliberate learning and discourses of learning online. *Language and Education, 26*(2), 139-150. doi:10.1080/09500782.2011.642880

Baumeister, R. F., & Leary, M. R. (1995). The need to belong: Desire for interpersonal attachments as a fundamental human motivation. *Psychological Bulletin, 117*(3), 497-529. doi:10.1037/0033-2909.117.3.497

Becker, B. E., & Luthar, S. S. (2002). Social-emotional factors affecting achievement outcomes among disadvantaged students: Closing the achievement gap. *Educational Psychologist, 37*(4), 197-214. doi:10.1207/s15326985ep3704_1

Beilock, S. L., Gunderson, E. A., Ramirez, G., & Levine, S. C. (2010). Female teachers' math anxiety affects girls' math achievement. *PNAS Proceedings of the National Academy of Sciences of the United States of America, 107*(5), 1860-1863. doi:10.1073/pnas. 0910967107

Bell, A. E., Spencer, S. J., Iserman, E., & Logel, C. E. R. (2003). Stereotype threat and women's performance in engineering. *Journal of Engineering Education, 92*(4), 307-312. doi:10.1002/j.2168-9830.2003.tb00774.x

Berndt, T. J., Hawkins, J. A., & Jiao, Z. (1999). Influences of friends and friendships on adjustment to junior high school. *Merrill-Palmer Quarterly, 45*(1), 13-41. doi:10.1353/ mpq.2004.0014

Berndt, T. J., & Keefe, K. (1995). Friends' influence on adolescents' adjustment to school. *Child Development, 66*(5), 1312-1329. doi:10.2307/1131649

Birmingham, D., & Calabrese Barton, A. (2014). Putting on a green carnival: Youth taking educated action on socioscientific issues. *Journal of Research in Science Teaching, 51*(3), 286-314. doi:10.1002/tea.21127

Bonney, R., Phillips, T. B., Ballard, H. L., & Enck, J. W. (2015). Can citizen science enhance public understanding of science? *Public Understanding of Science, 25*(1), 2-16. doi:10.1177/0963662515607406

Bouillion, L. M., & Gomez, L. M. (2001). Connecting school and community with science learning: Real world problems and school–community partnerships as contextual scaffolds. *Journal of Research in Science Teaching, 38*(8), 878-898. doi:10.1002/tea.1037

Bowen, G. M., & Roth, W. M. (2007). The practice of field ecology: Insights for science education. *Research in Science Education, 37*(2), 171-187. doi:10.1007/s11165-006-9021-x

Brown, B. B., Bakken, J. P., Ameringer, S. W., & Mahon, S. D. (2008). A comprehensive conceptualization of the peer influence process in adolescence. In M. J. Prinstein & K. Dodge (Eds.), *Peer influence processes among youth* (pp. 17-44). New York, NY: Guildford Publications.

Buhs, E. S., & Ladd, G. W. (2001). Peer rejection as antecedent of young children's school adjustment: An examination of mediating processes. *Developmental Psychology, 37*(4), 550-560. doi:10.1037/0012-1649.37.4.550

Buhs, E. S., Ladd, G. W., & Herald, S. L. (2006). Peer exclusion and victimization: Processes that mediate the relation between peer group rejection and children's classroom engagement and achievement? *Journal of Educational Psychology, 98*(1), 1-13. doi:10.1037/0022-0663.98.1.1

Butler, D. L. (1998). The strategic content learning approach to promoting self-regulated learning: A report of three studies. *Journal of Educational Psychology, 90*(4), 682-697. doi: 10.1037/0022-0663.90.4.682

Cappella, E., Kim, H. Y., Neal, J. W., & Jackson, D. R. (2013). Classroom peer relationships and behavioral engagement in elementary school: The role of social network equity. *American Journal of Community Psychology*, *52*, 367-379. doi:10.1007/s10464-013-9603-5

Chicago Botanic Gardens (n.d.). Budburst. Retrieved from https://budburst.org/

Cohen, E. G. (1994). Restructuring the classroom: Conditions for productive small groups. *Review of Educational Research*, *64*(1), 1-35. doi:10.3102/00346543064001001

Cohen, G. L., & Steele, C. M. (2002). A barrier of mistrust: How negative stereotypes affect cross-race mentoring. In J. Aronson (Ed.), *Improving academic achievement: Impact of psychological factors on education* (pp. 303-327). San Diego, CA: Academic Press. doi:10.1016/B978-012064455-1/50018-X

Cohen, G. L., Steele, C. M., & Ross, L. D. (1999). The mentor's dilemma: Providing critical feedback across the racial divide. *Personality and Social Psychology Bulletin, 25*(10), 1302-1318. doi:10.1177/0146167299258011

Coleman, J. S. (1961). *The adolescent society: The social life of the teenager and its impact on education.* Oxford, United Kingdom: Free Press of Glencoe.

Coleman, J. S. (1988). Equality and excellence in education. In H. J. O'Gorman (Ed.), *Surveying social life: Papers in honor of Herbert H. Hyman.* (pp. 376-392). Middletown, CT: Wesleyan University Press.

Connell, J. P., & Wellborn, J. G. (1991). Competence, autonomy, and relatedness: A motivational analysis of self-system processes. In M. R. Gunnar & L. A. Sroufe (Eds.), *Self processes and development* (pp. 43-77). Hillsdale, NJ: Lawrence Erlbaum Associates, Inc.

Cordova, D. I., & Lepper, M. R. (1996). Intrinsic motivation and the process of learning: Beneficial effects of contextualization, personalization, and choice. *Journal of Educational Psychology, 88*(4), 715-730. doi:10.1037/0022-0663.88.4.715

Cornell University (2018). The Lost Ladybug Project. Retrieved from http://www.lostladybug.org/

Corpus, J. H., McClintic-Gilbert, M. S., & Hayenga, A. O. (2009). Within-year changes in children's intrinsic and extrinsic motivational orientations: Contextual predictors and academic outcomes. *Contemporary Educational Psychology, 34*(2), 154-166. doi:10.1016/j.cedpsych.2009.01.001

Croizet, J. C., & Claire, T. (1998). Extending the concept of stereotype threat to social class: The intellectual underperformance of students from low socioeconomic backgrounds. *Personality and Social Psychology Bulletin, 24*(6), 588-594. doi:10.1177/0146167298246003

Cronje, R., Rohlinger, S., Crall, A., & Newman, G. (2011). Does participation in citizen science improve scientific literacy? A study to compare assessment methods. *Applied Environmental Education & Communication, 10*(3), 135-145. doi:10.1080/1533015X.2011.603611

Csikszentmihalyi, M., Rathunde, K. R., & Whalen, S. (1993). *Talented teenagers: The roots of success and failure.* New York, NY: Cambridge University Press.

Czopp, A. M., Kay, A. C., & Cheryan, C. (2015). Positive stereotypes are pervasive and powerful. *Perspectives on Psychological Science, 10*(4), 451-463. doi:10.1177/1745691615588091

Deci, E. L., & Ryan, R. M. (1985). The general causality orientations scale: Self-determination in personality. *Journal of Research in Personality, 19*(2), 109-134. doi:10.1016/0092-6566(85)90023-6

Deci, E. L., & Ryan, R. M. (2000). The "what" and "why" of goal pursuits: Human needs and the self-determination of behavior. *Psychological Inquiry, 11*(4), 227-268. doi:10.1207/s15327965pli1104_01

Dickinson, J. L., Shirk, J. L., Bonter, D. N., Bonney, R., Crain, R., Martin, J., . . . Purcell, K. (2012). The current state of citizen science as a tool for ecological research and public engagement. *Frontiers in Ecology and the Environment, 10*(6), 291-297. doi:10.1890/110236

Dika, S. L., & Singh, K. (2002). Applications of social capital in educational literature: A critical synthesis. *Review of Educational Research, 72*(1), 31-60. doi:10.3102/00346543072001031

Duke, N. N., Borowsky, I. W., & Pettingell, S. L. (2011). Adult perceptions of neighborhood: Links to youth engagement. *Youth & Society, 44*(3), 408-430. doi:10.1177/0044118X11402852

Dweck, C. S. (1986). Motivational processes affecting learning. *American Psychologist, 41*(10), 1040-1048. doi:10.1037/0003-066X.41.10.1040

Elmose, S., & Roth, W. M. (2005). Allgemeinbildung: Readiness for living in risk society. *Journal of Curriculum Studies, 37*(1), 11-34. doi:10.1080/0022027041000229413

Farmer, T. W., Estell, D. B., Bishop, J. L., O'Neal, K., & Cairns, B. D. (2003). Rejected bullies or popular leaders?: The social relations of aggressive subtypes of rural African American early adolescents. *Developmental Psychology, 39*, 992-1004. doi:10.1037/0012-1649.39.6.992

Farmer, T. W., Lines, M. M., & Hamm, J. V. (2011). Revealing the invisible hand: The role of teachers in children's peer experiences. *Journal of Applied Developmental Psychology, 32*(5), 247-256. doi:10.1016/j.appdev.2011.04.006

Funk, C., & Parker, K. (2018, Jan. 9). Women and men in STEM often at odds over workplace equity. Retrieved from http://www.pewsocialtrends.org/2018/01/09/women-and-men-in-stem-often-at-odds-over-workplace-equity/

Furrer, C., & Skinner, E. (2003). Sense of relatedness as a factor in children's academic engagement and performance. *Journal of Educational Psychology, 95*(1), 148-162. doi:10.1037/0022-0663.95.1.148

Galloway, A. W. E., Tudor, M. T., & Haegen, W. M. V. (2006). The reliability of citizen science: A case study of Oregon white oak stand surveys. *Wildlife Society Bulletin, 34*(5), 1425-1429. doi:10.2193/0091-7648(2006)34[1425:trocsa]2.0.co;2

Gest, S. D., Farmer, T. W., Cairns, B. D., & Xie, H. (2003). Identifying children's peer networks in school classrooms: Links between peer reports and observed interactions. *Social Development, 12*(4), 513-529. doi:10.1111/1467-9507.00246

Gonzales, P. M., Blanton, H., & Williams, K. J. (2002). The effects of stereotype threat and double-minority status on the test performance of Latino women. *Personality and Social Psychology Bulletin, 28*(5), 659-670. doi:10.1177/0146167202288010

Good, C., Rattan, A., & Dweck, C. S. (2012). Why do women opt out? Sense of belonging and women's representation in mathematics. *Journal of Personality and Social Psychology, 102*(4), 700-717. doi:10.1037/a0026659

Gray, S. A., Nicosia, K., & Jordan, R. C. (2012). Lessons learned from citizen science in the classroom. A response to "The future of citizen science." *Democracy & Education, 20*(2), Article 14. Retrieved from: https://democracyeducationjournal.org/home/vol20/iss2/14

Hamm, J. V., Farmer, T. W., Dadisman, K., Gravelle, M., & Murray, A. R. (2011). Teachers' attunement to students' peer group affiliations as a source of improved student experiences of the school social–affective context following the middle school transition. *Journal of Applied Developmental Psychology, 32*(5), 267-277. doi:10.1016/j.appdev.2010.06.003

Hamm, J. V., Farmer, T. W., Robertson, D., Dadisman, K. A., Murray, A., Meece, J. L., & Song, S. Y. (2010). Effects of a developmentally based intervention with teachers on Native American and White early adolescents' schooling adjustment in rural settings. *Journal of Experimental Education*, *78*(3), 343-377. doi:10.1080/00220970903548038

Harackiewicz, J. M., Barron, K. E., Tauer, J. M., Carter, S. M., & Elliot, A. J. (2000). Short-term and long term consequences of achievement goals: Predicting interest and performance over time. *Journal of Educational Psychology, 92*(2), 316-330. doi:10.1037/0022-0663.92.2.316

Harber, K. D., Gorman, J. L., Gengaro, F. P., Butisingh, S., Tsang, W., & Ouellette, R. (2012). Students' race and teachers' social support affect the positive feedback bias in public schools. *Journal of Educational Psychology, 104*(4), 1149-1161. doi:10.1037/a0028110

Harber, K. D., Stafford, R., & Kennedy, K. A. (2010). The positive feedback bias as a response to self-image threat. *British Journal of Social Psychology*, *49*(1), 207-218. doi:10.1348/014466609X473956

Harrison J. A., Underhill, L. G., & Barnard, P. (2008). The seminal legacy of the Southern African Bird Atlas Project. *South African Journal of Science, 104*, 82-84. doi:10.2989/salals.2008.26.2.5.569

Hiller, S. E. (2012). *The impact of a citizen science program on student achievement and motivation: A social cognitive career perspective*. (Dissertation thesis). Retrieved from Proquest Dissertations and Theses Global. (UMI Number 3547088)

Hiller, S. E., & Kitsantas, A. (2014). The effect of a horseshoe crab citizen science program on middle school student science performance and STEM career motivation. *School Science and Mathematics, 114*(6), 302-311. doi:10.1111/ssm.12081

Houseal, A. K., Abd, E. F., & Destefano, L. (2014). Impact of a student–teacher–scientist partnership on students' and teachers' content knowledge, attitudes toward science, and pedagogical practices. *Journal of Research in Science Teaching, 51*(1), 84-115. doi:10.1002/tea.21126

Hulleman, C. S., Godes, O., Hendricks, B. L., & Harackiewicz, J. M. (2010). Enhancing interest and performance with a utility value intervention. *Journal of Educational Psychology, 102*(4), 880-895. doi:10.1037/a0019506

Irwin, A. (1995). *Citizen science.* London, United Kingdom: Routledge.

Iyengar, S. S., & Lepper, M. R. (1999). Rethinking the value of choice: A cultural perspective on intrinsic motivation. *Journal of Personality and Social Psychology*, *76*(3), 349-366. doi:10.1037/0022-3514.76.3.349

Jeanpierre, B., Oberhauser, K., & Freeman, C. (2005). Characteristics of professional development that effect change in secondary science teachers' classroom practices. *Journal of Research in Science Teaching*, *42*(6), 668-690. doi:10.1002/tea.20069

Johnson, D. W., & Johnson, R. T. (1989). *Cooperation and competition: Theory and research.* Edina, MN: Interaction Book Company.

Johnson, D. W., & Johnson, R. T. (2005). New developments in social interdependence theory. *Genetic, Social, and General Psychology Monographs*, *131*(4), 285-358. doi:10.3200/MONO.131.4.285-358

Jollymore, A., Haines, M. J., Satterfield, T., & Johnson, M. S. (2017). Citizen science for water quality monitoring: Data implications of citizen perspectives. *Journal of Environmental Management, 200*, 456-467. doi:10.1016/j.jenvman.2017.05.083

Juvonen, J. (2006). Sense of belonging, social bonds, and school functioning. In P. A. Alexander & P. H. Winne (Eds.), *Handbook of educational psychology* (pp. 655-674). Mahwah, NJ: Lawrence Erlbaum Associates Publishers.

Karabenick, S. A. (2004). Perceived achievement goal structure and college student help seeking. *Journal of Educational Psychology*, *96*(3), 569-581. doi:10.1037/0022-0663.96.3.569

Kim, G., & Lee, H. (2017a). Effects of community-based SSI programs on promoting middle school students' understanding of issues and character and values as citizens: Focused on fine dust issues. *Journal of the Korean Association for Science Education, 37(*6), 911-920.

Kim, G., & Lee, H. (2017b). Perceptions of teachers, program instructors, and local experts on implementing community-based socioscientific issues programs. *Journal of the Korean Association for Science Education, 37*(3), 453-464.

Kim, Y., Lee, H., & Kim, J. (2013). Korean pre-service science teachers' belief on science teaching and learning and its evolution. *Journal of Science Education, 37*(1), 40-51. doi:10.21796/jse.2013.37.1.40

Kindermann, T. A. (2015). Peer group influences on students' academic motivation. In K. R. Wentzel & G. B. Ramani (Eds.), *Handbook on social influences on social-emotional, motivation, and cognitive outcomes in school contexts* (pp. 31-47). New York, NY: Routledge.

Krasny, M. E., Russ, A., Tidball, K. G., & Elmqvist, T. (2014). Civic ecology practices: Participatory approaches to generating and measuring ecosystem services in cities. *Ecosystem Services, 7*, 177-186. doi:10.1016/j.ecoser.2013.11.002

Ladd, G. W. (1990). Having friends, keeping friends, making friends, and being liked by peers in the classroom: Predictors of children's early school adjustment? *Child Development*, *61*(4), 1081-1100. doi:10.2307/1130877

Ladd, G. W. (2006). Peer rejection, aggressive or withdrawn behavior, and psychological maladjustment from ages 5 to 12: An examination of four predictive models. *Child Development*, *77*(4), 822-846. doi:10.1111/j.1467-8624.2006.00905.x

Ladd, G. W., Herald-Brown, S. L., & Reiser, M. (2008). Does chronic classroom peer rejection predict the development of children's classroom participation during the grade school years? *Child Development*, *79*(4), 1001-1015. doi:10.1111/j.1467-8624.2008.01172.x

Ladd, G. W., Kochenderfer, B. J., & Coleman, C. C. (1996). Friendship quality as a predictor of young children's early school adjustment. *Child Development*, *67*(3), 1103-1118. doi:10.2307/1131882

Lee, H. (2018). *What is SSI education?* Seoul, South Korea: Parkyoung Story.

Lee, H., & Yang, J. E. (2019). Science teachers taking their first steps toward teaching socioscientific issues through collaborative action research. *Research in Science Education*, *49*(1), 51-71. doi:10.1007/s11165-017-9614-6

Leff, S. S., Kupersmidt, J. B., Patterson, C. J., & Power, T. J. (1999). Factors influencing teacher identification of peer bullies and victims. *School Psychology Review, 28*(3), 505-517.

Lepper, M. R., Corpus, J. H., & Iyengar, S. S. (2005). Intrinsic and extrinsic motivational orientations in the classroom: Age differences and academic correlates. *Journal of Educational Psychology*, *97*(2), 184-196. doi"10.1037/0022-0663.97.2.184

Leung, B. P., & Silberling, J. (2006). Using sociograms to identify social status in the classroom. *California School Psychologist*, *11*, 57-61. doi:10.1007/BF03341115

Levy-Tossman, I., Kaplan, A., & Assor, A. (2007). Academic goal orientations, multiple goal profiles, and friendship intimacy among early adolescents. *Contemporary Educational Psychology, 32*(2), 231-252. doi:10.1016/j.cedpsych.2006.06.001

Linnenbrink-Garcia, L., & Patall, E. A. (2015). Motivation. In L. Corno & E. M. Anderman (Eds.), *Handbook of educational psychology* (3rd ed., pp. 91-103). New York, NY: Taylor and Francis Inc.

Lipstein, R., & Renninger, K. A. (2006). "Putting things into words": 12-15-year-old students' interest for writing. In P. Boscolo & S. Hidi (Eds.), *Motivation and writing: Research and school practice* (pp. 113-140). New York, NY: Kluwer Academic/Plenum.

Loevinger, J., & Blasi, A. (1991). Development of the self as subject. In J. Strauss & G. R. Goethals (Eds.), *The self: Interdisciplinary approaches* (pp. 150-167). New York, NY: Springer-Verlag.

Lorenz, A. (2016). *The influence of a citizen science project: Student attitudes, sense of place, and understanding of science practices* (Unpublished doctoral dissertation). University of Wyoming, Laramie, WY.

Marsh, H. W. (1990a). Causal ordering of academic self-concept and academic achievement: A multiwave, longitudinal panel analysis. *Journal of Educational Psychology*, *82*(4), 646-656. doi:10.1037/0022-0663.82.4.646

Marsh, H. W. (1990b). The structure of academic self-concept: The Marsh/Shavelson model. *Journal of Educational Psychology*, *82*(4), 623-636. doi:10.1037/0022-0663.82.4.623

Marsh, H. W. (1993). The multidimensional structure of academic self-concept: Invariance over gender and age. *American Educational Research Journal, 30,* 841-860. doi:10.3102/00028312030004841

Marsh, H. W., Trautwein, U., Lüdtke, O., Köller, O., & Baumert, J. (2005). Academic self-concept, interest, grades, and standardized test scores: Reciprocal effects models of causal ordering. *Child Development*, *76*(2), 397-416. doi:10.1111/j.1467-8624.2005.00853.x

Marx, D. M., & Goff, P. A. (2005). Clearing the air: The effect of experimenter race on target's test performance and subjective experience. *British Journal of Social Psychology*, *44*(4), 645-657. doi:10.1348/014466604X17948

Marx, D. M., & Roman, J. S. (2002). Female role models: Protecting women's math test performance. *Personality and Social Psychology Bulletin*, *28*(9), 1183-1193. doi:10.1177/01461672022812004

McDougall, P., & Hymel, S. (1998). Moving into middle school: Individual differences in the transition experience. *Canadian Journal of Behavioural Science / Revue Canadienne Des Sciences Du Comportement*, *30*(2), 108-120. doi:10.1037/h0085811

McIntyre, R. B., Paulson, R. M., & Lord, C. G. (2003). Alleviating women's mathematics stereotype threat through salience of group achievements. *Journal of Experimental Social Psychology*, *39*(1), 83-90. doi:10.1016/S0022-1031(02)00513-9

Meece, J. L., Anderman, E. M., & Anderman, L. H. (2006). Classroom goal structure, student motivation, and academic achievement. *Annual Review of Psychology*, *57*, 487-503. doi:10.1146/annurev.psych.56.091103.070258

Meece, J. L., & Miller, S. D. (2001). A longitudinal analysis of elementary school students' achievement goals in literacy activities. *Contemporary Educational Psychology*, *26*(4), 454-480. doi:10.1006/ceps.2000.1071

Midgley, C., Kaplan, A., Middleton, M., Maehr, M. L., Urdan, T., Anderman, L. H., &… Roeser, R. (1998). The development and validation of scales assessing students' achievement goal orientations. *Contemporary Educational Psychology*, *23*(2), 113-131. doi:10.1006/ceps.1998.0965

Moreno, J. L. (1934). *Who shall survive?: A new approach to the problem of human interrelations*. Washington, DC: Nervous and Mental Disease Publishing Co. doi:10.1037/10648-000

Mueller, M. P., Tippins, D., & Bryan, L. A. (2012). The future of citizen science. *Democracy & Education, 20*(1), Article 2. Retrieved from https://democracyeducationjournal.org/home/vol20/iss1/2/

Murphy, M. C., Steele, C. M., & Gross, J. J. (2007). Signaling threat: How situational cues affect women in math, science, and engineering settings. *Psychological Science*, *18*(10), 879-885. doi:10.1111/j.1467-9280.2007.01995.x

Nelson-Le Gall, S. A. (1985). Motive–outcome matching and outcome foreseeability: Effects on attribution of intentionality and moral judgments. *Developmental Psychology*, *21*(2), 332-337. doi:10.1037/0012-1649.21.2.332

Neuhauser, A., & Cook, L. (2016, May 17). 2016 U.S. News/Raytheon STEM Index shows uptick in hiring, education. *U.S. News*. Retrieved from https://www.usnews.com/news/articles/2016-05-17/the-new-stem-index-2016

Newman, C., Buesching, C. D., & MacDonald D. W. (2003). Validating mammal monitoring methods and assessing the performance of volunteers in wildlife conservation – "Sed quis custodiet ipsos custodies?". *Biological Conservation, 113*(2), 189-197. doi:10.1016/S0006-3207(02)00374-9

Newman, R. S. (2000). Social influences on the development of children's adaptive help seeking: The role of parents, teachers, and peers. *Developmental Review*, *20*(3), 350-404. doi:10.1006/drev.1999.0502

Nicholls, J. G. (1984). Achievement motivation: Conceptions of ability, subjective experience, task choice, and performance. *Psychological Review*, *91*(3), 328-346. doi:10.1037/0033-295X.91.3.328

Niemiec, C. P., & Ryan, R. M. (2009). Autonomy, competence, and relatedness in the classroom: Applying self-determination theory to educational practice. *Theory and Research in Education, 7*(2), 133-144. doi:10.1177/1477878509104318

Osborne, J. (2001). Testing stereotype threat: Does anxiety explain race and sex differences in achievement? *Contemporary Educational Psychology, 26*(3), 291-310. doi:10.1006/ceps.2000.1052

Otis, N., Grouzet, F. M. E., & Pelletier, L. G. (2005). Latent motivational change in an academic setting: A 3-year longitudinal study. *Journal of Educational Psychology, 97*(2), 170-183. doi:10.1037/0022-0663.97.2.170

Parker, J. G., & Asher, S. R. (1987). Peer relations and later personal adjustment: Are low-accepted children at risk? *Psychological Bulletin, 102*(3), 357-389. doi:10.1037/0033-2909.102.3.357

PARRISE (2014-2017). Retrieved from https://www.parrise.edu

Patall, E. A., Cooper, H., & Wynn, S. R. (2010). The effectiveness and relative importance of choice in the classroom. *Journal of Educational Psychology, 102*(4), 896-915. doi:10.1037/a0019545

Patrick, H., Anderman, L. H., Ryan, A. M., Edelin, K. C., & Midgley, C. (2001). Teachers' communication of goal orientations in four fifth-grade classrooms. *The Elementary School Journal, 102*(1), 35-58. doi:10.1086/499692

Patrick, H., Kaplan, A., & Ryan, A. M. (2011). Positive classroom motivational environments: Convergence between mastery goal structure and classroom social climate. *Journal of Educational Psychology, 103*(2), 367-382. doi:10.1037/a0023311

Penrose, D., & Call, S. M. (1995). Volunteer monitoring of benthic macroinvertebrates: Regulatory biologists' perspectives. *Journal of North American Benthological Society, 14*(1), 203-209. doi:10.2307/1467735

Purdie-Vaughns, V., Steele, C. M., Davies, P. G., Ditlmann, R., & Crosby, J. R. (2008). Social identity contingencies: How diversity cues signal threat or safety for African Americans in mainstream institutions. *Journal of Personality and Social Psychology, 94*(4), 615-630. doi:10.1037/0022-3514.94.4.615

Raddick, M. J., Bracey, G., Gay, P. L., Lintott, C. J., Cardamone, C., Murray, P., & . . . Vandenberg, J. (2013). Galaxy zoo: Motivations of citizen scientists. *Astronomy Education Review, 12*(1). doi:10.3847/AER2011021

Reis, H. T., Collins, W. A., & Berscheid, E. (2000). The relationship context of human behavior and development. *Psychological Bulletin, 126*(6), 844-872. doi:10.1037/0033-2909.126.6.844

Renninger, K. A., & Hidi, S. (2002). Student interest and achievement: Developmental issues raised by a case study. In A. Wigfield & J. S. Eccles (Eds.), *Development of achievement motivation* (pp. 173-195). San Diego, CA: Academic Press. doi:10.1016/B978-012750053-9/50009-7

Roseth, C. J., Johnson, D. W., & Johnson, R. T. (2008). Promoting early adolescents' achievement and peer relationships: The effects of cooperative, competitive, and individualistic goal structures. *Psychological Bulletin, 134*(2), 223-246. doi:10.1037/0033-2909.134.2.223

Roth, W. M. (2003). Scientific literacy as an emergent feature of collective human praxis. *Journal of Curriculum Studies, 35*(1), 9-23. doi:10.1080/00220270210134600

Roth, W. M., & Lee, S. (2004). Science education as/for participation in the community. *Science Education, 88*(2), 263-291. doi:10.1002/sce.10113

Ryan, A. M. (2000). Peer groups as a context for the socialization of adolescents' motivation, engagement, and achievement in school. *Educational Psychologist, 35*(2), 101-111. doi:10.1207/s15326985ep3502_4

Ryan, A. M., Kuusinen, C. M., & Bedoya-Skoog, A. (2015). Managing peer relations: A dimension of teacher self-efficacy that varies between elementary and middle school teachers and is associated with observed classroom quality. *Contemporary Educational Psychology, 41*, 147-156. doi:10.1016/j.cedpsych.2015.01.002

Ryan, A. M., & Patrick, H. (2001). The classroom social environment and changes in adolescents' motivation and engagement during middle school. *American Educational Research Journal, 38*(2), 437-460. doi:10.3102%2F00028312038002437

Ryan, A. M., & Pintrich, P. R. (1997). "Should I ask for help?" The role of motivation and attitudes in adolescents' help seeking in math class. *Journal of Educational Psychology, 89*(2), 329-341. doi:10.1037/0022-0663.89.2.329

Ryan, R. M., & Connell, J. P. (1989). Perceived locus of causality and internalization: Examining reasons for acting in two domains. *Journal of Personality and Social Psychology, 57*(5), 749-761. doi:10.1037/0022-3514.57.5.749

Ryan, R. M., La Guardia, J. G., Solky-Butzel, J., Chirkov, V., & Kim, Y. (2005). On the interpersonal regulation of emotions: Emotional reliance across gender, relationships, and cultures. *Personal Relationships, 12*, 145-163. doi:10.1111/j.1350-4126.2005.00106.x

SAFECAST (n.d.). Open environmental data for everyone. Retrieved from https://blog.safecast.org/

Schunk, D. H., Meece, J. R., & Pintrich, P. R. (2014). *Motivation in education: Theory, research, and applications* (4th ed.). Boston, MA: Pearson.

Senko, C., & Hulleman, C. S. (2013). The role of goal attainment expectancies in achievement goal pursuit. *Journal of Educational Psychology, 105*(2), 504-521. doi:10.1037/a0031136

Shah, H. R., & Martinez, L. R. (2016). Current approaches in implementing citizen science in the classroom. *Journal of Microbiology & Biology Education, 17*(1), 17-22. doi:10.1128/jmbe.v17i1.1032

Shih, M., Ambady, N., Richeson, J. A., Fujita, K., & Gray, H. M. (2002). Stereotype performance boosts: The impact of self-relevance and the manner of stereotype activation. *Journal of Personality and Social Psychology, 83*(3), 638-647. doi:10.1037/0022-3514.83.3.638

Shih, M., Pittinsky, T. L., & Ambady, N. (1999). Stereotype susceptibility: Identity salience and shifts in quantitative performance. *Psychological Science, 10*(1), 80-83. doi:10.1111/1467-9280.00111

Shim, S. S., Kiefer, S. M., & Wang, C. (2013). Help seeking among peers: The role of goal structure and peer climate. *The Journal of Educational Research, 106*(4), 290-300. doi:10.1080/00220671.2012.692733

Smith, E. P., Osgood, D. W., Caldwell, L., Hynes, K., & Perkins, D. F. (2013). Measuring collective efficacy among children in community-based afterschool programs: Exploring pathways toward prevention and positive youth development. *American Journal of Community Psychology, 52*, 27-40. doi:10.1007/s10464-013-9574-6

Sobieski, C., & Dell'Angelo, T. (2016). Sociograms as a tool for teaching and learning: Discoveries from a teacher research study. *Educational Forum, 80*(4), 417-429. doi:10.1080/00131725.2016.1207734

Sperling, E., & Bencze, J. L. (2010). "More than particle theory": Citizenship through school science. *Canadian Journal of Science, Mathematics and Technology Education, 10*(3), 255-266. doi:10.1080/14926156.2010.504487

Steele, C. M. (1997). A threat in the air: How stereotypes shape intellectual identity and performance. *American Psychologist, 52*(6), 613-629. doi:10.1037/0003-066X.52.6.613.

Steele, C. M., & Aronson, J. (1995). Stereotype threat and the intellectual test performance of African Americans. *Journal of Personality and Social Psychology, 69,* 797-811. doi:10.1037//0022-3514.69.5.797

Steele, C. M., & Aronson, J. (2000). Stereotype threat and the intellectual test performance of African Americans. In C. Stangor (Ed.), *Stereotypes and prejudice: Essential readings.* (pp. 369-389). New York, NY: Psychology Press.

Steinberg, L., & Silverberg, S. B. (1986). The vicissitudes of autonomy in early adolescence. *Child Development, 57*(4), 841-851. doi:10.2307/1130361

STEPWISE (n.d.). Retrieved from https://webspace.oise.utoronto.ca/~benczela/STEPWISE.html

Sutton, S. (2009*). Increasing science literacy among English language learners through plant phenological monitoring: A citizen science program at Sequoia and Kings Canyon National Parks*. (Master's thesis). Retrieved from Proquest Dissertations and Global Theses. (UMI Number 1473523)

The Great Sunflower Project (n.d.). Retrieved from https://www.greatsunflower.org/

Thoman, D. B., Smith, J. L., Brown, E. R., Chase, J., & Lee, J. Y. K. (2013). Beyond performance: A motivational experiences model of stereotype threat. *Educational Psychology Review, 25*(2), 211-243. doi:10.1007/s10648-013-9219-1

Thomas, A. E. (2017). Gender differences in students' physical science motivation: Are teachers' implicit cognitions another piece of the puzzle? *American Educational Research Journal, 54*(1), 35-58. doi:10.3102/0002831216682223

Tomasek, T. M. (2006). *Student cognition and motivation during the Classroom BirdWatch citizen science project.* (Dissertation thesis). Retrieved from Proquest Dissertations and Global Theses. (UMI Number 3242410)

Urdan, T. C. (1997). Examining the relations among early adolescent students' goals and friends' orientation toward effort and achievement in school. *Contemporary Educational Psychology, 22*(2), 165-191. doi:10.1006/ceps.1997.0930

Urdan, T. (2004). Predictors of academic self-handicapping and achievement: Examining achievement goals, classroom goal structures, and culture. *Journal of Educational Psychology, 96*(2), 251-264. doi:10.1037/0022-0663.96.2.251

Urdan, T. C., & Maehr, M. L. (1995). Beyond a two-goal theory of motivation and achievement: A case for social goals. *Review of Educational Research, 65*(3), 213-243. doi:10.3102/00346543065003213

Vansteenkiste, M., Zhou, M., Lens, W., & Soenens, B. (2005). Experiences of autonomy and control among Chinese learners: Vitalizing or immobilizing? *Journal of Educational Psychology, 97*(3), 468-483. doi:10.1037/0022-0663.97.3.468

Walker, C. O., Greene, B. A., & Mansell, R. A. (2006). Identification with academics, intrinsic/extrinsic motivation, and self-efficacy as predictors of cognitive engagement. *Learning and Individual Differences, 16*(1), 1-12. doi:10.1016/j.lindif.2005.06.004

Wallace, D., & Bodzin, A. (2017). Developing scientific citizenship identity using mobile learning and authentic practice. *Electronic Journal of Science Education, 21*(6), 46-71. Retrieved from http://ejse.southwestern.edu/article/view/17688

Wentzel, K. R., Barry, C. M., & Caldwell, K. A. (2004). Friendships in middle school: Influences on motivation and school adjustment. *Journal of Educational Psychology, 96*(2), 195-203. doi:10.1037/0022-0663.96.2.195

Wentzel, K. R., Battle, A., Russell, S. L., & Looney, L. B. (2010). Social supports from teachers and peers as predictors of academic and social motivation. *Contemporary Educational Psychology*, *35*(3), 193-202. doi:10.1016/j.cedpsych.2010.03.002

Wentzel, K. R., & Wigfield, A. (1998). Academic and social motivational influences on students' academic performance. *Educational Psychology Review*, *10*(2), 155-175. doi:10.1023/A:1022137619834

Wolters, C. A. (2004). Advancing achievement goal theory: Using goal structures and goal orientations to predict students' motivation, cognition, and achievement. *Journal of Educational Psychology*, *96*(2), 236-250. doi:10.1037/0022-0663.96.2.236

Yeager, D. S., Trzesniewski, K. H., Tirri, K., Nokelainen, P., & Dweck, C. S. (2011). Adolescents' implicit theories predict desire for vengeance after peer conflicts: Correlational and experimental evidence. *Developmental Psychology, 47*(4), 1090-1107. doi:10.1037/a0023769

Zeidler, D. L., Sadler, T. D., Simmons, M. L., & Howes, E. V. (2005). Beyond STS: A research-based framework for socioscientific issues education. *Science Education*, *89*(3), 357-377. doi:10.1002/sce.20048

In: Enhancing STEM Motivation through Citizen Science Programs ISBN: 978-1-53616-038-3
Editors: Suzanne E. Hiller and Anastasia Kitsantas

Chapter 5

EMPOWERING YOUNG PEOPLE THROUGH CO-CREATED SCIENTIFIC RESEARCH

Elizabeth A. C. Rushton[1,2,*] and Becky Parker[2]
[1]University College London, Institute of Education, London, UK
[2]The Institute for Research in Schools, London, UK

ABSTRACT

The purpose of this chapter is to describe the Institute for Research in Schools (IRIS), provide an overview of the range of IRIS citizen science programs and discuss two IRIS case study programs in detail, *Genome Decoders* and *Well World*. The conceptual and structural frameworks of each of these case study programs are documented, alongside with template materials including guidance on how to effectively plan research collaborations with schools and resources that support student agency in research. Five educational implications are discussed: (1) establish research collaborations that are well-planned and structured; (2) develop research collaborations where students have agency; (3) raise students' awareness of and aspirations towards STEM careers; (4) create research communication and dissemination opportunities for school students and finally, (5) enhance students' soft skills through participation in research. Areas of future research are identified.

Keywords: authentic research, school students, soft skills, STEM, STEM careers, teachers

The Institute for Research in Schools (IRIS, 2018) works with schools across England and Scotland, and supports individual schools in Brussels, Norway, Hong Kong, Dubai and Colombia. The context for the IRIS program described in this chapter is broadly the English STEM (science, technology, engineering and mathematics) education and careers landscape. However, the trends and themes observed in STEM participation are broadly representative of

* Corresponding Author's E-mail: erushton@researchinschools.org

the those found across the United Kingdom and beyond (Bøe, Henriksen, Lyons & Schreiner, 2011; Brown, Brown & Bibby, 2008; Gilbert et al., 2006; Nardi & Steward, 2003).

Student Participation in STEM Education

Reiss and Mutjaba (2017) suggested that low participation in STEM subjects in post-16 years education (Hodgen, Marks & Pepper, 2013; Royal Society, 2011) continues to be an area of concern for the U.K. government, as they identified a highly-trained STEM workforce as a crucial component of a globally competitive workforce (Department for Business, Innovation and Skills, 2009; Truss, 2014). Not only are STEM skills shortages an area of concern for businesses, the lack of engagement with data contributes to the disenfranchisement of young people in a society increasingly dominated by the use and manipulation of data. The Science and Technology Committee of the U.K. House of Commons highlighted the growth of the use of "big data" and "machine learning" in decision making in areas across public life, including the financial sector, the legal sector, the criminal justice system, healthcare and education and recognised that there is a need for "decisions by algorithms to be challenged, understood and regulated" (House of Commons Science and Technology Committee, 2018). Citizen science – the participation of the public in scientific research – typically involves the public in the collection of data or processing of large-scale datasets with a view to achieving both scientific (for the scientists) and educational (for the participants) outcomes. The role of citizen science programs in providing the general public with the ability to understand and shape socio-economic and political decisions has been recognised by some researchers, particularly in the area of local environment decision making (Ballard, Dixon, & Harris, 2017; Bonney, Phillips, Ballard, & Enck, 2016). As will be outlined in this chapter, IRIS gives young people a realistic understanding of science and data and enables them to develop skills so that they can contribute in a way that is valued and recognised by the wider science community.

Participation rates in STEM post-16 have been attributed to the demands of the English education system that requires young people to choose three or four subjects post-16 (Hodgen, Marks, & Pepper, 2013) but are also due to the high levels of student disaffection with science and mathematics courses for young people aged 11-16 years (Reiss & Mutjaba, 2017). Over the past two decades, a range of initiatives have been implemented to try and increase the number of students taking STEM subjects post-16 years. These have included changes to the STEM curricula (Cadwallader, 2018; Homer, Ryder, & Donnelly, 2013); STEM ambassadors (Gartland, 2014), and subject-specific continuing professional development for STEM teachers (Allen & Sims, 2017; Institute of Physics, 2017). The most recent figures from the U.K. Joint Council for Qualifications (2018) suggest that these initiatives have had some success, as increasing numbers of students continue with STEM subjects: in 2018 36.2% of all entries were STEM subjects, an increase from 28.0% in 2009, 33.4% in 2014 and 34.5% in 2017 (Joint Council for Qualifications, 2018). Reiss and Mutjaba (2017) highlight that, "It is difficult to apportion success to specific initiatives, given how many there have been" (p. 146). Male students are still more likely to continue with STEM post-16, accounting for 57% of STEM entries (Joint Council for Qualifications, 2018). Apart from biology, where there are more entries for female students than male (9.0% and 6.4% respectively), there are more entries from male students than female students in chemistry (male = 7.0%, female = 6.4%), mathematics (male = 16.2%, female = 8.6%), Further mathematics (male = 3.1%, female = 1.0%) and physics (male = 8.1%, female = 1.9%) (Joint Council for Qualifications, 2018).

STEM Career Education

Career education helps young people to better understand the relationship between educational goals and occupational outcomes. STEM courses are valued for careers such as medicine and engineering (Wang & Degol, 2013), but many students do not know what sort of jobs they could do in STEM beyond these two sub-sets of STEM careers (Bøe & Henriksen, 2013). Career choice is also linked to students' identification of their own ability and confidence levels, as when students are more confident that they can be successful in subjects, including mathematics and science, they are more likely to persist and continue to study STEM subjects (Wigfield & Eccles, 2002). In the context of a career education that has been described as weak (Reiss & Mujtaba, 2017), research has shown that many students make choices about their post-16 education early in their school careers, sometimes as early as 11 years (Tai, Liu, Maltese & Fan, 2006). Reiss and Mutjaba (2017) highlight the importance of embedding STEM careers education into STEM lessons so that students are aware of the range of career opportunities that are available to them if they continue with STEM post-16 years and the value and transferability of the skills they develop through further study in STEM subjects. Reiss and Mutjaba (2017) also suggest the important role that a parent and/or teacher can have in demonstrating to a young person that STEM subjects are worthwhile and that the young person is capable of continued success in STEM. IRIS provides the context for young people to understand the range, value and transferability of STEM subjects and careers and provides them with opportunities to genuinely contribute so that they can see that they are already successful in STEM. The following section sets out IRIS' vision and approach for science education for young people.

An Introduction to the Institute for Research in Schools

The vision of the Institute for Research in Schools (IRIS) is a science education where research is a key element of learning STEM such that both the teacher and student are contributing, valued members of the scientific community (Parker, Fox & Rushton, 2018). IRIS has three core aims:

1. To increase the number of young people and teachers who engage with STEM research at school in a way that raises aspiration, participation and attainment so that more young people, especially those from disadvantaged groups, continue with careers in STEM;

2. To establish within the teaching profession the model of the 'teacher scientist' as opposed to the science teacher so that a role within a school of Leader in STEM Research and Development is integral to the delivery of STEM Education in the school;
3. To enable teachers to be connected as key partners in STEM research with Universities, Industry and Research Councils, and so retain them in the profession.

Launched in March 2016, IRIS is a U.K.-based charity that works to provide an education where research is a key element of STEM (science, technology, engineering and mathematics) learning and where students work on genuine problems (Bloom, 2016; George, 2018; Parker, 2017; Parker, Fox, & Rushton, 2018) in collaboration with their teachers (Rushton & Reiss, 2019). In their study of research active science teachers, Rushton and Reiss (2019) found that teachers' participation in research with their students enhanced their professional development and increased job satisfaction. Rushton and Reiss (2019) suggest that providing teachers with the time and the opportunity to be active in research could be an alternative to U.K.-based teacher recruitment and retention initiatives that are focused on financial incentives. IRIS makes cutting edge research projects open to schools so that students, technicians and teachers can experience the excitement and challenge of science. IRIS makes data accessible to schools, provides teacher training and resources and loans scientific research equipment at no financial cost to the school. This approach to science education resonates with the concept of authentic learning, as students and teachers are contributing to knowledge by focusing on what is not already known, as part of a disciplined inquiry that has value beyond the classroom (Lombardi, 2007; Newmann, Marks, & Gamoran, 1996). Students and teachers collaborate with active researchers based in universities and industry, and IRIS supports schools in building research networks and provides access to data and experimental equipment. This is a social constructivist approach to learning, where students are supported by their peers, teachers and other collaborators, both to develop their understanding of science and to further science itself (Parker et al., 2018; Rushton & Reiss, 2019). In the first three years of IRIS' work, schools have been actively involved in STEM research, involving thousands of students (George, 2018). Collaborating with IRIS, school students have presented at symposia hosted by universities and academic institutions including the Wellcome Trust (London, United Kingdom), the Wellcome Genome Campus (Cambridgeshire, United Kingdom), the Rutherford Appleton Laboratory (Oxfordshire, United Kingdom) and the Physics Department at the University of Oxford (Oxford, United Kingdom). Students have also authored (Hatfield, 2010) and co-authored papers published in peer reviewed journals (Furnell, Shenoy, Fox, & Hatfield, 2018; Hatfield et al., 2019; Parker, Rushton, Thomas & Hatfield, 2019; Whyntie & Harrison, 2014; 2015). This is indicative of the capacity of young people to make a genuine contribution to furthering scientific knowledge and understanding in a way that has, to date, been undervalued (George, 2018), and this is perhaps the distinctive feature of the IRIS approach to citizen science. The following section considers the wider context of school-based citizen science programs, predominantly from the United Kingdom and the United States, recognising that there is a range of approaches and typologies of citizen science programs. (Shah & Martinez, 2016).

A Review of School-Based Citizen Science Literature

An Introduction to School-Based Citizen Science Programs

Having introduced the IRIS vision and approach for science education, it is important to set this initiative in the context of a growing body of literature that considers citizen science in educational settings, including schools and universities. This review will focus on programs that have an element that is specifically designed for, and/or implemented as a school-based initiative. It will consider the aims of school-based citizen science programs, how they are structured, the role of students and teachers, the evaluation of these programs and the challenges and opportunities identified in the literature.

The Purpose and Aims of School-Based Citizen Science Programs

Citizen science projects are increasingly being implemented in school settings which include children aged 5-18 years old. Shah and Martinez (2016) identify U. S. citizen science programs that are appropriate for children in three age groups. For example, students aged 5-11 years old participate in programs such as *Feederwatch* (Bonter & Cooper, 2012); students aged 11-14 years old participate in *Galaxy Zoo* (Lintott et al., 2008); and students aged 14-18 years old participate in *FoldIt* (Curtis, 2018). The purpose and aims of school-based citizen science programs can be placed into four broad categories: Firstly, programs that aim to increase students' scientific literacy; secondly, programs that aim to increase students' motivation for and engagement with scientific inquiry; thirdly, programs that increase student attainment and retention in STEM (science, technology, engineering and mathematics) subjects; and, lastly, programs that increase science capital and promote positive attitudes towards science-based careers.

Increasing scientific literacy through citizen science programs. Shah and Martinez (2016) suggested that the common goal for school-based citizen science programs, "is to motivate students' scientific inquiry, literacy and imaginations to examine the world around them at a young age" (p. 20). An evaluation of a citizen science project led by Trumbull, Bonney, Bascom and Cabral (2000) showed that 78% of participants identified that their scientific thinking and processing had improved because of their involvement in the project, and Cronje, Rohlinger, Crall and Newman (2011) and Woodcock et al. (2017) suggested that citizen science programs can improve participants' understanding of scientific methods, reasoning and research findings. Bonney et al., (2016) found that citizen science projects such as BirdSleuth (Bonney, Phillips, Enck, Shirk, & Trautman, 2014) refined students' hypothesis design and gave students a better understanding of the nature of scientific research. An evaluation of the GLOBE project (Butler & MacGregor, 2003) suggested that students who had participated in the project were more likely to use data effectively in decision making and had improved measurement skills than students, who had not been involved in the program (Bonney et al., 2016). Some citizen science programs have developed resources for teachers to explore student attainment and attitudes towards science. Perelló, Ferran-Ferrer, Ferré, Pou, and Bonhoure (2018) developed a rubric for teachers to assess the scientific, communication and ICT competencies and participation and motivation attitudes of their students who participated in one of five citizen science projects. The study involved 17 teachers and 547 students and suggests that the students developed "excellent" scientific competencies and improved their communication and ICT competencies (Perelló et al., 2018).

Increasing student engagement with scientific inquiry. Linked to the first aim of improvements in scientific literacy is the desire to increase students' engagement with scientific inquiry. Inquiry-based learning has been described as seeking, "to move beyond a pedagogy of instruction and recall of facts to enhance or produce a pedagogy of engagement and practice" (Edwards, McDonnell, Simpson, & Wilson, 2018, p. 197). Bell, Urhahne, Schanze and Ploetzner (2010) suggested that in inquiry-based learning, students are supported by their teachers to ask their own research questions, evaluate evidence and explain and disseminate results. Inquiry-based learning provides both undergraduate students (Coleman & Mitchell, 2014; Freeman et al., 2014; Powell & Harmon, 2014) and school students (Vitone et al., 2016) with the structured experiences necessary to develop and enhance scientific literacy. However, for many school students this way of approaching science is not part of their regular experience of curriculum science and they may, at first, find it challenging (Rushton & Reiss, 2019). Bonney et al. (2016) suggested that citizen science projects can provide a useful way of engaging students who might not otherwise be interested in science and scientific inquiry, especially when those projects have an overarching narrative that resonates with the student, for example environmental conservation, as in the GLOBE project (Bonney et al., 2016; Butler & MacGregor, 2003). The importance of a compelling narrative in providing the "hook" that increases student engagement and motivation for scientific inquiry has been a feature of IRIS projects including Genome Decoders (Parker et al., 2018; Rushton & Parker, 2019; Rushton & Reiss, 2019) and MELT.

Increasing student attainment and retention in STEM subjects. Increased scientific literacy and motivation to participate in scientific inquiry can have positive impacts upon student attainment and retention in STEM subjects (Trautmann, Shirk, Fee & Kransy, 2012), and many programs are intentionally designed to achieve educational goals that are linked to national curriculum standards (Bonney et al., 2016; Butler & MacGregor, 2003). Wieman and Gilbert (2015) suggested that citizen science programs can increase attainment and retention in school contexts, and this finding is replicated in the university setting, with reduced attrition rates for STEM undergraduate students who participate in research in the first year of their degree (Mervis, 2016). Mitchell et al. (2017) suggested that inquiry-based learning for both school and undergraduate students increases attainment as this approach to science education can "promote the retention of material by increasing deep thinking in students" (p. 2). An important feature of many citizen science programs is that they bridge the gap between university-based research and school educations and enable partnerships to develop that smooth this transition (Shah & Martinez, 2016). Developing this pathway through university and school partnerships can increase student retention in STEM subjects (Shah & Martinez, 2016). Evaluations of IRIS programs suggested that when school students engage with authentic research in partnership with university researchers, they develop both the academic experience and interpersonal skills that support a successful transition from school science to university research science (Parker et al., 2018).

Increasing science capital and positive attitudes towards careers in STEM. Students' aspirations towards science-based careers, attitudes towards studying science and STEM careers advice continues to be an important area of research (Archer et al., 2013; DeWitt & Archer, 2015; De Witt, Archer & Moote, 2018; DeWitt, Archer & Osborne, 2014; Reiss &

Mutjaba, 2017). Wallace and Bodzin (2017) suggest that students who participate in citizen science programs may develop more positive attitudes to science and STEM careers; early findings from the first-year evaluation of IRIS' Genome Decoders project support this. Shah and Martinez (2016) identify that citizen science programs can have both individual and community level outcomes and Jordan, Ballard, and Phillips (2012) suggest that the individual learning gains of participants can have a positive impact on the science capital and levels of trust of scientists at a community level. Positive attitudes towards careers in STEM can be encouraged when citizen science projects enable students to link the scientific skills they have gained with a capacity to make a positive contribution to science and the wider community, for example, through increasing environmental awareness (Ballard et al., 2017; Butler & MacGregor, 2003).

These four broad aims are not an exhaustive list of the goals identified by school-based citizen science programs; however, they do outline the key areas of focus. The structure and format of school-based citizens science programs also vary, including the ways in which programs are implemented and the roles of teachers and students.

Implementing School-Based Citizens Science Programs

When identifying four categories of citizen science programs, Bonney et al. (2016) used the term "curriculum-based" citizen science programs to describe programs that involve school students. It is important to note, however, that citizen science programs are delivered in both classroom-based, formal curriculum lessons and in informal settings including extra-curricular clubs (Bonney et al., 2016; Kolb, 2014). Both curricular and extracurricular projects involve students working, supported by teachers or other adults, to contribute to larger "parent" citizen science projects that are often linked to national curriculum standards and frameworks (Bonney et al., 2016). Although many have recognised the capacity of school-based citizen science programs to achieve the four aims linked to student development outlined earlier in this chapter, Calabrese Barton (2012) challenged the capacity of schools to provide contexts for citizen science programs that bring about social change. Calabrese Barton suggested community development and personal empowerment is only possible if the concept of place is recognised as a crucial component of a citizen science program. Calabrese Barton 2012) asks whether "citizen science" should be reframed as "citizens' science," placing importance on science expertise that is linked to societal change and located in a specific community (p. 3). Different typologies of citizen science programs have varying roles for the participants, and school-based citizen science programs also have varying roles for both the students and teachers who participate (Bonney et al., 2016).

The Role of Students in School-Based Citizen Science Programs

Research focused on school-based citizen science programs is predominately situated in ecological and environmental studies, which reflects the early trend in citizen science programs as a whole (Bonney et al., 2009). Ecological and environmental studies have considerations of place and agency at the centre of their research, which means they provide useful settings in which to consider the role students can take in school-based citizen science programs (Ballard et al., 2017). This is important when considering whether school-based citizen science programs provide young people with an opportunity to make a genuine contribution to scientific knowledge (Ottinger, 2010) and to lead a rigorous research project (Calabrese Barton, 2012). Research suggests that when young people have a leading role in the research process, they are better able to work collaboratively to ask and answer research questions (Ardoin, Castrechini, & Hofstedt, 2014; Cammarota & Fine, 2010). Ballard et al., (2017) and Harris and Ballard (2018) found that when students are given ownership of data collection and analysis, disseminating and communicating research findings, and providing guidance to their local community, young people are more likely to develop agency as they are positioned as experts. Ballard et al., (2017) also recognised that students will develop agency when they perceive that the research project they are contributing to is authentic, particularly when their work is associated with multiple aspects of the project work and is not solely linked to data collection that is then used by professional scientists. School-based citizen science projects that develop student agency include repeated experiences for students that generate or develop connections to a place and, provide students with explicit ways to contribute to authentic research science (Ballard et al., 2017). Students who participate in authentic, complex scientific research can contribute at a range of entry points and in a variety of ways (Sadler et al., 2010), from "dirty wet field work, to writing for the public, to digging into complex datasets, to talking to new people" (Ballard et al., 2017, p. 73). Sharing their findings with outside audiences is important for students to develop their identity as researchers, providing their expertise to a range of stakeholders who recognise the value of their work (Ballard et al., 2017; Harris & Ballard, 2018).

The Role of Teachers in School-Based Citizen Science Programs

The role of teachers in school-based citizen science programs is more varied, with some viewing a teacher in a practical, facilitatory role, whereas other programs promote a more collaborative approach of teachers working alongside their students. Shah and Martinez (2016) emphasised the role of teachers in managing students' workloads and providing financial support, as these are factors that have been previously recognised as decreasing motivators and limiting outcomes for students (Nov, Arazy & Anderson, 2011). Harris and Ballard (2018) suggest that the role of the teacher is to position their students as people who do science by encouraging them to take on meaningful roles, by posing open, critical questions and by supporting students to design, implement, evaluate and disseminate their own research projects. Harris and Ballard also recognised that teachers have an important role in emphasising the local and global context and importance of the research. By framing students' work as both part of a larger scientific endeavor, as well as an issue that is relevant to their own community, teachers can encourage motivation and foster engagement. In this model, teachers are "co-learners" who work with students to "figure out new understandings together" (Harris & Ballard, 2018, p. 36). As will be expanded upon in greater detail in the later sections of this chapter, teachers in IRIS programs work in partnership with students, collaborating in areas of authentic science research where the answers are not already known. This orientation positions both the students and their teachers (and technicians) as people who can contribute to a wider intellectual endeavor where they are "capable investigators" rather than disenfranchised automatons (Harris & Ballard, 2018, p. 36).

Challenges and Opportunities for School-Based Citizen Science

School-based citizen science projects encounter challenges that are experienced by citizen science programs in general, as well as those that are specific to their context. School-based citizen science programs are well placed to overcome some of the barriers facing wider citizen science programs, and these challenges and opportunities will be briefly considered here.

The dearth of evaluation and longitudinal studies of outcomes and impact is a challenge for citizen science programs as a whole. Curtis (2018) recognises that there is insufficient published research regarding the implementation of online citizen science in the classroom and that more longitudinal studies and evaluation of outcomes is needed to better understand how implementation of these programs in school settings can be improved. Although there has been some recent research focused on developing tools to allow teachers to assess the development of students' competencies, attitudes and motivation through participation in citizen science projects (Perelló, 2018), there is still much research needed to develop evaluation tools that accurately measure a range of outcomes and capture the duration of any impact observed (Curtis, 2018; Perelló, 2018). Pandaya (2012) has highlighted that citizen science programs that reach the largest audiences tend to appeal to white, affluent and highly-educated participants. School-based citizen science programs have the potential to reach underserved audiences, particularly if programs are implemented as part of a school-based curriculum because this ensures that they provide an educational opportunity and experience that is delivered to all students (Bonney et al., 2016; Curtis, 2018). Evaluation of IRIS programs suggested that school-based citizen science programs implemented as part of an extra-curricular activity can reach students from a range of socio-economic groups, but there is more work to be done to increase the participation of those from more deprived contexts.

School-based citizen science programs face challenges that are specifically connected to their context. These include practical issues of scheduling and availability of staff and student time to prepare for and carry out the program (Butler & MacGregor, 2003). The variability in global education systems, both in terms of the pedagogical frameworks used and the timing of the academic year, can provide additional challenges where programs rely on the collection of systematic and continuous data throughout the year (Butler & MacGregor, 2003). Teachers are a key part of the school-based context; programs need to include significant support and training for teachers if they are to be effective and focused on authentic science research (Bonney et al., 2016). Teaching in the United Kingdom remains a challenging context. In 2017, 80% of teachers surveyed reported that they were considering leaving the profession; this attrition is an increase from 50% in 2014 (Adams, 2018). The loss of experienced teaching professionals in England is keenly felt in science teaching, with English schools facing a severe shortage of science teachers (Migration Advisory Committee, 2016). In this challenging context it can be difficult for teachers to incorporate citizen science programs into curricular and/or extra-curricular activities due to a lack of financial and staffing resources and rapid staff turnover. Bonney et al. (2016) suggested that for many teachers it is impossible to adopt programs that are not specifically linked to national standards or curricular. IRIS has found that school leaders are more willing to embrace the IRIS approach to science education when they understand that outcomes for students include increasing both academic attainment (e.g., progression to university and/or science-based careers) and personal development and enrichment (e.g., increased confidence and resilience). This allows teachers the time to incorporate programs that may not generate immediate attainment gains (e.g., increased GCSE grades) but are focused on longer-term holistic growth (Rushton & Reiss, 2019).

Finally, school-based citizen science projects still encounter the perception that young people are not able to make a genuine contribution to scientific research and furthering understanding (George, 2018; Parker, 2017). IRIS citizen science programs seek to provide young people and their teachers with the opportunity to engage in scientific research and have their contribution to knowledge recognised and valued by the wider community through disseminating their findings in conferences (Rushton & Reiss, 2019) and peer-reviewed papers (Furnell et al., 2018; Hatfield, 2010; Hatfield et al., 2019; Parker et al., 2018, 2019; Whyntie & Harrison, 2014, 2015). IRIS programs work across a range of subjects and disciplines; these areas are outlined in the following section, before a more detailed description of four of the programs and their resources.

An Overview of IRIS Citizen Science Programs

IRIS currently supports schools to participate in research projects that cover different areas of STEM research, including physics (*CERN@school*, *Higgs Hunters*, *James Webb Space Telescope*), biology (*Genome Decoders*, *Well World*), chemistry (*Ionic Liquids*) and environmental sciences (*MELT*). Many of these projects require students to use technology to analyse large datasets.

Physics Research

IRIS physics projects are *CERN@school*, *Higgs Hunters* and *The James Webb Space Telescope project*. *CERN@school* lends the Timepix technology to schools and is a project led by Professor Becky Parker, first developed over ten years ago with Parker and her students. Over the last decade *CERN@school* has grown to include more than 300 schools and thousands of students. Resource materials and online support has been developed by IRIS and schools have used the technology both to support curriculum activities and provide stimulus for student research. Student work includes research on radiation levels during a solar eclipse and a payload launched in space (Furnell et al., 2018; Hatfield et al., 2019; Parker et al., 2019). Research projects with the Timepix technology have led to the development of the Langton Ultimate Cosmic ray Intensity Detector (LUCID), which has enabled students to develop an understanding of radiation in space and large-scale data analysis (Furnell et al., 2018).

Higgs Hunters provides students with the opportunity to work with physicists at the University of Oxford to analyse data from the Large Hadron Collider (LHC), CERN. The LHC is an engineering and scientific triumph, with thousands of people working together to gather and analyse data that will help us better understand the story of our universe. *Higgs Hunters* is a development of the *Zooniverse* citizen science project (Simpson, Page, & DeRoure, 2014), where 20,000 members of the public analysed data captured in the ATLAS experiment at the LHC, looking for the "Baby Higgs" particle. Through the *Zooniverse* project, over one million classifications were made and, *Higgs Hunters* gives students access to this dataset and the opportunity to find new particles and new aspects of physics. Students aged 16-18 years of age, from five different schools presented their findings at a 2-day conference to physicists from the University of Oxford, and their work is currently being collated as a "proceeding" from this event that will be shared with the wider Physics community.

The James Webb Space Telescope or *Webb* project is a partnership between IRIS and astronomers from the Science and Technology Facilities Council's (STFC) U.K. Astronomy Technology Centre (UK ATC). The research aim is to identify potential targets for the James Webb Space Telescope, which NASA has described as the world's premier science observatory for the next decade. Scheduled for launch on an Ariane 5 rocket from French Guiana in early 2021, *Webb* will help astronomers understand more about how planets, stars and galaxies formed and evolved. *The James Webb Space Telescope's* unprecedented infrared sensitivity will allow scientists to explore massive clouds of cosmic dust to better understand how the materials around and within us are formed and spread through space. In the *Webb* project, launched by IRIS in September 2018, students are analysing spectral data from a previous mission – the Spitzer Space Telescope – which has observed over thirteen thousand-point sources. Astronomers and education staff from the UK ATC have worked with IRIS to develop a classroom activity in which students examine and classify this rich collection of existing data. The ultimate goal is for students to collaborate with UK ATC astronomers on the selection of potential targets for *The James Webb Space Telescope* and the development of an observing proposal which makes the scientific case for pointing the huge space observatory at these objects. The resulting work will be used to not only identify potential targets for *The James Webb Space Telescope* but will also produce the first fully classified catalogue of these sources, which is expected to be of scientific value in its own right.

Biology Research

The two biology projects, *Genome Decoders* and *Well World*, are discussed in detail later in this chapter. In summary, *Genome Decoders* is a collaboration with scientists based at the Wellcome Sanger Institute, Cambridgeshire, United Kingdom where students are annotating the entire DNA of the human whipworm genome. The human whipworm is a parasite that causes a neglected tropical disease and the annotation could identify vulnerabilities that could contribute to future treatments or cure. *Well World* is a project based upon the work of school students from a school in Canterbury, United Kingdom who have explored the links between biodiversity and physical and mental well-being of young people.

Chemistry Research

IRIS' chemistry project is *Ionic Liquids*, where students explore the design, synthesis and application of ionic liquids. IRIS' role in *Ionic liquids* is to provide guidance on how to develop student research projects with external partners, developed Ionic Liquids in collaboration with students and teachers at other institutions. This partnership approach is different to the role IRIS has played in initiating and developing projects such as *CERN@school*, *Genome Decoders* and *MELT*.

Environmental Science Research

Monitoring the Environment, Learning for Tomorrow (*MELT*) is a project with two strands, "carbon footprint" which is primarily aimed at students aged 9-14 years old and "earth observation," aimed at students aged 14-18 years old. Through the "carbon footprint" strand, students can calculate the carbon footprint of their homes and schools, which provides baseline data for researching activities to reduce the carbon footprint of their school. The "earth observation" strand enables students to analyse images identifying glaciological features and iceberg calving episodes from images taken by the Sentinel-1 satellite. This analysis is shared with scientists based at the Centre for Polar Observation and Monitoring, U.K. These two research strands can be followed independently or jointly depending on the individual context of participating schools. *MELT* was launched in 2018 and to date, 30 schools are contributing their research in both strands of the project.

The following section gives detailed case studies of two IRIS programs. *Genome Decoders* outlines effective frameworks for school-based citizen science programs and *Well World* suggests template materials for supporting student agency in school-based citizen science programs.

Genome Decoders and *Well World* – Two Examples of IRIS Citizen Science Programs

Genome Decoders – Frameworks for School-Based Citizen Science Programs

Genome Decoders is a collaboration between IRIS and the Parasite Genomics team at the Wellcome Sanger Institute, the Wormbase team at EMBL-EBI (European Molecular Biology Laboratory-European Bioinformatics Institute) and the Wellcome Genome Campus Public Engagement team. All the scientific partners involved in this project are based at the Wellcome Genome Campus, Cambridge, United Kingdom, the site of world-leading organisations in the fields of genomics and computational biology. The public engagement team, part of Connecting Science, supports scientists from across the campus to engage public audiences with genome research. *Genome Decoders* is a research project that enables school students to annotate the estimated 15,000 genes of the Human Whipworm (*Trichuris trichuria*) genome. The whipworm is a parasite that causes *Trichuriasis*, a currently incurable neglected tropical disease (NTD) that impacts over 500 million people globally, mainly children living in the poorest parts of the tropics, including Africa, Asia and South America (Barda, Keiser & Albonico, 2015). Annotating the genome enables the identification of the genes that code for proteins important during infection and disease which can lead to identifying a treatment and or cure (Rushton et al., 2019). The overarching aim of *Genome Decoders* is to provide an opportunity for school students to contribute to an authentic genomics research project, specifically to support the

NTD research community, and to explore the impact of this on the students and their teachers and technicians. *Genome Decoders* was primarily focused at students aged 15-18 years of age.

Genome Decoders is founded upon a collaborative and co-created approach to citizen science, where students and their teachers co-create knowledge with scientists and other specialist partners. The learning goals and scientific goals are distinct but were established at the same time and are of equal importance. The learning goals are appropriate and achievable and are rooted in genuine scientific research that contributes knowledge that is recognised by the wider scientific community as valid, reliable, accurate and useful (Rushton & Reiss, 2019). Although scientists established the research questions for *Genome Decoders*, the development of the resources was a collaborative process between the students, teachers and scientists and, at the end of Year 1, has led to co-created out-puts, including students co-creating training videos at the end of Year 1 of the project. Genome Decoders has been of direct benefit to the researchers, and in 2018 the Genome Decoders Team won the Wellcome Genome Campus Engaged Team Prize (Connecting Science, 2018). *Genome Decoders* is a hybrid of the two frameworks outlined by Bonney et al. (2009) and Bonney et al. (2016). The following section sets out the structural framework of *Genome Decoders* during Year 1, including a detailed outline of the different development and implementation phases. This may prove a useful framework for others planning school-based citizen science programs that have both educational and scientific outcomes.

The Framework of Genome Decoders

The development and implementation phases of Genome Decoders are included in Appendix A. During the Development phases, school students contributed to the annotation of a genome using *Apollo*, a web-based genomic annotation editing platform (Lee et al., 2013). The web-based nature of the software meant that students were able to work at both home and school. To annotate a gene, students interpreted the sequencing data generated by Illumina and Pacific Biosciences sequencing platforms to identify gene-coding regions. Using this data, students were able to predict gene structure. During early development phases, scientists from both the Wormbase team and the Parasite genomics team established the annotation interface used by students to collaboratively annotate the genome. *Apollo* software is a tool generally used by post-graduate and post-doctoral researchers and professional curators (Dunn et al., 2019) and this, coupled with the use of complex language and concepts related to genome research, meant that training materials had to be developed to support students, teachers and technicians through the process.

Resources included an introductory booklet that set out the context of genomics and the key principles of gene structure and, written "getting started" guidelines to support participants with using the web-based software. These were supplemented with narrated walk-through tutorial videos that were made available on *YouTube*. A key feature of the development of these resources was the participation of school students who were involved in trialing and refining the training materials in the both the development and implementation phases. This meant that the experiences and perspectives of the students as collaborators was integral to the project from the outset. The importance of the role the students had in the project was also clearly communicated at the beginning of the project, with Dr. Matthew Berriman, Senior Group Leader, Wellcome Sanger Institute writing in the introduction to the student guide:

> With lots of volunteers looking at the genome, we should be able to find every gene and find plenty of new clues to what makes human whipworms different to other animals, and maybe some new vulnerabilities that can be attacked. I've always found it exciting to be the first person to see the basic building blocks of an organism. For the human whipworm, we haven't yet looked, so you'll be the first. Can we really work as a team to analyse every gene? I'm really excited to see what you'll uncover. (Berriman, 2017)

Implementation Phase

The project was officially launched in September 2017 at a day-long event at the Wellcome Genome Campus which was attended by students and staff from participating schools and featured presentations from the research scientists, public engagement workshops and an opportunity for the school students to engage with genome research and research scientists as well as introductory videos (The Institute for Research in Schools, 2017). Dr Julian Rayner, Director of Communicating Science at the Wellcome Genome campus said at the event:

> This project shows the enormous benefit of incorporating research into education. These genes are not simply a paper exercise, they are real genes, which no one has looked at before. The students will be the world experts on them and, will be contributing directly to our understanding of a major global pathogen. (Rayner, 2017)

School students who had been part of the development phase of the training materials presented their experiences to encourage and inspire other school students to participate. During the first implementation phase, students were recruited to the project through IRIS' network of schools (Appendix A). During this phase, 52 schools had registered, with over one thousand students allocated an individual avatar to use with the *Apollo* software. This contrasts with other citizen science projects which use platforms that are specifically created for citizen science projects, such as *Zooniverse* (Simpson et al., 2014). The use of *Apollo* software in *Genome Decoders* gives school students access to a research tool that they otherwise would not encounter until at least post-graduate studies. Each student using the software has a unique avatar allowing their participation and accuracy to be measured both at an individual and school level. This ability to electronically track and measure participation has been a common characteristic of previous contributory citizen science projects (Becker-Klein, Peterman & Stylinski, 2016; Bonney et al., 2009); however, the use of this software solely with school-age children is unique (Rushton & Parker, 2019).

During first and second implementation phases, students were given practice regions to explore and annotate, supported by the training materials and guidance. Teachers were connected to each other and the research scientists through an email group where they could ask questions and share their experiences. To increase the level of direct and specific feedback to students, a token system was introduced at the beginning of phase three of the implementation phase (Appendix A). During phase four of the implementation of *Genome Decoders*, the training materials and introductory booklet were reviewed, with feedback from students, teachers and technicians incorporated to improve the resources for the beginning of Year 2 in September, 2018. Two school students spent a week at the Wellcome Sanger Institute, working with the research scientists and contributing to the evaluation of the project. Phase 4 was also an opportunity to review school participation and to actively recruit new schools to participate in Year 2.

Appendix A sets out the different phases of development and implementation, with an indicative timeline and focus of activity for each phase. Broadly speaking this timeline should provide a useful guide for those working with school-based citizen science programs. However, it is important to note that every project will have variability in the length of each phase and the nature and focus of activity during each phase. Rushton and Parker (2019) set out the evaluation opportunities for citizen science programs that have the capacity to track individual and group activity and accuracy levels. Appendix A also indicates that there are opportunities for informal, formative evaluation of the program during the development and implementation phases of school-based citizen science programs like *Genome Decoders*. Appendix A shows that during the development and implementation phases, it is important that school-based citizen science programs develop activities and resources that support students to actively engage with the research project. Students need support to develop an awareness of, and interest in, research relevant to the project, and students should be encouraged where possible to identify their own research questions, plan their investigation and share their findings. IRIS' *Well World* project includes opportunities for students to create their own research questions and this project and supporting materials are outlined in the following section.

Well World – Resources to Support Student Agency in School-Based Citizen Science Projects

IRIS' *Well World* project developed from the work of school students (16-18 years of age) and their teachers, from a school in the United Kingdom. The students posed an initial research question, "Does biodiversity make us happy?" inspired by a wooded area of their school site, known locally as "The Orchard." Students and teachers were initially motivated to research the links between biodiverse areas and health and well-being to raise awareness and appreciation of the value of "The Orchard" within their school community. When developing their study, students considered research that demonstrates the positive role green space can have in promoting physical and mental well-being (Barton & Pretty, 2010; Burgess, Harrison, & Limb, 1988; Pretty, Peacock, Sellens, & Griffin, 2005; Wood, Hooper, Foster, & Bull, 2017). Green space includes a range of environments including national parks, public parks and domestic gardens (Kaplan, Kaplan, & Ryan, 1998). In the context of increasing global concern about the protection of the environment, including anthropogenic global warming, deforestation, soil degradation and water and air pollution (Millennium Ecosystem Assessment, 2006) and rising levels of child and adult physical inactivity and obesity (U.K. Department of Health, 2009), multi-institutional initiatives including *Green Gym* (Pretty et al., 2007) and *Blue Gym* (White, Pahl, Wheeler, Fleming, & Depledge, 2016) have sought to reconnect populations to their environments and promote positive health outcomes associated with physical activity in the natural environment. The term "green exercise" was first used by Pretty et al., (2005) and has been defined as "activity in the presence of nature" (Barton & Pretty, 2010 p. 3947).

Students designed an experiment to answer the question "Does biodiversity make us happy?" In the experiment they identified three areas in the school site that participants were asked to walk in for a similar distance and length of time. These areas included an indoor school area, the outdoor running track and the wooded area known as "The Orchard." Participants' physical well-being was measured before and after walking through pulse and blood pressure measurements and their mental well-being was assessed through the shortened Spielberger State Trait Anxiety Index (STAI) questionnaire (Marteau & Bekker, 1992). The students were supported in their research by an expert curator of insects at the Museum of Zoology and their biology teacher. Financial support was provided by a Royal Society Partnership grant, awarded in 2015. The Royal Society partnership grants have been established since 2000, and the scheme awards up to £3000 to schools or colleges to support students aged 5-18 years of age to buy equipment to run a STEM investigation project in partnership with a STEM professional (Royal Society, 2018). This grant enabled the purchase of field equipment to measure biodiversity levels, equipment to measure blood pressure and pulse rates and travel and related expenses to enable a scientist to visit the students and support their work measuring biodiversity in their school site. According to Turner (2018) in reference to the *Well World* project,

> Working to make greenspaces around schools more biodiverse and accessible to students has never been more important. We are living in a world where people are becoming more disconnected than ever from nature and projects like this are a fun and engaging way to counteract this problem. Helping young people to connect with local biodiversity and learn more about it is a vital first step to engaging them in conservation. (personal communication, June 25, 2018)

Students and teachers also recognised the valuable experience of the research project, according to the biology teacher:

> Taking part in this project has been incredibly rewarding both as a teacher but also as a biologist. It has been wonderful to watch successive sets of students discover the trials and tribulations of taking part in real research from planning to evaluation and experience the thrill of finding out something new. I am now able to reference our research in the classroom and pupils can experience the importance and relevance of experimental work on their doorstep. After 20 years in the classroom this experience has reminded me of my love of science, not just my love of teaching it. (Goodfellow, personal communication, February 27, 2018)

Two videos of the *Well World* project were produced by IRIS, with contributions from the students, teachers and academics involved in the project (Institute for Research in Schools, 2018). IRIS' key role in developing *Well World* beyond the initial school was to provide resources that would support students and teachers in a range of schools to develop their own research exploring the links between the environment and mental and physical well-being. These resources are shared in the following section and include:

- A guide to planning an investigation (Table 1)
- A summary report template (Appendix B)
- A journal club exercise for students (Appendix C)

Supporting Resources

Planning an investigation. Conversations with teachers highlighted the need for support in identifying the different stages of research and planning appropriate activities for each phase of research. This was particularly relevant for a project such as Well World, as the project introduced students and teachers to an area of research where they could determine their own research questions, rather than a project like Genome Decoders, where the research questions were established. Table 1 sets out five phases of research and suggests activities and research questions. Table 1 provides guidance on how to develop and implement a research project and could be used by teachers or simply given to a group of students to use. There are no suggestions of how long each phase should take so that teachers and students are able to incorporate the project into the available time they have; for example, an after-school club, a curriculum lesson or an individual investigation by a student may alter time requirements. Sources of information regarding ethics and risk assessments are signposted in this document so that as with the research, students and teachers are given agency to lead their own research.

Summary report template. Another area where students and teachers identified that they needed support was in writing-up their research findings, as this was an area where they felt they did not have the necessary skills or experience to communicate their research findings. As with Table 1, the report template (Appendix B) was created so that teachers could use it with their students or for students to use independently. The report template provides a scaffold that divides the task of writing research findings into smaller tasks (e.g., recording research questions, describing results, identifying next steps). This provides a template for students and teachers creating a summary of their research that can then be part of an endeavor that builds over different school years, enabling students and teachers to see that they are part of a longitudinal research effort. Writing a research summary can also encourage students and teachers to disseminate their findings more widely perhaps through a presentation in a school assembly or parents' evening, giving a talk or poster presentation at an IRIS conference, or preparing a paper for publication.

Journal club activity. To further underpin the concept that research should be informed by previous scholarship in the area, and to develop the academic literacy of students and teachers, a journal club activity is included in the supporting resources (Appendix C). This task is relatively simple and asks a student to read a suggestion abstract that is related to their research area and to work through the abstract using a set of questions. This encourages the reader to dissect the different components of research contained in an authentic example of research, providing them with a sense of how research planning evolves into published findings. This task enables the reader to consider their own research in the context of established scholarship, fostering a sense of being part of a research community that has developed from previous work whilst still retaining the idea that there is much the reader can contribute to research in this area. Holding a journal club as part of an informal, group activity can reduce some of the barriers that young people may face when accessing this type of material as working together they can decipher unfamiliar terminology, discuss the relevance of the research to their own lives and communities and develop a sense of working as team of researchers.

Table 1. A Guide to Planning an Investigation

Phase 1 – Identifying the key aspects of the investigation
- What aspect of this topic are you particularly interested in? - What has already been written about this topic? Does this research shape your area of interest? - Form a research question and break this down into sub-questions - What will you measure? How will you measure it? What equipment will you need? - Consider how you will analyse your datum. Will this have an impact on data collection?
Phase 2 – Permissions, Risk Assessments and Project Timetable
- Consider who will take part in your study. Who do you need to ask permission? Are you working with other students/schools? - Ensure that you work with a teacher to write a Risk Assessment for any investigations that you plan and ensure that you have permission from the appropriate member of school staff. - Ensure that you consider the ethical implications of your research. - Create project timetable. What are your key activities? Which order would be most sensible to carry them out?
Phase 3 – Data collection and recording
- How are you going to collect and record your data? How will you collate data between different groups? - Once you have finished the first phase of data collection, review the accuracy and validity of your datum? Are there any improvements you could make to the collection and recording?
Phase 4 – Data analysis and evaluation
- Analyse your data using the information you have collected and recorded, how does this relate to your research questions? Have you answered them? What further questions are raised? - What issues have you encountered in your investigation? How have you dealt with them? What should be done differently in another phase of investigation?
Phase 5 – Communicating your findings
- Have you completed the Summary Report sheet? - How else can you communicate your results? Could you present them to your peers as part of a seminar or assembly? Could you write a report of your findings to members of your community who might be interested (e.g. Headteacher, local university)? Could you share your report on your school website? - Can you write your research up in an academic poster or journal article?

The final section of this chapter considers the educational implications and future research directions of involving young people in genuine scientific research in light of the literature considered in this chapter and the examples presented from IRIS.

Educational Implications and Future Research Directions

1. Establish Research Collaborations That Are Well-Planned and Structured

Previous school-based citizen science programs have recognised some of the challenges of working within the frameworks of school academic years, the demands of external examinations and the workload and retention of teachers. If teachers and school students are to make a genuine contribution to research, they need to be involved in the planning of research collaborations so that they can develop projects with frameworks that will be easier to incorporate into school structures and curricula. School staffing and teaching timetables are often established 6 months ahead of the academic year. Therefore, if teachers are to be given time to facilitate school-based citizen science programs, the funding and/or staffing resources need to be identified with schools so that teachers can incorporate programs into staffing models as early as possible. Where possible, provisions should be made so that school-based citizen science programs are not the sole responsibility of one teacher but shared across a department or faculty. Citizen science programs should be supported by a member of the school senior management team, so that this work is incorporated into the school's approach to STEM learning and the development of their staff and students. This will support sustainable and continued participation in research projects that will often extend across more than one academic year. This is an important area to consider when planning research collaborations that incorporate schools.

2. Develop Research Collaborations Where Students Have Agency

Much of IRIS' work is to develop resources, frameworks and guidance so that teachers can support their students to develop agency within a research collaboration. Critiques of citizen science projects suggest that some projects do not give participants the opportunity to ask and answer research questions that are relevant to their lives and communities. In IRIS' citizen science projects, school students, teachers, researchers work together to identify questions, develop appropriate methods and resources and communicate their research. This agency is a key part of raising students' aspirations towards STEM careers, developing their self-efficacy, and soft skills including teamwork, resilience, and communication skills. For many students, having the opportunity to better understand the issues that face both their local communities (e.g., *Well World*) and global challenges (e.g., *Genome Decoders*, *MELT*) and to be part of research that seeks to provide solutions is a key source of motivation to initially participate and to continue through periods of challenge and difficulty.

3. Raise Students' Awareness of and Aspirations towards STEM Careers

School-based citizen science has been recognised as a way of reaching new and more diverse audiences. Certainly, schools provide an opportunity to increase participation in science through programs that target socio-economic groups or geographical regions that have reduced engagement in STEM careers. As has been described throughout this chapter, students who participate in IRIS citizen science programs experience active research science where they are a contributing member of a research collaboration. These projects may include opportunities to visit scientific institutions including CERN, Wellcome Sanger Institute (Cambridgeshire, United Kingdom), and Rutherford Apple Laboratory (Oxfordshire, United Kingdom). As is shown by Rushton and Parker (2019), these opportunities raise students' awareness of the nature of these institutions and the variety of STEM careers associated with research. Crucially, when students encounter these institutions, they do so not as visitors but as collaborators and partners, with a specific and valued role in a wider collaboration. This does much to increase the aspirations of a young person to continue with a STEM career because their involvement in STEM post-18 years of age would simply be an extension and continuation of an experience and role that they have already begun whilst at school. By involving young people as researchers, their aspirations towards STEM careers increases because they understand that science is an area where there is still much to learn, and their contribution is needed to advance understanding further. Much school-based citizen science has viewed the role of teachers as research managers, facilitators and funders, providing practical and financial support (where appropriate) for students. IRIS' experience of school-based research suggests that teachers have a more significant role, mentoring students through research and providing scientists with a critical understanding of the challenges and opportunities of working with young people in this way. IRIS continues to explore the experiences of teachers who participate in research with their students to better understand the role teachers play in school-based citizen science programs that have high quality scientific and educational outcomes.

4. Create Research Communication and Dissemination Opportunities for School Students

A key component of IRIS' approach is to provide school students the opportunity to share their research with their peers, their local communities and the wider research communities. Over the last five years, IRIS has supported conferences and symposia that bring students together to discuss their research in events focused around a specific research area or theme. In 2018-2019, IRIS will host five regional conferences in Exeter, London, Cambridge, York and Edinburgh where young people will have the opportunity give presentations and display research posters across the range of IRIS projects to an audience of researchers, students and teachers. This sharing of multi-disciplinary research through a regional conference will foster a sense of research across a wider community and potentially generate collaborations. IRIS continues to support students to disseminate their research through peer-reviewed publication. Young people working with IRIS can contribute to the design and creation of videos that introduce their research to a wider audience. For example, the introductory videos for *Well World* were co-created by the students as part of a team with researchers (i.e., IRIS staff, video producer, and director). These varied opportunities enable a greater range of young people to communicate their research, and high participation rates are linked to projects that have well planned opportunities for students to have agency, and to develop soft skills that are transferable to other areas of their lives.

5. Enhance Students' Soft Skills and Self-Motivational Beliefs through Participation in Research

If students are to develop soft skills, the school-based citizen science programs need to enable students to have agency and provide them with opportunities to communicate their research beyond their own research group. Students who participate in genuine research develop self-motivational beliefs including greater self-efficacy and interest in STEM. Through experiencing authentic research, students not only have the ability to communicate to a variety of audiences and increase their capacity to work collaboratively in a team, but they also work individually, develop resiliency and the ability to sustain their motivation and engagement when the work becomes challenging. Involving students in the evaluation of their research provides them further opportunities to think critically and be reflective about their work and their own performance. These skills (communication, collaboration), self-motivational beliefs (self-efficacy, interest), and constructs (resiliency) are sometimes difficult to measure and harder still to directly attribute to specific research projects. Rushton and Parker (2019) suggest some approaches to evaluation that could be useful in developing this aspect of school-based citizen science programs. Future development of school-based citizen science projects should consider how best to document the soft skills development of students (and indeed other participants) and the ways in which this development can be promoted as part of the educational outcomes of the program.

ACKNOWLEDGEMENTS

The authors would like to thank the school students, teachers and technicians, academic and industry partners who have worked with IRIS over the past three years. IRIS' work has been supported by grants from Welcome, the Garfield Weston Foundation, the UK Space Agency and the Royal Commission for the Exhibition of 1851.

Appendix A

THE DEVELOPMENT AND IMPLEMENTATION PHASES OF Genome Decoders

DEVELOPMENT PHASE

LENGTH: 6 MONTHS (MARCH - AUGUST 2017)

PHASE ONE
Length: 3 months (March - May 2017)

Focus of activity: Project aims and objectives identified; financial, time and staffing resources identified and allocated; project implementation plan written and agreed with partners.

PHASE TWO
Length: 3 months (June - August 2017)

Focus of activity: Planning for launch event; preparation and refining of written training materials e.g. subject brochure and background reading references; development of evaluation plan and tools; testing of training methodology with a small group of teachers and students; testing of training material with students from one school to refine ahead of main roll out.

IMPLEMENTATION PHASE (YEAR ONE)

LENGTH: 12 MONTHS (SCHOOL YEAR: SEPTEMBER 2017 - AUGUST 2018)

PHASE ONE
Length: 3 months (September - November 2017)

Focus of activity: Launch event in September 2017; recruitment and registration of schools; creation of video training materials; school training using training datasets; school engagement with evaluation baseline.

PHASE TWO
Length: 4 months (October 2017 - January 2018)

Focus of activity: student training; student annotations; student and teacher feedback regarding training materials and annotation process, as well as accuracy of work completed to date.

PHASE THREE
Length: 5 months (February 2018 - June 2018)

Focus of activity: student annotations; direct feedback from scientists to students; training for students, teachers and technicians via webinars.

PHASE FOUR
Length: 2 months (July - August 2018)

Focus of activity: school engagement with evaluation including surveys, focus groups and interviews; refinement of training materials by students and scientists; communication of research via conference presentations and research papers; planning for Year two implementation phase. Recruitment of new cohort of schools and additional students in existing schools to replace school leavers.

APPENDIX B

1 School name
2 School address
3 Teacher contact
4 Well World researchers (names)
5 Date
6 Description of biodiversity results. Please include details of each site, you could include a map and/or labelled photographs on a separate sheet.
7 Description of well-being results. Remember to include the results of both your physical and psychological measures. Is your school already involved in any well-being schemes? If so, what are they and what activities does the school undertake? E.g. mindfulness, green exercise

APPENDIX C

Exploring 'Green exercise' research

Green exercise = activity in the presence of nature

Using Google scholar, find the following article:

Thompson Coon, J., Boddy, K., Stein, K., Whear, R., Barton, J., & Depledge, M. H. (2011). Does participating in physical activity in outdoor natural environments have a greater effect on physical and mental wellbeing than physical activity indoors? A systematic review. *Environmental science & technology*, *45*(5), 1761-1772.

In pairs/small groups, read through the abstract of the paper and think about the following questions:

1) What is the focus of the study?
2) Why is this study important?
3) How does this study relate to other research/ ideas?
4) Which methods are used in the study?
5) What are the ethical considerations?
6) What are the key findings?
7) How could you develop this into a research question you could investigate?

REFERENCES

Adams, R. (2018, April 1). Vast majority of teachers considered quitting in past year – poll. *The Guardian*. Retrieved from https://www.theguardian.com/education/2018/apr/01/vast-majority-of-teachers-considered-quitting-in-past-year-poll

Allen, R. & Sims, S. (2017). *Do National STEM Learning Network professional development courses keep science teachers in the classroom?* Education Datalab. Retrieved from https://wellcome.ac.uk/sites/default/files/science-teacher-retention.pdf

Archer, L., DeWitt, J., Osborne, J., Dillon, J., Willis, B. & Wong, B. (2013). "Not girly, not sexy, not glamorous": Primary school girls' and parents' constructions of science aspirations. *Pedagogy, Culture & Society, 21*(1), 171-194. doi:10.1080/14681366.2012.748676

Ardoin, N. M., Castrechini, S. & Hofstedt, M. K. (2014). Youth–community–university partnerships and sense of place: two case studies of youth participatory action research. *Children's Geographies, 12*(4), 479-496. doi:10.1080/14733285.2013.827872

Ballard, H. L., Dixon, C. G., & Harris, E. M. (2017). Youth-focused citizen science: Examining the role of environmental science learning and agency for conservation. *Biological Conservation, 208*, 65-75. doi:10.1016/j.biocon.2016.05.024

Barda, B. D., Keiser, J., & Albonico, M. (2015). Human trichuriasis: Diagnostics update. *Current Tropical Medicine Reports, 2*(4), 201-208. doi:10.1007/s40475-015-0063-x

Barton, J., & Pretty, J. (2010). What is the best dose of nature and green exercise for improving mental health? A multi-study analysis. *Environmental Science & Technology, 44*(10), 3947-3955. doi:10.1021/es903183r

Becker-Klein, R., Peterman, K., & Stylinski, C. (2016). Embedded assessment as an essential method for understanding public engagement in citizen science. *Citizen Science: Theory and Practice, 1*(1), 8. doi:10.1177/0963662515607406

Bell, T., Urhahne, D., Schanze, S., & Ploetzner, R. (2010). Collaborative inquiry learning: Models, tools, and challenges. *International Journal of Science Education, 32*(3), 349-377. doi:10.1080/09500690802582241

Berriman, M. (2017, September 28). *The role of Genome Decoders in the fight against Neglected Tropical Diseases*. Speech given at the Genome Decoders launch, Wellcome Genome Campus, Cambridgeshire, England.

Bloom, A. (2016). Dr Becky Parker: Our 'amazing' universe can get students excited about studying science. *The Times Education Supplement*. Retrieved from https://www.tes.com/news/dr-becky-parker-our-amazing-universe-can-get-students-excited-about-science

Bøe, M. V., Henriksen, E. K., Lyons, T., & Schreiner, C. (2011). Participation in science and technology: Young people's achievement-related choices in late-modern societies. *Studies in Science Education, 47*(1), 37–72. doi:10.1080/03057267.2011.549621

Bøe, M. V., & Henriksen, E. K. (2013). Love it or leave it: Norwegian students' motivations and expectations for post-compulsory physics. *Science Education, 97*(4), 550-573. doi:10.1002/sce.21068

Bonney, R., Cooper, C. B., Dickinson, J., Kelling, S., Phillips, T., Rosenberg, K.V., & Shirk, J. (2009). Citizen science: a developing tool for expanding science knowledge and scientific literacy. *BioScience, 59*(11), 977-984. doi:10.1525/bio.2009.59.11.9

Bonney, R., Phillips, T. B., Ballard, H. L., & Enck, J. W. (2016). Can citizen science enhance public understanding of science? *Public Understanding of Science, 25*(1), 2-16. doi:10.1177/0963662515607406

Bonney, R., Phillips, T. B., Enck, J., Shirk, J., & Trautmann, N. (2014). Citizen science and youth education. *National Research Council Committee on Out-of-School Time STEM.* Washington, DC: National Research Council.

Bonter, D. N., & Cooper, C. B. (2012). Data validation in citizen science: A case study from Project FeederWatch. *Frontiers in Ecology and the Environment, 10*(6), 305-307. doi:10.1890/110273

Brown, M., Brown, P., & Bibby, T. (2008). "I would rather die": Reasons given by 16-year olds for not continuing their study of mathematics education. *Research in Mathematics Education, 10*(1), 3–18. doi:10.1080/14794800801915814

Burgess, J., Harrison, C. M., & Limb, M. (1988). People, parks and the urban green: A study of popular meanings and values for open spaces in the city. *Urban Studies, 25*, 455-473. doi:10.1080/00420988820080631

Butler, D. M., & MacGregor, I. D. (2003). GLOBE: Science and education. *Journal of Geoscience Education, 51*(1), 9-20. doi:10.5408/1089-9995-51.1.9

Cadwallader, S. (2018). *Assessing science practical skills at A level.* Ofqual blog. Retrieved from https://ofqual.blog.gov.uk/2018/05/10/assessing-science-practical-skills-at-a-level/

Calabrese Barton, A. M. (2012). Citizen (s') Science. A Response to "The Future of Citizen Science." *Democracy & Education, 20*(2), 1-4. Retrieved from https://democracy educationjournal.org/cgi/viewcontent.cgi?referer=https://scholar.google.co.uk/&httpsredir=1&article=1044&context=home

Cammarota, J. & Fine, M. (2010). *Revolutionizing education: Youth participatory action research in motion.* New York, NY: Routledge.

Coleman, J. S. & Mitchell, M. (2014). Active learning in the atmospheric science classroom and beyond through high-altitude ballooning. *Journal of College Science Teaching, 44*(2), 26-30. doi:10.2505/4/jcst14_044_02_26

Connecting science (2018). *Connecting science public engagement prizes. Retrieved* from https://publicengagement.wellcomegenomecampus.org/connecting-science-public-engagement-prizes

Cronje, R., Rohlinger, S., Crall, A. & Newman, G. (2011). Does participation in citizen science improve scientific literacy? A study to compare assessment methods. *Applied Environmental Education & Communication, 10*(3), 135-145. doi:10.1080/1533015X.2013.876302

Curtis, V. (2018). Realising the potential of online citizen science. In *Online Citizen Science and the Widening of Academia* (pp. 167-187). Palgrave Macmillan, Cham. doi:10.1007/978-3-319-77664-4_8

Department for Business, Innovation and Skills (2009). *Higher ambitions: The future of universities in a knowledge economy.* London, United Kingdom: Department for Business, Innovation and Skills.

DeWitt, J. & Archer, L. (2015). Who aspires to a science career? A comparison of survey responses from primary and secondary school students. *International Journal of Science Education, 37*(13), 2170-2192. doi:10.1080/09500693.2015.1071899

DeWitt, J., Archer, L. & Moote, J. (2018). 15/16-year-old students' reasons for choosing and not choosing physics at A Level. *International Journal of Science and Mathematics Education*, 1-17. doi:10.1007/s10763-018-9900-4

DeWitt, J., Archer, L., & Osborne, J. (2014). Science-related aspirations across the primary–secondary divide: Evidence from two surveys in England. *International Journal of Science Education, 36*(10), 1609-1629. doi:10.1080/09500693.2013.871659

Dunn N.A., Unni D. R., Diesh C., Munoz-Torres, M., Harris, N. L., Yao, E., . . . Lewis, S. E. (2019). Apollo: Democratizing genome annotation. *PLOS Computational Biology* 15(2): e1006790. doi.org/10.1371/journal.pcbi.1006790

Edwards, R., McDonnell, D., Simpson, I., & Wilson, A. (2018). Educational backgrounds, project design and inquiry learning in citizen science. In C. Herodotou, M. Sharples, & E. Scanlon (Eds.) *Citizen inquiry: Synthesising science and inquiry learning* (pp. 195-209). Oxford, United Kingdom: Routledge.

Freeman, S., Eddy, S. L., McDonough, M., Smith, M.K., Okoroafor, N., Jordt, H., & Wenderoth, M.P. (2014). Active learning increases student performance in science, engineering, and mathematics. *Proceedings of the National Academy of Sciences*, *111*(23), 8410-8415. doi:10.1073/pnas.1319030111

Furnell, W., Shenoy, A., Fox, E., & Hatfield, P. (2018). First results from the LUCID-Timepix spacecraft payload onboard the TechDemoSat-1 satellite in LEO. *Advances in Space Research 63*(5), 1523-1540. doi:10.1016/J.ASR.2018.10.045

Gartland, C. (2014). *STEM strategies: Student ambassadors and equality in higher education.* London, United Kingdom: IOE Press.

Gilbert, J., Bell, D., Black, P., Donnelly, J., Harlen, W., Ince, M., . . . Reiss, M. (2006). *Science education in schools: Issues, evidence and proposals.* London: TLRP. Retrieved from www.tlrp.org/pub/documents/TLRP_Science_Commentary_FINAL.pdf

George, M. (2018). Teachers want to be scientists too. *The Times Education Supplement.* Retrieved from https://www.tes.com/news/tes-profile-science-advocate-becky-parker

Goodfellow (2018, February 27). Interview with E. Rushton, Canterbury, United Kingdom.

Harris, E., & Ballard, H. (2018). Real science in the palm of your hand. *Science and Children*, *55*(8), 31-37. doi:10.2505/4/sc18_055_08_31

Hatfield, P. (2010). Using line intensity ratios to determine the geometry of plasma in stars via their apparent areas. *High Energy Density Physics*, *6*(3), pp. 301-304. doi:10.1016/j.hedp.2009.10.001

Hatfield, P., Furnell, W., Shenoy, A., Fox, E., Parker, B., Thomas, L. & Rushton, E. A. C. (2019). IRIS opens pupils' eyes to real space research. *Astronomy and Geophysics 60*(1) 1.22-1.24. doi:10.1093/astrogeo/atz046

Hodgen, J., Marks, R. & Pepper, D. (2013). *Towards universal participation in post-16 mathematics: Lessons from high-performing countries.* London, United Kingdom: Nuffield Foundation.

Homer, M., Ryder, J., & Donnelly, J. (2013). Sources of differential participation rates in school science: The impact of curriculum reform. *British Educational Research Journals*, *39*(2), 248–265. doi:10.1080/01411926.2011.635783

House of Commons Science and Technology Committee (2018). Algorithms in decision-making. *Fourth Report of Session 2017-19.* Retrieved from https://publications. parliament.uk/pa/cm201719/cmselect/cmsctech/351/351.pdf

Institute for Research in Schools (2108) Well World introductory videos. Retrieved from https://www.youtube.com/watch?v=moiWKVdn-Pg&index=2&list=PLX_O545Tok NiYCVMJP--AIW3d3jzuhIQx

Institute of Physics (2017). Improve gender balance. Reflections on the impact of interventions in schools. *An Institute of Physics Report.* Retrieved from http://www. iop.org/publications/iop/2017/file_69171.pdf

Joint Council for Qualifications (2018). *A-Levels.* Retrieved from http://www.jcq.org.uk/examinationresults/a-levels

Jordan, R. C., Ballard, H. L., & Phillips, T. B. (2012). Key issues and new approaches for evaluating citizen-science learning outcomes. *Frontiers in Ecology and the Environment, 10*(6), 307-309. doi:10.1890/110280

Kaplan, R., Kaplan, S., & Ryan, R. L. (1998). *With people in mind. Design and management of everyday nature.* Washington, DC: Island Press.

Kolb, D.A. (2014). *Experiential learning: Experience as the source of learning and development* (2nd ed). Upper Saddle River, NJ: Pearson Education.

Lee, E., Helt, G.A., Reese, J. T., Munoz-Torres, M. C., Childers, C. P., Buels, R. M., . . . Lewis, S. E. (2013). Web Apollo: A web-based genomic annotation editing platform. *Genome Biology, 14*(8), R93. doi:10.1186/gb-2013-14-8-r93

Lintott, C. J., Schawinski, K., Slosar, A., Land, K., Bamford, S., Thomas, D., . . . Murray, P. (2008). Galaxy Zoo: Morphologies derived from visual inspection of galaxies from the Sloan Digital Sky Survey. *Monthly Notices of the Royal Astronomical Society, 389*(3), 1179-1189. doi:10.1111/j.1365-2966.2008.13689.x

Lombardi, M. M. (2007). Authentic learning for the 21st century: An overview. *Educause Learning Initiative, 1*(2007), 1-12.

Marteau, T. M., & Bekker, H. (1992). The development of a six-item short-form of the state scale of the Spielberger State—Trait Anxiety Inventory (STAI). *British Journal of Clinical Psychology, 31*(3), 301-306.

Mervis, J. (2016). Genuine research keeps students in science. *Science, 352*(6291), 1266. doi:10.1126/science.352.6291.1266

Migration Advisory Committee (2016). *Partial review of the shortage occupation list: Review of teachers.* Migration Advisory Committee. Retrieved from https://www.gov.uk/government/consultations/migration-advisory-committee-mac-review-on-teacher-shortages

Millennium Ecosystem Assessment. (2006). *Ecosystems and human well-being.* Washington, DC: Island Press.

Mitchell, N., Triska, M., Liberatore, A., Ashcroft, L., Weatherill, R., & Longnecker, N. (2017). Benefits and challenges of incorporating citizen science into university education. *PLoS ONE 12*(11): e0186285. doi:10.1371/journal. pone.0186285

Nardi, E. & Steward, S. (2003). Is mathematics T.I.R.E.D? A profile of quiet disaffection in the secondary mathematics classroom. *British Educational Research Journal, 29*(3), 345–367. doi:10.1080/01411920301852

Newmann, F. M., Marks, H. M., & Gamoran, A. (1996). Authentic pedagogy and student performance. *American Journal of Education, 104*(4), 280-312. doi:10.1086/444136

Nov, O., Arazy, O., & Anderson, D. (2011). Dusting for science: Motivation and participation of digital citizen science volunteers. Paper presentation at the annual 2011 iConference (pp. 68-74), Seattle, WA.

Ottinger, G. (2010). Buckets of resistance: Standards and the effectiveness of citizen science. *Science, Technology, & Human Values, 35*(2), 244-270. doi:10.1177/0162243909337121

Pandya, R. E. (2012). A framework for engaging diverse communities in citizen science. *Frontiers in Ecology and the Environment, 10*(6), 314-317. doi:10.1890/120007

Parker, B. (2017). Why don't we let young people contribute to cutting edge science at school? *The Guardian*. Retrieved from https://www.theguardian.com/science/life-and-physics/2017/nov/03/why-dont-we-let-young-people-contribute-to-cutting-edge-science-at-school

Parker, B., Fox, E., & Rushton, E. A. C. (2018). *IRIS – promoting young peoples' participation and attainment in STEM and reigniting teachers' passion for science education. Impact 2.* Retrieved from https://impact.chartered.college/article/parker-iris-stem-students-teachers-participation-research/

Parker, B., Rushton, E. A. C., Thomas, L. and Hatfield, P. (2019). Transforming Education with the Timepix detector – ten years of CERN@school. *Radiation Measurements*. doi.org/10.1016/j.radmeas.2019.03.008

Perelló, J., Ferran-Ferrer, N., Ferré, S., Pou, T., & Bonhoure, I. (2018). High motivation and relevant scientific competencies through the introduction of citizen science at secondary schools. An assessment using a rubric model. In C. Herodotou, M. Sharples, & E. Scanlon (Eds.) *Citizen inquiry: Synthesising science and inquiry learning* (pp. 150-175). Oxford, United Kingdom: Routledge.

Powell, N. L., & Harmon, B. B. (2014). Developing scientists: A multiyear research experience at a two-year college. *Journal of College Science Teaching*, *44*(2), 11-17. doi:10.2505/4/jcst14_044_02_11

Pretty, J., Peacock, J., Hine, R., Sellens, M., South, N. & Griffin, M. (2007). Green exercise in the UK countryside: Effects on health and psychological well-being, and implications for policy and planning. *Journal of Environmental Planning and Management, 50*(2), 211-231. doi:10.1080/09640560601156466

Pretty, J., Peacock, J., Sellens, M. & Griffin, M. (2005). The mental and physical health outcomes of green exercise. *International Journal of Environmental Health Research, 15*(5), 319-337. doi:10.1080/09603120500155963

Rayner, J. (2017, September 28). *The opportunity provided by student involvement in Genome Decoders*. Speech given at the Genome Decoders Launch, Wellcome Genome Campus, Cambridgeshire, England.

Reiss, M. J. & Mujtaba, T. (2017). Should we embed careers education in STEM lessons? *The Curriculum Journal*, *28*(1), 137-150. doi:10.1080/09585176.2016.1261718

Royal Society (2011). *A "state of the nation" report on preparing the transfer from school and college science and mathematics education to UK STEM higher education*. London, England: The Royal Society.

Royal Society (2018). *Partnership Grants*. Retrieved from https://royalsociety.org/grants-schemes-awards/grants/partnership-grants/

Rushton, E. A. C., & Parker, B. (2019). Evaluating the impacts on young people and their teachers who participate in genuine scientific research whilst at school. In S. E. Hiller & A. Kitsantas (Eds), *Enhancing STEM motivation through citizen science programs*. Hauppauge, NY: Nova Science Publishers, Inc.

Rushton, E. A. C. & Reiss, M. J. (2019). From science teacher to 'teacher scientist': Exploring the experiences of research-active science teachers in the UK. *International Journal of Science Education*. doi:10.1080/09500693.2019.1615656

Sadler, T. D., Burgin, S., McKinney, L., & Ponjuan, L. (2010). Learning science through research apprenticeships: a critical review of the literature. *Journal of Research in Science Teaching 47* (3), 235–256. doi:10.1002/tea.20326

Simpson, R., Page, K. R., & De Roure, D. (2014). Zooniverse: Observing the world's largest citizen science platform. In *Proceedings of the 23rd International Conference on World Wide Web* (pp. 1049-1054). ACM.

Shah, H. R. & Martinez, L. R. (2016). Current approaches in implementing citizen science in the classroom. *Journal of Microbiology and Biology Education*, *17*(1), 17. doi:10.1128/jmbe.v17i1.1032

Tai, R. H., Liu, A. C. Q., Maltese, D. V., & Fan, X. (2006). Planning early for careers in science. *Science*, 312(5777), 1143-1144.

The Institute for Research in Schools (2017). *Introducing the human whipworm genome decoders project.* Retrieved from https://www.youtube.com/watch?v=xrlu13AoXcE&feature=youtu.be

Trautmann, N. M., Shirk, J., Fee, J., and Krasny, M. E. (2012). Who poses the question? Using citizen science to help K–12 teachers meet the mandate for inquiry. In J.L. Dickinson, & R. Bonney (Eds). *Citizen science: Public collaboration in environmental research* (pp. 179-190). Ithaca, NY: Cornell University Press.

Trumbull, D. J., Bonney, R., Bascom, D., & Cabral, A. (2000). Thinking scientifically during participation in a citizen-science project. *Science Education*, *84*(2), 265-275. doi:10.1002/(SICI)1098-237X(200003)84:2<265::AID-SCE7>3.0.CO;2-5

Truss, E. (2014). *Support for maths and science teaching.* Lecture given at Imperial College London. Retrieved from https://www.gov.uk/government/speeches/elizabeth-truss-on-support-for-mathsand-science-teaching

Turner, E. (2018). Interview featured in the *Well World* video. Retrieved from https://www.youtube.com/ watch? v=moiWKVdnPg&list=PLX_ O545TokNiYCVMJP--AIW3d3jzuhIQx&index=2

U. K. Department of Health (2009). *Be Active and Healthy. A Plan for Getting the Nation Moving*. London, United Kingdom.

Vitone, T., Stofer, K. A., Steininger, M. S., Hulcr, J., Dunn, R., & Lucky, A. (2016). School of ants goes to college: integrating citizen science into the general education classroom increases engagement with science. *Journal of Science Communication*, *15*, 1-24. doi:10.22323/2.15010203

Wallace, D. & Bodzin, A. (2017). Developing scientific citizenship identity using mobile learning and authentic practice. *Electronic Journal of Science Education*, *21*(6). doi:17688-69716-1-PB

Wang, M. T., & Degol, J. (2013). Motivational pathways to STEM career choices: Using expectancy–value perspective to understand individual and gender differences in STEM fields. *Developmental Review*, *33*(4), 304-340. doi:10.1016/j.dr.2013.08.001

White, M. P., Pahl, S., Wheeler, B. W., Fleming, L. E. F., & Depledge, M. H. (2016). The "Blue Gym": What can blue space do for you and what can you do for blue space? *Journal of the Marine Biological Association of the United Kingdom, 96*(1), 5-12. doi:10.1017/S0025315415002209

Whyntie, T., & Harrison, M. A. (2014). Simulation and analysis of the LUCID experiment in the Low Earth Orbit radiation environment. *Journal of Physics: Conference Series, 513*(2), p. 022038. Bristol, U.K.: IOP Publishing.

Whyntie, T. & Harrison, M. A. (2015). Full simulation of the LUCID experiment in the Low Earth Orbit radiation environment. *Journal of Instrumentation*, *10*(03), p.C03043. doi:10.1088/1748-0221/10/03/c03043

Wieman, C., & Gilbert, S. (2015). Taking a scientific approach to science education, part II – changing teaching. *Microbe, 10*(5), 203-207. doi:10.1128/microbe.10.203.1

Wigfield, A., & Eccles, J. S. (2002). The development of competence beliefs, expectancies for success, and achievement values from childhood through adolescence. In A. Wigfield, & J.S. Eccles (Eds). *Development of achievement motivation* (91-120). San Diego, CA: Academic Press.

Wood, L., Hooper, P., Foster, S., & Bull, F. (2017). Public green spaces and positive mental health–investigating the relationship between access, quantity and types of parks and mental wellbeing. *Health & Place*, *48*, 63-71. doi:10.1016/j.healthplace.2017.09.002

Woodcock, J., Greenhill, A., Holmes, K., Graham, G., Cox, J., Oh, E. Y., & Masters, K. (2017). Crowdsourcing citizen science: Exploring the tensions between paid professionals and users. *Journal of Peer Production*. Retrieved from https://researchportal.port.ac.uk/portal/files/8039057/COXj_2017_cright_JPP_Crowdsourcing_citizen_science_exploring_the_tensions_between_paid_professionals.pdf

SECTION III: CREATING EFFECTIVE COLLABORATIONS AMONG EDUCATORS AND CITIZEN SCIENCE PROGRAMS

In: Enhancing STEM Motivation through Citizen Science Programs ISBN: 978-1-53616-038-3
Editors: Suzanne E. Hiller and Anastasia Kitsantas

Chapter 6

NATURALISTS' PERSPECTIVES OF OUTDOOR LEARNING, ADOLESCENTS AS CITIZEN SCIENTISTS, AND STEM CAREER MOTIVATION

Suzanne E. Hiller*[*] *and L. Earle Reybold
George Mason University, VA, US

ABSTRACT

The purpose of this qualitative case study is to illuminate effective educational practices from the perspective of naturalist-educators well-versed in working with adolescents in authentic, outdoor learning activities, such as citizen science. Fifteen expert naturalists were recruited from urban, suburban, and rural areas to participate in semi-structured interviews. The authors identified three emergent themes using a constant comparative analysis approach: (a) foundational aspects of outdoor learning, including scientific observation skills, (b) instructional strategies for promoting self-motivational beliefs and cognitive growth, and (c) the influence of these types of programs on student affect. A cross comparison of the data and these themes revealed an overarching central theme of the influence of outdoor programs, such as citizen science, on student identity formation and STEM career pathways. Implications of this study extend to the goals of informal and formal educators, researchers, and stakeholders in developing and implementing field based activities for students.

Keywords: affect, citizen science, scientific observation skills, self-efficacy, STEM career motivation

Distinct from many science-oriented occupations, naturalists serve a unique role in the conservation, preservation, and educational efforts on multiple scales (local, regional, continental, and global). Merriam-Webster (2018) defines a naturalist as "a student of natural history." Master naturalists provide valuable insight into data analysis and interpretation within

[*] Corresponding Author's E-mail: shiller@gmu.edu

natural environments, recommendations for wildlife management, education initiatives for the general public including school, club, and interest group presentations, as well as interpretive walks (EnvironmentalScience.org, 2018). A compelling contribution of naturalists within local communities is to provide a platform which includes individuals in scientific practices so that the general public has insight into scientific findings (Merendlender, Crall, Drill, Prysby, & Ballard, 2016).

A specific activity which involves the public in scientific research and often relies on educational expertise of naturalists is citizen science. Citizen science initiatives are dependent on volunteer contributions to collect large volumes of data on environmental/ biological/ societal factors (Crall, Newman, Stohlgren, Holfelder, Graham, & Waller, 2011). Examples of citizen science programs constitute studying the interaction between urban rainfall and impervious surfaces, contrails, pollinating insects, and garden hedgehogs, as some examples (deVos, Leijnse, Overeem, & Uijlenhoet, 2017; Fowler, Whyatt, Davies, & Ellis, 2013; Roy, Baxter, Saunders, & Pocock, 2016; Williams, Stafford, & Goodenough, 2015).

This joint effort between professional and amateur scientists provides greatly needed resources in garnering large amounts of data, offsetting research costs, and contributing to the general body of knowledge related to science topics. For the hobbyist, participating in citizen science programs increases science knowledge, enhances environmental attitudes, and enables individuals to interact with professional scientists (Bonney, Ballard, Jordan, McCallie, Phillips, Shirk, & Wilderman, 2009a; Devine-Wright, Devine-Wright, & Fleming, 2004; Meinhold & Malkus, 2005). For school-age learners, there is evidence that citizen science programs impact motivation and achievement (Creel, Nite, Almarri, Shafik, Mari, & Al-Thani, 2017; Hiller & Kitsantas, 2014; Wallace & Bodzin, 2017). In their role as educators, naturalists often provide an invaluable service as a bridge in this dynamic; these experts maintain validity of the research study while training individuals in scientific observation skills and data collection procedures to collect viable data points (Hiller & Reybold, 2011).

From the perspective of a learner involved in outdoor experiences, such as citizen science programs, an equally beneficial aspect of the work of a naturalist is the training individuals receive through guidance and modeling while conducting fieldwork. This type of interaction is central to enhancing fine grained skills required for scientific inquiry in and out of the classroom and serves as a catalyst to shape an individual's identity formation as a scientist (Eberbach & Crawley, 2009; Hiller & Kitsantas, 2015). In other words, through new social/academic affiliations, learners begin to view themselves as members of the scientific community as they begin to consider new career pathways (Bombaugh, 2000).

Recently, research has focused on the impact of exposure to field experts on adolescents' achievement and STEM career motivation, yet the benefits of teaching and learning in the outdoors have been recognized for a much longer period. For instance, over 100 years ago, Hocking, the Principal of the Melbourne Continuation School, outlined the benefits of nature-based instruction for educational purposes:

> Movement and colour are found there, adaptation of structure to environment is learnt by the sea-side or in the tea-tree glade, . . . The glory of sunrise and of sunset, the grandeur of the mountain range purpled by distance, the profusion of the fern gully, the sparkle and music of the waterfall, the majesty of the gathering thunderclouds, cannot be brought into the school-room. Out of doors there is access to appropriate conditions. Data must be collected, and a general familiarity with things cultivated . . . [T]hat the larva of the Buprestis is destructive to

trees means nothing when told, but to see the larvae at work gives meaning to the statement. (Hocking, 1907, p. 234)

In the last century, the way modern advances have shaped naturalistic studies would not have been predictable, yet the diverse and fluctuating variables associated with outdoor learning continue to be a viable instructional platform today.

Albeit technological innovations have enhanced capabilities of field studies through the use of readily available devices, probes, sensors, simulations, and online programs, the study of nature affords poignant educational experiences, particularly coupled with field expert training. Replicating techniques by field experts and collecting data develops mastery experiences, heightens individuals' perceptions about their capabilities in the science realm, extends scientific process skills, and reshapes vocational identities (Bombaugh 2000; Eberbach & Crowley, 2009; Hiller & Kitsantas, 2014).

Working with learners in outdoor settings provides a unique context for developing science-oriented skill sets. Today, as in the past, naturalists who serve as educators are aware of the multiple benefits of outdoor instruction. The insights of naturalists, with the dual function as educator, are of great use in understanding how to encourage individuals to engage in authentic science practices. Although, citizen science research often centers on the impact of adult volunteers, this work addresses the needs of students 10-25 years of age, a developmental stage coined as adolescence (Steinberg, 2015).

The purpose of the current work is to present the collective voice of naturalist educators who work with adolescents from a three-prong perspective: (a) key aspects of training adolescent learners; (b) effective instructional techniques for preparing students for activities such as citizen science; and (c) the positive benefits of these types of programs on self-efficacy, affect, identify formation and STEM career motivation. In this work, we first provide an overview of protocol training and scientific observation skills as essential components of citizen science programs to situate the current study. Next, we outline the basis for this study from a theoretical perspective, particularly in terms of developing student skills sets and self-motivational beliefs. Then we detail a qualitative case study rooted in the viewpoint of naturalists who work as educators preparing students with refined skill sets, which are applicable to outdoor learning and citizen science programs.

Theoretical Perspective

Our work follows the supposition that outdoor education provides unique opportunities to construct knowledge. In particular, the ability to collect data based on multisensory processes enables students to develop knowledge acquisition in a way that diverges from a traditional classroom setting. Rooted in this perspective, we frame this work according to social cognitive career theory (Lent, Brown, & Hackett, 1994). Social cognitive career theorists reference the impact of interventions on self-motivational beliefs, such as self-efficacy and interest, when considering individual career paths. Moreover, social supports and barriers are central to assessing potential inequities and limitations within educational endeavors. These factors are implicit in studies of science career motivation for middle, secondary, and collegiate students (Hiller & Kitsantas, 2014; Lent, Miller, Smith, Watford, Lim, & Hui, 2016; Navarro, Flores, & Worthington, 2007; Patrick, Care, & Ainley, 2011).

We contend that naturalists who guide students in science inquiry are reflective of imperative social supports within a social cognitive career framework. For through modeling and feedback, this type of training has the potential to advance learners from novice to expert levels of performance (Eberbach & Crowley, 2009; Hiller, 2017; Hiller & Kitsantas, 2015). As scientific observation skills are the impetus for data interpretation (Cartwright, 1989), and with the required component of protocol training in citizen science programs, trained naturalists are scaffolding supports for learners. Within this dynamic, how the individual perceives their capabilities in science-oriented tasks (self-efficacy) in coordination with an expert mentor, is a highly potent predictor of achievement and motivation (Bandura, 1997), particularly in science education contexts (Britner & Pajares, 2006; Cleary & Labuhn, 2013; DiBenedetto & Zimmerman, 2010; Hiller, 2017; Hiller & Kitsantas, 2016). Further, sustained interest through repeated exposure (Hidi & Ainley, 2008), is the hallmark of motivation. Figure 1 highlights the dynamics between scientific observation skills, self-efficacy, affect, and identity formation as a keystone of motivation. Instruction in scientific observation skills (counting, classifying, collecting, and measuring), as described by Eberbach and Crowley (2009), are necessary skills sets for scientists. Students who develop mastery experiences in the use of these strategies are likely to have heightened self-efficacy.

Self-efficacy, a term which describes a learner's perceptions about their capabilities in a specific area, is a pivotal self-motivational belief. Students with strong self-efficacy are better able to readjust their learning style and exhibit resiliency when setbacks occur (Cleary, 2018; Cleary & Labuhn, 2013). This construct is subject/task defined, a foundational aspect of achievement and originates from four sources including mastery experiences, vicarious experiences (influence of mentors, teachers, models), social persuasion (influence of peers), and negative physiological states, such as anxiety. Although not considered a direct source of self-efficacy, Bandura (1997) posits that affect is instrumental in steering self-efficacy.

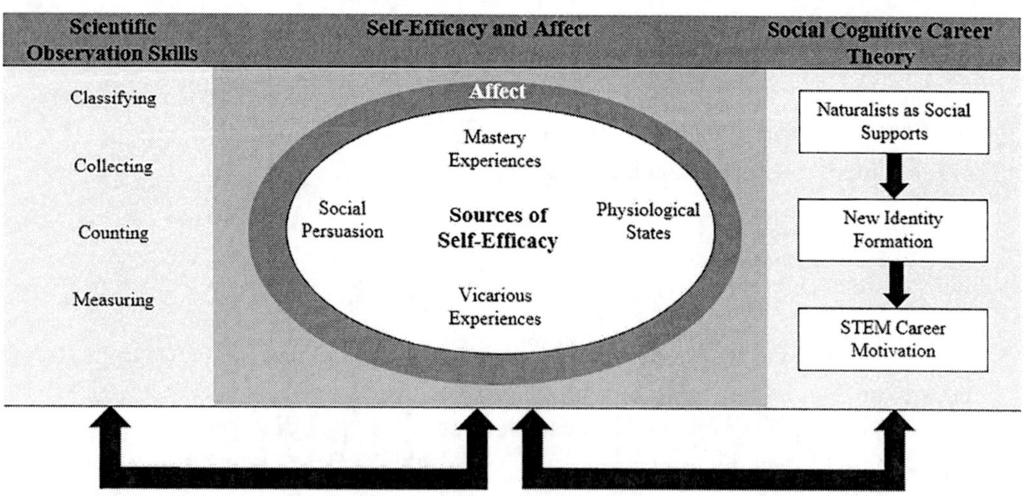

Figure 1. The dynamics of scientific observation skills, self-efficacy, affect, and STEM career motivation in a citizen science context.

Both positive and negative affect correlate with self-motivational beliefs (Roeser, Peck, & Nasir, 2006). For example, someone with positive affect will exhibit interest and pleasure in learning, thereby enhancing overall self-efficacy. Conversely, negative affect may lead to avoidance of activities and manifest as frustration, fear, or anger (Carver, Avivi, & Laurenceau, 2008). Affect is reflective of an individual's mood and emotion with the first being a more enduring tendency and is integral to attitude formation (Manstead, 2002). Emotion, in contrast, is generally a strong response, which demands an individual's immediate attention. Moreover, a lack of affect can have a detrimental impact on reasoning (Bjørnebekk, 2008). For naturalists, often interested in attitudinal changes in regards to the environment, affect as it relates to motivational influences, such as self-efficacy and interest, is of the greatest importance.

Identity formation describes student reflections on themselves as individuals and their roles as community members. Societal, biological, and cognitive factors coupled with previous experiences shape a learner's perceptions of self (Crocetti, 2017). Identity formation as the basis for STEM career motivation is another component addressed in Figure 1. From a social cognitive career theory standpoint, the naturalists serve as social supports well-positioned to alter students' self-perceptions as capable scientists (Bombaugh, 2000; Eberbach & Crowley, 2009; Hiller, 2012). Over the course of several experiences or one with a profound impact, students may exhibit STEM career motivation through academic activities, outside projects/interests, and course selection. In the course of the progression of scientific observation skills, increased self-efficacy and affect, learners have the potential to build an identity as a scientist and heighten STEM career inclinations. Navigating this interaction between scientific observation skills, increased self-efficacy, positive affect, and changing identity formation, naturalists are influential motivators for bolstering student potential. The following literature review outlines research that addresses scientific observation skills, self-efficacy, and affect as positive outcomes of well-implemented citizen science programs for adolescents which in turn can influence an individual's identity formation and STEM career motivation.

Protocol Training, Scientific Observation Skills, and Citizen Science

An innate component of citizen science programs, which serves as a conduit between naturalist and learner, is protocol training. In order to maintain validity in data analysis and interpretation, it is essential that scientists convey a systematic approach in collecting data; clear guidelines are foundational in guiding scientific observation skills. Often, this training rests on the work of field experts who model, guide, and evaluate an individual's data collection techniques. When professional scientists include children and adolescents in field work, maintaining validity becomes even more pronounced as some individuals are critical that children's observation skills and data collection approaches are not as accurate as adults (Hiller & Reybold, 2011). Straightforward protocol training offsets these types of concerns as students have been found to collect suitable data for scientific studies (Hiller, 2012; Pocock & Evans, 2014). Snäll, Kindvall, Nilsson, and Pärt (2011) recommend using a check-list of factors and questions used prior to citizen science programs to account for training and quality issues. Elements to be considered include the type of data, the quality of species identification, duration of data collection, and regional considerations among other issues.

The goals of each scientific study are contextually embedded, thereby determining the nature of protocol training which can vary in terms of geographical/logistical requirements, level of field expert support, level of citizen scientist involvement, and complexity of data collection requirements. For example, in a citizen science study of the general health of the

Great Barrier Reef, the scientific research design had some unique specifications in that volunteers were part of the snorkel tourism industry in Australia and limited to 20 min intervals to assist with data collection. This parameter mandated that comprehensive data collection occurred in an expedited manner. In addition, protocol training required that volunteers were able to collect data three times a day in the same location within less than 20 min for each cycle (Beeden, et al., 2014).

Trained ranger citizen scientists learned to locate shallow waters, such as lagoons, reef crests, slopes, and/or reef flats. Specific and well-defined protocol related to using fin kicks to locate starting points and coordinate with volunteers on the boat above to identify an initial GPS fix to center the survey. Scuba divers surveyed the area while completing a thorough inventory to assess species of microalgae present, the coral types under study, and evidence of disease, predation, and damage to the coral. Simultaneously, snorkel divers swam the designated 5 m circumference above to maintain a comprehensive view of the area under investigation. In addition, photographs were taken to assess environmental impacts on the corals from several vantage points (Beeden, et al., 2014).

Citizen science programs such as the Great Barrier Reef study, exemplify how concerted volunteer efforts are essential and valuable resources to large scale studies. Clearly specified protocol design provides the groundwork for rigorous data collection and subsequent interpretation. While citizen scientists are of great support to this type of research, there are substantial educational benefits to individuals engaged in scientifically oriented activities in conjunction with professional scientists who serve as models in cognitive learning.

Within classroom settings, students often collect data prescribed during daily activities. However, limited time is typically allocated for refining scientific observation skills. Authentic learning experiences, such as citizen science, are intrinsically tied to the quality of volunteer observations. In this regard, the benefits are mutually beneficial for the research team and learner; citizen scientists who have refined scientific observation skills are more likely to provide usable findings while student volunteers enhance their science-oriented skill sets for advanced academic learning. Further, students flourish when their work is valued and that their efforts have some visible merit (Chawla & Kushing, 2007).

Scientific observation skills are tools of scientists which are continuously refined during a cyclical process and often shaped by cultural influences and societal demands (Tracy, 2018). These influences illuminate the role of citizen science programs in the current global context steered by growing needs and supported with advanced technological resources. With skilled training in data collection from field experts in the natural world, the information gathered can be useful in analysis particularly in terms of identifying large scale patterns of animal/plant populations and other environmental factors (Bonney et al., 2009a; Harrison, Underhill, & Barnard, 2008).

A major theme in informal natural science learning settings and citizen science programs specifically relates to the role of scientific observation skills in data collection (Eberbach & Crowley, 2009). Observation skills used for scientific investigations during field work greatly influence the development of scientific findings (Cartwright, 1989) as the use of a "five senses approach" creates an opportunity for enriched learning through taxonomic comparisons, determining cause and effect relationships, comparing collected field data to scientific theories, and drawing a connection to human activity and the natural world (Auer, 2008). Often overlooked as instinctive, scientific observation skills require concerted cognitive effort. As Tracy (2018) explained, "When we consider scientific observations—those paragons of a purportedly objective gaze—we find, in fact, that they are often complex, contingent, and distributed phenomena, much like human vision itself" (p. 242). From this viewpoint, scientific observations differ from general observations in terms of a coordinated effort between physiological aspects of sensory input, conceptual understanding, content knowledge, science inquiry knowledge, the use of scientific tools, and refined data collection skills (Ballantyne & Packer, 2009; Eberbach & Crowley, 2009; Kim & Park, 2018).

As potential mentors, professional scientists distinguish themselves by synthesizing observation skills with an in depth understanding of classification distinctions. Expert scientists not only engage in the use of observations, but their perceptions are intertwined with comparisons based on a taxonomic system (Eberbach & Crowley, 2009). When scientists create collections of natural specimens, the activity is an example of the interconnectedness of scientific observation skills, documentation, and distinctions between specimens. As such, activities centered on identification, classification, measurement, and the maintenance of collections all encompass scientific observation skills (Eberbach & Crowley, 2009).

While scientific observations skills are crucial in the realm of professional scientists, for students, this skill is not always clearly defined and highlighted within science curricula. Naturalistic studies may offset deficiencies by aiming to reinforce, extend, and promote students' science foundations, particularly in terms of scientific observation skills. Scaffolding experiences with teacher/mentor/guide feedback is critical in the formation of high levels of development (Zimmerman, 2000) and is applicable to incorporating scientific observation skills within program guidelines. Via mentorship of fine-grained and nuanced observations, individuals incorporate sensory perceptions into an organizational structure which becomes part of long-term memory (Mayer & Wittrock, 2006). For citizen scientists, whose role has recently become more integral in research design formulation and data interpretation through collaborative projects, advanced scientific observation skills are even more impactful.

The complexities of informal learning require specific training in scientific observation skills, according to Eberbach and Crowley (2009). Training students in scientific observation skills is most effective when learners are guided through a series of practical experiences coupled with reflective questions and naturalist modeling (Eberbach & Crowley, 2009; Hiller & Kitsantas, 2015). From this perspective, the training of scientific observation skills should emphasize a progression in three steps that proceeds from novice to mastery level: noticing, distinguishing, and recognizing. In the higher-level stages, students recognize more specific features and learn to discount information, which is not useful in identification and comparisons (Eberbach & Crowley; Hiller & Kitsantas, 2015).

This type of coordination between scientific observation skills, content understanding, and documentation has been exemplified in research studies of citizen science programs. A research study which incorporated protocol training, scientific observation skills, and joint efforts with community members and scientists centered on the study of monarch butterflies. Through this program, teachers and students worked as teams while benefitting from instruction and modeling from professional researchers. Further, student/teacher teams applied protocol parameters, scientific observation skills, documentation processes, and background knowledge to contribute to a citizen science survey on butterflies known as the *Monarch Larva Monitoring Project*. Collaborations between high school students, science teachers, and scientists during butterfly data collection, resulted in an increased level of content knowledge development and self-efficacy for teacher and student teams (Jeanpierre, Oberhauser, & Freeman, 2005). Similarly, Ballantyne and Packer (2009) found that programs were most effective when students used scientific observation skills to raise questions, compare differing points of view, develop personal value and belief systems and draw conclusions. These types of activities offer students the opportunity to develop advanced scientific observation skills through a modeling process.

The need for well-developed scientific observation skills are not exclusive to citizen science programs. One study within formal classrooms highlighted the negative impact of limited support in the development of scientific observation skills, which can hamper overall student progress. Chinn and Malhotra (2002) targeted instruction on extending scientific observation skills as necessary for overall conceptual understanding of science topics. In a series of experiments designed to test the impact of accurate scientific observation skills on cognitive development, Chinn and Maholtra identified setbacks in student progress in stages. Notably, for the learner, conceptual change was stifled primarily at the initial stage of observation, which greatly altered later stages of development. Distinguished by these findings, students with limited initial experiences in scientific observation skills were not able to advance in scientific conceptual understanding to their greatest potential. This study exemplifies how developing high levels of expertise in scientific observation skills are dependent on repetitive exposure, modeling, and feedback (Chinn & Maholtra, 2002) and extend to citizen science educational activities (Hiller & Kitsantas, 2015).

Not only is well-defined instruction in scientific observation skills essential for cognitive development, but students benefit from understanding the basis for the use of scientific observation skills. For example, students must be aware that personal perceptions, experiences, and background knowledge shape the way individuals interpret their observations. In order to assist students to learn about the nature of scientific observations, Kim and Park (2018) created an instructional program as an intervention based on 10 statements designed from the input of scholars in educational, psychology, and philosophy fields. Statements which reflected the tenets of scientific observations included, "Given observation is often possible only for a person who has specific scientific knowledge," "The observer often observes selectively with a specific purpose or expectation," and "Observations can vary with changes in the surrounding context or condition" (p. 866).

Tenth graders ($N = 110$) in Korea engaged in the intervention program and then completed assessments on their understanding of the nature of scientific observations. Following activities centered on training the nature of scientific observations, students completed a measure to gauge their level of agreement with corresponding statements about the nature of scientific observations. Following the intervention, participants were found to reach a general consensus about the nature of scientific observations. In other words, students had developed a consistent and concise understanding of the nature of scientific observations based on training within the classroom. From this work, Kim and Park (2018) concluded that teaching students about the nature of scientific observations is feasible within school programs. This work exemplifies the possibilities of training students on scientific observation skills which extend to outdoor, authentic learning opportunities. Helping students to become metacognitively aware of the processes and nature of scientific observation skills is essential in developing fine-grained skill sets; this awareness is the basis of data collection and interpretation for scientists with high levels of expertise (Cartwright, 1989; Eberbach & Crowley, 2009; Hiller & Kitsantas, 2015).

The studies outlined in this section highlight the implications of encouraging scientific observation skills and the deficits that may develop if students do not receive explicit instruction in the processes and understanding of the nature of scientific observations. As scientific observation skills are fundamental in the work of citizen scientists, programs such as these for adolescents provide authentic experiential learning in support of school wide instruction.

Most recently, not only have professional researchers included adolescents in research studies, but a growing number of educators/researchers are noting positive outcomes for learners. While professional researchers use of technology on field sites, apps, and online programs adds to the growing amount of data, these technological resources do not replace the role of highly skilled individuals who identify nuances of understanding through scientific observations (Young, 2016). For this reason, learners thrive from multiple experiences to utilize scientific observation skills in collaboration with field experts and scientists. Most notably, for those students who are highly self-efficacious in the use of scientific observation skills, there are far reaching ramifications in terms of achievement and STEM career motivation.

Self-Efficacy

As a dominant predictor of achievement across a variety of domains, high levels of self-efficacy are necessary to extend academic performance (Bandura, 1997). The effect of self-efficacy is relevant across science-oriented topics, including outdoor learning contexts, environment education initiatives, and citizen science programs (Devine-Wright, et al., 2004; Hiller & Kitsantas, 2014; 2016; Sutton, 2006; Tomasek, 2009). In a seminal study, Britner and Pajares (2006) pinpointed the influence of the sources of self-efficacy on overall self-efficacy and science achievement. Middle school students ($N = 319$) completed measures reflecting the sources of self-efficacy and achievement. Of the four sources of self-efficacy, mastery experiences had the greatest impact on science self-efficacy. Moreover, as mastery experiences, vicarious experiences, and social persuasion experiences improved, negative manifestations of physiological states, such as anxiety diminished. Given the influence of self-efficacy on achievement, Britner and Pajares cited the pivotal role of improving student self-efficacy to foster science career trajectories.

From the perspective of environmentally-oriented studies, students' self-efficacy for shaping environmentally friendly behavior and responsibility beliefs are primary objectives for training individuals. For example, children instructed in informal natural learning settings with expert modeling have higher self-perceptions of their ability to control global warming and have higher responsibility beliefs (Devine-Wright, et al., 2004). Heightened self-motivational beliefs extend to influence student achievement and goals as well as demonstrated in a study of high school students in environmental science programs. Meinhold and Malkus (2005) measured self-efficacy, attitudes, content knowledge, goals, and behaviors with 848 high school students. The results indicated that students with high levels of self-efficacy and content knowledge had strong attitudes about stewardship and exhibited tendencies towards promoting conservation efforts. Given the findings from these studies, children who experience instruction in natural science learning conditions have higher levels of self-efficacy and marked responsibility in terms of protecting the environment when coupled with explicit instruction.

Self-efficacy is central to an array of outdoor learning programs, including citizen science. As noted previously, scientific observations skills are central to citizen science activities. For this reason, Hiller and Kitsantas (2016) developed and validated a scale known as the Citizen

Science Self-Efficacy Scale (CSSES). Tested with 248 eighth grade students, this measure was found to strongly correlate with the sources of self-efficacy as well as a performance measure. Although designed within the context of a horseshoe crab citizen science program, the CSSES serves as a model for citizen science researchers interested in measuring self-efficacy for scientific observation skills in a range of citizen science programs (Hiller & Kitsantas, 2016), which could be applicable to diversified instructional programs related to outdoor and/or environmentally oriented programs.

Incorporating measures which reveal student self-motivational beliefs, such as self-efficacy and interest, are mainstays in examining the potential of citizen science programs on student achievement and career paths. In addition, understanding the impact of access to naturalists on student affect and identity formation further advances pedagogical frameworks.

Outdoor Learning and Affect

For naturalists, a primary concern is developing student confidence (self-efficacy) and a positive emotional response to being outdoors. Students who are self-efficacious and have positive associations with natural settings are more likely to become interested and develop sustained engagement in outdoor activities. Notably, affect influences both cognitive development and motivation; this interaction forms an inseparable triad (Ainley, 2006). In terms of outdoor learning and citizen science, establishing positive affect is necessary to encourage cognitive growth and sustained interest in naturalistic activities. Rooted in this initiative, well-designed citizen science programs have the potential to spur knowledge acquisition, sustained engagement, scientifically oriented skill sets, and attitudes and behaviors as affective components (Bonney et al., 2009b).

Hidi and Ainley (2008) claimed, "The development of interest as a predisposition results from continued exposure to and reengagement with content and is an outcome of positive affect, increased knowledge, and stored value" (p. 85). These repeated experiences are particularly influential when the framework builds on a contextually interesting topic or on the general interests of an individual. In turn, increasing levels of interest and exposure to these specific topics will further enhance an association reflective of positive affect (Ainley, 2006).

The interconnectedness of affect and cognition and the combined influence on attitude results in the way individuals evaluate events, people, or objects. For example, Wilson (2007) studied the role of emotional responses and decision-making on individual choice on environmental issues. Participants ranked conservation management problems from wildlife disease, deer population, and trail erosion with numeric scales ranging from 1-7 ranking their emotional reactions towards the problems. Wilson concluded that participants made decisions related to conservation efforts not necessarily on cognitive functions but through emotional responses. Understanding how individuals respond to natural settings and environmental issues on an emotional level is important in steering learners' developmental experiences.

Outdoor settings may elicit a variety of reactions. In addition to emotional response, sensory stimulus further influences affect. Notably, naturally derived fragrances in outdoor settings have been linked to promoting the affective states of mood, alertness, and calmness (Weber & Heuberger, 2008). Further, increasing affect has been linked to stronger content knowledge gains. For instance, environmental programs which focused solely on developing content knowledge were not as successful at altering conservation behavior without an affective component (Cachelin, Paisley, & Blanchard, 2009). For students in the outdoors, a variety of experiences rooted in scientific observations and sensory inputs has the potential to create positive associations through affect (Ballantyne & Packer, 2009). When stakeholders are aware of the benefits of sensory stimuli and affect on learner's responses, a new arsenal of instructional practices become readily available to educators.

In a novel citizen science program designed as a rehabilitation platform for individuals with challenging memory deficits, participants navigated through a polluted canal to identify foreign objects. Sixty engineering students volunteered to participate in the intervention study which targeted the affective states of long-term excitement, short term excitement, meditation, frustration, and engagement. While participants generally found the activity fun and engaging, the work was difficult which led to some frustration. The implications of this citizen science program are that these types of citizen science programs can foster scientific literacy, self-efficacy, and affect (Palermo, Laut, Nov, Cappa, & Porfiri, 2017). Notably, this program used citizen science as a backdrop for use with a specific subset of students who may benefit from memory deficit training. The work reflects the potential ways that citizen science augments educational practices to meet diversified individual needs while incorporating affect as an influential construct.

Identity Formation and STEM Career Motivation

Previous sections in this literature review have outlined critical skill sets (scientific observations skills), self-motivational beliefs (self-efficacy and interest), and emotional constructs (affect) integral to outdoor learning, including citizen science activities. When these components are successfully integrated within instructional programs, not only do students make gains in terms of achievement, there is potential for alterations in their identity as scientist and the possibility of igniting STEM career paths (Bombaugh, 2000).

Identity formation in terms of vocational choices stems from the use of sensory perceptions in field work (Ballantyne & Packer, 2009; Eberbach & Crowley, Hiller & Kitsantas, 2014). Collaboration with scientists through citizen science projects impacts children's career paths in terms of shaping their identity as capable scientists (Bombaugh, 2000).

Bombaugh's supposition states that early exposure to participation in field work builds a basis for future participation in scientific research, and for some individuals, a professional career. To utilize this educational platform to encourage career motivation, it is noteworthy that the impact of citizen science programs revolve around the specific requirements of each type of citizen science program (Cronje, Rohlinger, Crall, & Newman, 2011). However, from an educational perspective, the defining feature of any of these programs is the role of field experts in modeling science process techniques, imparting content knowledge, and guiding volunteers through protocol in collecting data whether in the field or online.

As a reflection of the potential of working with field experts, one study of a citizen science programs was situated in the Great Smoky Mountains Institute at Tremont. Researchers concluded that field experts enhanced the learning opportunities of children by adapting protocol requirements, creating team roles in terms of collecting and recording data, encouraging additional research, and working with children in outdoor sites during data collection (Prysby, Oberhauser, & Petersen, 2007).

With the inclusion of naturalists as instructors, and with scientific observation skills as a cornerstone of student learning, Hiller and Kitsantas (2014) sought to extend the understanding of citizen science programs on adolescent achievement and motivation through a quasi-experimental study situated in the study of horseshoe crab populations. Eighth grade students ($N = 86$) participated in a comparison or treatment group studying horseshoe crabs as part of the state curriculum. Both groups of students received the same classroom training; however, the treatment group went to a beach to receive protocol training on collecting data on horseshoe crabs for a professional biologist's exploratory study.

The outcomes of this study highlighted that the citizen science intervention positively influenced the self-motivational beliefs of self-efficacy, interest, and outcome expectations. Mastery experiences, vicarious experiences, and social persuasion experiences were positively impacted by the experience with social persuasion from other classmates having strong main effects. A strong implication from this study was that path analysis revealed that interest and self-efficacy covaried to influence science performance on a content measure and overall career goals (Hiller & Kitsantas, 2014).

In a successive study rooted in career motivation and achievement, Wallace and Bodzin (2017) analyzed ninth grade students in a plant phenology citizen science program. The quasi-experimental research design centered on the impact of an authentic phenology citizen science program using a MobiLAP approach as the intervention, whereas the control group studied within the school setting. Specifically, Wallace and Bodzin examined whether the citizen science program influenced student identify formation, interest in contributing to citizen science programs, and STEM career motivation. Through data analysis, the outcome of the study mirrored the work of Hiller and Kitsantas (2014) in that student interest, self-efficacy, and science identity formation heightened following the citizen science program. Wallace and Bodzin concluded that given the effects of the short exposure to the citizen science programs, students would benefit from prolonged experiences.

Fundamentally, expert modeling and support provides the impetus to develop students' scientifically oriented skill sets, reshape identities, and encourage STEM career options. In fact, outdoor interventions impact students by providing social support, academic engagement, group identity, active citizenship, connectedness to the world, and a problem-based curriculum. In addition, multisensory experiences enhance student development (Ballantyne & Packer, 2009). With the many positive benefits associated with outdoor learning, and the intrinsic role of field experts as instructional guides, the perspective of naturalists in terms of effective training strategies is most prescient. The purpose of this study is to examine field experts' perceptions of effective environmental education practices, particularly as related to citizen science. This study illuminates effective pedagogy for individuals interested in incorporating informal natural science learning within academic programs. In their hinged capacity as field experts and educators, uniquely trained naturalists offer foundational information for administrators, practitioners, and policymakers committed to improving outdoor learning. The following research question guided inquiry:

- How do citizen science and outdoor educational activities promote scientific observation skills, self-motivational beliefs, identity formation, and STEM career motivation?

METHOD

As educators and researchers, we believe knowledge is both personal and contextual (Guba & Lincoln, 2005). Individuals understand the world based on their social experiences and personal interpretations (Patton, 2015) and so to understand the impact of citizen science and other outdoor learning activities for children, we focused this study on the perspectives of naturalist educators actively engaged in these activities. Our goal was to examine the value of these activities from the perspective of those who use them regularly; thus, all aspects of the research design and methods choices stemmed from this aim, particularly selection of participants (Reybold, Lammert, & Stribling, 2013).

Setting and Participant Selection

Initially, we identified three local naturalists using purposive selection (Patton, 2015) based on two criteria: professional field experts and trained naturalist volunteers. Early in the study, we became concerned with the limitations of our sample; these participants were from the same organizational affiliations, geographical locations, educational backgrounds, and ethnic/racial groups. One of the issues surrounding participation in naturalist career training programs is the lack of diversity based on racial/ethnic backgrounds; European Americans are more prominent in naturalist occupations (Merendlender, et al., 2016).

Our selection process was a combination of network and chain techniques, but familiar concerns dogged our intentions. The initial sample was primarily European American and male. Our challenge was whether to report out findings and account for homogeneity, or intentionally redirect our selection criteria to advocate for diversity in the profession. Wanting to understand diverse experiences and perspectives, we employed a network selection method adjusted for sex/gender, race/ethnicity, and geographical region (Patton, 2015) to contact a broader range of individuals. Also, we added one additional criterion—educator. These changes in selection strategy allowed us to account for perspectives associated with urban, suburban, and rural programs across the region; they also focused our data more specifically on pedagogy. We identified 16 participants across a 270-mile region that spanned the northeast region of the United States. We subdivided the regions based on the convenience referral system by Geographical Regions A (suburban), B (urban and suburban), and C (rural). For the present study, 15 individuals participated in interviews as logistical difficulties prevented one individual from engaging in the study. The interview locations were based on individual preference and occurred outdoors, in cafes, in hotel lobbies, and at work locations.

Participants represented federal agencies, non-profit organizations, and private organizations; they all were environmental educators with extensive training in scientific fields such as botany, ichthyology, entomology, anthropology, and/or geology. Figure 2 shows the progression of participant interviews based on initial gatekeepers and referrals to agencies. Initial interviewing occurred in a suburban region known as Geographical Region A. Participants A10 and A7 made critical referrals to Geographical Region B which was comprised of three urban areas and one suburban area. The naturalists in Geographical Region B were from diverse backgrounds which offset initial concerns about participant selection. A contact with a third gatekeeper in Geographical Region C (a rural area) resulted in an additional three participants. Overall, 50% of participants were male and 50% were female. The diversity of the study mirrored the limited diversity in the naturalist field based on racial/ethnic backgrounds with 73% European American, 20% African American, and 7% Latino.

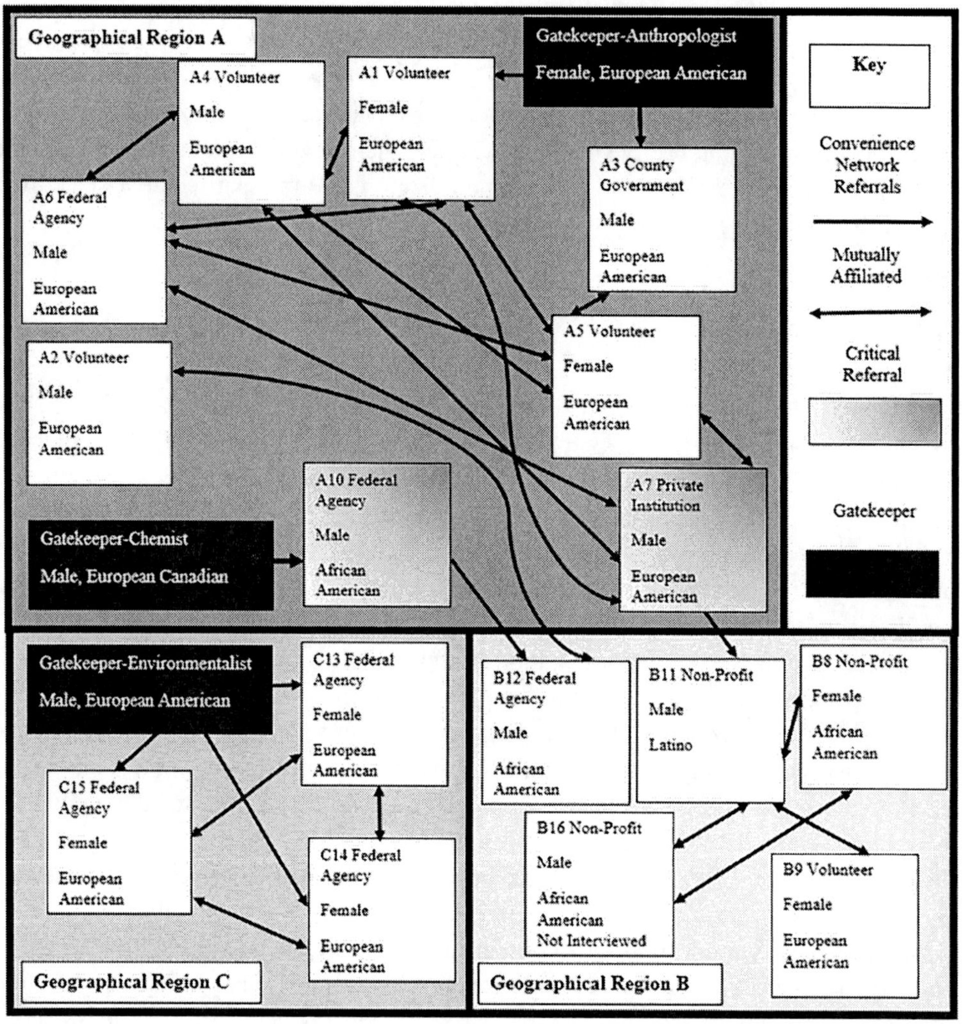

Figure 2. Graphic representation of environmental educator referrals.

Another factor in the study was that participants selected in Geographical Region A had more mutual connections in the naturalist community than other participants. This difference was likely due to the greater distances between participants in Geographical Regions B and C as well as from the researchers.

Data Collection, Coding, and Data Analysis

Because of our focus on participant perspectives, we conducted in-depth interviews as suggested by Patton (2015). The purpose of the semi-structured interviews was to highlight the perspectives of naturalists on effective outdoor educational practices. The interview guide, which was used to steer the discussion, included questions such as, "What are some ways of knowing or learning in the outdoors? What types of skills should children be taught in the outdoors? How do children benefit from citizen science programs?" Just as we refocused our participant selection criteria after initial analysis, we adjusted our interview guide across data collection to explore ideas specific to a program or person or to narrow our focus on naturalist educators' perspectives as shown in Appendix A. This semi-structured interview approach allowed us to identify themes shared by participants, but not at the expense of following up on stories specific to any one participant.

We analyzed interview transcripts using three techniques to assist in the development of subsequent interview questions. Multiple coding techniques are particularly useful in exploratory studies when research questions and data collection foci are in development (Saldaña, 2009). The first analysis technique we applied to the initial three interviews was the constant comparative approach (Corbin & Strauss, 1990; Walker & Myrick, 2006); this method examines data using three phases of coding: open, axial, and selective. Within the open coding phase, we applied three distinct coding methods—Descriptive Coding, *In Vivo* Coding, and Values Coding (Saldaña, 2009) to generate new questions for subsequent interviews, further develop emerging codes, and check our interpretations along the way. Table 1 provides an overview of how to differentiate between the three types of coding processes based on the writing of Saldaña.

Table 1. Coding Systems According to Saldaña

Type of coding system	Quote	Analysis
Descriptive Coding is a categorization system that analyzes transcripts in a broad way.	"Descriptive Coding leads primarily to a categorized inventory, tabular account, summary, or index of the data's content" (p. 72).	Categorize data by applying a word or short phrase such as a descriptive noun. (p. 70)
In Vivo Coding centers on a participant's perspective.	*In Vivo* Coding, "honors the participant's voice . . . *In Vivo* coding is one of the methods to employ during grounded theory's initial coding" (p. 74).	Focus on words or phrases in the transcripts which need to be highlighted due to repetition or emphasis. (p. 75)
Values Coding reflects participants' perspectives based on values, beliefs, and attitudes.	"A value is the importance we attribute to oneself, another person, thing, or idea" (p. 89). "An attitude is the way we think and feel about oneself, another person, thing, or idea" (p. 89). "A belief is part of a system that includes our values and attitudes plus our personel knowledge, experiences, opinions, prejudices, morals, and other interpretive perceptions of the social world" (pp. 89-90).	First code units into values, attitudes, and beliefs. Then categorize the units. Reflect on the categories as interrelated ideas. (p. 91)

For the remaining 12 participant transcripts, we applied two forms of coding; these forms included Values Coding and Descriptive Coding. As transcriptions were coded, participant statements were group based on a classification system of whether these statements were beliefs, attitudes, or values to highlight the participants' voice. As data from interviews were grouped into three main groupings of attitudes, beliefs, and values, we refined open coding within each grouping by employing Descriptive Coding (Saldaña, 2009). This process enabled us to code transcripts based on noun labels in order to build a solid foundation for the axial phase.

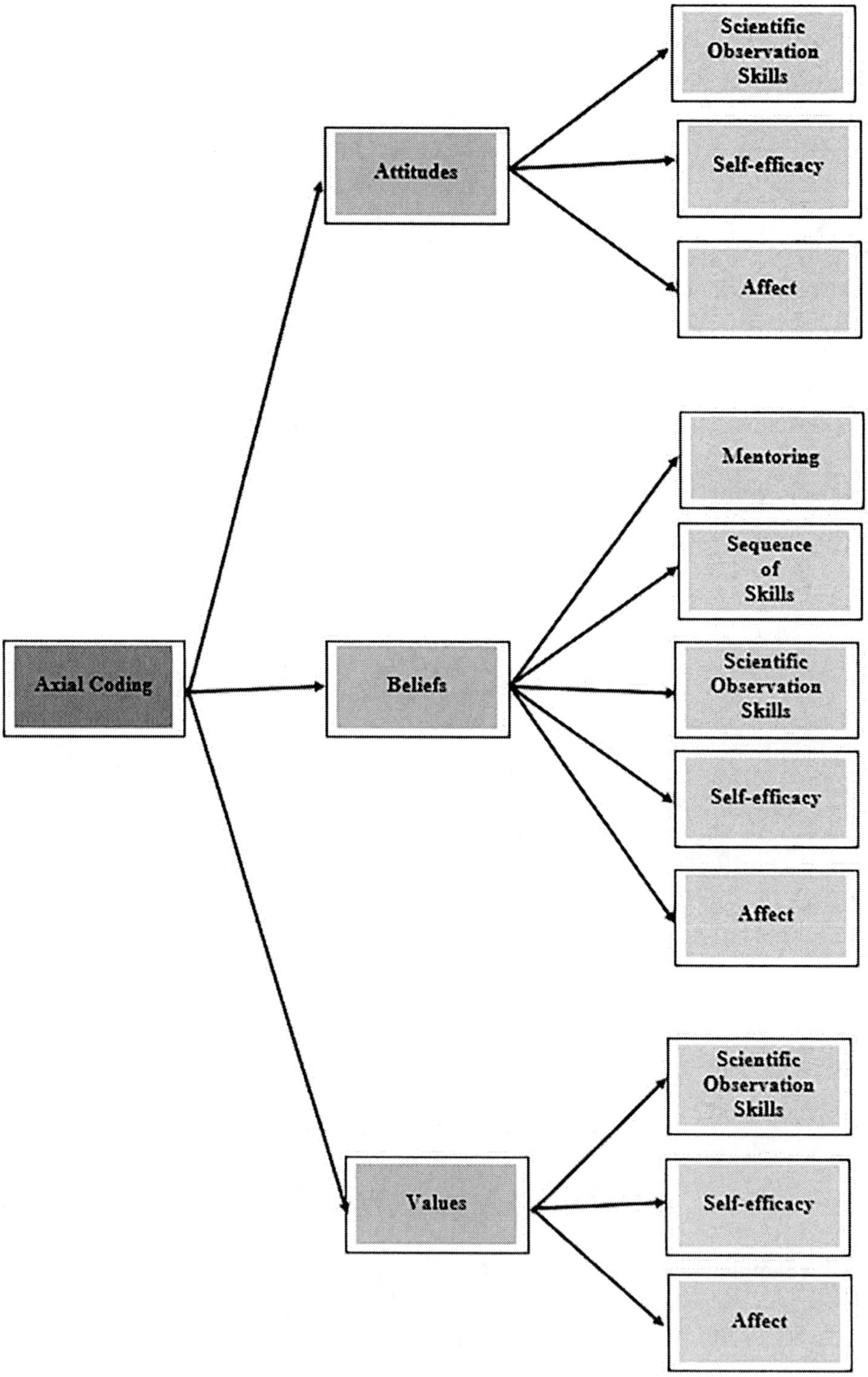

Figure 3. Axial coding categories.

Axial Coding

Based on the working definitions of attitudes, beliefs, and values in the open coding phase and Descriptive Coding, participant responses were grouped into categories within each of the three domains. Within the attitudes section, three categories included scientific observations, self-efficacy, and affect. In the beliefs section, five categories which emerged were scientific observations, self-efficacy, affect, sequence of skills, and mentoring. In the values section, the identified categories included curriculum, scientific observation skills, self-efficacy, and affect as seen in Figure 3.

Through the development of categories within the attitudes, beliefs, and values coding sections, emergent themes developed. Categories from the three major domains of attitudes, beliefs, and values were grouped together in order to develop the emerging themes of *scientific observation skills*, *self-efficacy*, and *affect* as shown in Figure 3. The repetition of these categories within each of the three domains is indicative of the great emphasis naturalists placed on the themes of scientific observation skills, self-efficacy, and affect, particularly with educational practices geared towards citizen science. In the last iteration of data interpretation, an overarching central theme of *identity formation and STEM career motivation* was developed based on the three emergent themes as shown in Figure 4.

As shown in Figure 4, the central them of *identity formation and STEM career motivation* was drawn from across the three initial emergent themes (scientific observation skills, self-efficacy, and affect). These three critical areas stressed by naturalists ultimately shape the learner's identity formation as a member of a scientific community and influence an individual's future STEM career pathway.

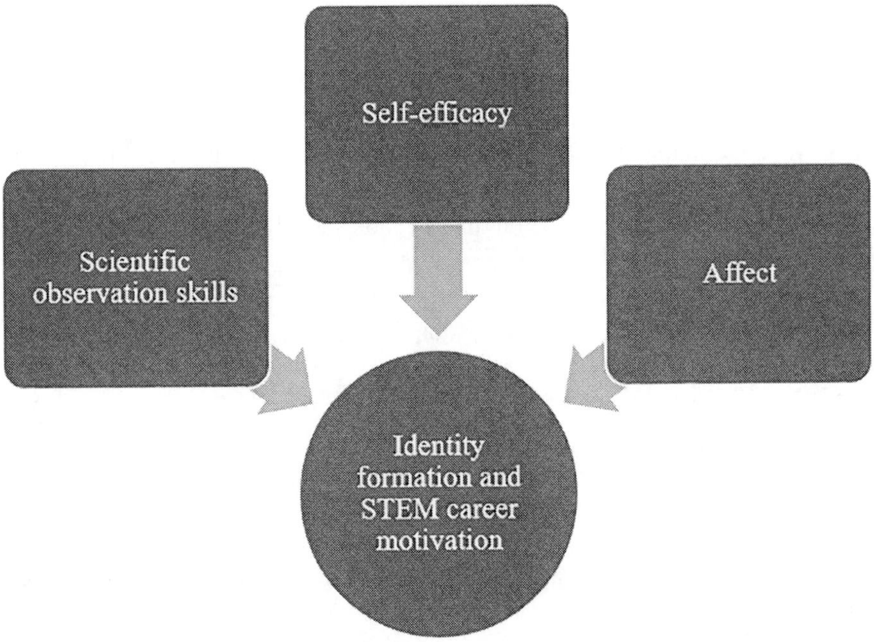

Figure 4. Emergent themes and central theme.

RESULTS

The following discussion centers on the emergent themes of scientific observation skills, self-efficacy, and affect. Next, we draw across these emergent themes to highlight the central them of identity formation and STEM career motivation. In line with current literature and based on the naturalists' experiences with training children in the outdoors, we highlight these themes as essential to fostering effective outdoor practices, particularly in tandem with citizen science.

Scientific Observation Skills

By far, the importance of scientific observation skills surfaced throughout the interviews regardless of the naturalists' backgrounds. The types of skills to be emphasized had striking similarities throughout. According to William, students needed to be trained to be "hyper visual." The range of activities rooted in sensory input and scientific study included collecting, classifying, counting, comparing, and measuring. As Kevin described,

> It's about interface with the resource . . . There are many limitations on the input of sensory stimuli in the classroom. The ability to smell, touch, hear, even taste elemental aspects of nature can only be replicated indoors in poor fashion. And how would you observe the interactions of living things in the classroom? Through a video? Out the window? (Hiller, 2012, p. 1)

The development of scientific observation skills begins with simple exposure to the outdoors. Dion described walking with students in a city park and observing 25 living things within an hour. When working with students' scientific observation skills as a sequence of skills, Dion recommended taking "everything in small steps." Acclimating students to an unfamiliar setting started the sequence of events in developing scientific observation skills. According to these naturalists, establishing comfort levels was an underpinning of the ability for students to develop in depth scientific observation skills. Central to the discussion of the progression of these skills was the infusion of affect and self-efficacy as a result of the informal learning experiences. From Kyle's perspective,

> Sometimes people will get just immediately intimidated by something so we come from that affective side. You don't have to . . . be smart to do this. You have to have a nose. If you have a nose, you're in the game. You got eyes, you're in the game. (Kyle)

To further refine scientific observation skills, the naturalists recommended activities which focused on binary classifications. For example, Eileen found that having students measure where impervious surfaces occurred was an activity that reached many children as a basis for a binary classification system. Comparing and contrasting natural specimens or events were useful in developing affective and cognitive skills.

Some naturalists had a concern that requiring students to identify species too early in their development could trigger some negative responses (affect). For Cary, an expectation of younger children would be to know the difference between an insect and an arachnid. However, very detailed identification is more applicable to mature students or those who have a specific interest. From Jerry's perspective, the level of identification should be steered by the objectives. In his experience, some children became motivated in further study by identifying the Latin name of dragonflies. A commonality among all of the naturalists was that classification and identification skills were context specific and based on the needs of the individual.

Dion advocated for a sequence of steps where children were encouraged to feel comfortable in the outdoors with interpretive activities and basic observational games. His preference was to work with students through their schools over several years. Scientific observation skills developed during this time period with activities such as monitoring oysters through counts, size measurements, and analysis of interaction with other living organisms. Comparably, students extended their scientific observation skills through seining. The students who walked through their neighborhood everyday form a new appreciation for their environment. In Dion's words, "It continues to build. So, I can see that it is opening up their eyes . . . With that opportunity alone, it can kind of take them elsewhere as opposed to the norm."

In general, the field experts cited a variety of experiences, which were indicative of the lack of experience in scientific observation skills. Several naturalists noted that they had worked with students who were not able to differentiate between horses and cows. As Gene indicated,

> Taking them out in the woods . . . can be very traumatic for some of them . . . They've never seen animals. I've had the debate with trips taken out with kids; horses-cows. No cows-horses. Kids don't know the difference between cows and horses. (Gene)

Christina discussed how after planting a garden with a student, she was told that the girl had never planted anything that wasn't in a plastic cup. Other naturalists described students ducking at the noise of woodpeckers, which they mistook for gunfire or running in fear at the sight of some deer. In the view of the field experts, consistent exposure to the outdoors is the basis for forming in depth scientific observation skills. The inability to work in informal settings leads to unexpected responses and limited awareness of environmental issues.

Certainly, the field experts repeatedly referred to a need to set a protocol for methods in collecting data and training by a mentor. Wanda indicated, "There has to be a formal protocol because then your science is not going to hold water. It's not going to be considered really professional or sound science."

In many cases, the field experts believed that students' scientific observation skills surpass that of adults and, as such, have the ability to contribute to professional databases with training. Eileen indicated that in her line of work, data contributions from student citizen scientists would be helpful. However, she had the impression that in the scientific community, input from children was not considered valid. She claimed that when she discussed children's citizen science projects to scientists, the response had been, "You know you can't use those data. They won't be reliable." Eileen perceived that children's contributions were not considered to be accurate enough for professional research. In her view, the data could be adequate with very specific protocols and training. From Wanda's vantage, "There's 60,000,000 of those jokers out there. And so, let's build some programs to involve them in protecting their own environmental health."

Self-Efficacy

For the field experts, the development of scientific observation skills led to increased self-efficacy. Wanda described the sequence as moving from observation to training, then training on a particular interest, and then training on responsibility and ownership of individual behavior. Repeated exposure to work in the outdoors leads to increased motivation and self-efficacy. Cindy described working with one boy in a park who did not appear to be engaged. After the activity was through, she realized that he had been identifying trees with a field guide during an interpretive walk.

In another instance, Christina described the experience of working with students to build a boardwalk through a student developed garden. In their journaling of the experience, she relayed how important the experience was to the children because of the sense of responsibility. Many of the children wrote that they were concerned that people would fall on their boardwalk if it was not well constructed. Not only were the students connected to nature, but they were very excited that they had been given such immense responsibility. This type of connection to the environment and the community, bolstered student self-efficacy through a sense of responsibility.

In order for students to develop self-efficacy, the field experts are central figures in mentoring students. Jerry indicated that in this type of mentoring relationship, the naturalist should be an adult who makes the students feel comfortable and safe. The mentor should be energetic and help the students discover things for themselves. Many of the naturalists were involved in collaboration with schools and the general public with interpretive walks and facilitating activities and projects. The participants viewed these types of activities as opportunities to extend themselves academically, collaborate on research, and contribute to the community.

Participants identified student participation in outdoor activities as a vehicle for developing self-efficacy. Dion cited the need for new experiences to broaden the range of career options for students in the future. For Kevin, one of the greatest advantages to working outdoors was:

> Ownership. You know for a child to feel that their contribution . . . in some element of knowledge or some contribution in action. When you can stand back and look at the fruits of your labors. You can see the tangible products of your labors. There are few things that bring more satisfaction than that. (Kevin)

Citizen science, in which students contribute to professional databases, is one activity several of the naturalists had worked on with children. Examples of citizen science projects ranged from firefly studies, vernal pools, fish populations, impervious surfaces, and plant phenology. The majority of the participants viewed citizen science as a natural extension in the process of developing scientific observation skills, self-efficacy, and affect. Cindy claimed, "I think citizen science programs are the bomb. They are such a phenomenal way for children to really become involved and help our environment in a way that is very satisfactory to them." Christina felt the experience of citizen science would give insight into potential careers for children. Other benefits included reinforcing the scientific method, connecting the students to the natural world, and increasing student self-efficacy.

Affect

Current educational practices place great emphasis on performance-based assessment, yet by contrast, naturalists emphasized affective growth. A consistent concern was that focus on performance-based assessments detracted from environmental programs. This attention to promoting affective growth through environmental experiences is relevant as there is an integral relationship between affective and cognitive states in terms of processing information (Manstead, 2002). For the naturalists, attitude development over repeated exposures to the outdoors was a central consideration. As Kyle indicated when describing the experiences of students from urban environments,

> What I notice a lot of the times is what lacks is that piece that brings it to relevancy for the kids which falls back on the affective learning. There's a valid reason why these kids don't like trees. Because they've got other things to think about. (Kyle)

Naturalists in the study cited repeated exposure to experiential learning outside in terms of changes in individuals' perceptions towards the environment. Wanda indicated that outdoor experiences created long term attitudinal changes towards studying in the outdoors. She noted that many children have a strong dislike for science in schools because the instruction is not based in outdoor settings. Without outdoor instruction in formal settings, Wanda feared that there would be a "further degradation" of attitudes related to stewardship. From a conservation effort perspective, Cindy stated,

> We can't ask children to love something they have no knowledge of. We can't ask them to save something they have no relationship with or connection with. We can't continue to let them think that we could save . . . the whales if they write a check. (Cindy)

Dion found from past experiences that students working with oysters and learning of their role as filter feeders developed a greater appreciation and connection with nature. Exposure to this type of natural study, according to Dion, spurs a passion for the natural sciences. Experiences in the outdoors can reshape personal perspectives in terms of interconnections with the natural world. According to Kevin,

> If you have the opportunity to sit and watch some critter do something that you've never seen it do, or even something that is striving for its own survival, it kind of helps put your life, your existence, your importance or lack thereof into perspective. (Kevin)

Identity Formation and STEM Career Motivation

In the course of working with students in the outdoors, emergent themes revealed that naturalists centered on increasing scientific observation skills, self-efficacy, and affect. Drawing across these emergent fields, findings indicated a central theme was to promote identity formation and STEM career motivation. From the perspective of the naturalists, scientific observation skills, self-efficacy, and affect were critical foundational aspects for working with students in the outdoors. With this framework in place, for some students changes in identity formation as a scientist and encouragement in STEM career motivation becomes an outcome of experiential learning.

In fostering new identities as scientists, naturalists offer support by way of tangible resources, continuity in relationships, and modeling. For example, naturalists typically relayed that they initially focused on whether students were dressed properly, had appropriate materials, and knew safety procedures. Christina indicated that modeling all aspects of scientific behavior was crucial for shaping student identity and providing positive experiences.

> We do run into the fact that do they have the proper clothing for coming out into a situation like that. And so, we had hip wader boots for them . . . even though some of the children were very short. (Christina)

Whether engaged in outdoor activities or citizen science, providing materials not only encouraged participation, but the tools needed by scientists are instrumental in molding student identities as scientists. Cindy claimed,

> There's something about giving a child a clipboard that immediately gives them this sense of being someone other than who they are. Now they're going to be an explorer or a scientist or researcher. And it brings on this whole new persona. So, I always give them a clipboard and pens . . . I think citizen scientry has the same kind of effect. When children realize that they're actually collecting data, it just really makes them feel like they're involved and important and doing something worthwhile. (Cindy)

Even with the advent of technological tools to collect data, access to field sites and corresponding practices was irreplaceable for Edward. Although he uses technological tools to take measurements in his own environmentally oriented professional work, he insisted that all individuals he mentors have experiences doing field work. From his perspective, it is not possible to understand the task fully without knowledge of the setting and corresponding tools first hand.

An equally imperative concern for naturalists interested in encouraging identity formation is establishing consistent, positive relationships with students. When referring to students from low economic areas in urban centers, Lisa stressed the need for continuity and reliability over the course of the year through school-naturalist collaborations.

> And trust is big for them . . . for some people they're so used to people coming in and saying they're going to do stuff and not do it . . . So just collectively, they're used to people not following through. So for you to say, 'I'm going to see you four times,' and you see them four times. By the third time, you've got a bond with them . . . It helps with the conditioning, the reconditioning of the relationship . . . their confidence in themselves. (Lisa)

In this case, Lisa highlighted how building strong relationships with students as a mentor encouraged self-efficacy in terms of trying new outdoor related tasks. In this regard, mentorship becomes what Christina coined as "a social thing . . . not just connecting to nature." In this dynamic Kevin suggested "There's far too much evidence that you can change . . . the negative posture of a student if you will, with the positive experiences of outdoor based education."

Cindy viewed citizen science programs embedded in school-based environmental programs as an important way to enhance student identities.

> I'm always recommending that schools become part of a citizen science program. . . You know there's a sense of independence. We learn a lot of things in the outdoors. But when you're outdoors exploring nature, there's a sense of independence. There's a sense of being in control, being the alpha dog, being a scientist . . . It's just you and that butterfly and you and that tree. (Cindy)

By providing the basic necessities and support structure to work in the outdoors, students become relevant members of a scientific team functioning not as one individual but as a group contributor. Gene described the process in these terms:

> We take it through that stage again here identify an issue or a problem, do an action, and take it through . . . completion. In between, we have to do our research, we have to work together, we have to get the hypothesis, see if it works, maybe try a few different avenues, maybe fail a couple of times. (Gene)

When students are enveloped in the behavior and practices of scientists, learners have a greater understanding of scientific activity and potential career paths. Citizen science is one activity cited by naturalists to encourage student participation in STEM. As Edward stated in reference to an article about citizen science and engineers, "They're more likely to become engineers . . . That's probably just one example of something that would make them feel more empowered." Christina viewed citizen science as a bridge between skills sets (such as scientific observation skills) taught in school settings and potential STEM career paths. "It gives them an introductory, actually gives them a little insight into a career . . . And it does build on skills development . . . if it's identification skills, or critical thinking skills, or measurement skills."

Lisa highlighted the lack of exposure to nature as a barrier to potential career paths, particularly for individuals in crowded urban areas. "I think when people lost their connections to nature, it just naturally lost the connection as that as a possible career option." From Lisa's perspective, programs such as citizen science are an inspiring way to promote STEM careers for children in low-income urban areas. For these children, citizen science

> gives them one more thing to believe in and hope for in their lives . . . There's a lot of people who don't make connections between subjects in school and careers, so anytime you can touch on a career track . . . I think there's potential. (Lisa)

The naturalists in this study, regardless of geographical location, cited the benefits of outdoor learning in connecting individuals to the community. When students learn how to conduct field work, and the importance of this type of endeavor, students acquire the skills sets and knowledge to protect local resources as community volunteers or professional scientists. In fact, Nate indicated that in terms of some of his students' career aspirations, "Quite a few of them have a desire to go further in the field and actually several have gone on to be environmental lawyers."

Regardless of the level of inspiration drawn from the outdoor experiences described by the naturalists, working in collaborative/mentoring relationships has the potential to change student perceptions and career pathways. What these types of experiences have to offer as Kevin described is that ". . . every child regardless of their age is looking for the same thing . . . We're looking for the affirmations that I'm valuable, and what I do is valuable. My contribution is valuable."

Discussion

In this study, we identified three emergent themes: the impact of outdoor learning via mentoring opportunities on scientific observation skills, self-efficacy, and affect from the perspective of naturalists. Based on these three emergent themes, an overarching theme ensued in that the support of scientific observation skills, self-efficacy, and affect ultimately steers identity formation and STEM career formation. In addition, through an analysis of the interview data, we identified a predominant proposition; participants believe the most important aspect of reaching children is through simply having them exposed to the outdoor environment, and the essential skill to emphasize is scientific observation skills.

Although there were many similarities between the interviews of the participants, there were two areas in which there were some alternate views. In the first case, some participants viewed the development of scientific observation skills as a progression of activities. This view would mirror the suggested progression outlined by Eberbach and Crowley (2009) and extended by Hiller and Kitsantas (2015) which begins with novice and moves to expert through modeling and interaction with a mentor. Some of the naturalists viewed citizen science programs as an end stage in the phases of development. Other naturalists viewed the training of scientific observation skills to be interwoven and that citizen science programs and various skills could be embedded throughout the observation training. The interpretation of the sequence of skills was individually based, and there were no commonalities identified by gender, race, geographic regions, educational backgrounds, or group affiliations.

All of the naturalists agreed that the first step in training students in the outdoors was to help them to become more comfortable and to be safe which manifests as positive affect. There was a broad difference between the needs of students from city environments as compared to suburban and rural counterparts. Where the students from outside the city tended to be more concerned with ticks and snakes as described by the naturalists, students from city environments were afraid of being outside and were uncomfortable with trees and animals based on negative experiences in neighborhood settings. As such, each naturalist tailored instruction to meet the needs of a specific population within a given context.

Previous environmentally oriented studies have found a strong affective component to learning in the outdoors (Ballantyne & Packer, 2009). In an Austrian study of the impact on olfactory stimulus on mood, researchers found that natural scents in the outdoors regardless of visual processes impacted the affective state of participants (Weber & Heuberger, 2008). Similarly, the naturalists in the current study cited scientific observation skills as integral to the development of affective and cognitive learning in outdoor settings. The findings of this study highlight naturalists' perceptions of the magnitude of scientific observation skills as well as the notion that although children are competent in protecting the environment, opportunities to demonstrate these capabilities are not always available. Citizen science and outdoor learning help to eradicate this deficiency by forming collaborations with the scientific community, developing fine grained scientific skills, and promoting scientifically oriented careers.

The focus on scientific observation skills mirrors national standards in terms of scientific inquiry. In these guidelines, children are to be encouraged to understand the nature of science, develop skills to become independent inquirers about the natural world, and form dispositions to use the skills, abilities, and attitudes associated with science (National Academy of Science, 2018). Educational policies and curricula centered on promoting environmental education as part of formal education should consider outlining specific goals to target scientific observation skills. Similarly, collaborative efforts with field experts, such as citizen science programs, promote affective and cognitive development and serve to augment formal educational standards.

Educational Implications

This research highlights key themes within outdoor learning settings, such as training individuals to engage in citizen science activities. The current study reveals implications for both practical applications in educational settings as well as research design. Educational programs should target the development of scientific observation skills, student self-efficacy through mastery experiences, and positive affect through repeated exposure to authentic tasks. For example, whether in a formal classroom or informal natural learning setting, providing students with access to specimens or natural objects and classification keys provides repetitive experiences. Naturalists and teachers provide an additional layer of support by highlighting aspects of the object students might miss (i.e., veins in insect wings, patterns in seastars, shapes in arthropods).

Although it is not always possible to work with students outside during the formal school structure, teaching scientific observation skills is possible through lessons with natural specimens and online citizen science resources. A collection of natural specimens provides ample opportunities for this type of instruction. Students may count, collect, classify, and measure easily available specimens found in their natural habitat. With these items, teachers may incorporate scientific observation skills in topics such as biodiversity, biomes, and taxonomic levels. Drawing games where some students describe the key features of a specimen while another student draws at the board is yet another engaging way to encourage scientific observation skills.

Onsite citizen science programs extend this type of instructional approach where students apply their learned skills for authentic research. In this regard, students may help to identify species in a variety of online programs, for example. Most recently, many citizen science programs include online feedback from scientists on volunteer contributions adding to the scaffolding effect.

Stakeholders interested in encouraging outdoor learning can assist schools by providing funding and collaborative opportunities with naturalists. These experts who visit schools can draw on students' scientific observation skills by having them identify animals not only visually but by listening to sounds from nature and identifying corresponding organisms.

Further, naturalists who meet with teachers and students to discuss ways to maximize outdoor learning spaces provide an entirely new dimension to learning within school settings. An even more enriching experience is when students meet naturalists at their school site and then conduct experiential activities in outdoor settings following training with experts. In this regard, field experts are a powerful resource for promoting children's' growth through collaborative outdoor experiences.

From a research design perspective, studies centered on the impact of citizen science programs on adolescent achievement and STEM career motivation should extend beyond self-motivational beliefs (e.g., self-efficacy and interest) to include affect. Researchers focused on assessing the impact of citizen science programs might consider including measures related to scientific observation skills, self-efficacy, and affect while examining the interaction among these effects to reflect the input of naturalists from this study.

Naturalists dedicate their life's work to engaging students in a way that will promote sustained motivation. In fact, citizen science programs are inherently based on volunteers who are interested in contributing to professional research (Pandya & Dibner, 2019). This intrinsic component of citizen science creates a dilemma in citizen science research for adolescents. When teachers incorporate citizen science within classroom instruction as a requirement, students are not volunteering. And for researchers gauging self-motivational beliefs such as interest, assessing interest levels of volunteers may skew the results as participants typically have high levels of interest (Hiller, 2012; Hiller & Kitsantas, 2014). In this case, researchers may have difficulty assessing the impact of citizen science on interest for students in general. The juxtaposition between citizen science rooted in volunteer activity and the need to measure interest levels for a variety of diverse learners, is a unique predicament for citizen science researchers concerned with adolescent self-motivational beliefs, science achievement and STEM career motivation.

Limitations and Future Directions

Some limitations of this study relate to participant selection. Although great efforts were made to include varying perspectives, additional research on different regional areas and varying racial/ethnic backgrounds may shed alternate views on educational training. In particular, a wider selection of participants from all the major geographical regions of the United States would provide helpful comparisons. Moreover, we recognize that initial gatekeepers as well as critical references influenced the narrative of this work as their referrals ultimately selected the participants included in this study.

In addition, two broad perspectives of the naturalists may be incongruent with other factions in the environmental community. First, since the issue of recognition of student data collection for professional databases is sometimes controversial, research examining the perceptions of scientists in regards to student input may help inform the literature base on how to establish more cohesive partnerships between schools and the scientific community.

On-site administrators play an important role in ensuring that students meet standards while promoting environmental education opportunities. For this reason, a similar study examining administrative perceptions of the impact of environmental programs within the school would potentially afford the opportunity to see if the attitudes, beliefs, and values of naturalists are in alignment with school officials. Further, emerging state environmental literacy plans and state curricula should be examined for explicit directives on promoting scientific observation skills.

Conclusion

The natural world, saturated with opportunities for varied experiences, affords a unique context for student development. As an educational springboard, scientific observation skills are the impetus for the development of strong mastery experiences, self-efficacy, positive affect, and for some students, a career pathway.

This work outlined the benefits of citizen science programs, particularly in informal natural science learning settings, as a means to strengthen scientific observation skills, self-efficacy, and affect. In turn, these positive benefits offer the possibility of creating science-oriented identities and future career options. The intertwining of these constructs has been further examined and supported through the perspectives of field experts. Without repeated access to well-defined learning experiences, students are constricted in moving beyond novice skill sets and levels of understanding. It is the field expert, through modeling, who mediates a more advanced form of learning and development.

For the naturalists in this study, developing scientific observation skills and self-efficacy within environmental contexts is essential to children's' affective and cognitive development. Cindy, in describing a canning experience with a student said,

> I knew she'd never seen canning. But what happened shocked me . . . What I wasn't prepared for was when we went to the orchard . . . We got to the fields where we were going to do our picking. She stopped and I said, "What's wrong?" and she said, "I've never seen fruit growing on trees." And I cried because how can it be this way? We can't raise children like that.

For many individuals, this scenario is unimaginable. Yet, field experts indicated consistently that this lack of exposure to the natural world is the reality for many children in the United States. Citizen science programs and the complementary experiences of working with naturalists to do field work, offers an authentic learning opportunity which magnifies student scientific observation skills, self-efficacy, and affect. Most notably, naturalists trained as educators are essential community members in the endeavor to not only maximize academic learning and STEM career motivation but to reshape childhood learning experiences in the outdoors.

APPENDIX A

Semi-Structured Interview Guide

How do naturalist programs empower children?

Do children sometimes become teachers during environmental experiences?

In what ways do students contribute to research?

What impact do these contributions have on students?

Have you been aware of incidences where students had a political impact on the environment?

Is credibility of environmental programs a key component?

How do students establish and maintain credibility?

In what ways are students responsible for environmental programs?

How does learning in the outdoors differ from the school setting?

What are key characteristics children need in environmental education programs?

Is technology compatible with outdoor learning?

Would you approach outdoor instruction using a sequence of skills?

APPENDIX A (CONTINUED)

What are the benefits of informal learning?

How important is database use in constructing knowledge outdoors?

Have you heard of the term "citizen scientist?"

Are citizen scientist programs a secondary curriculum?

If students benefit from learning observation, identification, and community impact skills, would a citizen science program complement these goals?

Should citizen science be taught in schools?

Are children capable of being citizen scientists?

How can field experts support environmental education programs?

What role should databases play in environmental education programs?

Have you known of children becoming professional scientists who participated in environmental education programs?

What are barriers to environmental education programs?

REFERENCES

Ainley, M. (2006). Connecting with learning: Motivaton, affect and cognition in interest processes. *Educational Psychology Review, 18*(4), 391-405. doi:10.1007/s10648-006-9033-0

Auer, M. R. (2008). Sensory perception, rationalism, and outdoor environmental education. *International Research in Geographical and Environmental Education, 17*(1), 6-12. doi:10.2167/irgee225.0

Ballantyne, R. & Packer, J. (2009). Introducing a fifth pedagogy: Experience-based strategies for facilitating learning in natural environments. *Environmental Education Research, 15*(2), 243-262. doi:10.1080/13504620802711282

Bandura, A. (1997). *Self-efficacy: The exercise of control.* New York, NY: W. H. Freeman and Company.

Beeden, R. J., Turner, M. A., Dryden, J., Merida, F., Goudkemp, K., Malone, C. & Birtles, A. (2014). Rapid survey protocol that provides dynamic information on reef conditions to managers of the Great Barrier Reef. *Environmental Monitoring and Assessment, 186*, 8527-8540. doi:10.1007/s10661-014-4022-0

Bjørnebekk, G. (2008). Positive affect and negative affect as modulators of cognition and motivation: The rediscovery of affect in achievement goal theory. *Scandinavian Journal of Educational Research, 52*(2), 153-170. doi: 10.1080/00313830801915788

Bombaugh, R. (2000). From citizen scientist to engineers. *Journal of Professional Issues in Engineering Education and Practice, 126*(2), 64-68. doi:10.1061/(asce)1052-3928(2000)126:2(64)

Bonney, R., Ballard, H., Jordan, R., McCallie, E., Phillips, T., Shirk, J. & Wilderman, C. C. (2009a). Public participation in scientific research: Defining the field and assessing its potential for informal science education: A CAISE inquiry group project. Retrieved from Center for Advancement of Informal Science Education (CAISE) website: http://informalscience.org/sites/default/files/PublicParticipationinScientificResearch.pdf

Bonney, R., Cooper, C. B., Dickinson, J., Kelling, S., Phillips, T., Rosenberg, K. V. & Shirk, J. (2009b). Citizen science: A developing tool for expanding science knowledge and science literacy. *BioScience*, *59*(11), 997-984. doi:10.1525/bio.2009.59.11.9

Britner, S. L. & Pajares, F. (2006). Sources of science self-efficacy beliefs of middle school students. *Journal of Research in Science Teaching*, *43*(5), 485-499. doi:10.1002/ tea.20131

Cachelin, A., Paisley, K. & Blanchard, A. (2009). Using the significant life experience framework to inform program evaluation: The Nature Conservancy's Wings and Water Wetlands Education Program. *The Journal of Environmental Education*, *40*(2), 2-14. doi:10.3200/joee.40.2.2-14

Cartwright, N. (1989). *Nature's capacities and their measurements*. Oxford, United Kingdom: Clarendon Press.

Carver, C. S., Avivi, Y. E. & Laurenceau, J. P. (2008). Approach, avoidance, and emotional experiences. In A. J. Elliot (Ed.), *Handbook of approach and avoidance motivation* (pp. 385-397). New York, NY: Psychology Press.

Chawla, L. & Kushing, D. F. (2007). Education for strategic environmental behavior. *Environmental Education Research*, *13*(4), 437-452. doi:10.1080/13504620701581539

Chinn, C. A. & Malhotra, B. A. (2002). Children's responses to anomalous scientific data: How is conceptual change impeded? *Journal of Educational Psychology*, *94*(2), 327-343. doi:10.1037/0022-0663.94.2.327

Cleary, T. J. (2018). *The self-regulated learning guide: Teaching students to think in the language of strategies*. New York, NY: Routledge.

Cleary, T. J. & Labuhn, A. S. (2013). Application of cyclical self-regulation interventions in science-based contexts. In H. Bembenutty, T. J. Cleary, & A. Kitsantas (Eds.), *Applications of self-regulated learning across diverse disciplines: A tribute to Barry J. Zimmerman* (pp. 89-124). Charlotte, NC: Information Age Publishing, Inc.

Corbin, J. & Strauss, A. (1990). Grounded theory research: Procedures, canons, and evaluative criteria. *Qualitative Sociology*, *13*, 3-21. doi:10.1007/bf00988593

Crall, A. W., Newman, G. J., Stohlgren, T. J., Holfelder, K. A., Graham, J. & Waller, D. M. (2011). Assessing citizen science data quality: An invasive species case study. *Conservation Letters*, *4*, 433-442. doi:10.1111/j.1755-263X.2011.00196

Creel, B., Nite, S., Almarri, J. E., Shafik, Z., Mari, S. & Al-Thani, W. A. (2017, June). Inspiring interest in STEM education among Qatar's youth. Paper presentation at the annual conference of the American Society for Engineering Education. Columbus, OH (Paper ID # 20731)

Crocetti, E. (2017). Identity formation in adolescence: The dynamic of forming and consolidating identity commitments. *Child Development Perspectives, 11(2), 145-150.* doi:10.1111/cdep.12226

Cronje, R., Rohlinger, S., Crall, A. & Newman, G. (2011). Does participation in citizen science improve scientific literacy? A study to compare assessment methods. *Applied Environmental Education & Communication*, *10*, 135-145. doi:10.1080/1533015x. 2011.603611

Devine-Wright, P., Devine-Wright, H. & Fleming, P. (2004). Situational influences upon children's beliefs about global warming and energy. *Environmental Education Research*, *10*(4), 493-506. doi:10.1080/1350462042000291029 deVos, L., Leijnse, H., Overeem, A. & Uijlenhoet, R. (2017). The potential of urban rainfall monitoring with crowdsourced automatic weather stations in Amsterdam. *Hydrology and Earth Systems Science*, *21*, 765-777. doi:10.5194/hess-21-765-2017

DiBenedetto, M. K. & Zimmerman, B. J. (2010). Differences in self-regulatory processes among students studying science: A microanalytic investigation. *The International Journal of Educational and Psychological Assessment*, *5*, 2-24. doi:10.1016/ j.lindif.2013.04.004

Eberbach, C. & Crowley, K. (2009). From everyday to scientific observation: How children learn to observe the biologist's world. *Review of Educational Research*, *79*(1), 39-68. doi:10.3102/0034654308325899

EnvironnmentalScience.org. (2018). What does a naturalist do? Retrieved from https://www.environmentalscience.org/career/naturalist.

Fowler, A., Whyatt, D., Davies, G. & Ellis, R. (2013). How reliable are citizen-derived scientific data? Assessing the quality of contrail observations made by the general public. *Transactions in GIS*, *17*(4), 488-506. doi:10.1111/tgis.12034

Guba, E. G. & Lincoln, Y. S. (2005). Paradigmatic controversies, contradictions, and emerging influences. In N. K. Denzin & Y. S. Lincoln (Eds.), *The handbook of qualitative research*, (3rd ed., pp. 191-216). Thousand Oaks, CA: SAGE Publications, Inc.

Harrison, J. A., Underhill, L. G. & Barnard, P. (2008). The seminal legacy of the Southern African Bird Atlas Project. *South African Journal of Science*, *104*, 82-83.

Hidi, S. & Ainley, M. (2008). Interest and self-regulation: Relationships between two variables that influence learning. In D. H. Schunk and B. J. Zimmerman's (Eds.), *Motivation and self-regulated learning*, (pp. 77-110). New York, NY: Routledge.

Hiller, S. E. (2012). *The impact of a citizen science program on student achievement and motivation: A social cognitive career perspective*. (Dissertation thesis). Retrieved from ProQuest Dissertations & Theses Global. (UMI Number 3547088)

Hiller, S. E. (2017). *Mastering science with metacognitive and self-regulatory strategies: A teacher-researcher dialogue of practical applications for adolescent students. Education in a competitive and globalizing world* (Vol. 2). Hauppauge, NY: Nova Science Publishers, Inc.

Hiller, S. E. & Kitsantas, A. (2014). The effect of a horseshoe crab citizen science program on student science performance and STEM career motivation. *School Science and Mathematics Journal*, *114*(6), 302-311. doi:10.1111/ssm.12081

Hiller, S. E. & Kitsantas, A. (2015). Fostering student metacognition and motivation in STEM through citizen science programs. In A. Peña-Ayala's (Ed.) *Metacognition: Fundaments, Applications, and Trends*, (Vol. 76, pp. 193-221). Cham, Switzerland: Springer.

Hiller, S. E., & Kitsantas, A. (2016). The validation of the citizen science self-efficacy scale (CSSES). *The International Journal of Environmental and Science Education*. doi:10.12973/ijese.2016.405a

Hiller, S. E. & Reybold, L. E. (2011, October). *Field expert's perceptions of scientific observations in the natural world*. Paper presentation at the annual research symposium of the North American Association for Environmental Education, Raleigh, NC.

Hocking, J. (1907). Field naturalists' work: It's value to teachers. *The Victorian Naturalist*, *23*, 234.

Jeanpierre, B., Oberhauser, K. & Freeman, C. (2005). Characteristics of professional development that effect change in secondary science teachers' classroom practices. *Journal of Research in Science Teaching*, *42*(6), 668-690. doi:10.1002/tea.20069

Kim, S. & Park, J. (2018). Development and application of learning materials to help students understand ten statements describing the nature of scientific observation. *International Journal of Science and Mathematics Education, 16*, 857-876. doi:10.1007/s10763-017-9823-5

Lent, R. W., Brown, S. D. & Hackett, G. (1994). Toward a unifying social cognitive theory of career and academic interest, choice, and performance. *Journal of Vocational Behavior, 45*, 79-122. doi:10.1006/jvbe.1994.1027

Lent, R. W., Miller, M. J., Smith, P. E., Watford, B. A., Lim, R. H. & Hui, K. (2016). Social cognitive predictors of academic performance in engineering: Applicability across gender and race/ethnicity. *Journal of Vocational Behavior, 94*, 79-88. doi:10.1016/j.jvb.2016.02.012

Manstead, A. (2002). The role of affect in the formation and change of attitudes. *Marketing for Sustainability*, 202-212.

Mayer, R. E. & Wittrock, M. C. (2006). Problem solving. In P. A. Alexander & P. H. Winne (Eds.), *Handbook of educational psychology*, (pp. 369-389). New York, NY: Routledge.

Meinhold, J. L. & Malkus, A. J. (2005). Adolescent environmental behaviors: Can knowledge, attitudes, and self-efficacy make a difference? *Environment and Behavior, 37*(4), 511-531. doi:10.1177/0013916504269665

Merendlender, A. M., Crall, A. W., Drill, S., Prysby, M. & Ballard, H. (2016). Evaluating environmental education, citizen science, and stewardship through naturalist programs. *Conservation Biology, 30*(6), 1255-1265. doi:10.1111/cobi.12737

Merriam-Webster Dictionary. (2018). *Naturalist.* Retrieved from https://www.merriam-webster.com/dictionary/naturalist

National Academy of Sciences. (2018). *National science education standards.* Retrieved from http://www.nap.edu/openbook.php?record_id=4962&page=R1

Navarro, R. L., Flores, L. Y. & Worthington, R. L. (2007). Mexican American middle school students' goal intentions in mathematics and science: A test of social cognitive career theory. *Journal of Counseling Psychology, 54*(3), 320-335. doi:10.1037/0022-0167.54.3.320

Palermo, E., Laut, J., Nov, O., Cappa, P. & Porfiri, M. (2017). Spatial memory training in a citizen science context. *Computers in Human Behavior, 73*, 38-46. doi:10.1016/j.chb.2017.03.017

Pandya, R. & Dibner, K. A. (2018). Learning through citizen science. Enhancing opportunities by design. A consensus study of the National Academies of Science, Engineering, and Medicine. National Academic Press: Washington, D. C. Retrieved from https:// www.nap.edu/ catalog/ 25183/ learning-through-citizen-science-enhancing-opportunities-by-design? utm_ source= NASEM+News+and+Publications& utm_ campaign= 1ea596a1e6-NAP_ mail_ new_ 2018-11-06& utm_ medium= email&utm_ term= 0_ 96101de015- 1ea596a1e6- 104793173&goal= 0_ 96101de015- 1ea596a1e6-104793173&mc_cid=1ea596a1e6&mc_eid=9a43192017

Patrick, L., Care, E. & Ainley, M. (2011). The relationship between vocational interests, self-efficacy, and achievement in the prediction of educational pathways. *Journal of Career Assessment, 19*(1), 61-74. doi:10.1177/1069072710382615

Patton, M. Q. (2015). *Qualitative evaluation and research methods* (4th ed.). Thousand Oaks, CA: SAGE Publications, Inc.

Pocock, M. & Evans, D. M. (2014). The success of the Horse-Chestnut Leaf-Miner, *Cameraria ohridella*, in the UK revealed with hypothesis-led citizen science. *PLoS ONE*, *9*(1), e86226. doi:10.1371/journal.pone.0086226

Prysby, M., Oberhauser, K. & Petersen, C. (2007, November). Citizen science: Active, inquiry and service learning for environmental educators. Paper presented at the annual conference for the North American Association for Environmental Education, Virginia Beach, VA.

Reybold, L. E., Lammert, J. D. & Stribling, S. M. (2013). Participant selection as a conscious research method: Thinking forward and the deliberation of 'emergent' findings. *Qualitative Research*, *13*(6), 699-716. doi: 0.1177/1468794112465634

Roy, H. E., Baxter, E., Saunders, A. & Pocock, M. J. (2016). Focal plant observations as a standardized method for pollinator monitoring: Opportunities and limitations for mass participation citizen science. *PLoS ONE*, *11*(5), e0155571. doi:10.1371/journal.pone.0155571

Roeser, R. W., Peck, S. C. & Nasir, N. S. (2006). Self and identity processes in school motivation, learning, and achievement. In P. A. Alexander and P. H. Winne (Eds.), *Handbook of educational psychology*, (2nd ed, pp. 391-424). New York, NY: Routledge.

Saldaña, J. (2009). *The coding manual for qualitative researchers*. Thousand Oaks, CA: SAGE Publications, Inc.

Snäll, T., Kindvall, O., Nilsson, J. & Pärt, T. (2011). Evaluating citizen-based presence data for bird monitoring. *Biological Conservation*, *144*, 804-810. doi:10.1016/j.biocon.2010.11.010

Steinberg, L. (2015). *Age of opportunity: Lessons from the new science of adolescence*. Boston: Houghton Mifflin Harcourt Publishing Company.

Sutton, S. (2009). *Increasing science literacy among English language learners through plant phenological monitoring: A citizen science program at Sequoia and Kings Canyon National Parks*. (Master's thesis). Retrieved from Proquest Dissertations and Global Theses. (UMI Number 1473523)

Tomasek, T. M. (2006). *Student cognition and motivation during the Classroom BirdWatch citizen science project*. (Dissertation thesis). Retrieved from the Proquest Dissertations and Global Theses. (UMI Number 3242410)

Tracy, G. (2018). Learning to see. *American Scientist*, *106*(4), 242-249. doi:10.1511/2018.106.4.242

Walker, D. & Myrick, F. (2006). Grounded theory: An exploration of process and procedure. *Qualitative Health Research*, *16*(4), 547-559. doi:10.1177/1049732305285972

Wallace, D. & Bodzin, A. (2017). Developing scientific citizenship identity using mobile learning and authentic practice. *Electronic Journal of Science Education*, *21*(6), 46-71.

Weber, S. T. & Heuberger, E. (2008). The impact of natural odors on affective states in humans. *Chemical Senses*, *33*, 441-447. doi:10.1093/chemse/bjn011

Williams, R. L., Stafford, R. & Goodenough, A. E. (2015). Biodiversity in urban gardens: Assessing the accuracy of citizen science data on garden hedgehogs. *Urban Ecosystem*, *18*, 819-833. doi:10.1007/s11252-014-0431-7

Wilson, R. S. (2007). Balancing emotion and cognition: A case for decision aiding in conservation efforts. *Conservation Biology*, *22*(6), 1452-1460. doi:10.1111/j.1523-1739.2008.01016.x

Young, M. T. (2016). Technology and technique: The role of skill in the practice of scientific observation. *Perspectives on Science*, *24*(4), 396-415. doi:10.1162/posc_a_00213

Zimmerman, B. J. (2000). Attaining self-regulation: A social cognitive perspective. In M. Boekart, P. R. Pintrich, and Zeidner, M. (Eds.), *Handbook of self-regulation*, (pp. 13-39). New York, NY: Academic Press.

In: Enhancing STEM Motivation through Citizen Science Programs ISBN: 978-1-53616-038-3
Editors: Suzanne E. Hiller and Anastasia Kitsantas

Chapter 7

STUDENT MOTIVATION AND TEACHER/SCIENTIST COLLABORATION IN CITIZEN SCIENCE PROGRAMS: MAKING SCIENCE LEARNING FUN AND RELEVANT

Maria K. DiBenedetto **[*]** ***and Dale H. Schunk***
The University of North Carolina at Greensboro,
Bryan School of Business and Economics, NC, US

ABSTRACT

This chapter discusses the power of citizen science programs to motivate student learning by building self-efficacy beliefs on how to conduct authentic scientific research. We describe research on motivation and self-efficacy in science education followed by research on the importance of teacher/scientist collaborations in science learning. This is continued by a discussion on key aspects of successful citizen science programs, and we present two examples of successful programs used in schools throughout the United States and globally. Throughout the chapter we emphasize the importance of active learning and how citizen science programs spark, excite, and motivate students and their teachers as they collect data for scientific research. We discuss how citizen science programs have the capacity to transform learners into agents of change, citizens who care about the Earth - and we conclude the chapter with educational implications and recommendations for future research.

Keywords: learner self-efficacy for conducting research, motivating students to learn and citizen science programs, self-efficacy and citizen science programs, teacher/scientist collaboration in citizen science programs

---

[*] Corresponding Author's E-mail: mkdibenedetto@uncg.edu

In the 2015 PISA (Program for International Student Assessment) report, approximately 540,000 tenth grade students from 72 countries were assessed on reading, mathematics, and science (OECD, 2018). The United States ranked 25th in science achievement with little change in ranking since 2006. Yet the United States has produced more Nobel Laureates in science (265), followed by the second most awarded recipients of 84 from the United Kingdom (The Nobel Prize, n.d.). These data are confusing. Why are the students of the United States, which heavily values education and produces many prestigious scientists, scoring well below the top performers? And what is being done to help promote science achievement in the United States?

The United States National Science Foundation (NSF) is a government organization that focuses on science education and research. In its vision statement: "A Nation that is the global leader in research and innovation" (NSF, 2018, p. 8), NSF emphasizes the importance of creating leaders in research and education. NSF provides financial support to educators who are developing and instituting STEM (science, technology, engineering, mathematics) initiatives in schools throughout the country. Critical to the success of these programs is student motivation to participate and engage in learning, and teachers' willingness to participate and learn new strategies for teaching and implementing science education.

One of the most exciting and impactful ways to foster science motivation and learning is with citizen science programs (CSPs). CSPs provide opportunities for scientists, educators, and students to collaborate on "real" research; to generate new knowledge; to interact with people globally; and to potentially transform learners into motivated, thoughtful and caring individuals of our Earth (Bela et al., 2016). CSPs are the focus of this chapter.

Our chapter begins with a discussion of motivation research in the science classroom using Bandura's (1986) social cognitive theory. Social cognitive theory presents an agentic view of the learner. Agency includes having a sense of control over one's learning and believing in one's capability, or *self-efficacy* to plan and implement actions to reach goals (Bandura, 1997). By helping students develop a sense of personal agency, educators can positively influence learners' self-efficacy for science learning and thus their motivation to do so.

This section is followed by a discussion of research on teachers' collaborating with scientists to support self-efficacy and learning in the classroom. Collaboration with scientists offers a way for educators to provide opportunities for mastery experiences for their students in science learning (Hiller & Kitsantas, 2014). Successful experiences at mastering a task foster a sense of control, which helps build self-efficacy for future similar tasks (Bandura, 1986; Zimmerman & DiBenedetto, 2008) and can make learning enjoyable and relevant.

Following the research on self-efficacy and teacher collaboration, we describe two CSPs. Students Discover is a project funded by NSF to improve STEM education in middle school students by providing them with opportunities to engage in authentic scientific practices. It involves a partnership between scientists from North Carolina State University and the North Carolina Museum of Natural Sciences and includes collaborations among teachers, students, scientists, and district leaders to implement citizen science lessons in the classroom (Students Discover, 2018). We highlight one specific CSP in the Students Discover program -Ant Picnic- and describe how it promotes students' self-efficacy and teachers' collaboration with scientists. This is followed by a discussion of Global Learning and Observations to Benefit the Environment (GLOBE), a CSP that began on Earth Day in 1994. It has since expanded to include over 30,000 K-12 schools in 121 countries (The GLOBE Program, 2018a). GLOBE is aimed at assessing the rate and impact of climate change on the environment and is sponsored by prestigious organizations such as the National Aeronautics and Space Administration

(NASA), the National Oceanic Atmospheric Administration (NOAA), NSF, the U.S. State Department, and the Youth Learning as Citizen Environmental Scientist organization. In addition to describing the GLOBE CSP, we discuss how it fosters self-efficacy beliefs, collaborations among teachers throughout the world, and STEM education.

We then summarize the connections between these programs and self-efficacy and teacher collaborations and provide educational implications. The chapter concludes with recommendations for future research.

RESEARCH ON MOTIVATION IN SCIENCE EDUCATION

Self-Efficacy Sources

Social cognitive theory emphasizes the role of capability beliefs (self-efficacy) in motivation learning, and achievement (Bandura, 1986, 1997). Self-efficacy is a key motivation variable that is hypothesized to influence choice of activities and the goals one sets, the amount of effort put into reaching these goals, and the degree to which one persists in the face of challenges and obstacles. Bandura (1997) postulated that students gain self-efficacy beliefs from four sources: social persuasion, vicarious experiences, physiological indexes, and mastery performances.

Social persuasion refers to encouragement from others to complete a task (Bandura, 1997; Schunk & DiBenedetto, 2016). Telling students, "You can do it!" can build self-efficacy beliefs. When students know others believe in them, they may develop self-efficacy to engage in a task. For example, science students may be fearful to conduct lab dissections, but when a teacher explains the procedure and encourages them to try, they are conveying a belief in students' capabilities to perform the dissection.

Vicarious experiences refer to observing others model activities, which the observer would not have been able to perform on his or her own without having seen how it is done first. For example, science teachers will often model how to conduct an experiment before the students begin working on the lab. Learners obtain information from their observations, such as how-to problem-solve when confronted with challenges, how to carry out the steps of the scientific method, and how to persist when results are not immediate. Research shows that when learners see models who are competent or similar to themselves their self-efficacy to perform similar tasks increases (Schunk, 2012).

Physiological indexes refer to physical and emotional states learners experience when attempting tasks (Bandura, 1997). Experiencing what it feels like to hold test tubes and to be able to calculate data can provide students with cues about future successes and failures (Schunk & DiBenedetto, 2016). For example, students who experience jubilation at successfully mixing chemicals or correctly estimating the effects of water pollution on plant growth will likely feel self-efficacious in future experiments involving similar protocols.

Mastery experiences are powerful indicators of competency and build self-efficacy (Bandura, 1997; Zimmerman & DiBenedetto, 2008). Students who experience success in performing and completing tasks are likely to feel efficacious, whereas students who experience failures may anticipate poor performances in similar future academic tasks. A student who builds a solar car that out-races all of her classmates' cars is likely to feel more competent in her capability to build other similar models.

Self-Efficacy Research

Research has examined the role of the four hypothesized sources of self-efficacy to determine if they make an independent contribution to students' science self-efficacy. In a study with middle school students (Britner & Pajares, 2006), 319 participants completed a survey assessing self-efficacy sources for learning science, along with other variables such as science self-concept, grade self-efficacy, self-regulated learning, and anxiety. As predicted, each of the sources of science self-efficacy correlated with one another, with science self-efficacy, and with science grades. Science self-efficacy was the strongest predictor of grades, with mastery learning experiences contributing most of the variance to the regression. These findings are important because they support Bandura's (1997) tenet that mastery experiences build feelings of competence.

Little research exists on whether science self-efficacy beliefs change over time (DiBenedetto & Bembenutty, 2013) and whether mastery experiences can help sustain students' motivation for learning science as students advance in grade levels. Lofgran, Smith, and Whiting (2015) conducted a study on students' self-efficacy beliefs as they transitioned from elementary to middle to high school. Participants were 522 students in grades 6-9 who completed surveys assessing their self-efficacy for learning science. Results indicated that self-efficacy beliefs gradually declined as they reached the upper grades. While students' motivational beliefs were not significantly different from each other in the lower grades, the self-efficacy for each grade level below ninth was significantly different from ninth graders' self-efficacy. This finding suggests that students' motivation for learning science declines with development.

Other research has demonstrated changes in self-efficacy within a science course. In a study examining changes in self-efficacy over the course of a semester in intermediate college level Biology courses, students' self-efficacy declined from the original assessments taken at the start of the semester to the end of the term (DiBenedetto & Bembenutty, 2013). Declining motivation over time will likely not foster future selection of science courses or science careers and can result in lower science achievement. The studies above suggest that by engaging students in mastery learning experiences using instructional approaches that promote opportunities for authentic learning, science self-efficacy may not only be sustained but also influence students' future plans for taking additional science courses or choosing careers in science (Britner & Pajares, 2006).

Collaboration Between Teachers and Scientists

Science teachers are often not aware that they are building self-efficacy through classroom lessons such as laboratory activities that provide students with the opportunity to experience mastery (DiBenedetto, 2018). While these lab experiments may motivate students, they provide hands-on learning experiences that often take an artificial approach: The problem is laid out, students know the exact steps to carry out the experiment, and there are uniform conclusions among all students participating in the lab (Shah & Martinez, 2016). In addition, there are time constraints (Hiller & Kitsantas, 2014), as they are usually conducted within science classrooms during the school day and in-between other content area classes. Although these approaches to teaching science in schools help students learn about the scientific method, they do not replicate the many nuances and challenges that arise when conducting real scientific research (Hiller & Kitsantas, 2014). They also do not provide students with authentic mastery experiences in which they can engage in true scientific inquiry, interpret and analyze data, and draw scientific conclusions – skills necessary for building self-efficacy for participating in science (Shah & Martinez, 2016).

NSF has invested in projects to improve STEM education and research in formal and informal settings (NSF Research on Learning in Formal and Informal Settings (DRL), 2018). The focus of these projects has been to encourage collaboration among scientists and educators and they view themselves as "a catalyst for change – advancing theory, method, measurement, development, and application in STEM education" (para. 1). This highlights the importance of science teachers collaborating with scientists, engineers, and officials from museums, parks, and recreation facilities to provide students with opportunities to learn about how science inquiry takes place, how to generate questions and hypotheses, and how to conduct research, collect and analyze their own data, and interpret findings (Shah & Martinez, 2016).

More traditional interactions between teachers and museums have been centered on providing educators with resources through organized field-trips, summer workshops for teachers, and traveling science trunks filled with artifacts, lesson plans, and materials needed for simple labs (Summers, 2004). Each of these activities is designed to complement existing curriculum rather than change the ways in which educators teach and students learn or promote opportunities for genuine scientific inquiry experiences. More recent research, however, has shifted the focus from factual content-based information and predetermined step-by-step protocols to exploring the unknown and challenging students to think critically, analyze, and explore different possibilities in informal settings outside of the classroom (Shah & Martinez, 2016). These types of learning environments are more likely to stimulate interest, curiosity, motivation, and include at-the-moment science research (Faria, Guilherme, Gaspar, & Boaventura, 2015).

Faria et al. (2015) conducted a study in Lisbon that was part of a larger project aimed at improving students' scientific knowledge and understanding of life on the ocean's seafloor. The researchers worked with a public-school teacher and her 25 fourth graders to learn about the deep sea. The study involved a session where the students were read a short story about the adventures of three children who discovered objects at the beach and connected this adventure to a pioneering scientist who believed in deep sea life. This was followed by two trips to two different museums where students learned about various exhibits, searched for artifacts, and responded to questions on worksheets. The study concluded with the students completing a final task of writing a diary about the scientist, his nautical yacht, scientific equipment, and findings, as well as conducting research on the characteristics of deep-sea fish species. The results from this study showed that students developed higher interest, motivation, and perceptions about learning science. In addition, the teacher reported that she learned more about the subject content, the students and what motivates and excites them to learn, and the need to create more opportunities for exploration and research.

The above study highlights how collaborations between scientists and teachers can heighten students' interests and motivation about learning science, along with providing educators valuable information about students and a deeper understanding of content. Benefits also include access to the scientific community, participation in actual scientific research, and engagement in authentic, inquiry-based learning (Houseal, Abd-El-Khalick, & Destefano, 2014).

In a study funded by NSF (Houseal et al., 2014), researchers examined the effects of content knowledge gain, attitudes and pedagogical practices among teachers and content knowledge gain and perceptions of scientists on students who worked with park rangers and researchers at Yellowstone National Park. The experimental group consisted of teachers and students engaged in science activities that included investigating topics of their own interests, using tools typically used by rangers and scientists to conduct measurements and collect data, and presenting data to audiences beyond the classroom. Teachers in the experimental group received professional development training from university researchers, educational specialists, geologists, and rangers. The goals of the professional development training were: to increase teacher content knowledge on how to make connections between the Mammoth Hot Springs and early life on Earth, coral reefs, and life on other planets; to help teachers discover ways to make the content more accessible to students by having them involved in actual data collection; and to have teachers participate in the field experiences in the same way their students would. The comparison group of teachers did not receive the professional development, had less instructional time dedicated to the topics, had shorter field experiences, and used the same instruments for data collection before and after the expeditions.

Findings revealed that teachers and students in the experimental group had significant increases in content knowledge, and students' attitudes and perceptions towards science and scientists also improved. In addition, teachers learned strategies to transform their pedagogical practices, to motivate students to enhance their interest, and to actively engage them in learning science. The evidence suggests the effectiveness of having educators and students participate in collaborative opportunities with scientists and researchers within authentic contexts. An important additional benefit of collaboration among the teachers, scientists, and rangers resulted after teachers examined the research protocol prior to beginning the experiment. Using their knowledge of the students, they made recommendations that made it easier for the students to collect data. These modifications provided more accurate and reliable measures than would have been obtained if the original protocol were followed. Thus, not only do educators and their students benefit from these partnerships, but they also open the door to enhancing the quality of research for the scientists.

While the above studies do not specifically address self-efficacy and mastery learning, they show that by building students' sense of confidence at conducting scientific research in real-life contexts, students are gaining experience and exposure to the dynamic and fluid nature of conducting true scientific research. Social cognitive theory postulates that these positive mastery learning experiences promote self-efficacy for engaging in similar science tasks in the future. In addition, students' direct exposure to models such as scientists, researchers, and science professionals (i.e., rangers) provide students with insights into possible career options.

TAKING IT A STEP FURTHER: CITIZEN SCIENCE PROGRAMS

Citizen science has been defined as "public engagement in scientific research with collaboration with professional scientists" (Shah & Martinez, 2016, p. 17). It involves research that contributes to scientific knowledge and understanding, learning, and environmental stewardship (Kobori et al., 2015). In the United States, CSPs were historically done by amateur volunteers (naturalists) who would explore the wilderness for new species. CSPs have typically been associated with counting birds, amphibians, and butterflies in certain areas, or monitoring lilac and honeysuckle flowering. Findings from CSPs have provided data helping scientists understand important topics such as species diversity, climate change, pollution, invasive species, and land-use change.

Key Aspects of CSPs

While CSPs provide scientists with a volume and wide-breadth of data that a single or small group of scientists would not have been able to obtain, it is important for participants to follow a standard protocol to ensure the validity of the data collected (Tredick et al., 2017). Tredick et al. developed a rubric containing six criteria for evaluating CSPs, which can be taken into consideration when developing a CSP. The first criterion is to examine the stakeholders such as scientists, conservationists, and educators and the resources available to conduct the research. This involves an understanding among members participating in the CSP of the actual protocol or steps involved in collecting the research and the funds necessary for things such as training, materials, and technology. The second criterion is whether the CSP has defined goals and objectives. This includes the recruitment and training of volunteers by educating them on the importance of the research as well as the timeframe for data collection. Methods, which encompasses design and implementation of the data collection, is the third criterion for a sound CSP. The methods employed are particularly important because participants, such as teachers and students who do not have prior training on how to conduct authentic scientific research, need to be carefully trained on how to follow a series of steps to ensure the validity of the data collected. The protocol should include simple, easy-to-follow directions. The remaining three criteria are focused on data entry, storage, and analysis; reporting and disseminating the findings; and evaluating the outcomes of the research and its overall contribution to science. Teachers and their students may not be involved in the development of a CSP or in evaluating the outcomes of the research, but as participants of the CSP, they are often responsible for the data entry. Teachers often include extension activities, which involve having students analyze what they observed and collected, thus building and refining students' scientific inquiry skills, enhancing mastery experiences for conducting research, and increasing self-efficacy for making authentic scientific observations (Hiller & Kitsantas, 2014).

Critical to the protocol of a CSP program is training students on how to make systematic observations. Research shows that students can be trained to make scientific observations and that these differ from typical everyday observations (Eberbach & Crowley, 2009). Erberbach and Crowley (2009) differentiate between novice, intermediate, and expert observations. Novice observations might entail noticing that birds are different from other animals. Intermediate observations include noticing specific features such as wings that are used by birds to fly. Expert or scientific observations are more sophisticated in that they involve ignoring irrelevant features while stimulating related knowledge. For example, noting that sea birds (birds that spend a significant time at sea) differ from marine or shorebirds in how they use the ocean (birds that rely on the ocean for food but spend most of their time on land). These authors also discuss a framework in which students may be taught how to change the ways in which they make everyday (novice) observations become more powerful and productive (expert) observations. Through question asking, classroom discourse, and generating interest, students may experience a change or *transform* the ways in which they look at the world. The following section describes how CSPs can be used to create such a transformation in learners.

Recent research has focused on the need for *transformative* changes to take place in science learners (Wals & Dillon, 2013). A transformative change refers to a change in one's view of the world and how one interacts with the world (Bela et al., 2016). This is different from learning content for the sole purpose of understanding and performing well on assessments. An example of a transformative change is when students who witness plastic contents in the bodies of seagulls and turtles become emotionally moved such that they sign up for a 2-week beach litter removal project, in addition to reducing their own plastic purchases and waste. The transformation results in an action or behavioral change. One of the tenets of social cognitive theory is that learning is a process that occurs socially before becoming internalized (Schunk, 1999). Through CSPs, students interact with educators, scientists, and researchers as they learn how to engage in hands-on science activities. These social activities result in internalizing feelings of self-efficacy and competence. In their roles as "scientists," students begin to experience changes or transformations about the ways in which they think of themselves and their capability to do science (Wals & Dillon, 2013).

Research has supported the impact of CSPs on middle school and high school students' motivation and interest (Hiller & Kitsantas, 2014). In a study conducted by Hiller and Kitsantas (2014), 86 eighth-grade students participated in a horseshoe crab CSP to examine the effects on career goals and social cognitive theory constructs such as sources of self-efficacy, choice goals, outcome expectations, and interest. Students in both the treatment and comparison groups received instruction on how to conduct measurements and collect data on horseshoe crabs. The treatment group, however, also conducted fieldwork at the beach with naturalists and collected data for a professional biologist on the horseshoe crabs. Pre- and post-test measures revealed that there was a positive treatment effect on all social cognitive constructs for students who participated in the CSP. In particular, citizen science students outperformed the comparison group on academic achievement mastery experiences, and self-efficacy for conducting scientific observations. These results demonstrate the power of CSPs to transform students' competency beliefs, interest in learning and engaging in the scientific process which may ultimately influence subsequent choices in taking science courses or pursuing science careers.

Ant Picnic CSP

Although not widely known, there are several CSPs that are thriving around the world, educating teachers and students on how to participate as scientists and collect data needed and then used by scientists globally. Students Discover is one such program focused on linking middle school teachers and students to STEM learning outside of the classroom. The main goal of the NSF program is to develop a model for engaging students in STEM education that involves hands-on authentic learning experiences (Branon, Booth Freeman, Bryant, Townsend, & Fabre, 2018). The project has been on-going for three years with collaboration between teachers, district administrators, institutional partners, and scientists who together develop lesson plans to "participate in cutting-edge research through data collection, analysis, and digital communication" (Branon et al., 2018, para. 8).

Ant Picnic is a CSP in the Students Discover program that was developed out of research at the Rob Dun Lab at North Carolina State University (Students Discover: Ant Picnic, 2018). Scientists asked questions such as which ant species are found where, what food preferences ants have to understand more about ants, and their role in their existing ecosystems. Learning more about ants can provide researchers with information about diet preferences, environmental conditions, and the impact of climate change in different parts of the world. In addition, students learn about the basic food groups and how macromolecules (protein, carbohydrates, nucleic acids, and lipids) are essential for good nutrition.

The Ant Picnic lesson is clearly outlined on the Students Discover website. Although students of all ages can participate, the lessons work particularly well with younger learners who do not need to change classes because of the timeframe for data collection. Information about materials needed for the lesson plan is provided such as an instructional video, information on how teachers can prepare students in advance of data collection, and how students carry out the lesson. Additional information includes the link between this CSP and the North Carolina State Standards; worksheets on learning and understanding food; and how to collect, record, and submit the data online.

Figure 1. Image of card used to help initiate interest and motivation. Students Discover (2018). Ant picnic. Retrieved from http://studentsdiscover.org/lesson/ant-picnic/

Researchers from North Carolina State University work closely with teachers and students by educating and training them to conduct scientific research. They, along with artists, developed a number of "fun" and learner-friendly handouts that are distributed to teachers and students to help motivate them. Figure 1 provides an example of educational material which is humorous and informative. By using puns and misspellings, cute illustrations, and the formal names of the ants, students' interests are likely to be sparked, along with their motivation to learn more about the project and ants.

Figures 1 and 2 describe steps involved in carrying out the science lab. The Ant Picnic lesson begins with a warm-up activity about ants and how annoying they can be when on a picnic as they are often trying to eat picnickers' food and snacks. The teacher then introduces students to diet and nutrition by having them make lists of the foods they have just eaten and to place these items into food categories.

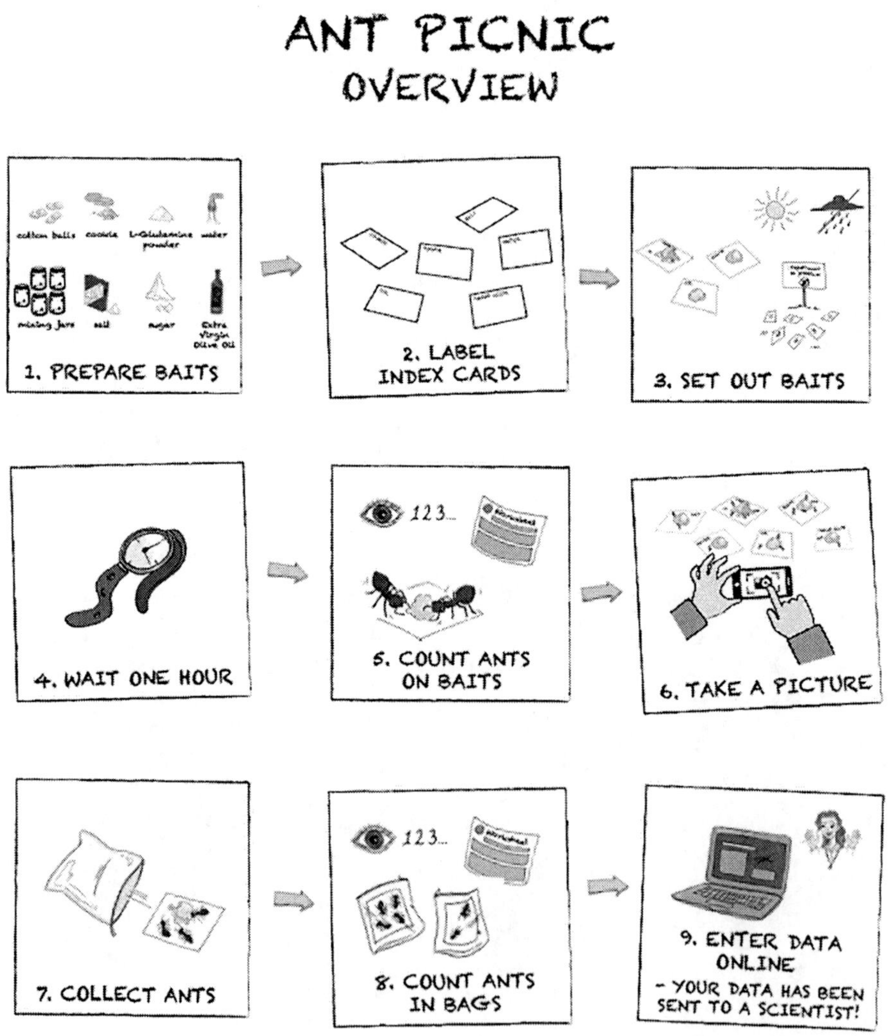

Figure 2. Overview of the steps involved in conducting the data collection. Sorger, P. M., & Derouin, P. (n.d.) Ant picnic activity. Retrieved from https://drive.google.com/file/d/1S39aXSexmIiau WE2rEA2eay5IvkrJLv4/view

This is followed by explaining to students that the purpose of the lab is to discover information about the types of ants in the school's ecosystem and the types of foods they eat. In the second step, students are placed into groups of 3-6 students who, with the teacher, prepare six "baits" (amino acids, cookies, sugar, olive oil, salt, and water). Five baits (excluding the cookies) are mixed with water in a jar and one cotton ball is placed in each jar to soak up the solution. Five index cards are prepared with the name of each bait and then teachers and students take a "quiz." The quiz, Are you ready for Ant Picnic? provides a review of whether the materials needed for the activity are ready (See Figures 3 and 4).

Teachers: Are you ready for Ant Picnic?

Take this quiz to self-check if you are ready to do the Ant Picnic experiment with your students.

IF YOU RECEIVE 16 POINTS THEN YOU ARE READY TO GO.

* Required

How many experiments do I plan to set up with my students? i.e. How many groups will I put students in? * 1 point

- ○ Just one - I'll set the experiment up as a demo
- ○ I'll put them in groups of 2 - that means more experiments but my students can handle it
- ○ I'll put them in groups of 4 - each group can set up one experiment
- ○ I'll put them in groups of 6 - so each group member is responsible for one food type
- ○ I'll put students in larger groups (6+)

Do I have all ingredients ready to make an Ant Picnic? Select all that apply. * 1 point

- ☐ Extra Virgin Olive Oil - about 1 cup
- ☐ Amino Acids: L-Glutamine powder - 3 1/2 tablespoons (50 g)
- ☐ Sugar - 3 1/2 tablespoons (50 g)
- ☐ Salt - 1/2 teaspoon (2.5 g)
- ☐ Pecan Sandies cookies (or a different shortbread cookie if you are at a nut-free school)
- ☐ Access to water (tap water is ok)

Figure 3. Students Discover (n.d.) Excerpt of the teacher's quiz. Ant picnic. Retrieved from https://docs.google.com/forms/d/e/1FAIpQLSfSmb5RP6DMvpDm2eW37QV6HEKVf92AAWxMXGs0uCgY42cv8w/viewform

Students: Are you ready for Ant Picnic?

Take this quiz to find out if you are ready to participate in a real science experiment.

You need 13 points to pass this quiz.
When you are done, take a screenshot of the score page and submit it to your teacher.

* Required

How many food types are we testing in the Ant Picnic experiment? * 1 point

- ○ 1
- ○ 3
- ○ 5
- ○ 6

Which of the following is NOT a food type we are testing as part of the Ant Picnic experiment? * 1 point

○ Extra Virgin Olive Oil

○ Salt

Figure 4. Students Discover (n.d.). Excerpt of the student's quiz. Ant picnic. Retrieved from https://docs.google.com/forms/d/e/1FAIpQLSdZkRoTmdRWFSQb32vw3JwUVqOEmZRbCtGnF5Yed75Fx0fzyQ/viewform

At the experimental site, students place a cotton ball on the corresponding index card, crush the cookie and place it on the sixth corresponding card, then record the start time, location, and weather conditions of the experiment. They return to their classrooms for one hour before going back to the experimental site. When they are back at the site, they carefully count the number of ants on each card, turn the index cards over to count any ants on the underside of the card, and record this information on their worksheets. Students are instructed then to take a picture of each bait card, even if there are no ants on the card. The cards with the ants and their cotton balls are placed in individual sealed bags and students then return to their classrooms. In their classrooms they count the number of ants three times without opening the bags and record the average number of ants per bait on their worksheets. The final step occurs when the teacher enters students' data and pictures onto a website where the information is sent to scientists and incorporated with data collected from others around the world.

Extension activities are also available, such as the *Ant Data Analysis,* where students can use the data collected from the Ant Picnic on the Common Online Data Analysis Platform (CODAP). Using CODAP, students enter their findings, analyze tables and graphs, and compare their data to existing global data sets. A second extension activity includes the *Ant ID Activity* where students collect odorous house ants using sugar and send them to scientists who examine their DNA. In this activity students use microscopes to learn how to identify the parts of ants and how to euthanize, mount, and create their own ant collection. The *Spirit Ant Activity* is an engaging lesson where students learn that ants have "personality characteristics" (for example, see Figure 5). Upon learning about the different types of ants and their characteristics, students make a poster comparing their own personalities to the ants they identify with. The *Tree of Life Activity* helps students learn about different species and how they are connected to one another. Students learn about classification systems, create posters and explore a phylogenic tree of life. The *Pipe Cleaner Ant Activity* is another hands-on lesson where students learn the names and body parts of ants by creating models of ants using pipe cleaners. Students may create their models based on a specific species of ants such as those learned about in the *Spirit Ant Activity* by adjusting the size, color, and shape of the model to match their ant species of choice.

Each of these activities provides students with hands-on learning opportunities in a fun and engaging way that fosters scientific knowledge, not just about content, but also on the *process* of science. They transform students into scientists who collect data that become part of a global pool and identify, organize, and relate it to themselves. These activities motivate students by building their sense of mastery and efficacy as they realize they are capable of participating in true science – data that they collect are part of a global data collection pool that is being used to generate our understanding of the world. Students can learn about the overall findings of their projects, and they may generate their own research questions for future studies. Research supports this impact on motivation and academic achievement on middle school students who engage in CPSs (Hiller & Kitsantas, 2014).

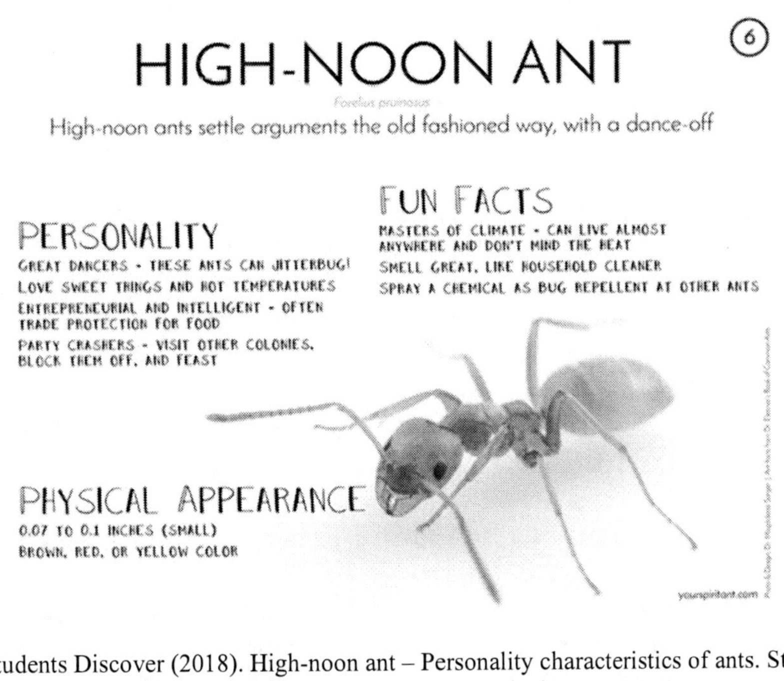

Figure 5. Students Discover (2018). High-noon ant – Personality characteristics of ants. Students' ant picnic. Retrieved from http://studentsdiscover.org/lesson/ant-picnic

Global Learning and Observations to Benefit the Environment CSP

Another CSP is Global Learning and Observations to Benefit the Environment (GLOBE), which consists of several international CSPs (The GLOBE Program, 2018a). Its mission is to get the public, teachers, and students from around the world involved in doing real science – science that is focused on understanding climate change, the environment, and the Earth. GLOBE views students as true scientists because they are generating hypotheses, making observations, and collecting data. Teachers, students, and citizens can become involved in a variety of projects such as the GLOBE Mosquito Project targeted at understanding the mosquitos that carry the Zika virus (The GLOBE Program, 2018b).

In the GLOBE Mosquito CSP, students learn about the life cycle of mosquitos and how they are vectors, or insects that carry diseases. Due to warming climates and changes in land use, disease carrying mosquitos have been found in regions typically not associated with having them. The GLOBE Mosquito Project is focused on identifying where mosquitos are breeding and then eliminating the breeding sites to prevent future outbreaks. Millions of people die each year from mosquito-borne diseases; CSPs help scientists identify where the mosquitos are found and participate in eliminating the breeding sites (Kohl & Clark, 2017).

On the GLOBE website, there is a video showing a Brooklyn College professor talking to students about participating in the Mosquito Project. She describes it as a way to collect data and to have fun while doing it. Teachers and students download the Global Observer App which contains two protocols: GLOBE clouds (used to record data on the sky and clouds) and the GLOBE Mosquito Habitat Mapper. These protocols allow citizen scientists to collect information on hundreds of thousands of mosquitos and to provide these data to scientists in NASA.

Middle, secondary, and higher education teachers interested in participating in this CSP are trained by GLOBE on disease spreading mosquitos and on how to use the app, as well as how to safely and correctly collect the data. Students collect water samples from places where water collects such as ponds, canals, and standing objects (i.e. buckets, bird feeders). To collect larvae, students use a kit that includes items such as a turkey baster to collect the water, an inexpensive microscope to examine the presence of larvae in the water sample, latex gloves, and plastic bags to transport the larvae samples. Teachers are trained on a mosquito larvae protocol, which includes background information to be provided to students on local mosquitoes and on how to identify each mosquito's genus and species; on ways to safely collect the larvae; on deciding whether to collect samples from containers (i.e. plant pots, vases, discarded tires) or non-containers; natural habitats, such as marshes, ponds, and puddles; and on ways to accurately obtain the measurements using the microscopes (for example, not counting tadpoles or small fish). In addition to providing students with protocol worksheets for recording their findings and describing the water conditions, students are encouraged to keep field notes as scientists do and to take pictures of the experimental site. Upon completion of the data collection, students make scaled field maps of the sampling sites using a handout available on the website.

If students have downloaded the Mosquito Habitat Mapper, they attach a small macro lens to their iPhones, Androids, tablets, or laptops, which allows them to record their data and location, to take pictures of the larvae, and to upload them to the GLOBE database (Kohl & Clark, 2017). The website provides detailed information on the steps as described below in addition to educational material about mosquitoes and larvae. (Figure 6 shows the steps outlined in the Mosquito Habitat Mapper tutorial found online and created by staff at NASA.)

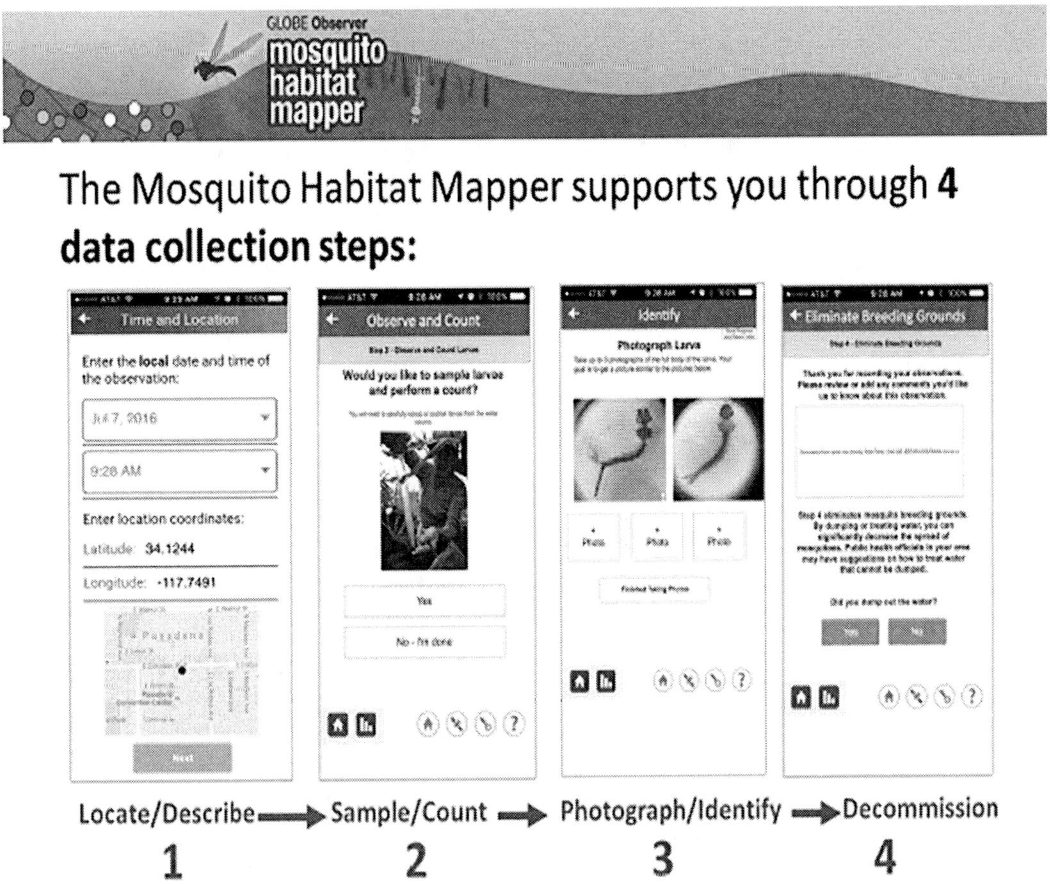

Figure 6. Slide from the mosquito habitat mapper tutorial. GLOBE flyer on the Air Quality Program. The GLOBE Program (n.d.). Mosquitoes training. Retrieved from https://observer.globe.gov/training/mosquitoes

In the first step, students identify breeding sites, which are typically found near sites of standing water. They then open the app which will automatically download the latitude, longitude of where the students are located in addition to the date and time. Students are also provided with a map to verify their location. A convenient feature of the Mapper is that the app may be used regardless of whether Wi-Fi is available. Once Wi-Fi becomes available, the information will automatically be sent to a database, which collects information from around the world. Students are now ready to begin collecting samples of water that contain the larvae. The GLOBE tutorial provides specific instructions on the sampling method depending on the sampling tools the students are using. Students are recommended to collect at least five samples of water to ensure that larvae are present in at least one of the samples. The protocol instructs students to wait before each sample collection to ensure the larvae have risen to the surface of the water. This is because each time the water is disturbed, the larvae swim beneath the surface to hide; they rise again to the top to get oxygen to breathe. Students place samples into plastic bags leaving air for them to breathe.

In the second step, students are observing the larvae under a microscope or using the macro lens which can be attached to their mobile devices and counting the number of larvae found. Students may use a precise number count or estimate. The GLOBE tutorial offers additional information on how the larvae may look different, which is due to the developmental stage of the larvae in the sample. Step 3 involves photographing and identifying the larvae. Students pour a part of the water sample onto a white tray or plate and, using a dropper or spoon, obtain one larva to place on the plate. The instructions warn students not to obtain too much water or the larva will swim, making it difficult to photograph and examine. Students position the larva using a problem or toothpick to examine it using the microscope or macro lens. They identify the type of larva using a local mosquito larva key to the Mosquito Habitat Mapper app, take a picture and submit the photograph along with the identification by following the instructions on the app. The final step involves eliminating the breeding site. This can be something as simple as dumping a container of water onto the ground or tipping over a tire that has become filled with water.

The relatively new Mosquito Habitat Mapper app provides scientists at NASA, teachers and their students with immediate information on mosquitos and their potential breeding sites from around the world. In addition to the tutorials found on the GLOBE website, there are games, such as *Zika Zap Bingo*, that students can play to learn more about the life cycle of mosquitos and their breeding habitats.

Along with the Mosquito Project, the GLOBE website contains information about many opportunities for schools around the world to become involved in CSPs. In their *Stars and STEM Stories: News and Events* site, GLOBE highlights stories written about several of the projects dating back to 1997. One recent, exciting news article focused on a high school in Queens, New York for students who are deaf (Giarratano, 2018). Students and teachers were asked by a scientist from the NASA Langley Research Center if they would be interested in participating in a CSP called GLOBE's Air Quality Program. The program involves making observations and collecting data on the sky in order to help understand and track air pollution - tiny particles called aerosols - used in middle and secondary schools. Figure 7 shows a flyer used by GLOBE to recruit schools globally to get involved in the project. GLOBE also recommends that students use the flyer to connect with scientists who are studying aerosols and their effects on climate.

One teacher at the Lexington School for the Deaf, for example, embraced this as an opportunity to help her students who struggle with English as a second language and who rely on visuals to learn and communicate. The teacher, her assistant, and students collaborated with a research scientist from NASA Langley on how to use GLOBE's atmosphere protocols and Calitoo sun photometers to obtain measurements on the sun's light intensity that may be blocked by a number of causes such as aerosols, volcanic ash, or clouds. Students also recorded observations on the characteristics of the sky such as color, clouds, and visibility to determine air quality. A datasheet is also available on the GLOBE website to record information about cloud identification.

Students worked together in small groups or teams and each student had a different role on each team doing science work such as converting the local time to the universal time zone [Converted Universal Time (UTC)] and taking barometric pressure readings. Each student took notes on their observations of the sky and compared notes with one another and either entered their measurements on an online data form or a mobile app described further below.

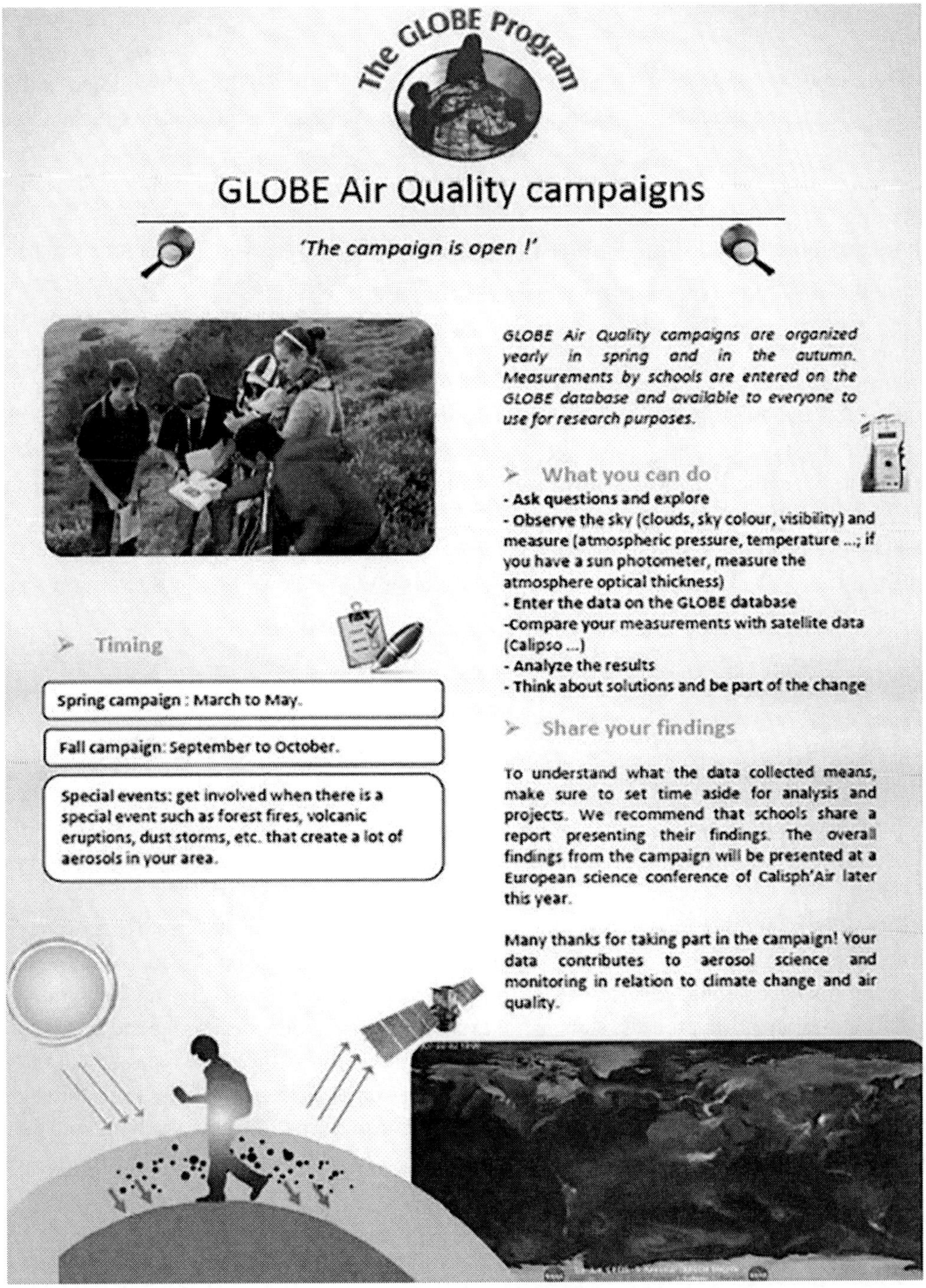

Figure 7. GLOBE flyer on the air quality program. The GLOBE Program (n.d.). Instruments using for air quality campaign. Retrieved from https://www.globe.gov/web/air-quality-campaign/overview/instruments

The teacher described the students' enthusiasm, interest, and motivation to learn more about the Earth and the impacts humans have on Earth (Giarratano, 2018). A particularly meaningful event occurred when students participated in a video conference with scientists and engineers from NASA Langley, one of whom was deaf. This scientist spoke with the students about being an engineer for NASA Langley for over 25 years and about hard work, persistence, and being proud of being deaf. One can infer that these students were transformed, motivated, excited to learn and contribute to science. They not only learned the content but realized the impact humans have on the environment when they pollute. In addition, they observed successful and enthusiastic role models (a form of vicarious learning), one of whom was similar to themselves. This engineer encouraged students to feel good about themselves and told them they are capable of becoming scientists or engineers in the future, a form of social persuasion (Bandura, 1997).

In addition to the mastery experiences of collecting measurements and using the data to understand air quality, students were also encouraged by a group of scientists, one who shared the same disability. As indicated earlier in this chapter, mastery and vicarious experiences and social persuasion are effective sources of self-efficacy (Schunk & DiBenedetto, 2016). CSPs offer students and teachers opportunities to become more confident through persuasion by teachers and scientists, to learn vicariously from models, to experience physiological and emotional sensations of success and competency, and to engage in mastery events, which taken together, transform students by building self-efficacy for engaging in scientific research. As seen from the preceding example, students with disabilities - as well as those without - can benefit from educational opportunities such as those in CSPs.

Educational Implications

The material summarized in this chapter makes it clear that participation in CSPs can have beneficial effects on motivation and learning of students and educators. A clear implication is that educators should seek out more opportunities to involve their classes and schools in these programs.

But aside from this implication, CSPs have a number of features that can be incorporated into the regular instructional program. CSPs are designed to help people do science rather than just learn science. Doing science means being behaviorally, cognitively, and emotionally engaged in the scientific process, such as by exploring phenomena, designing and conducting experiments, and compiling data. This suggests a move away from the static types of laboratory experiments typically found in science classes to addressing more real-world issues. Motivation is apt to be enhanced when learners and teachers believe that what they are studying may have an impact on their environments. This perceived sense of value or importance is critical for motivation.

Doing science taps directly into the idea espoused in this chapter that interest and motivation benefit when learners experience a sense of self-efficacy for being able to perform well. Self-efficacy has been shown to relate positively to students' choice of activities, effort, and persistence, all of which help to sustain motivation over long periods. Sustaining motivation is critical because significant scientific problems are unlikely to be resolved in a 50-minute class period.

Advances in technology mean that learners today can be digitally connected with others in multiple sites. Such data sharing also helps to promote interest and motivation because learners receive vicarious self-efficacy information that helps to substantiate the importance of what they are engaged in. In short, there are many opportunities for learners to become engaged in the process of science, which are apt to have benefits on their interest, motivation, and learning.

Recommendations for Future Research

From a social cognitive perspective, CSPs have enormous potential for increasing student motivation for learning science. Students' participation in these programs builds self-efficacy for learning science, has the capability of making the learning enjoyable for both students and teachers, teaches valuable and authentic real-world lessons, and empowers students to contribute to the making of scientific knowledge. While CSPs have been around for decades, traditionally they have not been used in educational settings and with younger students. More research is needed on students in different grade levels to examine the relationship between social cognitive theory and learning in CSPs.

Additional research is needed to understand the mechanisms in which science self-efficacy increases through CSPs, not only for students, but also for educators. Teachers often struggle with how to make science learning meaningful and engaging while using the Common Core State Standards and the Next Generation Science Standards to guide instruction (Greene, Anderson, O'Malley, & Lobczowski, 2018; White & DiBenedetto, 2018). Teachers may feel limited in their time and ability to work with scientists, naturalists, or national park rangers to participate in a CSP. Research is needed to demonstrate the effectiveness and potential impact of these programs on achievement, student interest, and how they can map against state mandated standards along with providing teachers with the tools needed to generate CSP lesson plans.

Artificial intelligence is increasingly becoming an area of interest in education (Popenici & Kerr, 2017). As students at younger ages are exposed to immediate access to the Internet, phones that provide answers to questions, computer games where players can compete against other players in other parts of the world, and cars that drive themselves, there will be more demand on learners' agility to adapt to technology advances that extend beyond the classroom. CSPs offer students the opportunity to learn the scientific process and the ways that the measurements they obtain become part of a universal collection of data. These data are used to understand a breadth of issues related to our Earth such as invasive species, infectious diseases, ailing ecosystems, biodiversity, pollution, climate change, and how robots can be used to diagnose and remove cancer cells. Research is needed on how CSPs can be used to teach topics in the physical sciences, such as engineering, physics, and chemistry. These are fields in which U. S. students lag behind other countries. Providing students with the tools to participate in scientific inquiry in an authentic, fun, and engaging way may foster not only a love for life-long learning but also an avid interest in science that could last their lifetimes.

References

Bandura, A. (1986). *Social foundations of thought & action: A social cognitive theory.* Upper Saddle River, NJ: Prentice-Hall, Inc.

Bandura, A. (1997). *Self-efficacy: The exercise of control.* New York, NY: Freeman.

Bela, G, Peltola, T, Young, C. J., Balazs, B, Arpin, I., Pataki, G., Hauck, J., . . . Bonn, A. (2016). Learning and the transformative potential of citizen science. *Conservation Biology,* 30(5), 990-999. doi:10.1111/cobi.12762

Branon, S., Booth Freeman, S., Bryant, L., Townsend, L., & Fabre, M. (2018). *The Students Discover Project partnerships and citizen science in North Carolina classrooms.* Connected Science Learning, 4. Retrieved from http://csl.nsta.org/2018/04/the-students-discover-project/

Britner, S. L., & Pajares, P. (2006). Sources of science self-efficacy beliefs of middle school students. *Journal of Research in Science Teaching,* 43(5), 485-499. doi:10.1002/tea.20131

DiBenedetto, M. K. (Ed.). (2018). *Connecting self-regulated learning and performance with instruction across high school content areas.* Dordrecht, The Netherlands: Springer.

DiBenedetto, M. K., & Bembenutty, H. (2013). Within the science pipeline: Self-regulated learning and academic achievement among college students in science classes. *Learning and Individual Differences,* 2, 218-224. doi:10.1016/j.lindif.2012.09.015

Eberbach, C., & Crowley, K. (2009). From everyday to scientific observation: How children learn to observe the biologist's world. *Review of Educational Research,* 79(1), 39-68. doi:10.3102/0034654308325899

Faria, C., Guilherme, E., Gaspar, R., & Boaventura, D. (2015). History of science and science museums: An enriching partnership for elementary school science. Science & Education: *Contributions from History, Philosophy and Sociology of Science and Mathematics,* 24(7-8), 983-1000. doi:10.1007/s11191-015-9773-7

Giarratano, R. (2018). Students at Lexington School for the Deaf in New York (USA) participate in GLOBE Air Quality Campaign – Discover their power to contribute to science in a significant way. *Stars and STEM Stories.* Retrieved from: https://www.globe.gov/news-events/globe-stars

Greene, J. A., Anderson, J. L., O'Malley, C. E., & Lobczowski, N. G. (2018). Fostering self-regulated science inquiry in physical sciences. In M. K. DiBenedetto (Ed.), *Connecting self-regulated learning and performance with instruction across high school content areas* (pp. 163-184). Dordrecht, The Netherlands: Springer.

Hiller, S. E., & Kitsantas, A. (2014). The effect of a horseshoe crab citizen science program on middle school student science performance and STEM career motivation. *School Science and Mathematics,* 114(6), 302-311. doi:10.1111/ssm.12081

Houseal, A., Abd-El-Khalick, F., & Destefano, L. (2014). Impact of a student-teacher-scientist partnership on students' and teachers' content knowledge, attitudes toward science, and pedagogical practices. *Journal of Research in Science Teaching,* 51(1), 84-115. doi:10.1002/tea.21126

Kobori, H., Dickinson, J. L., Washitani, I., Sakurai, R., Amano, T., Komatsu, N., …Miller-Rushing, A. J. (2016). Citizen science: A new approach to advance ecology, education, and conservation. *Ecological Research,* 31, 1-19. doi:10.1007/s11284-015-1314-y

Kohl, H. R., & Clark, A. (2017). *National Science app tackles mosquito-borne disease.* GLOBE Implementation Office. Retrieved from https://www.globe.gov/news-events/globe-news/newsdetail/globe/nasa-citizen-science-app-tackles-mosquito-borne-disease

Lofgran, B. B., Smith, L. K., & Whiting, E. F. (2015), *Science self-efficacy and school transitions. School Science & Mathematics,* 115(7), 366-376. doi:10.1111/ssm.12139

National Science Foundation, (2018). *Building the future: Investing in discovery and innovation - NSF strategic plan for fiscal years* (FY) 2018 – 2022. Retrieved from https://www.nsf.gov/pubs/2018/nsf18045/nsf18045.pdf

National Science Foundation, (2018). *Research on learning in formal and informal settings* (DRL). Retrieved from https://www.nsf.gov/ehr/drl/about.jsp

OECD (2018). PISA 2015 Results in focus. Paris: OECD. Retrieved from https://www.oecd.org/pisa/pisa-2015-results-in-focus.pdf

Popenici, S. A. D., & Kerr, S. (2017). Exploring the impact of artificial intelligence on teaching and learning in higher education. *Research and Practice on Technology Enhanced Learning*, 12(1), 2-13. doi:10.1186/s41039-017-0062-8

Schunk, D. H. (1999). Social-self interaction and achievement behavior. Educational Psychologist, 34(4), 219-227. doi:10.1207/s15326985ep3404_3

Schunk, D. H. (2012). Social cognitive theory. In K. R. Harris, S. Graham, & T. Urdan (Eds.), APA educational psychology handbook: *Theories, constructs, and critical issues* (Vol. 1, pp. 101-123). Washington DC: American Psychological Association.

Schunk, D. H., & DiBenedetto, M. K. (2016). Self-efficacy theory in education. In K. R. Wentzel & D. B. Miele (Eds.), *Handbook of motivation at school* (2nd ed., pp. 34-54). New York, NY: Routledge.

Shah, H. R., & Martinez, L. R. (2016). Current approaches in implementing citizen science in the classroom. *Journal of Microbiology and Biology Education,* 17(1), 17-22. doi:10.1128/jmbe.v17il.1032

Sorger, P. M., & Derouin, P. (n.d.) *Ant picnic activity.* Retrieved from https://drive.google.com/file/d/1S39aXSexmIiauWE2rEA2eay5IvkrJLv4/view

Students Discover (2018). *Ant Picnic.* Retrieved from http://studentsdiscover.org/

Summers, S. (2004). Museums as resources for science teachers. *Science Scope,* 27(9), 27-28.

The Nobel Prize, (n.d.). Nobel prizes and laureates. Retrieved from https://www.nobelprize.org/prizes/

The GLOBE Program: A Worldwide Science and Education Program (2018a). Retrieved from https://www.globe.gov/

The GLOBE Program: A Worldwide Science and Education Program (2018b). About mosquitoes that transmit the Zika virus. Retrieved from https://www.globe.gov/web/globe-mosquito-project/overview/about-the-project

The GLOBE Program (n.d.). *Instruments using for air quality campaign.* Retrieved from https://www.globe.gov/web/air-quality-campaign/overview/instruments

Tredick, C., Lewison, R., Deutschman, D., Hunt, T., Gordon, K., & Von Hendy, P. (2017). *A rubric to evaluate citizen-science programs for long-term ecological monitoring. Bioscience*, 67(9), 834-844. doi:10.1093/biosci/bix090

Wals, A. E. J., & Dillon, J. (2013). Conventional and emerging learning theories: Implications and choices for educational researchers with a planetary conscious. In R. B. Stevenson, M. Body, J. Dillon, & A. E. J. Wals (Eds.), *International handbook of research on environmental education (*pp. 253-261). New York, NY: Routledge.

White, M. C., & DiBenedetto, M. K. (2018). Self-regulation: An integral part of standards based education. In D. H. Schunk & J. A. Greene (Eds.), *Handbook of self-regulation of learning and performance* (2nd ed., pp. 208-222). New York, NY: Routledge.

Zimmerman, B. J., & DiBenedetto, M. K. (2008). Mastery learning and assessment: Students and teachers in an era of high stakes testing. *Psychology in the Schools,* 45(3), 206-216. doi: 10.1002/pits.20291

In: Enhancing STEM Motivation through Citizen Science Programs ISBN: 978-1-53616-038-3
Editors: Suzanne E. Hiller and Anastasia Kitsantas

Chapter 8

Science Educators' Perspectives on Learning to Utilize Citizen Science in our Teaching

Lacey Huffling[1,*], Heather Scott[1], Regina Collins[2], Britt Gantt[3], Hayward Johnson[4] and Melissa Weeks[5]

[1]Georgia Southern University, GA, US
[2]Trinity Christian School, GA, US
[3]Jenkins County Middle School, GA, US
[4]Swainsboro High School, GA, US
[5]Pickens County High School, GA, US

Abstract

This chapter explores how six science educators (two middle school science teachers, two high school science teachers, and two postsecondary science teacher educators) engaged in self-study of their own practice in regard to incorporating a citizen science learning experience into their classrooms. Using a collaborative self-study, qualitative data were collected through self-reflections. We analyzed for themes in regard to our perceptions of student learning and teacher identity development to better understand what affordances and constraints we had in common to inform our practice. Four themes emerged for perceptions of student learning: student interest, collaborative learning, authentic practice, and scientific observation skills, while three themes emerged for teacher identity development: working through inexperience, observing student perceptions, and examination. Overall, the citizen science learning experience increased engagement for all three levels of students (middle, high, and postsecondary) from the educators' perspectives, provided access to authentic scientific practices, helped educators engage in new teaching practices, and developed our identity formation as science educators.

Keywords: authentic practice, citizen science, collaborative learning, collaborative self-study, scientific observation skills, student interest, and teacher identity development

[*] Corresponding Author's E-mail: lhuffling@georgiasouthern.edu

As science educators, we are committed to providing our students with opportunities to experience authentic science practices whenever possible, particularly through collaborative group work as a means of increasing our students' science interest and literacy. In this regard, we came together and designed a learning experience to introduce our students to ecological concepts covered in our state standards, through scientific inquiry, scientific practices, and citizen science research defined as "partnerships between those involved with science and the public in which authentic data are collected, shared, and analyzed" (Jordan, Ballard, & Phillips, 2012, p. 307). As we worked together to implement this project, we viewed "scientific literacy as a property of collective activity rather than individual minds" (Roth & Lee, 2002, p. 33). Our goal was to provide students opportunities to understand why scientists were asking the general public to help with their research, to contribute to citizen science research, to analyze data collected, and to communicate their results to their peers. To this end, we engaged in a self-study to examine this learning experience as our students became citizen scientists.

Literature Review

Citizen science research has expanded in the past couple of decades (Bonney, Phillips, Ballard, & Enck, 2016) due to technological advancements that have enabled global participation in scientific data collection. Thus, scientists are leveraging the general public's interest in science to collect and share data creating a global network to examine population trends, invasive species, phenology shifts, species range changes, and species richness and density (Cooper, Dickinson, Phillips, & Bonney, 2007; Dickinson, Zuckerberg, & Bonter, 2010). Dickinson and colleagues also highlighted how citizen science has the potential to aid in understanding human motivation to engage in environmental monitoring as data can be gathered on the participants themselves. Given how citizen science engages the general public in authentic scientific practices, such as observation and data collection, citizen science within formal education settings can be used to introduce students in the scientific practices and cross-cutting concepts embedded in the Next Generation Science Standards (NGSS Lead States, 2013; Koomen, Rodriguez, Hoffman, Petersen, & Oberhauser, 2018).

When coupled with formal education, citizen science has the potential to be an effective science learning experience for students if educators give time for students to grapple with the science behind the study, and then to use their newfound understandings to develop their research questions (Hiller & Kitsantas, 2015). In fact, in her dissertation exploring how three middle school teachers used the popular *eBird* citizen science project within their classrooms, Tomasek (2006) discovered that students' participation in the citizen project engaged them in question generation above and beyond what the teachers normally experienced with their students. Tomasek (2006) also discovered student involvement with citizen science affords students the opportunities to grapple with the parts of science that are usually provided through guided inquiry. Having to analyze which data they need to collect or how to refine their question does not often happen in the science classroom. In their study of over 1,500 college students, Mitchell, Triska, Liberatore, Ashcroft, Weatherill, and Longnecker (2017) found students' environmental engagement increased significantly after participating in citizen science projects. In addition, "students subsequently reported being more careful to avoid errors in their own data collection and making greater efforts to contribute records that were useful for future scientific research" (para. 1).

In their survey of 1000 schools with teachers who received GLOBE training and 500 schools with teachers who regularly submitted data to the project, Penuel and Means (2004) found student contributions were higher and data collected was more reliable when teachers were supported in project implementation and data collection through collaboration with other educators. Citizen science participation has also been discussed as a means to increase youths' science learning, environmental stewardship, connection to place, and civic engagement (Bonney et al., 2016; Calabrese Barton, 2012; Krasny, Russ, Tidball, & Elmqvist, 2014).

Given these benefits to students, the potential synergy of citizen science research and formal education is vast, yet there is a need to better understand how teachers themselves view citizen science and how they are leveraging it in their classrooms (Bracey, 2018). Recently, Koomen and colleagues (2018) examined citizen science data collection and science fair project development for four middle school students in one teachers' classroom. They discovered that citizen science engaged students in authentic science practices while providing opportunities for the teacher to serve dual roles of mentor and master. These dual roles could provide teachers with opportunities to further develop their science teacher identities as other studies have found teachers' pedagogical content knowledge (PCK) can expand with engagement in citizen science (Mueller & Tippens, 2012; Scheuch et al., 2018). Thus, our study aims to examine our science teacher identity development in implementing a citizen science learning experience through a collaborative self-study.

THEORETICAL PERSPECTIVE

Our study is informed by constructivist and socio-cultural perspectives which illuminate our meanings and perceptions through personal inquiry and socially-constructed knowledge of our practice (Samaras & Freese, 2009). Thus, the purpose of our self-study is to better understand our meanings and perceptions of our teaching and our students, past experiences and experiences teaching using citizen science projects. In particular, we chose to focus on our teacher identity development as science educators. For the purpose of our inquiry, we define science teacher identity development as a process by which one is recognized by self and others (Gee, 2005; Luehmann, 2007) as working toward teaching scientific observation skills through collaborative and authentic practice. By analyzing our individual reflections, we can begin to understand how citizen science projects can help sustain, encourage, or constrain our science teacher identity development.

Positioning Our Study

We engaged in a collaborative self-study, a collaboration of educators to examine their own teaching practices (Gallagher, Griffin, Ciuffetelli Parker, Kitchen, & Figg, 2011) and of our individual attempts to implement citizen science learning experiences in our classrooms. We elected to make this self-study collaborative in nature given the research on communities of practices and how engaging in such endeavors can provide opportunities for professional collaboration, community development, and professional growth (Gallagher et al., 2011; Hannigan, Raphael, White, Bragg, & Clark, 2016). Our research questions were:

1. What are the affordances of citizen science projects for our teacher identity development?
2. How does teaching with citizen science affect our perceptions of student learning?

METHOD

Participants and Setting

We were six science educators: two middle grades teachers, two high school teachers, and two postsecondary science educators.

Participant one. I have taught for 15 years teaching students from Grades 5-12. I particularly enjoy teaching middle school because it affords me the opportunity to make a difference in the lives of individuals during a time where their identity is forming. Some of my goals for my students include problem solving with a variety of approaches, understanding and explaining why they choose a method, and engaging them in metacog-nitive strategies.

Participant two. My school is 100% free and reduced lunch (Georgia Department of Education, 2018) . . . I conducted this learning experience during my first-year teaching. When we started discussing citizen science projects in my graduate science methods class, my first thought was that it was a way to bring science alive to my school. Using projects that get the students thinking about their community and their ecosystems compared to other ecosystems around the world really makes Life Science real for these students. It also gave a way for some students to virtually visit some of these places that they have heard of in their history and science classrooms.

Participant three. My school is a Title I school. The school is 87.6% free and reduced lunch (Georgia Department of Education, 2018). I teach Biology and Environmental Science, and I was in my second year of teaching when I taught this learning experience. My goals for my students are to open up their minds, connect with them, and leave a lasting impression. I am often surprised at how little my students are exposed to the world outside of their own community. I strive to not only expose them to the real world but to also help them understand the processes that control how things work. My favorite part about teaching is getting to know my students. The connections I make help me be a better teacher as well as allow me to have a positive impact on my students' lives.

Participant four. My school serves four local communities. The school is free and reduced lunch for 41% of the students (Georgia Department of Education, 2018). I have been a science educator for 15 years. I implemented our citizen science learning experience with two Biology classes. One of my goals as a science teacher is to help my students see the relevance of science in their everyday lives and the importance it holds in their futures.

Participant five. I have been a science educator for 13 years working in both formal and informal settings. One of my goals when working with preservice and in-service teachers is to introduce them to ways to engage their students in science learning that will not only expose their students to scientific practices and reasoning but also equip them to explore and observe their local communities.

I implemented this learning experience in a graduate science methods course for initial licensure in middle grades or secondary education and in a professional development (PD) workshop for science and math high school teachers. There were three Caucasian students (two males and one female) in the methods course and 62 teachers in the PD. Of the 62 teachers, all taught in urban Title I schools.

Participant six. I have been a science educator for 16 years. My goal in working with preservice teachers in elementary and middle grades education is to foster a passion for teaching science that engages a natural curiosity among children. I want them to use multiple approaches to teach science that encourage students to explore the natural world around them.

I used this citizen science learning experience in two Integrated Life/Earth Science courses for elementary and middle grades education preservice teachers as part of an undergraduate education degree. This is a required course for all of the students enrolled, and most take the course while they are sophomores or juniors (generally 19-21 years of age, though occasionally nontraditional students, too). The overall goal of this course is to increase science content knowledge while increasing pedagogy. The courses meet twice a week for 16 weeks of combined life and earth science content, which follows a lab/lecture approach. Almost every day of class has some hands-on or inquiry-based investigation for students to apply new content knowledge.

Study Design

Self-study is a relatively new field of research (Zeichner, 2007) and has the goal of assessing one's own practice and reframing it (Schuck & Russell, 2005). Dinkelman (1999) also argued that self-study promotes reflective teaching practice, and Russell and Korthagen (1995) ascertained that self-study can increase the professionalization of teachers. Self-study enables educators to guide their own development while creating a supportive group of critical friends (Schuck & Russell, 2005; Gallahger et al., 2011). For the purpose of our study, we defined a collaborative self-study as "questioning practice with the support of colleagues, and framing, assessing, and reframing [our] practice within the context of broader educational aims" (Samaras & Freese, 2009, p. 11). We collected qualitative data through self-reflections, and we analyzed these self-reflections for themes to better understand what affordances and constraints we had in common to help us reframe our practice. Given our research was a voluntary collaborative self-study, our project was IRB exempt, but we have used pseudonyms in order to maintain confidentiality.

Procedures

Our study originated with me (Alexa) learning about *WildCam Gorongosa* (Zooinverse, n.d.) at the North American Association for Environmental Education (NAAEE) 2014 conference. Over the past three years, I have worked to introduce preservice teachers and colleagues to this resource. Desiring to better understand how to help teachers leverage citizen science research in their classrooms and to analyze my own practice of introducing others to citizen science, I approached my science education colleague, Bonnie. After brainstorming with Bonnie on ways we could analyze our practice, we approached Catrina, Denis, Edward, and Francis to see if they wanted to join us in a collaborative self-study. Once the others also expressed interest, we collaborated over email, phone, and in person to further design our study.

Selection of the projects. We noted the three major citizen science categories set forth by a recent report funded by the National Science Foundation:

> (a) contributory projects, which are scientist designed, with the public often relegated to simple data collection; (b) collaborative projects, which are scientist structured, with citizens given freedom to refine project design, analyze data, or communicate findings; and (c) co-created projects, which are fully democratized, with the public actively engaging with scientists though all steps of the scientific process. (Bonney et al., 2009a, p. 17)

Table 1. List of Zooniverse Wildlife Camera Project

Site chosen	Description
AmazonCam Tambopata	Over 100 remote cameras monitor the forests between Refugio Amazonas and the Tambopata Research Center.
Canid Camera	Monitor wildlife species in young forests.
Cheetahs of Central Namibia	Monitor wildlife in Greater Waterberg Landscape, across wet season and dry season, near to and away from human settlements.
Cleveland Metroparks	Monitoring plots to monitor urban areas.
Michigan ZoomIN	Monitor mammals around Michigan lakes.
Numbat Discovery	Monitor Numbats and utilize this to ensure their survival.
Parasite Safari	Monitor watering holes and control sites at Ol Pejeta Conservancy and Mpala Research Centre in Laikipia County, Kenya.
Snapshot APNR	Monitor The Associated Private Nature Reserves (APNR) in South Africa.
Snapshot Grumeti	Hundreds of camera traps in Singita Grumeti Game Reserve monitor Africa's most elusive wildlife species.
Snapshot Serengeti	Hundreds of camera traps in Serengeti National Park, Tanzania, monitor Africa's most elusive wildlife species.
Western Montana Wildlife	Monitor marked eagles and eagle use of private lands by placing camera traps on carcasses throughout the Bitterroot Valley.
Wildwatch Kenya	Monitor giraffes and other animals that share the habitat in conservation field sites.

In analyzing the three categories, we decided for our first attempt to engage in citizen science, we would select contributory projects as we had limited time left in the school year and did not feel we had adequate time to engage in an outdoor schoolyard project that was geared more toward collaborative or co-created projects. We also chose online citizen science projects as our first adventure into teaching with citizen science as we all had access to the required technology, and we knew our students would be engaged with the technology. We selected wildlife camera projects as we were all teaching some aspect of ecology in our courses and knew with past teaching experiences that student engagement with wildlife photographs is typically high.

In searching for wildlife camera projects to join, we reviewed a couple of different platforms and decided to use projects located on Zooniverse (n.d.) as projects on there had similar formats and functions. We liked Zooniverse because we discovered we could create a class account, and all our students could login with this account. In looking at the wildlife projects on Zooniverse, we looked for projects that were less than 85% complete to ensure they would not be completed by the time we implemented the project in order to ensure our students would have the opportunity to contribute. We also wanted projects that provided additional data our students could record on a separate spreadsheet. The additional data were date, time, temperature, and moon phase. Finally, we looked for sites in different countries around the world as well as different biomes. Table 1 provides a list of The Zooniverse projects we used with a brief description of the project and which of us used it.

Developing our teaching modules. Once we selected the type of citizen science research we would be engaging in together, we developed a general model for implementing the citizen science learning experience, which we based on Bonney et al.'s (2009b) general model for developing a citizen science project. Table 2 shows the steps we outlined for developing our common learning experience.

We decided that each person would determine the timeframe for implementing our learning experiences with the wildlife camera sites since our class meeting times and students' interests varied. However, we did all use the Howard Hughes Medical Institute (HHMI) Biointeractive module on niche partitioning (n.d.) to introduce this learning experience to our students. This free resource uses images for students to observe and collect data and simulates an example of how and why scientists use images and observations to answer ecological questions. This was important as scientific observation skills were key in order for our students to be successful in their data collection. The HHMI resource also allowed for differentiation between middle grades and high school as the middle grades teachers only used the first module while the high school teachers used both modules in the Click and Learn.

The other aspect of the project that was similar is the way in which most of our students collected data and communicated their results through a scientific research poster. We developed a common template (Appendix B) and rubric (Appendix C) for these scientific posters. Each team of students observed and identified photographs to collect data. The data collection included communication with peers while classifying organisms. When data collection was complete, each student team developed a research question, analyzed data and created a research poster to share their results with their peers. Figure 1 shows one example for a possible imp

Table 2. Model for Developing Common Learning Experience with Citizen Science

Steps taken
1. Educator forms a group of critical friends (Schuck & Russell, 2005).
2. Educator chooses citizen science projects (Bonney et al., 2009).
3. Educators develop, test, and refine educational materials (Bonney et al., 2009).
4. Students develop their own question from data collected (Gray, Nicosia, & Jordan, 2012).
5. Students analyze and interpret data (Bonney et al., 2009).
6. Students disseminate results through a scientific poster (Bonney et al., 2009).
7. Educators measure outcomes through self-reflections (Bonney et al., 2009).

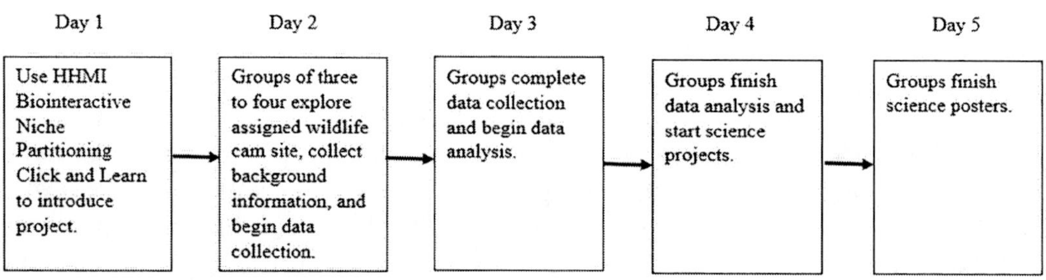

Figure 1. Sample 5-day implementation schedule.

Protocol training. Given the need for data collection consistency across participants in citizen science projects, we began with whole class instruction with a sample project on the classroom screen. Each teacher explained that each project was set up with a similar platform, so that the headings at the top of the page (e.g., ABOUT, CLASSIFY and COLLECT) could be used for each group to locate project information for their background research, determine progress for a designated site and determine where to start collecting. The initial HHMI module that was used to introduce the project also provided some initial practice with classification and data collection. We paired the students in partners for observation verification. We taught students how to use the site's classification feature to narrow down their choices and to use the built-in field guides when they had difficulty distinguishing between certain species. In addition, we shared with students how the photo identifications were verified using an algorithm that required so many volunteers to enter the same species identification. This helped them to understand how even if they did enter a specimen incorrectly that the scientists had set up checks and balances in the data collection.

We designed our own data collection protocols (Appendix D) for students to collect additional data for their groups. We created laminated cards for each student group with a screen shot of a photograph example and section of the data sheet on one side and images of the moon phases on the other. The photograph example used red arrows to indicate where students could find specific data needed for the data sheet. The phases of the moon were numbered so that students could enter a number into the data sheet to indicate the moon phase that matched the image. This reference card provided clear visuals to clarify for students exactly what was needed for their group's data sheet. We also had students rank their level of identification confidence on our class datasheet, so we could monitor how the students were feeling throughout the process.

Engaging in our community of practice. Throughout this study, we supported and encouraged each other. Most of our communication was via a group email due to work schedules and location constraints. We shared insights into how lessons were going or other resources we were adding to aid in content understanding. I (Alexa) checked in either by phone or in person prior to the others starting their unit of instruction to troubleshoot questions and to share insights I had gathered from the group. I also checked in when the others were in the midst of their teaching as a means of encouragement and support. At the conclusion of our implementation, we each shared our reflections and had the opportunity to read and provide comments to each other. We also made notes of what we would like to change for future classes, as we all plan to continue to use this in our classrooms.

Data Collection

As part of the self-study process, we each wrote a reflection based on our experiences implementing the citizen science wildlife camera projects and our perceptions of how students engaged and learned during the experience. We developed some common questions to address (i.e., What was the greatest affordance or best thing about teaching this lesson? How was student engagement during the project? How was it compared to other projects you have done in the past?). These questions focused on describing the classes in which we implemented the project, assessing our perceptions of student learning, and our prior experiences with citizen science to make our reflections semi-structured (see Appendix A for additional questions considered); however, the actual teaching reflections were left open-ended so that we could capture what was most important to us in regard to reflecting on our own practice (Loughran, Hamilton, LaBoskey, & Russell, 2004).

Data Analysis

We (Bonnie and Alexa) conducted a two-phase coding of our six reflections. The first phase was a priori coding with the parent codes of perceptions of student learning and discussions of teacher identity from our research questions (Stemler, 2001). Next, we used opening coding (Glaser & Strauss, 1967; Strauss & Corbin, 1998) to analyse the two parent codes further. From the open coding, four themes emerged for our perceptions of student learning: student interest, collaborative learning, authentic practice, and scientific observation skills. For teacher identity development, three themes emerged: working through inexperience, observing student perceptions, and examination of teaching practice as shown in Table 3.

Table 3. Qualitative Data Analysis of Reflections

Phase 1: A priori coding	Phase 2: Open coding and theme development	Example excerpts
Our perceptions of student learning	Student interest	"Students sought out working on the projects of their own volition and requested to work it, which was a first."
	Collaborative learning	"What struck me most during this class was how my students and I moved toward being colleagues. . ."
	Authentic practice	"Using citizen science in the classroom encouraged collaboration, reinforced science skills and connected content to the real world."
Discussions of teacher identity	Scientific observation	"Students were able to determine fur patterns and colors with consistency."
	Working through inexperience	"I had wanted to implement it in my classroom for some time, but I was not familiar enough with it to try it."
	Observing student perceptions	"I saw an increase in the amount of engagement from students who had been difficult to work with before. Some of these students showed an amount of excitement I had not seen in them all year."
	Examination of teaching practice	"It really opened my eyes to the purpose of wildlife studies and gave me a new perspective on the importance of content teaching."

Validity

Member checking was used to analyze our results (Creswell, 2012). Our parent codes were agreed upon by Bonnie and Alexa and presented to the other participants for agreement in that they aligned with our research questions. Emerging themes from the reflections were agreed upon by Bonnie and Alexa. These themes were connected back to the previously reviewed literature for alignment with best practices and confirmation of our findings. Member checking occurred when colleagues (Catrina, Denis, Edward, and Francis) were asked to review the results for verification of accurate wording and interpretation of their reflections. This helped to diminish researcher bias and to ensure we had captured other's perspectives as accurately as possible.

TEACHERS' SELF REFLECTIONS

Throughout the project, we each wrote personal reflections of the process. Prior to any one of us starting, we compiled a list of questions that would guide our self-reflection (see Appendix A). These prompts helped us to each attend to some of the same areas of focus while also providing an awareness of teaching anomalies. We each compiled a summary of our reflections.

Reflection One

Prior to this learning experience, I had never heard of citizen science. I was unaware of this resource, but once I heard of it, I realized citizen science research would be an excellent way to introduce my students to the research process. I was also intrigued as it allowed students to not only collect data for the scientists but to also compile a data set they could use to construct their own research questions. This created a great practice run before we headed into science fair projects where students design their own experiments and run them.

The overall implementation of the project went well. I used this project to prepare and introduce students to a different science project than a traditional science fair project that I have done in years past. Students were more successful with this project with regard to understanding scientific concepts (e.g. data analysis and transforming data) and their ability to independently comprehend and complete each step as they progressed through the project successfully. It enabled me to include components of collaboration amongst students, tiering based on how students interpreted the research, and facilitating the production of a scientific poster reflecting current practice in the field.

While the most daunting aspect of the planning process was the inclusion of the Google Classroom, this had immediate benefits of student collaboration in groups, instant dissemination of information and its scheduling, and the manipulation of student viewing versus editing the documents. I was able to formatively assess students' work as it was ongoing and receive reflections electronically versus in writing. In the past, I would have needed to wait for these to be handed in which would have eaten into instruction or group work time. In addition, I was able to have the students' complete data collection in class and collaborate on the writing at home. In the past, this process has been reversed for traditional science fair projects. It allowed me to help the students understand what the numbers meant, relationships between them, related trends, and how to transform the data into graphs. I was also able to discuss the purpose of the project and how it related to real-world contexts versus leaving these components to student interpretation outside of class.

Besides the Google Classroom, there were several aspects of this process that I found beneficial and would recommend for the future. For example, the Niche Partitioning Module 1 was instrumental in assisting the students in understanding how to take, analyze, and discuss data in preparation for the project. This module afforded them opportunities to become familiar with academic vocabulary, transform data from a table into graphs, relate data to the biome, and introduce them to determining data trends. I also found writing data down in a hard copy was helpful to the students in order for them to enter it into the computer at a later time, rather than switching back and forth between the Webcam site and the Google data sheet. Students found the laminated cards with phases of the moon and the example of data collection to be quite helpful and often requested them during the course of the project. Finally, utilizing Google Sheets for data collection was beneficial to teach students how to create graphs.

Student engagement with this project was exceptional. This was not simply a draw to the technology but to the science project itself. Students sought out working on the project of their own volition and requested to work on it, which was a first. This even occurred in their free time and during other class periods. Some students even completed more photograph identifications than initially required. For the first time, there were no complaints from parents or students during the implementation of the project. Students understood the goal and what they were supposed to complete, which reduced initial confusion. Their overall product met the objectives for the project and assisted them in preparation for their next science project and poster. The only complaints heard were with regard to collaboration within the group in terms of cooperation between members.

I would definitely use citizen science in the future. I believe it introduced the scientific process requiring a more formal scientific reasoning, writing, and practices within a whole versus its parts. It incorporated aspects of observation, data collection, data analyses, and data reporting while relating the purpose to research and how to determine trends in data. It resulted in increased discussion amongst the students as they collaborated on these ideas. Given the high level of student interest, I think it would be a great way to reward early finishers as a means of tiering.

My views on scientific practices have changed considerably through this process. Prior to this project, the only other time all the elements of the scientific process were engaged at one time was through a traditional science fair project. Citizen science has opened my eyes to real-world applications using a more current method of display through scientific posters. Students are afforded the opportunity to contribute to the scientific process through real-world science projects with other scientists. This places a higher intrinsic value on what they are accomplishing in that their product helps other scientists with their discoveries. It becomes a project where what the students do matters to the environment and the animals that live there versus simply completing a project for a grade.

Reflection Two

Prior to the implementation of this learning experience, I had very little knowledge or experience with citizen science projects. I have not participated in a citizen science experiment throughout my school time or my free time. The first time I heard about it was in my master's science methods. As a class, we explored the wildlife cam projects to see if it would work in a classroom setting. It was very entertaining. Ultimately a few of the students in the master's class decided that they wanted to implement it into their classroom. I used the same citizen science project in all my classes, which included gifted students and inclusion students.

The main reason I wanted to implement this project was because of how universally accessible it was. I implemented the project towards the end of the school year. This meant that my students had a working base knowledge of everything we would be looking at in the project. It tied up a big portion of the year and brought science alive. The students were able to participate in a real-life scientific experiment. To the students, it was just identifying animals. However, while identifying, the students also collected data on time, temperature, moon cycle, etc. All of the information was displayed on each picture for the students. This meant that all students were able to collect data as well. We worked with the data as a class, but all students were able to assist in gathering that data.

First, I will start with what went wrong for me in implementing this citizen science project into my classroom. All of the work I did with the citizen science project up to the implementation of it into my classroom was done on Chromebooks and laptops. In my classroom my students are 1:1; however, all they have access to an iPad generation 2. These iPads are seven years old with many new models and updates currently out. I worked with one iPad myself and was easily able to get the citizen science project to run smoothly. When I showed the students what to do, about half of the iPads did not handle the assignment well.

The first problem was that some of the assignment links did not work on all of the iPads. I could not figure out why some iPads ran them fine while some would not display them at all. Even in some of the links, the iPads would force close the Google applications we were using (Docs, Sheets, Maps, etc., . . .) and say that they were not up to date. Overall the first day went pretty poorly with a lot of technical issues with the iPads. I updated them all after school that day to make sure they were all up to date. After I made sure they were up to date, I got on two of the iPads to make sure that they worked. Four of the five links worked for both of the iPads. I eliminated the one link that was not working. The next day I tried again with the students. Again, the iPads were mainly just crashing and not working.

After that day, I booked the media center for the rest of the week. When I went to the media center the next day with the first class, I told them all to login. Our media center was out of date as well. Half of the computers did not work. On top of that, logging in took about 15 minutes for most of the students. The media center did not seem to be able to handle 20 students logging in at the same time.

Most of my troubles were with technology. This past year was the last year we would be in the current school building. Since they were moving into a new school, and it was the end of the year, technical problems were not priorities. I had to work around a lot of frustrating setbacks, but eventually I did get it to work relatively well for most students.

When it was going well, I was very happy with what the students were reporting. Since the project was mainly an identification-based project, with correct animals being input to the sites, the students were getting first-hand experience with observation. I was asked wonderful questions by students who would see animals that looked similar to, but not quite the same as the pictures of animals provided. The students were able to determine fur patterns and colors with consistency. When it was more complicated than that, the students were able to compare horn sizes and shapes. They were also able to click between body shapes (for example, a squirrel is a very different shape than a dog). Using all of these small identifiers together, the students were able to narrow down most of the animals to a choice between two or three.

During the days that it worked for everyone, I paid special attention to how gifted students, "average" students, and my inclusion students differed in their observation. I was very happy to see that there really was not a big difference. The gifted students asked more detailed questions when they did ask, but overall, I did not notice a significant difference in how they went about the identifications. The other biggest thing I picked up on was the amount of engagement throughout the classes. All of the students were actively contributing to these projects without being reminded to do so. They did not get off topic either. Even when their conversations strayed from the immediate task, they stayed within science and observation of animals. I heard stories from students who would identify an animal, then have a story to tell their partner about an interaction with, or fact about, that animal. I saw an increase in the amount of engagement from students who had previously been difficult to teach. Some of these students showed an amount of excitement I had not seen in them all year. They wanted to identify the animals and to input data. The students were excited to have all of their data be a part of something we did as a class. The inclusion students were also able to complete the project with little help from the parapro in the room. They were able to see specific details for identifying animals. They were also able to input data just like everyone else.

Overall, I was very happy with how the project went. I did have quite a few technical problems at the start and throughout, but even with the problems I felt like the students got a worthwhile learning experience. I felt like all students were able to participate evenly and enjoyed the identification process as well as the data input. Unfortunately, with all of the setbacks we did not get to finish the project. I would have loved to see the data used by the students to make a claim.

Reflection Three

I taught three classes using the citizen science learning experience. They were usually engaged in what I was teaching and were always looking for relevance and asking questions. My second class was very task-oriented. I could give them instructions, and they would get straight to work. They were not a group who would ask a lot of questions, but I knew they were always processing information. My third class was full of personality. They were always very engaged, and even if they weren't fully paying attention, they always got something out of the lesson. They liked to have fun and learn at the same time.

I had only heard of citizen science from other teachers and professors. I had heard of projects they had done in their classrooms and thought it was pretty cool. I had wanted to implement it in my classroom for some time but was not familiar enough with it to try it yet.

Teaching using citizen science was a great experience. I did not know what to expect going into the project, which was scary, but I got to enjoy the experience and learn like my students were able to. It really opened my eyes to the purpose of wildlife studies and gave me a new perspective on the importance of the content I am teaching. The students were pretty receptive to citizen science, and many found it intriguing. Using citizen science in the classroom encouraged collaboration, reinforced science skills, and connected content to the real world. There were definitely a few things that I had to consider and problems that needed to be addressed, but the overall outcome of using citizen science in the classroom was a success.

One benefit of using citizen science in the classroom is the amount of collaboration involved. I put my students in groups of two or three, and they had to work together to collect data. One of the group members was responsible for looking at the citizen science project pictures and telling the other group member the information from the picture. The other group member was responsible for recording the data on a spreadsheet. The whole class was collaborating on a single spreadsheet on which each group was responsible for their designated cells. The small group and whole class collaboration opportunities helped everyone get involved and reinforced the idea of working together for a common goal.

Not only was there collaboration within the classroom, but there was also collaboration with researchers and with other citizens around the world. There are numerous citizen science projects. My classes helped collect data for *AmazonCam*, *Cheetahs of Central Namibia*, and *Snapshot Serengeti*. My students and I found it exciting to help these research teams around the world. The fact that we were working with real people collecting real data made the students feel purposeful. When asked during class why citizen science is beneficial for high school students, students indicated it was beneficial because it helps them learn about the science project. They thought it was a good experience for them as it did not just help them, but it also helped the scientists who started the project collect their data. The project helped them learn more about the animals and the Amazon. It was also exciting for the students to know that they were part of the body of citizens working to collect the data. The percentage complete feature on the website, which indicates how much of the project had been completed, was engaging for the students. In fact, the day after one of my classes finished collecting data, the project's status changed to 100% complete. The students were able to see they had a first-hand part in completing the project, which was really cool. So much collaboration is important for science, and citizen science is definitely a great representation and tool to demonstrate that.

Another benefit of using citizen science in the classroom is the science skills it can reinforce and give meaning to. At the beginning of school, I give my students the opportunity to express their curiosity and show me some inquiry skills. There are usually some students who are more curious than others, but most students get engaged when they are trying to figure out a problem or answer a question. This is even more apparent when the students are interested in the topic or see a purpose in what they are trying to figure out. The way I used citizen science in my classroom gave students the opportunity to do exactly this. Once all the data were collected in our spreadsheet, each group looked through the data and came up with some research question that could be answered or explored using the data. For some students, especially the lower level ones, I had to help guide them in picking a question. Next, students transformed the data to make it more specific to the question they were trying to answer. Then, they created graphs and drew conclusions using the graphs. The students did not even realize it, but they were doing science. They made observations, collected data, developed research questions, transformed data, created graphs, drew conclusions, and shared their information. Citizen science helped engage students in doing science, which is what all science teachers should be striving to achieve.

Arguably, the best benefit of using citizen science in the classroom is how relevant it can be regarding the topics covered in life sciences. Students often get lost in the details of what they are learning in science. There are so many vocabulary words, theories, and abstract concepts, so it is always good whenever teachers can bring relevance to lessons. Citizen science does that by helping students understand not only why it is important to know these concepts but also how they can be used to benefit ecosystems as well as humans. Students noted citizen science helped them know what role animals play in the ecosystem. It helps us better understand what professional scientists are trying to show us. In an introduction activity, I was able to address concepts such as community ecology, niche partitioning, and DNA technology. These were all concepts we had learned throughout the class, so it was great to see students connecting the content to the real world. During their data collection and analysis, they were able to use content vocabulary and skills to address different situations. Throughout my education, I learned a lot of details about various things, and I did not realize how important each detail was until I was able to reflect and connect all the dots. Citizen science is a tool that can help students connect the dots and see how things they have learned are important.

When I was introduced to citizen science, I was a little skeptical. It sounded interesting, but I did not know how to implement it in the classroom. Upon exploration, I discovered there are many ways to incorporate citizen science into lessons. Those who are interested should definitely spend some time exploring to get a feel for what can be done. I learned a lot from exploring it and using it in my classroom. Moreover, my students learned a lot. They were engaged in collaboration, science skills, and content, while using technology. I will definitely be using it in my classroom again.

Reflection Four

I implemented our citizen science learning experience with two classes. In both classes, students were placed in groups of three or four. While the students' abilities would have all been considered "on grade level", the groups still varied in academic ability, behavior, and motivation levels. We are 1:1 with Chromebooks and use Google Classroom daily. This allowed me to assign groups and allow access to the Slides, Sheets, and Docs easily to the members of each group.

I had very little experience with citizen science before implementing this project. Two years ago, I introduced the idea to a class during a classification unit. We used HHMI's *WildCam Gorongosa* project. There was no research, hypothesis, or data collection or analysis. I only used the site to show them and let them gain experience identifying the animals in the wildcam photos. They enjoyed the experience, but it was only introductory and lacked depth.

I wanted to introduce my students this past year to something a little different from what we had been doing all year. I used this as an end of the year project. It was after end of course testing, so there was no new material to be taught in my curriculum. Too often, the end of the school year is filled with "fluff" to keep students entertained more than engaged. In their opinions, we are "done" so we should just be having fun. The first thing they want to know when they finish testing is that we aren't doing anything else for the rest of the year. Much to their surprise, and initially to their displeasure, we started this project. I wanted to do something that would keep them engaged, be relevant, and give them an opportunity to do something new while contributing to science.

At the onset, there was a little resistance to this project from a few of my students. Most thought it was pretty cool we were doing something that had a bigger picture, so to speak. I explained that they were taking part in a citizen science project and would be helping scientists around the globe. I also explained that they weren't the only students taking part in this and that I would be contributing some of the experience to be included in a book about citizen science. This hooked many of them because they could already tell it was not going to be mindless busy work. At the same time, it made some uneasy because they were hoping for mindless busy work. We began the project by completing HHMI's Niche Partitioning module. I realized that in the future I will need to make some adjustments to my lessons on niches. They struggled a little more with this activity than I had expected. We worked through it together and discussed other questions that could be asked about an ecosystem like this one.

After completing the Niche Partitioning activity, we moved on to our own research of various wildcam sites. Each group was assigned a different wildcam, and they spent some time familiarizing themselves with the locations, projects, research teams, and ecosystems involved. We had a few "practice rounds" identifying and seeing how each worked and what information was available on each picture. I showed them how to log in so their data collection would be counted toward the project. Students divided up the work and research and set out to complete their projects. Most groups did well collaborating with one another and working together to identify and collect data. I was able to walk around and check in with each group, assisting as necessary. I noticed one student was not logged in to the Zooniverse site and when I pointed it out, he thought I wanted them to log in to Google Classroom. I explained that if he wasn't logged in to the site, his identifications would not be counted toward the project. In explaining this, nearby students realized that this was "real." One girl expressed that she didn't know that it really counted, and now she'd take it more seriously. I realized that they thought if it didn't really mean anything, they didn't have to take it seriously. However, if real scientists were using their data, they'd better do it correctly. After this conversation, I noticed more students taking the task more seriously. There were some students that had not put much effort into most projects or assignments throughout the year but jumped into this project willingly and even stepped up as a leader in their group to make sure the identifications were done correctly. One of the groups was working on their identifications and became visibly frustrated so I went to check on them. They said they weren't getting any new pictures and didn't understand why. When I looked at the site, I noticed that it said the project was completed. These students definitely thought it was cool that they were probably the ones that completed it with their identifications.

There were some struggles throughout the learning experience, and some of that was timing. I want to do this project or a similar one next year, and I'd much rather start the year with it than end. There were delays because of network limitations due to testing. There were end of the year interruptions that were out of my control. Groups struggled to finish their posters before we ran out of time so I know I didn't get their best work. Also, it became apparent that much of their science work has always been surface level. While identifying and collecting data was easy and dare I say, fun, the analysis of data, asking good research questions, and drawing conclusions was more difficult. Being the end of the year, they didn't have much left and just wanted to be done. On the other hand, I also realized that if this was something they were more comfortable with or if it was a routine they were more accustomed to, it would not have been such a struggle for them. It is definitely an area that I can improve in my lessons in the future.

Something that I think stands out the most about this is that when the science is real, they are more engaged and take things more seriously. Students do not want fluff; they want substance and relevance. It was important to them that their work was being used as a part of something bigger. For this reason, I'll definitely use citizen science in the future. I'll also use it from the beginning and revisit the concept throughout the year. Now that I have gone through the process, I am more comfortable with it. I think fear is something that holds teachers back sometimes. If I were talking to a colleague about it, I'd probably stress that they should at least try it. They might be surprised at how well their students do and the level of engagement that is possible. I know I was.

Reflection Five

Prior to this self-study, I had participated in several citizen science projects. As a teacher, I had conducted phenology and lichen studies with my students. As a doctoral student, I planned and implemented a lizard population study at Camp Rockfish in Rockfish, NC (The Herp Project, 2018a). I also helped high school students contribute data to other Herp Project research studies (The Herp Project, 2018b), the Carolina Herp Atlas (2017), the Herp Mapper (n.d.), and Anole Annals (2018). As a teacher educator, I have contributed to numerous projects on Zooniverse. Though I have used citizen science in multiple ways, I still feel I have a lot to learn, and I had been contemplating conducting a self-study for a while.

I was the first one to teach the citizen science lessons. After completing some readings about citizen science, my methods students were interested in learning more about projects they could implement in their classes. Two of my students were in their first year of teaching middle school, so I geared these lessons toward the middle school age range. My students and I walked through these lessons during one of our class meetings. We began by going through the HHMI Click and Learn. We analyzed the two different modules and discussed the pros and cons of each module and how they aligned or did not align with State Standards.

Next, we reviewed the teaching module we had designed for this study. My students were completely engaged in the photo identification, and we noted right away that it was probably best to put students in partners during data collection to help facilitate the excitement of finding an interesting photograph. My students mentioned that working with a partner would allow students to have someone instantly validate their "cool" find. They also noted that on some of the sites having someone else verify what they were seeing helped to increase their confidence level, and they mentioned how this could lead to discussions on accuracy and precision with data collection. In fact, the students were so engaged that we lost track of time and ended up going past class dismissal time. When I pointed this out and told them they were dismissed, my students still chose to stay and identify animals. The next week, all three students reported back that they had continued to identify animals at home after class. My two first year teachers also ended up using the sites with students, and one even decided to join our self-study to implement the entire project and reflect on it.

The HHMI Click and Learn introduction felt rather rushed. I believe in the future, I would rather assign students the Click and Learn prior to class and have them come to class ready to discuss this resource, or I might assign it after class and have them reflect on how it could help introduce a project like the wildcam one. Either course of action would allow more time for them to reflect on the resource as well as brainstorm ways they might use it in the future.

What struck me most during this class was how my students and I moved toward being colleagues as we analyzed the developed materials and discussed how we could facilitate implementation of the project. My students openly offered suggestions and critiques, which prior to that point in the semester, they had not. I could also tell they were able to visualize using the lessons with their students, which I wonder if this increased their comfort with offering suggestions as opposed to other classes where they seem overwhelmed by the lessons or the teaching methodologies presented. I left this class wondering how I could leverage this type of learning in future classes. This resonated even further with me when the students reached out asking me to share the resources, so they could tweak them for their classes. It could be that my students, who all had biology degrees, were super engaged because they loved the science behind the citizen science projects. However, I am left wondering if the joint analysis of materials and brainstorming on how to improve them afforded beginning teachers an opportunity to engage in reflective teaching practices as colleagues rather than students and professor. To further explore this, I plan to create a similar project using astronomical photographs to see if I can tease apart content interest versus accessibility of the method's lesson itself.

The way in which I taught this learning experience during the PD was different as I was working with adult learners who were in-service teachers and had less time with them. I walked the teachers through the resources we had designed, let them experience the HHMI Click and Learn and Zooniverse, and then I shared with them student example posters from my colleagues.

Overall, the teachers were also highly engaged and were surprised that they were unaware of Zooniverse. I did not receive one question or overhear a comment along the lines of my students cannot do this, which is typically the case when I have taught PDs in the past. Instead, I heard more comments of "My students will love this," "I can't wait to share this with my grade level," and "Can we get a copy of these materials?" They were more resistant to the idea of students creating science posters, but when I shared with them examples from middle school and high school students of varying abilities from different counties across the state, they became much more interested in this type of assessment, especially as we discussed how the sections of the poster aligned with new state standards and was authentic work scientists would do.

With the teachers at the PD, I felt more like an instructor than a colleague as I did with my methods students. This was not my intent as I see PD as a collaborative learning experience because as educators we all have expertise to share. It could be that citizen science was so new to the teachers (not one of the 62 had ever heard of citizen science) they were trying to understand this new field of research being introduced to them, or it could have been that my time was limited so I focused more on breadth than depth. Needless to say, I left contemplating how I could facilitate a PD experience similar to my methods class as my desire is to engage colleagues in collaboration.

These teaching experiences also reinforced to me the necessity of relationship building and how trust is essential to true collaboration. With my methods students, I had worked with them for over eight weeks, and I had taught them the semester prior in another class. They knew me, and I knew them. They shared with me their struggles, their victories, and their frustrations. We worked on unit plans, conference presentations, and traveled together. With the teachers, I did not know them. I did not have time to invest in a relationship prior to working with them. Though I did have them answer questions such as how many years they had taught, what subjects they taught, what grade levels, etc., the format did not facilitate relationship building. To them, I was a stranger who had some information on teaching science. As I go into a new school year, I plan to continue introducing others to citizen science, but I also find that I am recommitting to and reevaluating how I go about building relationships with others.

Reflection Six

Prior to this teaching experience with citizen science, I was aware of a few citizen science projects, such as projects that record the amount and type of trash collected on beaches and adding to the database of spawning horseshoe crabs. I had also participated in the *WildCam Gorongosa* project with my own children to identify a number of pictures and organisms. They enjoyed it immensely, but I had not actually used a citizen science project formally in the classroom setting.

My group of students were 61 preservice teachers in either their sophomore or junior year of a bachelor's degree in education. Most were elementary education majors, but a handful were middle grades majors.

On Day 1 of the project, I provided Chromebooks for each student and my goal was to walk through Module 1 of the HHMI Click and Learn, introduce the project and have groups of students explore a variety of preselected citizen science sites. My first error was having each student begin with a Chromebook at the start of the lesson and logging in to the HHMI site. Since I needed to model the process, it would have worked best to have the whole class walk through the Click and Learn together using my SMARTboard and responding to questions individually in writing. However, once they had a Chromebook in front of them and got started with the Click and Learn, they immediately dispersed at different rates. This would not have been a problem, except that they actually needed a fair amount of technology assistance, and I was the only instructor in the room to move around to 30 different students. Fairly quickly, I realized that to allow everyone to get through the background information at the same time I would need to use the SMART board, so I had them all close the Chromebooks for the time being. This approach worked much better. We finished up the Click and Learn and I opened the Google Doc to talk through the initial instructions for the project.

Student groups were determined by the lab group table where they were sitting, and once they were ready to explore the various preselected sites, each student reopened the Chromebooks and designated assignments for exploring the sites. Since I did not have access to Google Classroom, I used a Google Doc open on the teaching station for each group to sign up for a site and to list group members and their Gmail addresses to add them to groups for sharing documents. For the remainder of class, they began researching about their project site using the template questions to gather background information.

On Day 2 of the project, I reserved a fairly large computer lab with a cart of Chromebooks. This gave each group room to separate from other groups and work collaboratively. I walked from group to group, assigning "pairs" to work together. Within each pair, one person would click through the photos on the project site, and the other person would enter data in the shared Google Sheet. I was surprised that I had to help them understand how multiple people could efficiently enter data in the same spreadsheet without typing over one another's work. The simplicity of asking one pair to start on row two, while another group started on row 16, did not occur to them. Once they got used to this process, they all moved pretty quickly through identifying at least 15 organisms. Our initial plan had been for me to use the data that they were collecting on that day and pull into a spreadsheet to model sorting and analyzing data. Because groups moved at different paces, etc., it would have been better to have a set of canned data just to model the process with. I will make sure that is the case next time. Also, I had to take a few minutes to explain how they would need to clean up the data in their sheets, by making sure that all cells were filled, and worded the same way. For example, spelling words the same way and being consistent with capitalization would insure that the sorting feature actually worked. Again, these were things that never occurred to us as instructors to explain to the students, but time to clean up the sheet was needed to make the next step feasible. We also added dropdown selections in the datasheet to help with this. Since I had the opportunity to teach two classes back to back with this process, the second class went more smoothly, which is usually case, because I knew which stumbling blocks to warn them about! The end of this class period had all groups with data collected, and the groups divided to work on the poster template and data analysis/production of graphs/charts. Any work left to finish up the poster template or data analysis, needed to be done as homework outside of class. Again, this is the great part of using Google! All of the group members had access to it once we left class, and as an instructor, I could use the version history feature to make sure that all students in the group contributed to the final product.

On Day 3, the groups reviewed their poster and made any final edits. Then each group presented to the rest of the class. They gave a brief overview of the actual research project and then discussed their chosen research question, results, and analysis. Most groups received a question or two from the class following the presentation. At the conclusion of all presentations, I handed out reflection questions for students to evaluate their perspective on participating in a citizen science project. I was encouraged by the reflective responses of many students who acknowledged the multifaceted benefits of this project - expanding their view and awareness of citizen science as it relates to their content knowledge of various world habitats, etc. and the amazing growth they had in the use of Google applications for participating in a group project. Many also indicated that they could see themselves using citizen science in their own future classrooms and that they hoped their administrators would appreciate this as a learning experience that provides a real context for students and puts meaning into assignments.

RESULTS

In analyzing our self-reflections, we aimed to address what were the affordances of teaching a citizen science learning experience for our teacher identity development and our students' perceptions of learning science. In the qualitative analysis of our reflections, we employed a two-phase coding of our six reflections. The first phase was a priori coding of teaching identity and perceptions of student learning (Stemler, 2001), while the second phase was open coding (Glaser & Strauss, 1967; Strauss & Corbin, 1998) of the two parent codes, giving rise to seven themes.

Teacher Identity Development

Teacher identity development was selected as a priori code as we knew that engaging in a collaborative self-study would give us an opportunity to have recognition from ourselves and our colleagues (Gee, 2005) as moving toward expertise in using citizen science in our classroom. Our second round of coding revealed three themes within the teacher identity excerpts: working through inexperience, observing student perceptions, and examination of teaching practices.

Working through inexperience. One aspect four of us mentioned was our inexperience using citizen science. Inexperience has often been demonstrated to be a hinderance for teachers in regard to implementation of inquiry-based projects (Windschiti, 2003). For instance, Edward wrote, "I had wanted to implement it [citizen science] into my classroom for some time but was not familiar enough with it to try it yet." The collaborative nature of our project allowed us to push through our own insecurities, as we knew that others were also working through their own inexperience. In addition, two of us were the only science teachers in our grade level. Having the support of other science education colleagues was a new experience and helped to alleviate some of the fear we felt in attempting to teach with citizen science.

Observing student perceptions. Each of us noted our observations of students' perceptions during the learning modules. Denis noted how he "saw an increase of engagement from students who had been difficult to work with before. Some of these students showed an amount of excitement I had not seen in them tall year." Catrina observed that "students sought out working on the citizen science projects of their own volition. ... This even occurred during their free time." The fact that we each compared and contrasted our students' perceptions with their previous perceptions confirmed our decision to engage in self-study as it indicated we were all engaged in reflecting on our students' perceptions of science learning prior to our participation in this study (Samaras & Freese, 2009).

Examination of teaching practices. The other theme which emerged when we considered our teacher identities was how we each reflected on our teaching practices by acknowledging setbacks we faced and setting goals for the future. "My views on scientific practices have changed considerably through this process" (Catrina). Frances compared her prior experience with citizen science two years earlier as only introductory and lacking depth, and she noted that in her future teaching she wanted to introduce citizen science earlier in the year and keep revisiting the projects. Our analyses of our own teaching practices align with the self-study goals of openness and reframing (Hamilton, Pinnegar, Russell, Loughran, & LaBoskey, 1998).

Our Perceptions of Students' Science Learning

At the beginning of this collaborative self-study, we all acknowledged that one of our reasons for joining this study was to engage students in authentic scientific practices, and we discussed how reflecting on student perceptions often drive our teaching practices. Thus, we came to the conclusion that one focus of all our reflections would be on how students perceived the citizen science learning experience. Four themes emerged through the open coding of this a priori parent code: student interest, collaborative learning, authentic practice, and scientific observation skills.

Student interest. Swarat, Ortony, and Revelle (2012) found that student interest in school science increased based on the activity students engaged in, which aligns with what all six of us noted as student interest levels dramatically increased with the citizen science learning experience. For instance, Alexa shared how her methods' students stayed after class was dismissed to continue data collection, while Denis discussed how, "All of the students were actively contributing to these projects without being reminded to do so." Students were fascinated with the photographs and motivated by how their data were directly contributing to a larger scientific project. As Hiller and Kitsantas (2014) found in their research on middle school students engaged in citizen science student, interest was higher when students collected data rather than simply learning about horseshoe crab ecology in the classroom.

Collaborative learning. Another theme that emerged across our experiences was the collaborative learning that occurred amongst groups and classes. As Edward noted in his reflection, the collaboration also extended to the researchers who designed each wildlife camera project. Each of us also noted that the level of collaboration was different than past times we had engaged our students in collaborative efforts. The students more readily shared the workload, engaged in conversation around science, and generated thought-provoking questions and observations (Hiller & Kitsantas, 2015; Tomasek, 2006). Catrina observed, "It resulted in increased discussion amongst the students as they collaborated on these ideas." This also indirectly indicates possible science identity formation as students were recognizing their own contributions as well as being recognized by their peers as being capable of contributing to science (Carlone, Haun-Frank, & Webb, 2011).

Authentic practice. Throughout the process, our students were engaged with scientific research. Repeatedly, we found that students were motivated to participate in authentic scientific work. The attention to detail was greater for students as they recognized their work would be contributed to an outside source which supports other citizen science research (Mitchell et al., 2017). First, we noted how the real world connections and authentic science research seemed to increase student engagement as well as their commitment to use precision for accurate data collection. In each case, students commented on how they took the data collection more seriously because it mattered and was important for scientific discovery. This is similar to what Calabrese Barton and Tan (2010) observed when students worked on authentic science issues within their community. In some ways, this project afforded students opportunities to engage their critical science agency (Basu, Calabrese Barton, Clairemont, & Locke, 2009) by (a) gaining a deeper understanding of content, (b) seeing themselves as having expertise in a field of science, and (c) developing their science identities by seeing themselves capable of contributing to science.

Scientific observation skills. The final theme each of us observed was the scientific observations skills students were able to use while completing the learning experience. Observation of the photographs allowed students to notice minute details and compare/contrast similar species, which brought about questions and conversations among the students. The data collection, analysis, and communication of results helped students better understand the process and nature of science. This supports the argument set forth by Gray, Nicosia, and Jordan (2012) in regard to engagement in citizen science promoting nature of science development in students. Denis noted how he did not observe differences between his identified inclusion and gifted students ability to observe and provide species identification. Francis shared that "some students had not put much effort into most projects or assignments throughout the year but jumped into this project willingly and even stepped up as leaders in their groups to make sure the identifications were done correctly."

Discussion

The use of citizen science increased the engagement level for students at three different levels of science education (middle school, high school, and postsecondary). Such projects are wonderful ways to authentically introduce and reinforce scientific practices. Students at each level had opportunities to observe, identify, classify, collect, predict, analyze, and communicate through participation with citizen science in collaboration with their peers. Not only did the students find this to be a refreshing twist to the way they had been participating in science classes, but the realization that their data contributed to a growing body of knowledge (valuable to an ecosystem somewhere outside of their school and managed by "real" scientists) validated the experience for them. The level of precision that was exhibited shifted with the accountability of having to login to a site and identify organisms from real wildlife cameras.

In addition, the end result of the project allowed students the experience of producing posters at the level a researcher would use at a formal poster session. The production of the posters encompassed several valuable educational skills that teachers sometimes struggle to incorporate in their curriculum and produce quality results, such as data collection and analysis, generation of tables or graphs to depict results through the use of Google Sheets, writing/editing their written work and their peers', and communication through presentation of the end product. Many of these skills may be individually covered in a lab report or class project, but this learning experience allowed students to incorporate all of these at one time by following the guided structure and templates provided. This initial scaffolding provides a foundation for future research in science classes as each of the science practices can be referenced and then advanced.

Limitations

Given the nature of this research, limitations exist as to how generalizable the findings are to other teachers. Even more of a challenge was our newness to self-study itself, and how to ensure our reflective practices were adequate. Our study is also limited in its ability to quantify student science learning success in regard to student participation in this project. Yet much of what we learned concerning student perceptions of learning has also been cited in the research literature that informed this study (Barton, 2012; Bonney et al., 2016; Krasny et al., 2014; Penuel & Means, 2004). Most importantly, this study was grounded in teacher perceptions and experiences of using citizen science in their classrooms, an approach some scholars find lacking in the literature on citizen science education (Bracey, 2018).

Future Research

Participating in this project and seeing the level of interest that citizen science generated motivated several of the teacher participants to think about how we could extend this project. For instance, one of the middle grade's teachers would like to use this same model to create a guiding framework using Google Docs to help students participate in citizen science projects and collect data as well as utilizing astronomy projects that could be used in her earth science class. Some of the questions that were generated by classroom students during the participation could also be elaborated on through individual or small-group work. The university faculty would like to set up wildlife cameras on the university campus to monitor bioactivity on campus. Once preservice teachers have participated in class with the citizen science project and assisted with the wildlife cams on campus, they may see the value of setting up such cameras on school grounds to generate some long-term data of bio activity at their future school sites.

Educational Implications

The greatest constraint experienced was technology-based as the citizen science projects we chose were all online projects, so we needed access to computers and the internet. However, this is also where we served as critical friends to each other as we readily shared the triumphs and challenges we had. For instance, Bonnie emailed everyone to let them know that capitalization and spelling mattered for sorting on the Google Sheets. Bonnie and Alexa also went to Catrina's classes and helped with technology implementation, as her students were the least experienced with technology. We also made sure to emphasize that the implementation of this project was determined by each of us and that our lessons, approaches, and technology use did not need to look the same. Denis, having experienced the most setbacks due to technology issues, felt he had not really succeeded until we debriefed with him and reminded him that it was not necessary to do the science posters and if his students learned, then he had achieved much with the learning experience. Having a group of fellow educators we could reach out to help make this constraint less volatile to our analysis of our teaching practice than if we had engaged in this endeavor by ourselves. Knowing others experienced setbacks and had to work through obstacles encouraged us to persevere.

Given our first attempts to use citizen science research with our students, we would highly encourage all science teachers to use citizen science, but there are a few caveats. First, make sure the necessary technology is available to access the projects or the data collection tools. Second, an introduction activity is a good idea to explain what citizen science is and give students an opportunity to practice collecting data. Third, have a plan for students to actually do something with the data being collected besides simply contributing to the dataset the scientists are compiling, such as developing their own question and presenting their findings in a research poster or presentation. When planning a project with citizen science, it is also a good idea to be familiar with collaboration tools such as Google Docs, Slides, Sheets, etc. Just like with any lesson, the more one plans, the better the outcome.

APPENDIX A

Teacher Reflection Questions

1) Describe how you implemented the lessons. What was the sequence, overall flow? How did you introduce the idea of citizen science?
2) What was the greatest affordance or best thing about teaching this lesson?
3) What was the biggest hindrance or greatest constraint with teaching this lesson?

Appendix A (Continued)

4) How have your views on incorporating scientific practices (observation, data collection, data analyses, and data reporting) into your teaching changed?
5) How was student engagement during the project? How was it compared to other projects you have done in the past?
6) What types of questions did students ask? What did students want to explore with the data?
7) If you were going to help a colleague implement this lesson, what would you tell them?
8) What will you change the next time you teach this?
9) Will you use citizen science in the future within your classroom? If so, why? If not, why?

Appendix B

Scientific Poster Template

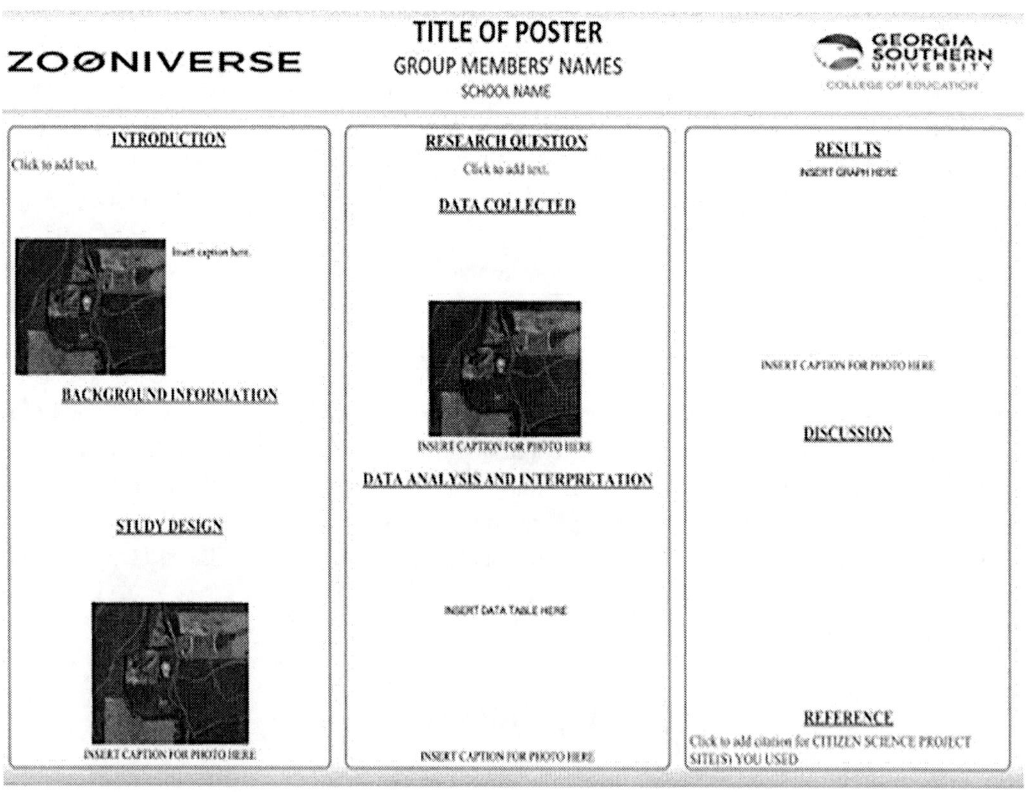

APPENDIX C

Scientific Poster Rubric

Concerns Areas that need work	Criteria Description reflecting achievement of mastery level of performance	Advanced Evidence of exceeding standards
	Introduction ❑ Site location/physical features ❑ Biome; found using *Bioviewer*	
	Background Information ❑ What are they trying to find out? Why was this citizen science project started? ❑ Discuss the research team: who are they? ❑ Represented institutions? ❑ Results so far? ❑ How much progress has been made?	
	Study Design ❑ Description of citizen science project (what are citizens being asked to do?) ❑ Description of how photographs were analyzed by your group ❑ Screenshot of study	
	Data Collected ❑ How many pictures were analyzed? ❑ Data collected from each picture ❑ Example image included with caption	
	Data Analysis and Interpretation ❑ Data transformed in a way that matches question ❑ Identify trends, etc. ❑ Data table included with caption ❑ Data table appropriately labeled	
	Research Question ❑ What question does your team have about the data?	
	Results (graph) ❑ Results displayed appropriately (bar or line graph, etc.) ❑ Graph contains all elements to clearly read (labels, title, etc.) with caption	
	Discussion ❑ Explains what results mean ❑ Ties results to at least one big idea ❑ Ties results to site location, physical features, or biome. ❑ Ties results to overall goal of citizen science project	
	Reference ❑ Citation is given for citizen science project	
	Reflection ❑ Turned in with poster ❑ What did I learn? ❑ What questions do I have now?	
	Poster Development ❑ Clear, concise writing ❑ Scientific vocabulary throughout ❑ Graphics in focus and relevant ❑ Poster had been proofed and little to no mistakes appear	
Unacceptable	**Acceptable**	**Target**

Total Score: _______________

Additional Comments:

APPENDIX D

Data Collection Protocol Card Side A

Photograph Example

Zooinverse (n.d.). Wildcam Gorgonosa on Zooniverse platform. Retrieved from https://www.wildcamgorongosa.org/#/

Data Table Example for Above Photograph

Research site	Date of photo	Time of day	Temperature (C)	Phase of Moon	Type of habitat	Animal identified	How many present?	Behavior observed	Trophic level	ID Confidence (1- not very; 5 - very)
Ol Pejeta Conservancy	10/09/2016	9:21 AM	25	3	Watering hole	Elephant	3	Drinking	Herbivore	5

Data Collection Protocol Card Side B

Moon Phase Chart

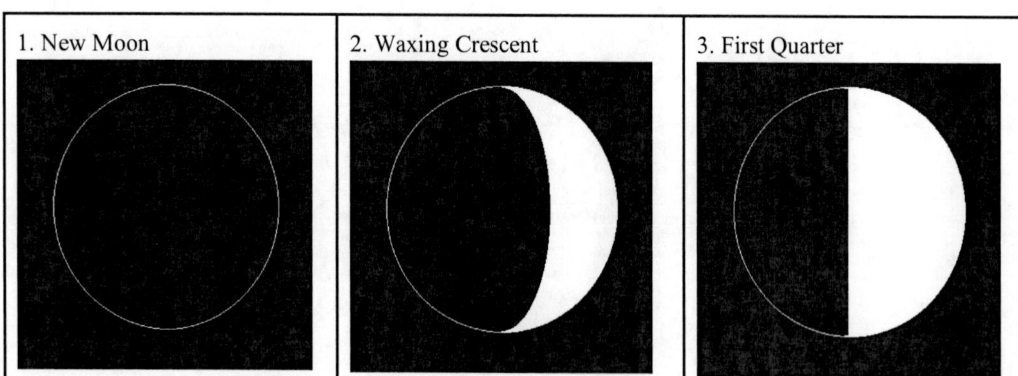

APPENDIX D (CONTINUED)

4. Waxing Gibbous 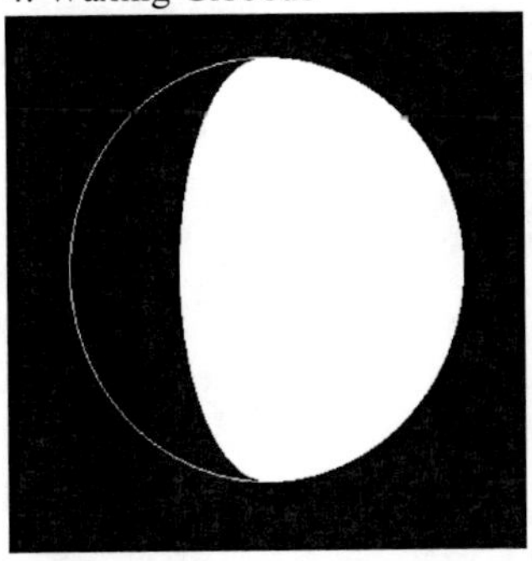	5. Full Moon 	6. Waning Gibbous 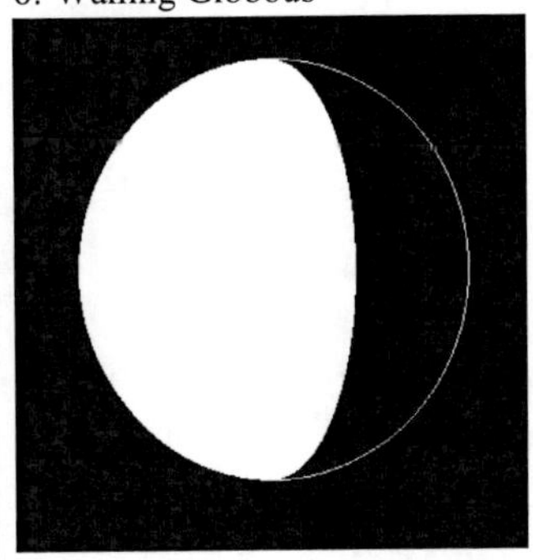
7. Third Quarter 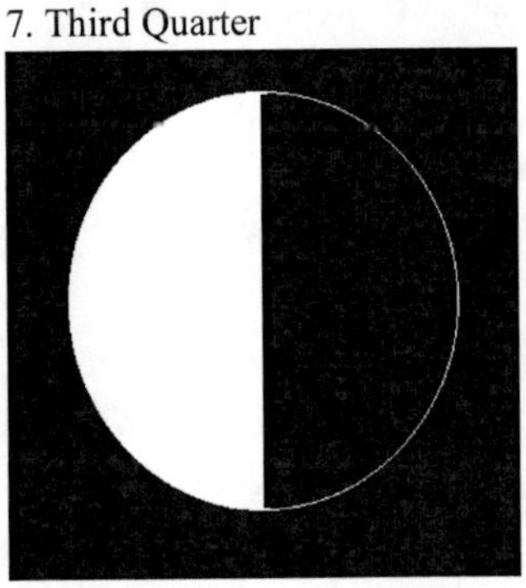	8. Waning Crescent 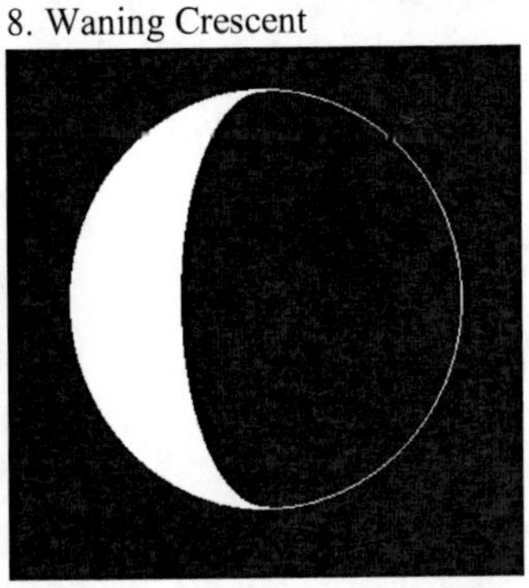	Select the moon phase most similar and record the number on your data sheet.

U. S. Naval Observatory; Astronomical Applications Department (n.d.). Phases of the moon illuminated. Retrieved from https://aa.usno.navy.mil/faq/docs/moon_phases.php.

REFERENCES

Anole Anals (2018). Retrieved from http://www.anoleannals.org

Basu, S. J., Calabrese Barton, A., Clairmont, N., & Locke, D. (2009). Developing a framework for critical science agency through case study in a conceptual physics context. *Cultural Studies of Science Education*, *4*(2), 345–371. doi:10.1007/s11422-008-9135-8

Bonney, R., Ballard, H., Jordan, R., McCallie, E., Phillips, T., Shirk, J., & Wilderman, C. C. (2009a). *Public participation in scientific research: Defining the field and assessing its potential for informal science education.* Center for Advancement of Informal Science Education. CAISE Inquiry Group Report

Bonney, R., Cooper, C. B., Dickinson, J., Kelling, S., Phillips, T., Rosenberg, K. V., & Shirk, J. (2009b). Citizen science: A developing tool for expanding science knowledge and Scientific literacy. *Bioscience, 59,* 977–984. doi:10.1525/bio.2009.59.11.9

Bonney, R., Phillips, T., Ballard, H., & Enck, J. (2016). Can citizen science enhance public understanding of science? *Public Understanding of Science, 25*(1), 2–16. doi: 10.1177/0963662515607406

Bracey, G. (2018). *Teaching with citizen science: An exploratory study of teachers' motivations and perceptions*. (Dissertation thesis). Retrieved from the Proquest Dissertations and Theses Global. (UMI Number 10791145)

Calabrese Barton, A. M. (2012). Citizen(s') science. A response to "The future of citizen science." *Democracy & Education, 20*(2), Article 12. Retrieved from https://democracyeducationjournal.org/home/vol20/iss2/12

Calabrese Barton, A., & Tan, E. (2010). We be burnin'! Agency, identity, and science learning. *Journal of the Learning Sciences, 19*(2), 187–229. doi:10.1080/10508400903530044

Carlone, H. B., Haun-Frank, J., & Webb, A. (2011). Assessing equity beyond knowledge- and skills-based outcomes: A comparative ethnography of two fourth-grade reform-based science classrooms. *Journal of Research in Science Teaching, 48*(5), 459–485. doi:10.1002/tea.20413

Carolina Herp Atlas (2017). Project overview. Retrieved from www.carolinaherpatlas.org.

Cooper, C. B., Dickinson, J., Phillips, T., & Bonney, R. (2007). Citizen science as a tool for conservation in residential ecosystems. *Ecology and Society, 12*(11). doi:10.5751/es-02197-120211

Creswell, J. (2012). *Educational research planning conducting and evaluating quantitative and qualitative research* (4th ed.). Boston, MA: Pearson

Dickinson, J. L., Zuckerberg, B., & Bonter, D. N. (2010). Citizen science as an ecological research tool: Challenges and benefits. *Annual Review of Ecology, Evolution, and Systematics, 41*(1), 149–172. doi:10.1146/annurev-ecolsys-102209-144636

Dinkelman, T. (1999). Self-Study in teacher education: A means and ends tool for promoting reflective teaching. In *Annual Meeting of the American Educational Research Association*. Montreal, Quebec, Canada.

Gallagher, T. L., Griffin, S. M., Ciuffetelli Parker, D., Kitchen, J., & Figg, C. (2011). Establishing and sustaining teacher educator professional development in a self-study community of practice: Pre-tenure teacher educators developing professionally. *Teacher and Teaching Education, 27*(5), 880-890.

Gee, J. P. (2005). *An introduction to discourse analysis: Theory and method (*2nd ed.). New York, NY: Routledge.

Georgia Department of Education. (2018). *Free and reduced lunch data.* Retrieved from http://www.gadoe.org/Pages/Home.aspx

Glaser, B. G., & Strauss, A. L. (1967). *The discovery of grounded theory. Strategies for qualitative research*. Chicago, IL: Aldine.

Gray, S. A., Nicosia, K., & Jordan, R. C. (2012). Failures of engagement: Lessons learned from citizen science in the classroom. *Democracy & Education, 20*(1), 1-5. doi:10.1080/1533015x.2012.777224

Hamilton, M. L., Pinnegar, S., Russell, T., Loughran, J., & LaBoskey, V. (1998). *Reconceptualizing teacher practice: Self-study in teacher education.* London, United Kingdom: Falmer Press.

Hannigan, S., Raphael, J., White, P., Bragg, L. A., & Clark, J. C. (2016). Collaborative reflective experience and practice in education explored through self-study and arts-based research. *Creative Approaches to Research, 9*(1), 84-110.

HerpMapper (n.d.). Retrieved from https://www.herpmapper.org

HHMI Biointeractive (n.d.). *Niche partitioning and DNAmetabarcoding.* Retrieved from https://www.hhmi.org/biointeractive/niche-partitioning-and-dna-metabarcoding

Hiller, S. E., & Kitsantas, A. (2015). Fostering student metacognition and motivation in STEM through citizen science programs. In A. Peña-Ayala's (Ed.) *Metacognition: fundaments, applications, and trends* (Vol. 76, pp. 193-221). Cham, Switzerland: Springer.

Hiller, S. E., & Kitsantas, A. (2014). The effect of a horseshoe crab citizen science program on middle school student science performance and STEM career motivation. *School Science and Mathematics, 114*(6), 302-311. doi:10.1111/ssm.12081

Jordan R. C., Ballard H. L., & Phillips T. B. (2012). Key issues and new approaches for evaluating citizen-science learning outcomes. *Frontiers in Ecology and the Environment, 10*, 307–309. doi:10.1890/110280

Koomen, M. H, Rodriguez, E., Hoffman, A., Petersen, C., & Oberhauser, K. (2018). Authentic science with citizen science and student-driven science fair projects. *Science Education, 102*(3), 593-644. doi:10.1002/sce.21335

Loughran, J. J., Hamilton, M. L., LaBoskey, V. K., & Russell, T. (Eds.). (2004). *International handbook of self-study of teaching and teacher education practices* (Vol. 53). Springer. doi:10.1017/CBO9781107415324.004

Mitchell, N., Triska, M., Liberatore, A., Ashcroft, L., Weatherill, R., & Longnecker, N. (2017). Benefits and challenges of incorporating citizen science into university education. *PLoS ONE 12*(11): e0186285. doi:10.1371/journal.pone.0186285

Mueller, M. P., and D. J. Tippins (2012*) Citizen science, ecojustice, and science education:* Rethinking an education from nowhere. In B. J. Fraser, K. Tobin, & C. J. McRobbie (Eds.), *Second international handbook of science education* (pp. 865–882). Dordrecht, Netherlands: Springer.

NGSS Lead States. (2013). *Next generation science standards: For states, by states.* Washington, DC: National Academy Press.

Office of Institutional Research: Enrollment Management. (2016). *College of Education: Enrollment by race group.* Retrieved from http://em.georgiasouthern.edu/ir/student-enrollment-data/

Penuel, W. R., & Means, B. (2004). Implementation variation and fidelity in an inquiry science program: An analysis of GLOBE data reporting patterns. *Journal of Research in Science Teaching, 41*(3), 294-315. doi:10.1002/tea.20002

Roth, W. M., & Lee, S. (2002). Scientific literacy as collective praxis. *Public Understanding of Science, 11*(1), 33-56. doi: 10.1088/0963-6625/11/1/302

Russell, T., & Korthagen, F. (Eds.). (1995). *Teachers who teach teachers: Reflections on teacher education*. London, United Kingdom: Falmer.

Samaras, A., & Freese, A. (2009). Looking back and looking forward: A Historical overview of the self-study school. In C. Lassonde, C.A. Galman, & S. Kosnik (Eds.), *Self-Study research methodologies for teacher educators* (pp. 3-19). Rotterdam, Netherlands: Sense Publishers.

Scheuch, M., Panhuber, T., Winter, S., Kelemen-Finan, J. Bardy-Durchhalter, M. & Kapelari, S. (2018). *Butterflies & wild bees: Biology teachers' PCK* development through citizen science. *Journal of Biological Education, 52*(1), 79-88, doi:10.1080/00219266.2017.1405530

Schuck, S., & Russell, T. (2005). Self study, critical friendship, and the complexities of teacher education. *Studying Teacher Education, 1*(2), 107-121. doi:10.1080/17425960500288291

Stemler, S. (2001). An overview of content analysis: Practical assessment. *Research & Evaluation, 7*, 137-146.

Strauss, A., & Corbin, J. (1998). *Basics of qualitative research,* Second edition. Thousand Oakes, CA: SAGE Publications, Inc.

Swarat, S., Ortony, A., & Revelle, W. (2012). Activity matters: Understanding student interest in school science. *Journal of Research in Science Teaching, 49*(40), 515-537. doi:10.1002/tea.21010

The Governor's Office of Student Achievement. (2018). Report Card. Retrieved from https://gosa.georgia.gov/report-card.

The Herp Project (2018a). *Lizards.* Retrieved from https://theherpproject.uncg.edu/curriculum.

The Herp Project (2018b). *Curricula.* Retrieved from https://theherpproject.uncg.edu/curriculum.

Tomasek, T. M. (2006). *Student cognition and motivation during the Classroom BirdWatch citizen science project.* (Dissertation thesis). Retrieved from the Proquest Dissertations and Theses Global. (UMI Number 3242410)

U. S. Naval Observatory; Astronomical Applications Department (n.d.). *Phases of the moon illuminated.* Retrieved from https://aa.usno.navy.mil/faq/docs/moon_phases.php

Windschiti, M. (2003). Inquiry projects in science teacher education: What can investigative experiences reveal about teacher thinking and eventual classroom practice? *Science Education, 87*(1), 112-143. doi: 0.1002/sce.10044

Zeichner, K. (2007). Accumulating knowledge across self-studies in teacher education. *Journal of Teacher Education*, *58*(1), 36–46. doi: 10.1177/0022487106296219

Zooinverse (n.d.). *Wildcam gorongosa.* Retrieved from https://www.wildcamgorongosa. org/#/

In: Enhancing STEM Motivation through Citizen Science Programs ISBN: 978-1-53616-038-3
Editors: Suzanne E. Hiller and Anastasia Kitsantas

Chapter 9

SCIENTIFIC DISCOVERY IN THE SCIENCE CLASSROOM VIA CONTRIBUTIONS TO CITIZEN SCIENCE

Carie Cardamone[1,*] and Bethany Cobb Kung[2]
[1]Tufts University, MA, US
[2]The George Washington University, DC, US

ABSTRACT

In this chapter, we describe an intervention designed to engage undergraduates enrolled in general education Astronomy with citizen science. Students often consider themselves lacking in scientific skills and unmotivated by inauthentic "cook-book" laboratory experiences. Citizen science provides a unique mechanism to integrate authentic scientific research into a course, bypassing barriers such as limited instructor time and background knowledge of students. We present a citizen science assignment designed to encourage students' intrinsic motivation and to be transparent to the learners. The assignment is based within the internet citizen science platform, The Zooniverse. Over the course of a month, students individually contributed to their selected Zooniverse project while reflecting on the experience. Then in groups, they explained the scientific purpose of the project and reflected on the impact of their contributions. After a trial implementation, we observed students perceiving themselves as capable of contributing to science and an increased depth in their understanding of the process of science. We reflected on the components of the assignment that were particularly key to its effectiveness and suggest that this assignment could be adapted for inclusion in many different undergraduate or K-12 curricula.

Keywords: assignment, astronomy, citizen science, general education, mindset, motivation, reflection, transparency

---

* Corresponding Author's Email: Carie.Cardamone@tufts.edu

Students enter our undergraduate classrooms with many preconceived notions about their capacity for learning science. Every student's prior exposure to science is unique, including both positive and negative experiences (Osborne, Simon & Collins, 2003). Unfortunately, previous bad experiences with science can lead some students to pessimistically conclude that they are simply "bad at" science.

This phenomenon – of students considering themselves incapable of "doing science" – is particularly endemic within general education undergraduate science classrooms. General education science courses vary widely in topic and include disciplines, such as biology, chemistry, physics, geology, astronomy, in addition to cross-disciplinary material. But these disparate subjects bely the fact that most of the students taking these courses have something profound in common: they have already made the decision (either consciously or unconsciously) that they *are not and never will be* scientists. In fact, while many general education courses are informally labeled "introductory" courses, this terminology is misleading since the vast majority of students will not be continuing on to higher-level science classes (Rowe et al., 2015). Indeed, these introductory courses represent the end of formal science learning for many students, not the beginning.

Almost every college or university requires that all students take a semester or two of science, regardless of their major; hence the need for general education science courses (Gaff, 1983; Warner & Koeppel, 2009). A common thread in general education science requirements is that science is interwoven into our modern society. Knowingly or unknowingly, we are all "consumers" of science on a daily basis. Therefore, citizens and leaders of the 21st century require a basic understanding and appreciation of the process of modern science, including its limitations. Even if students will not pursue science as a profession, they need basic science literacy to navigate their lives.

When students label themselves as "not a science person," they are often falling prey to a "fixed mindset" believing that scientific ability is a commodity that one is naturally born with (or not), rather than a set of skills that can be developed through study and practice (Dweck, 2008, 2016). This fixed mindset, combined with previous adverse science experiences, may cause students to view scientists as "others" to which they cannot relate. This way of thinking is, unfortunately, reinforced in the popular media. In TV and movies, scientists are often reduced to stereotypes (Chambers, 1983; Haynes, 2003; Long et al., 2010): lab-coated and brilliant but also awkward and antisocial. Even when scientists are cast in a positive light, their portrayal can be problematic if their genius seems completely foreign and unobtainable. The term "scientist" is often shorthand for "someone who can do anything." For example, a character who is an astrophysicist is also an expert in engineering, biology and chemistry. Stereotypical and unrealistic portrayals of scientists may be particularly problematic in the case of women and other historically-underrepresented minorities (Steele, 2010; Wyer, Schneider, Nassar-McMillan, & Oliver-Hoyo, 2010), given that the image of the scientists has almost always been limited to a white male swirling strange chemicals in a beaker.

How then can we move students toward a growth mindset about science and convince them that everyone is capable of understanding and even potentially contributing to science? To move them from seeing a scientist as an "other" to seeing a scientist as "self?" The goal is not to convince every student to become a professional scientist but to demonstrate to students that science is not entirely foreign, abstract or unknowable. Likewise, we want to encourage students to understand that science is not magical – that it has imperfections and uncertainties

– so that they can develop a healthy sense of skepticism. How can we ask our students to become, even in a limited way, scientists themselves? Enter citizen science.

CITIZEN SCIENCE

Citizen science projects create a venue for professional scientists and volunteers to work together to discover new knowledge and advance science. Science has a long history of particularly eager amateurs individually making significant contributions. For example, Mendel, now considered the father of modern genetics, was a monk whose work was initially ignored by the scientific community (De Castro, 2016). In contrast, modern citizen science projects are typically designed and maintained by professional scientists, but require the contributions of many individual citizens to enable new research. Typically, no special knowledge or skills are required of citizen science volunteers, though participants may have to take part in a brief training to prove that they understand the work. For example, they may have to successfully complete a task on a known set of sample data. The brain's hardwired ability to detect patterns and the goodwill of interested people are both key to the power of citizen science.

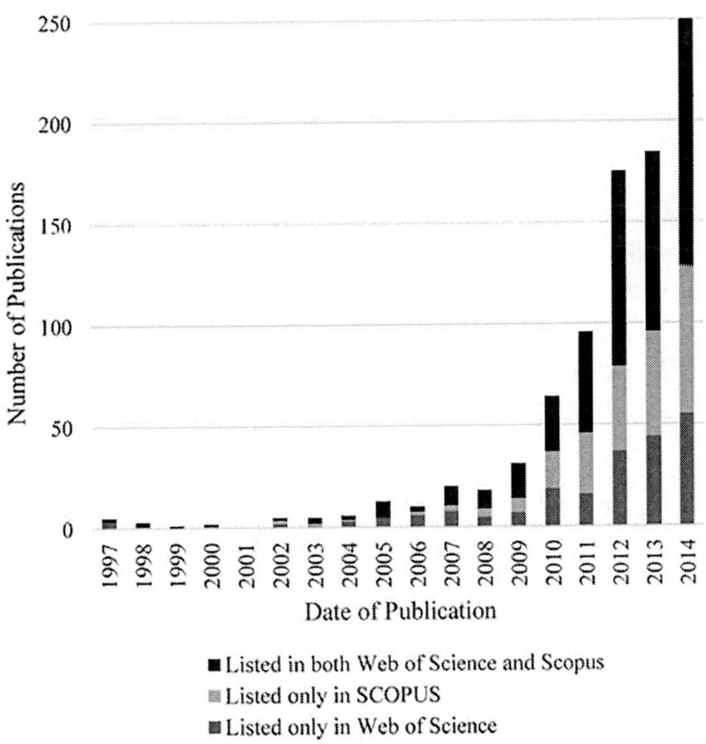

Figure 1. This chart, Figure 3 from Follett & Strezov (2015), shows the growth of published peer-reviewed articles on citizen science. Articles were collected via Web of Science and Scopus, peer-reviewed literature databases, and via searches for articles containing the topic "citizen science." Reprinted with permission.

Early examples of modern citizen science projects include bird, butterfly and frog counts led by organizations such as the National Audubon Society (n.d.), the North American Butterfly Association (2017) and the Association of Zoos and Aquariums (2018). Since 2006, the *Globe at Night* project (Scistarter, 2018a) has crowdsourced measurements of light pollution around the globe. Citizens have also helped scientists determine how proteins fold (Foldit, 2018).

Over the last decade, the proliferation of internet-connected devices has enabled citizen science to flourish. Anyone, anywhere, can now take part in citizen science if they have a laptop or smartphone with internet access. As a result, the number of scientific research projects that incorporate significant contributions from participants not formally trained in the sciences has expanded rapidly (Bonney et al., 2014). Citizen science projects typically involve either the collection of new scientific data, including observations of the natural world (e.g., Sullivan et al., 2014; The Cornell Lab of Ornithology, 2018) or the analysis of existing data that require either classification or interpretation (e.g., Fortson et al., 2011; Smith, Lynn, & Lintott, 2013; Zooniverse, n.d.). Citizen scientists have now contributed to hundreds of new scientific research articles published in journals across a dozen fields of study (Bonney et al., 2014; Follett & Strezov, 2015; Watson & Floridi, 2018). Figure 1 (from Follett & Strezov, 2015) shows the recent and rapid increase in the number of peer-reviewed publications related to citizen science. Citizen science is now regularly credited with facilitating important scientific discoveries.

Learning Science through Citizen Science

Although the primary motivation behind these citizen science projects is enabling new scientific discoveries, they also provide unique educational opportunities. Of note is the fact that citizens who participate in these projects have benefited from increased scientific knowledge (Bonney et al., 2014), and some studies have shown participation in citizen science can increase science literacy and positive attitudes toward science (Price & Lee, 2013). This suggests that citizen science might naturally have a place in the general education science classroom (Cardamone & Lobel, 2016; Kridelbaugh, 2016).

Citizen science activities that give students the opportunity to make a hands-on and valuable scientific contribution are aligned with research-based educational best-practices (Freeman et al., 2014; Olson & Riordan, 2012). Authentic participation in scientific research has been shown to be a powerful tool for student learning and engagement as it encourages self-identification with the act of doing science and stimulates student curiosity (Graham, Frederick, Byars-Winston, Hunter, & Handelsman, 2013). In particular, research experiences have been shown to increase student persistence and achievement, especially for students from historically underrepresented backgrounds (Espinosa, 2011; Hernandez, Schultz, Estrada, Woodcock, & Chance, 2013; Jones, Barlow, & Villarejo, 2010; Seymour, 2002; Seymour & Hewitt, 1997; Villarejo, Barlow, Kogan, Veazey, & Sweeney, 2008).

Developing authentic research projects for the introductory classroom from scratch is often too time consuming and resource intensive for an individual instructor to undertake. While many large sets of interesting data exist, they are often impractical for introductory students to use without significant individual mentoring because they require specialized knowledge and computer skills to analyze. Moreover, designing questions to investigate new and interesting areas requires a deep background in the discipline and area of study. Fortunately, the advent of modern citizen science means that there exists an abundance of available scientific research projects that are developed, vetted and maintained by researchers in a variety of disciplines. Undergraduate faculty can quickly and easily integrate authentic scientific research into an introductory science classroom using citizen science.

The Zooniverse

The Zooniverse represents one of the oldest, largest and most popular collections of "people powered research" projects (Zooinverse, n.d., see Figure 2). Citizen science projects on this platform are peer reviewed before launch and are designed to engage volunteers without any specialized knowledge or background, to produce measurable results, to encourage interactions between volunteers and scientists, and to inform the volunteers about all scientific results (Simpson, Page, & De Roure, 2014; Smith, et al., 2013). While the online interface and the specific scientific projects have changed and evolved since the release of the original *Galaxy Zoo* project in 2007, the platform's fundamental objectives have remained constant over time. By presenting volunteers with authentic research data and providing simple tools and instructions, volunteers can annotate data to produce scientifically useful results. For example, in the Zooniverse's *Snapshot Serengeti* project, participants are shown images taken from cameras set up by the researchers in the Serengeti. The cameras are meant to capture several images each time an animal moves in front of the sensor, though they sometimes trigger when no animal is present. If a participant does see an animal within a set of images, they identify it based on a menu of over 50 different types of animals such as buffalo, wildebeest, monkeys, and zebra. The project even includes a set of options to help narrow down the type of animal based on its general shape, color/pattern, and horn/tail shape. Researchers have used the resulting data to study the behavior of lions, hyenas, and cheetahs in the study area (Swanson, Arnold, Kosmala, Forester, & Parket, 2016).

The Zooniverse has over a decade's worth of experience with building and maintaining citizen science projects. Today, over one hundred projects have been implemented, spanning a diverse range of subjects from the humanities to the sciences – allowing instructors to target a variety of subjects, including applied topics like medicine, nature or climate (Zooinverse, n.d.). We selected this citizen science platform for use in our classrooms because each project includes extensive information about the project's background science, as well as forums for interaction with the project scientists and other volunteers, and blogs providing ongoing information about scientific results. The Zooniverse interface is also easy-to-use and is consistent between the many different projects. However, there are many other citizen science projects with a variety of different topics and contributions from the citizens that could be integrated into formal educational settings depending on the disciplinary focus and instructor-specific objectives for students learning. For example, Scientific American keeps an interactive database of ongoing projects as does SciStarter (Scientific American, 2018; SciStarter, 2018b).

Motivation

Under the expectancy-value theory (Atkinson, 1957; Eccles, 1983), motivation requires that individuals both attach some sort of personal value to a task and expect that they will succeed at that task. Students who have negative attitudes toward their personal scientific proficiency will, therefore, lack the motivation to do science because they believe that they will fail.

Fortunately, because citizen science projects have been designed to be accessible to volunteers without any specialized knowledge, even students with strong doubts about their own scientific capabilities can feel comfortable exploring and participating in these projects. To help motivate students, the history and purpose of citizen science should be explained before students dive into the task. In particular, it should be emphasized that citizen science has been designed to allow *non-scientists* to contribute to science in a meaningful way (Fortson, 2011; Franzoni & Sauerman, 2014).

When it comes to citizen science, the "value" side of the expectancy-value theory is trickier. In its purest form, citizen science is by definition an intrinsically-motivated activity since volunteers are moved to take part purely for their own internal interest or satisfaction. By bringing citizen science into the undergraduate classroom, faculty members are removing that aspect of personal choice. This adds an extrinsic motivational factor in the form of a grade that will be assigned, or pressure from peers to participate in the activity. While intrinsic motivation has been shown to result in "high-quality learning," extrinsic factors can sometimes work against learning because they cause students to resent or resist the task at hand (Ryan & Deci, 2000). As Ryan and Deci (2000) noted,

> because many of the tasks that educators want their students to perform are not inherently interesting or enjoyable, knowing how to promote more active and volitional (versus passive and controlling) forms of extrinsic motivation becomes an essential strategy for successful teaching. (p. 55)

Surveys of people who have voluntarily participated in citizen science show that the volunteers are primarily driven by the desire to contribute to real scientific research; they wish to help scientists accumulate knowledge (Evans et al., 2005; Raddick et al., 2010; Raddick et al., 2013). Some participants are so dedicated to this goal that they seek out or even build communities of like-minded peers (Fortson et al., 2011). These investigations clearly indicate that most people taking part in citizen science perceive their contributions as genuinely of benefit to science.

Students, when required to take part in citizen science, are unlikely to enter into such projects with the same frame of mind as volunteers. We hope, however, that as students participate over time, they will move from seeing the task as only having an extrinsic benefit (e.g., a grade) to having an intrinsic motivation in that they will feel good about being able to contribute to science. In bringing citizen science into the undergraduate classroom, we hope to give our students a taste of participating in real, modern science research and that they will also perceive their contributions as useful and beneficial to science.

Notably, the authentic nature of science in citizen science strongly contrasts with the inauthentic nature of many introductory science laboratory experiences. In a typical lab, students follow step-by-step instructions to reproduce already well-known results. While this type of entry-level lab might be a first step for students who plan to continue on in the sciences, other students might see rote and unoriginal exercises as nothing more than "busy-work." Even if students participate in more independent research experiences in a class context, those types of assignments must necessarily be limited in scope due to constraints of expertise, time, and equipment. Students are well-aware that such assignments are not "real science" because their work will only be seen by their instructor and possibly their peers – such work does not contribute to generalizable scientific knowledge. Hopefully, students recognize and respond positively to the authentic scientific contributions they can make through citizen science.

By tapping into student motivation, class activities designed to incorporate student participation in citizen science projects are potentially powerful ways to engage students in learning science and improve both their attitudes toward and performance in science. Motivation is a precursor to student engagement (emotional/affective, behavioral, and cognitive), which is a critical predictor of students' learning, persistence, and interest in a subject (Blumenfeld, Kempler & Krajcik, 2006; Christenson, Reschly, & Wylie, 2012). Moreover, enhancing student engagement has been shown to significantly enhance educational performance for all students but in particular, those who are at risk of failure. Hands-on activities, for example, are the most impactful in retaining students traditionally-underrepresented in STEM fields (Chang, 2002). Overall, many techniques for enhancing student engagement have been shown to be effective, including instructional approaches promoting student-student interactions, encouraging discussion and supporting the expression of the students' own viewpoints, as well as those that encourage students to construct meaning beyond the classroom (Christenson et al., 2012). Participation in citizen science is directly hands-on and because of its real-world implications, moves students to thinking beyond the classroom.

Designing a Course Citizen Science Assignment Assignment Overview

Citizen science provides a platform with which students in general education science courses can potentially improve or repair their personal scientific self-confidence and come to see science in a more positive light. To this end, we designed a Zooniverse-based citizen science assignment for Introductory Astronomy courses at two universities in the Eastern region of the United States. In each course, the assignment represented approximately 10 hours of work outside of class time, and 5-20% of a student's final grade. While this assignment was initially designed to fit within our Introductory Astronomy curricula, it can easily be modified for use within a diverse set of introductory courses, as the Zooniverse includes projects relevant to a variety of subjects (see Figure 2).

The assignment was broken down into two sections: an individual component and a group work component. The full text of the assignment is provided (see Appendix A) as is the rubric (see Appendix B) used to evaluate student work and determine their assignment grade. A detailed explanation of the assignment design and assignment components follows. The full text of the assignment and the associated grading rubric are provided in the appendices.

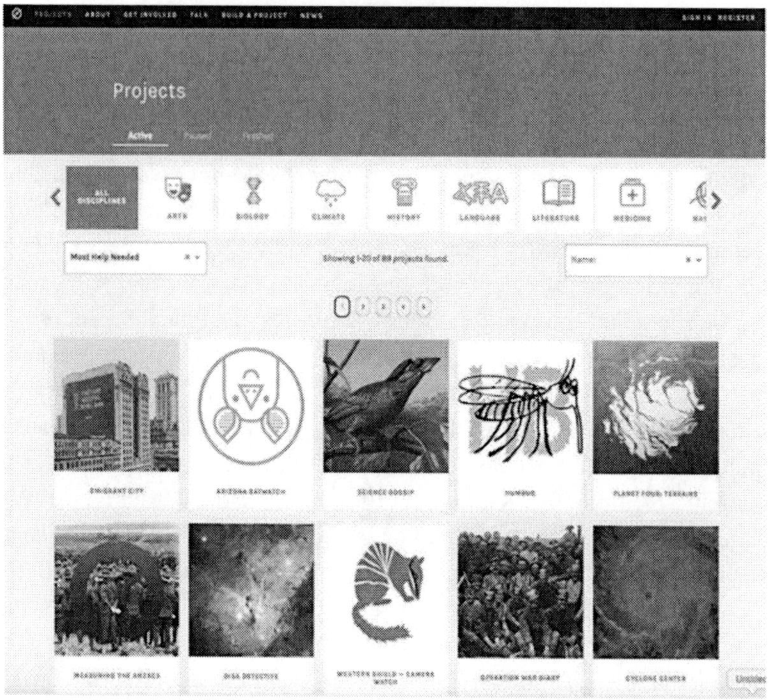

Figure 2. The Zooniverse projects page in October 2018, showing a variety of active citizen science projects. Zooinverse (n.d.). Retrieved from https://www.zooniverse.org/.

Design Philosophy - Transparency and Motivation

Faculty members often assume that their students understand the purpose of assignments and how to complete those assignments, but this is not always the case. Not every student is familiar with the "unwritten rules" of college. In transparent assignments (University of Nevada, 2018; Winkelmes, 2013), the purpose, tasks, and criteria for success are all explicitly defined. The purpose of the assignment should communicate, in jargon-free language, what students are meant to be learning through completion of the assignment. The listed tasks lay out what actions a student needs to take to complete the assignment and provides a clear task order when appropriate. Finally, clear criteria for success are provided. In this case, at the very beginning of the assignment period students were provided with the rubric (see Appendix B) that was to be used to grade their final work. For additional transparency, annotated examples of previous student work can be provided so that students can see exactly what form a successfully completed assignment might take.

Transparent assignments are particularly beneficial to students who are unfamiliar with college success strategies, including underrepresented and first-generation students (Winkelmes, Bernacki, Butler, Zochowski, Golanice, & Weavil, 2016). While our most capable and independent students may not require transparent assignments to achieve success, transparency can ensure that all students start on a more level playing field and is necessary when creating an inclusive environment in which all students can learn.

The assignment was designed to target students' intrinsic motivation. Transparency, by providing a clear purpose, can tap into intrinsic motivational factors as it helps students recognize that their own learning will benefit from the assignment. Because students are only likely to be motivated if they believe that they can be successful on a task, we were careful to introduce this assignment as a collaborative effort between professional scientists and the wider public, which required of the students no particular science aptitude beyond curiosity.

Providing opportunities for students to make choices and self-direct their learning can also enhance intrinsic motivation by giving students a greater sense of autonomy (Ryan & Deci, 2000). Therefore, the assignment incorporated choice by allowing students to select the Zooniverse project that they found most interesting. By selecting the project of their choice, students gain a sense of ownership and agency over the assignment (Christenson et al., 2012).

In addition to independent work, the assignment was designed to include a significant component of group work. By working together on a commonly chosen citizen science project, students may increase their sense of connectedness to that project. Teamwork also provides opportunities for students to engage in cooperative learning strategies.

Finally, a key component of the assignment design was the requirement that students relate the tasks they were doing while participating in citizen science to the project's ultimate scientific goals, and how those goals would advance collective knowledge. Relating student activities to the real-world taps into another source of intrinsic motivation.

Intervention: The Citizen Science Assignment Assignment Part One: In-Class Introduction and Individual Contributions

We introduced this assignment to our classes by framing citizen science as a collaborative effort between professional scientists and the wider public. We emphasized that citizen science is designed to enable new scientific research and discoveries and potentially has the side benefit of improving public understanding of science and the scientific process.

Students were then given some time to explore the Zooniverse website and the available projects. Our Introductory Astronomy students were restricted to selecting projects in either the space or physics categories. The only other limitation placed on project selection was the requirement that any selected project be less than 50% complete, to ensure that each project would remain up and running during the full period of the assignment. For the individual component, students contributed to a Zooniverse project over a period of approximately four weeks. This work was completed primarily outside of class time, although the assignment was introduced during class time and occasional in-class work time was also provided.

Over the assignment month, students were asked to visit and contribute to their Zooniverse project a minimum of eight separate times and to create a log noting each visit's date and time and a brief reflection on each visit. Visits did not have to take longer than 5-10 minutes, though some students spent more time than required when they were enjoying the task. In theory, students could have easily completed this component to the assignment during otherwise "dead" spaces in their schedule (between other classes, on the bus), especially if they had a smartphone and chose to participate in a project with mobile access. Some students opted to evenly space their visits throughout the month, completing two to three visits a week. Others chose to work intensely on the project over a few weeks, visiting daily. We required the visits to be spaced out in time to simulate how volunteers typically take part in a citizen science project and to allow students time between visits to begin researching the science behind their projects so that they could more fully grasp what they were actually accomplishing by participating in the task. While the assignment could easily be modified to require more visits, we recommend that students be required to make no fewer than about five visits. The student reflection logs showed that it took most students a few visits to become accustomed to the task and to gain confidence in their ability to successfully complete that task.

After each visit, students logged the visit date, the start and stop times of the visit, as well as writing a brief personal reflection on their experience. At one institution, students were asked to keep a digital record using the journal feature of the Blackboard course management system. At the second institution, students kept this journal either in a physical notebook or in electronic form simply using word processing software (e.g., Google Docs/Word). For the reflections, they were asked to note if they encountered anything particularly interesting or confusing. Reflections were only required to be a sentence or two and while some students kept their logs quite brief, paragraph length reflections were fairly common as well. Some students also used the logs to note their personal feelings about the project or task and to record images of interesting or exciting objects that they had encountered. We incorporated this journaling component into the assignment both to make certain that students were making multiple visits to their project and to ensure that they were thinking critically about the tasks in which they were engaged.

As a final step in the individual work, students took a screenshot of their Zooniverse recent projects webpage, which listed the total number of classifications that they had contributed to their project. By submitting this screenshot, they proved that they had been actively contributing to their citizen science project over that month for a total of at least 50 classifications. Most students easily exceeded this number during their visits.

Assignment Part Two: Group Video

After students selected their citizen science project, they formed groups of three or four students who had selected the same project on which to work. A few students preferred to work in partnerships or as individuals, which was also allowed. The groups worked together to try to understand and then articulate the basics of the science behind the citizen science project to which they were contributing – hence the need for all the students in a single group to be contributing to the same project. They also compared their own personal reactions to the projects, some of which proved quite common. By requiring students to investigate the science and to reflect on their participation, we meant to ensure that students were engaged with this material at a high cognitive level.

The final deliverable for the group work component of this assignment was a creative video presentation in which the group demonstrated their understanding of their particular Zooniverse project and in which they described their cumulative reflections on the task. Thanks to the proliferation of smartphones, most students had easy access to a video-enabled device. Some groups produced straightforward videos of themselves talking to the camera, while others used open-source web-based tools to make animations (e.g., Powtoons) or combined audio or text narration with slideshow images (e.g., PowerPoint). One group of students even filmed part of their video during a visit to the National Air and Space Museum.

Regardless of the chosen format, each video needed to answer four key questions about the students' participation in the citizen science project:

1. What is the scientific purpose of the citizen science project in which you participated?
2. What was the specific task (or tasks) that you did as a participant in this citizen science project?
3. Provide a *personal* reflection on the task: Was the task easy or hard? Was the task interesting or not? What did you enjoy the most and the least about the task? Did you encounter anything surprising? Confusing?
4. How did *your* participation help the citizen science project meet its science goals? How do the citizens' contributions to citizen science actually advance scientific knowledge?

A grading rubric (Appendix B) was provided to students at the onset of the assignment. The rubric included a points breakdown designed to clarify how students would be assessed on both the individual and group components of the project. The two components were each worth 50% of the final assignment grade in order to underscore their equal importance for student learning. For the group component, the rubric was designed to help guide the creation of the group video by emphasizing that students must address the four key questions and produce a clear and creative video, as well. When all materials were submitted, students received grades from the instructor using this detailed analytic rubric to provide feedback.

Faculty Reflection on the Assignment

We have utilized this Zooniverse-based citizen science assignment in several courses over a number of semesters in classes ranging from about 15-60 students. The students who have taken part in this assignment are primarily "traditional" undergraduates (18–22 years of age), who are not planning on majoring in a STEM field and who are enrolled in introductory science primarily to fulfill their general education graduation requirements.

In order to examine whether the assignment is meeting our pedagogical goals, such as tapping into student interest and intrinsic motivation and demonstrating to students that they are capable of contributing to real science, we reflect here on student responses to the assignment. Reflecting on teaching methods and students' experiences is a fundamental tool in the practice of the scholarship of teaching (Boyer 1990; Glassick, Huber, & Maeroff, 1997). Applying Schön's practice of Reflection-on-Action helps us to develop more effective practice as teachers (Calkins & Harris, 2017; Schön, 1992; Shulman, 1999).

Our reflection on this assignment was based primarily on the work completed and submitted by students. Specifically, the text students wrote as individuals in their journals/logs and the text, audio or visuals included in their group videos. The student log entries, in particular, provided unforeseen insight for each faculty member into their students' experiences with the assignment. Our reflection is also based on spontaneous discussions with students that occurred during class time or office hours. In order to maintain student confidentiality, we provide here our general impressions, rather than specific student quotes.

Of course, these reflections are based on *our* qualitative interpretations of what the students wrote or said and may be biased either because students only made positive comments in an attempt to "please" the instructors (who they knew would be grading their work) or because our reading of the journals are influenced by the fact that we assigned this work in the hopes that students would respond well to the assignment. That said, we believe that what the students wrote were mostly authentic reactions. Some entries were short and to the point, while others were much more extensive. The tone of many comments was neutral, but others were distinctly positive or negative. Finally, students frequently expressed interest and excitement in the assignment during informal (and ungraded) discussions with us during class time or office hours. In general, we believe that what students wrote in their logs and said in their videos support the effectiveness of this assignment in that the students took ownership of the citizen science task, demonstrated intrinsic motivation, and often expressed pride in their own contributions to science.

The vast majority of students fully completed the individual component, contributing to their project on at least eight separate occasions and completing a sufficient number of classifications. We found that providing several reminders over the assignment period and/or encouraging students to input project visits into their planners/calendars at the start of the assignment period was important, since the freedom to complete the visits at any time and the quick nature of the visits meant they were easy to overlook among other class work. While students sometimes commented that the visits themselves could be tedious, as we will describe below, there was no indication that the students found the multiple visits particularly onerous because the visits could be short and the log entries brief.

The group videos presented no more than the standard group assignment challenges, such as finding time in busy schedules for the group to meet and, of course, some unbalanced team member contributions. That said, while many of the videos submitted were relatively simple almost all were at least partially effective in answering the four key questions. A few of the videos were even extremely creative and beautifully produced.

Looking back at students' responses to this assignment, there are a number of key takeaways. In particular, when reading through student journals it became clear that the assignment contained two particularly critical components: requiring students to be able to articulate the scientific purpose of the citizen science and encouraging multiple student reflections on their participation in citizen science. This assignment would not have been effective without both of these components.

Thanks to the group video portion of the assignments, students were obligated to explore and understand, at least on a basic level, the scientific purpose behind the Zooniverse project to which they had contributed. What we found was that, in many cases, student journal reflections initially expressed confusion, uncertainty and often frustration with the citizen science task. After a few visits, the tone of the journal entries would shift to reflect student confidence in the specific task; in fact, it was extremely common for students at this point to remark that the task had become boring or tedious because of its repetitive nature. However, the most important shift would come when students started actually working on the assignment video. As students began to research the science behind their projects, their journal entries began to connect the tasks they were completing on the Zooniverse website with the actual science the project was hoping to accomplish. Happily, most student groups were able to articulate in their final videos the scientific purpose of their citizen science project, at least on a general level. This was critical because, in their journals, students only began to clearly describe the value and importance of their contributions once they understood the scientific purpose of their project.

An example of this progression can be seen in the groups that participated in the Zooniverse project, *Backyard Worlds: Planet 9* (n.d.). In this project, citizen scientists are shown night sky images that are the result of subtracting one image from a second image so that only differences between the two images remain. Alas, images rarely subtract perfectly leading to many image *artifacts* – basically meaningless errors – in the subtracted frames. The citizen scientists are tasked with differentiating between these artifacts and real objects that have shifted position between images. These moving objects are of interest because they might represent a new asteroid, comet or even planet. Initially, many students taking part in this project had no idea how to interpret what they were shown. Only after further research did the students' journal entries begin to reflect that they understood the purpose of the citizen science task. Had students not been prompted to research the background of the science by an extrinsic motivational factor (e.g., the assignment grade), they might never have sought out the information that helped them understand the task and connect it with the actual scientific research. Only once students understood the task and purpose of their citizen science project did they begin to express in their journals pride in having contributed to scientific knowledge, even in just a small way.

Working through these difficulties also gave students a mini-lesson in growth mindset. Over the course of the month, they basically all went from confused to competent in their chosen projects. The assignment directly demonstrated that with effort they could achieve understanding of the material and make a real contribution to science.

Reflection also proved to be an extremely important component of this assignment. Both the journal entries and the student videos included deep commentary on the students' experiences with citizen science through the assignment's duration. Many comments indicated that they found the work interesting; however, students also found the repetitive nature of the citizen science tasks to be tedious or boring at times.

It was in these reflections that we began to see evidence of students taking ownership of their chosen citizen science project, including expressions of pride in having contributed to "real" scientific research. As discussed previously, students are unlikely to have started the assignment particularly appreciating citizen science. In fact, most students were not even familiar with the concept of citizen science. In their reflections, however, they spontaneously began to write about the value they saw in citizen science, even when they found the tasks repetitive.

Interestingly, student journal narratives also frequently expressed significant discomfort with the "messiness" or "incompleteness" of the citizen science data. This might suggest that working with real, and often imperfect data, in the Zooniverse actually made students warier of scientific uncertainty. While it would be unfortunate if students became mistrustful of scientific data through this project, developing a healthy level of skepticism is beneficial given how science can be misrepresented in the media.

Such statements and realizations show that students are perhaps grasping for the first time some aspects of the scientific process, which includes the fact that not all progress is exciting or sexy and that scientific data can be messy and include errors and uncertainties. This is an achievement toward basic scientific literacy.

Student reflection was also vital because, in their journals and videos, students were obliged to use "I" statements to describe the process of science: "I classified," "I measured," "I categorized," etc. Through the Zooniverse citizen science platform, they were directly contributing to the creation of scientific knowledge. In describing their project, therefore, they had no choice but to become central to the narrative. This forced students to move from just imagining the scientist as simply "other" to acknowledging their own scientific capabilities. This is not to say that they suddenly pictured themselves as equal to the scientists who conceived of and built those citizen science projects, and no one (that we know of) suddenly expressed a desire to switch to a STEM major – but that was never our goal. We simply hoped to convince students that they are capable of contributing to science and feel that the assignment succeeded in that goal.

Finally, in some cases, students wrote that they were likely to continue participating in citizen science after the completion of the assignment. This indicates that they had moved from being entirely externally motivated to being intrinsically motivated, and we hope that some of our students do continue to make an impact through citizen science.

Educational Implications and Future Research

Based on our experience using this assignment in our classes, we believe that citizen science can provide students with a positive scientific experience that may help them to see themselves as capable of contributing to science, even in just a small way. We are also left with the impression that such an assignment helps students to appreciate the hard work that is required in science and to recognize that science is sometimes messy or imperfect (in other words, subject to uncertainties). For many of the students participating in this intervention, these classes represent the last time that they will formally interact with science. While it is unlikely that participating in citizen science in the classroom will cause a student to change his or her major, providing non-STEM majoring students with both a positive and realistic impression of science will help them to be better consumers of science in their lives.

Our investigation is currently limited by small class sizes. A larger sample of students completing pre and post surveys targeting students' attitudes towards science will allow for a statistical assessment of the impact of the experience. A larger sample might also allow analysis of any gender or other population differences in the experience of citizen science. To understand these results, it may also be necessary to measure student attitudinal changes that occur over the course of a month (the approximate time of the assignment) in a similar introductory science class in which no citizen science participation was included. This would help to isolate the effect of the assignment from any other course work.

Finally, we are sharing our assignment with faculty members in other STEM disciplines and with K-12 teachers. Additional research into the impact that this assignment might have on different age groups, or on students in different disciplines, could also provide interesting results. The use of citizen science with younger students is particularly compelling because if students have positive scientific experiences in K-12, they might be encouraged to pursue professions in STEM fields.

Conclusion

Citizen science projects, which have proliferated over the last decade, provide a unique opportunity for inclusively engaging students in introductory general education science classes. Citizen science projects give students direct, hands-on experience with research, tied to real scientific discoveries, without the need for any specialized knowledge or skills, which could inhibit participation. In this chapter we have presented an assignment designed to prompt intrinsic motivation, and through transparent design, engage students of diverse backgrounds in participation with scientific research.

Reflecting on our students' responses to this assignment, we noticed that while students are initially challenged by the citizen science tasks as they learn more about the science, they often come to appreciate that they can make a real contribution. For most students, working briefly on a citizen science project was a positive scientific experience. This type of assignment can easily be included in many different types of undergraduate level courses and could be modified for K-12 level, as well. Therefore, we propose citizen science as a valuable resource for instructors to bring real-world scientific investigations into their classrooms.

APPENDIX A

Full text of the "Contributions to Science" Assignment

Contribution to Science Assignment

Purpose

Citizen Science is a new movement creating collaborations between professional scientists and the wider public. These collaborations enable new scientific research and discoveries and improve public understanding of both science and the scientific process. Everyone can participate in Citizen Science, as the projects are designed so that no specialized knowledge or skills are required. Citizen Science harnesses the unique power and creativity of the human brain – unmatched by even the best computers – to recognize patterns and anomalies. In this activity, you will contribute directly to ongoing scientific research by participating in a Citizen Science Project. In doing so you will gain an understanding of and appreciation for a unique method of modern science.

Task

You will be responsible for *independently* contributing to a Citizen Science Project via "The Zooniverse" (www.zooniverse.org) over a period of 3 to 4 weeks (you will keep a log/journal of your work, as described below). You will then work with a group to understand and articulate the scientific purpose of the project and to produce a short (approximately 5 minute) video presentation that demonstrates your understanding by answering the key questions listed below. You can use any video method that you want. You can film yourself, or make an animation, or provide narration over still images, etc. – be creative!

Classification Log

You are required to visit your Zooniverse project and contribute "classifications" *a minimum* of 8 times over the period of the assignment. Each visit need only be 10 – 15 minutes in length. Typically, you would be able to classify 5 to 15 objects (or more, as you grow more

Appendix A (Continued)

skilled) during that time. Each visit must occur on a different day – ideally you will work on your project 2 to 3 times a week over the project period. Keep a simple digital log of your work using your "*Assignment Journal*" in Blackboard. Each time you visit the Zooniverse project, create a *new* journal entry to log your visit. In that journal entry, list the date and (approximate) work start/stop times **and** also include a *brief* reflection on each visit, being sure to note if you encounter anything interesting or confusing (a few sentences is sufficient). Once you have completed all your visits, take a screenshot of your "stats" page (go to "Profile" → "Your Stats"). Make sure your username, the phrase "you have made X classifications to date," and the project icon under "Recent projects" is visible in this screenshot (you may need to zoom out in your web browser). Over at least 8 visits, you should have classified a minimum of 50 objects (though most students can easily classify many more). Upload this screenshot into your journal (using the Insert/Edit Image function) as your final journal entry.

Video Presentation Key Questions

1. What is the scientific purpose of the Citizen Science Project in which you participated – in other words, what are the overarching science goals that the project is trying to achieve? What specific question (or questions) are the scientists seeking to answer? How will answering those questions advance scientific knowledge?
2. What was the specific task (or tasks) that you did as a participant in this Citizen Science Project? Show an example of this task. Explain why humans are better suited for performing this task than computers.
3. Provide a *personal* reflection on the task: Was the task easy or hard? Was the task interesting or not? What did you enjoy the most and the least about the task? Did you encounter anything surprising? Confusing? [Be sure include reflections that incorporate the personal experiences from *every* member of your team, especially if your experiences with the task differed significantly from one another.]
4. How did *your* participation help the Citizen Science Project meet its science goals? (In other words, how do the citizens' contributions to Citizen Science actually advance scientific knowledge?)

Assignment Overview

- **Step #1**: In class, you will select a project and a group (of 1 to 4 students). You may want to spend some time beforehand looking at the projects available at www.zooniverse.org in either the "Space" or "Physics" categories. The project you chose must be open for contributions (sort by "Most Help Needed") and must *be less than 50% complete* (so that the project doesn't end before you have time to make your contribution).
- **Step #2:** Add to your planner/calendar a minimum of 8 visits to your Zooniverse project or decide on a regular schedule of work (many projects can be worked on via your smartphone, so you might plan on working on the project during your commute or between other class times). Schedule a time to meet with your group to discuss the answers to the "key questions" and how you will make your video presentation (what

style of video, what hardware/software you will use, what everyone will specifically contribute, etc.). Finally, schedule a time when the group will work together to complete the video!

- **Step #3:** Visit and contribute classifications to your Zooniverse project at least 8 times. Each time you complete classifications, record the date and start/end times in your project journal on Blackboard. Include a brief reflection on your participation, being sure to note if you encounter anything interesting or confusing. Remember, each visit must be on a different day. (*You can't wait until the last minute to complete your classifications.*)
- **Step #4:** Once you have visited and contributed to your Zooniverse project a minimum of 8 times, take a screenshot of your "stats" page. (The "stats" page takes a few minutes to update after you have completed observations, so wait at least a half hour after completing your last classification to take a screenshot.) Upload this screenshot into a final entry in your project journal.
- **Step #5:** Work with your group to complete the video presentation. Look carefully over the rubric to ensure that your video satisfies all the criteria for success including creativity! Be sure your presentation is well organized and clear.
- **Step #6:** Share your video with your instructor through the assignment link on Blackboard.

Criteria for Success

The project will be graded via a rubric (see below). Each student will receive an *independent* score for the first section of the rubric (Individual Classification/Logs) and a *group* score for the other criteria of the rubric. The project is worth a total of 60 points.

APPENDIX B

Grading Rubric for Contributions to Science Assignment

Required Content	***Description of Requirement***	**Excellent:**	**Satisfactory:**	**Poor:**	**Failing:**
Individual Classifications and Log/Screenshot	*Visit and contribute to your Zooniverse project at least 8 times throughout the project period (approximately 2 to 3 visits per week), spending at least 10–15 minutes or so per visit. Visit dates/times and reflections must be recorded in a log and a screenshot of your final stats must be included [if no screenshot is submitted, your classification # = 0].*	8+ visits and at least 50 classifications	5 – 7 visits and least 50 classifications	2 – 4 visits or fewer than 50 classifications	only 1 visit (regardless of classificatio n #)
		15 points	10	5	0
		Log includes detailed reflections (e.g. interesting / surprising objects)	Log includes reflection, but the reflection lacks detail or is poorly written	Log is missing personal reflections	No log is submitted
		15 points	10	5	0

APPENDIX B (CONTINUED)

Required Content	***Description of Requirement***	**Excellent:**	**Satisfactory:**	**Poor:**	**Failing:**
Clarifies the scientific purpose of the project.	*What are the overarching science goals that the Citizen Science Project is trying to achieve? What specific questions are the scientists seeking to answer? How will answering those questions advance scientific knowledge?*	Outstanding discussion of scientific purpose; clear connections to advancing knowledge	Scientific questions and purpose are identified	Discussion of scientific purpose unclear or underdevelop ed	Not addressed
		10 points	7	3	0
Identifies specific tasks	*What was the specific task (or tasks) that you did as a participant in this Citizen Science Project? Show an example of this task. Explain why humans are better suited for performing this task than computers.*	Outstanding discussion of project task, clearly presented with an example demonstration	Project task described, with an example demonstration	Discussion of specific tasks unclear or underdevelop ed or no example	Not addressed
		5 points	4	2	0
Reflects on participation task	*Reflect on the task. Was the task easy or hard? Was the task interesting or not? What did you enjoy the most and the least about the tasks? Did you encounter anything surprising? Confusing?*	Detailed reflections incorporate personal experiences from every member of your team	Reflections clearly address a variety of perspectives towards the task	Reflections on participation unclear or underdevelop ed	Not addressed
		5 points	4	2	0
Connects individual contributions to broader project goals	*How do the citizens' contributions to Citizen Science actually advance scientific knowledge?*	Presents a clear and complete connection between participation and the project's science goals	Presents some connection between participation and the project's science goals	Connections unclear or underdevelop ed	Not addressed
		5 points	4	2	0
Project demonstrates creativity and clarity	*Final A/V presentation demonstrates creativity in design and is well-organized and clearly communicates your knowledge.*	Presents information in a way that is enjoyable to watch; well organized, each key question is clearly answered	Presents information in a way that is enjoyable to watch; some minor flaws in organization	Presents information in a way that is lacking in creativity or includes major flaws in organization	Confusing presentation, no creativity, demonstrate s little thought or time spent on it
		5 points	4	2	0
Overall Assignment Grade		**60 points**			

REFERENCES

Association of Zoos and Aquariums (2017). Retrieved from www.aza.org

Atkinson, J. W. (1957). Motivational determinants of risk-taking behavior. *Psychological Review*, *64*(6, Pt.1), 359–372. doi:10.1037/h0043445

Backyard worlds: Planet 9 (n.d.) Retrieved from https://www.zooniverse.org/projects /marckuchner/backyard-worlds-planet-9

Blumenfeld, P. C., Kempler, T. M., & Krajcik, J. S. (2006). Motivation and cognitive engagement in learning environments. In K. Sawyer (Ed.), *The Cambridge handbook of the learning sciences* (2nd ed., pp. 475–88). New York, NY: Cambridge University Press.

Bonney, R., Shirk, J. L., Phillips, T. B., Wiggins, A., Ballard, H. L., Miller-Rushing, A. J., & Parrish, J. K. (2014). Next steps for citizen science. *Science*, *343*(6178), 1436-1437. doi:10.1126/science.1251554

Boyer, Ernest L. (1990). *Scholarship reconsidered: priorities of the professoriate.* Lawrenceville, NJ: Princeton University Press. Retrieved from https://eric.ed.gov/?id=ED326149

Calkins, S., & Harris, M. (2017). Promoting critical reflection: An evaluation of the longer-term impact of a substantial faculty development program. *The Journal of Faculty Development, 31*(2), 29-36.

Cardamone, C., & Lobel, L. (2016). Using citizen science to engage introductory students: From streams to the solar system. *Journal of Microbiology & Biology Education*, *17*(1), 117-119. doi:10.1128/jmbe.v17i1.1082

Chambers, D. W. (1983). Stereotypic images of the scientist: The draw-a-scientist test. *Science Education*, *67*(2), 255-265. doi:10.1002/sce.3730670213

Chang, J. C. (2002). *Women and minorities in the science, mathematics and engineering pipeline. ERIC Digest.* Retrieved from ERIC database. (ED386327)

Christenson, S. L., Reschly, A. L., & Wylie, C. (2012). *Handbook of research on student engagement.* New York, NY: Springer Science & Business Media.

De Castro, M. (2016). Johann Gregor Mendel: Paragon of experimental science. *Molecular Genetics & Genomic Medicine*, *4*(1), 3-8. doi:10.1002/mgg3.199

Dweck, C. S. (2008). *Mindsets and math/science achievement.* New York, NY: Carnegie-IAS Commission on Mathematics and Science Education. Retrieved from https://lincs.ed.gov/professional-development/resource-collections/profile-534

Dweck, C. S. (2016). *Mindset: The new psychology of success* (2nd ed.). New York, NY: Penguin Random House.

Eccles, J. (1983). Expectancies, values, and academic behaviors. In J. T. Spence, (Ed.), *Achievement and achievement motives: Psychological and sociological approaches* (pp. 75-146). San Francisco, CA: W. H. Freeman.

Espinosa, L. (2011). Pipelines and pathways: Women of color in undergraduate STEM majors and the college experiences that contribute to persistence. *Harvard Educational Review*, *81*(2), 209-241. doi:10.17763/haer.81.2.92315ww157656k3u

Evans, C., Abrams, E., Reitsma, R., Roux, K., Salmonsen, L., & Marra, P. P. (2005). The neighborhood nestwatch program: Participant outcomes of a citizen-science ecological research project. *Conservation Biology*, *19*(3), 589-594. doi:10.1111/j.1523-1739.2005.00s01.x

Foldit (2018). Solve puzzles for science. Retrieved from fold.it/portal

Follett, R., & Strezov, V. (2015). An analysis of citizen science based research: Usage and publication patterns. *PLOS ONE*, *10*(11), e0143687. doi:10.1371/journal.pone.0143687

Fortson, L., Masters, K., Nichol, R., Borne, K., Edmondson, E., Lintott, C., Wallin, J. (2011). Galaxy Zoo: Morphological classification and citizen science. *ArXiv:1104.5513 [Astro-Ph]*. Retrieved from http://arxiv.org/abs/1104.5513

Franzoni, C., & Sauermann, H. (2014). Crowd science: The organization of scientific research in open collaborative projects. *Research Policy* *43*(1), 1-20. doi:10.1016/j.respol. 2013.07.005

Freeman, S., Eddy, S. L., McDonough, M., Smith, M. K., Okoroafor, N., Jordt, H., & Wenderoth, M. P. (2014). Active learning increases student performance in science, engineering, and mathematics. *Proceedings of the National Academy of Sciences*, *111*(23), 8410-8415. doi:10.1073/pnas.1319030111

Gaff, J. G. (1983). *General education today: A critical analysis of controversies, practices, and reforms* (1st ed.). San Francisco, CA: Jossey-Bass Inc Pub.

Glassick, C. E., Huber, M. T., & Maeroff, G. I. (1997). *Scholarship Assessed: Evaluation of the Professoriate. Special Report.* San Francisco, CA: Jossey Bass Inc.

Graham, M. J., Frederick, J., Byars-Winston, A., Hunter, A.-B., & Handelsman, J. (2013). Increasing persistence of college students in STEM. *Science*, *341*(6153), 1455-1456. doi:10.1126/science.1240487

Haynes, R. (2003). From alchemy to artificial intelligence: Stereotypes of the scientist in Western literature. *Public Understanding of Science*, *12*(3), 243-253. doi:10.1177/ 0963662503123003

Hernandez, P. R., Schultz, P. W., Estrada, M., Woodcock, A., & Chance, R. C. (2013). "Sustaining optimal motivation: A longitudinal analysis of interventions to broaden participation of underrepresented students in STEM": Correction to Hernandez et al. (2013). *Journal of Educational Psychology*, *105*(4), 1025-1025. doi:10.1037/a0034254

Jones, M. T., Barlow, A. E. L., & Villarejo, M. (2010). Importance of undergraduate research for minority persistence and achievement in Biology. *The Journal of Higher Education*, *81*(1), 82-115. doi:10.1080/00221546.2010.11778971

Kridelbaugh, D. M. (2016). The use of online citizen-science projects to provide experiential learning opportunities for nonmajor science students. *Journal of Microbiology & Biology Education*, *17*(1), 105-106. doi:10.1128/jmbe.v17i1.1022

Long, M., Steinke, J., Applegate, B., Knight Lapinski, M., Johnson, M. J., & Ghosh, S. (2010). Portrayals of male and female scientists in television programs popular among middle school-age children. *Science Communication*, *32*(3), 356-382. doi:10.1177/ 1075547009357779

National Audobon Society (n.d.). Retrieved from www.audubon.org

North American Butterfly Association (2017). Retrieved from www.naba.org

Olson, S., & Riordan, D. G. (2012). *Engage to Excel: Producing one million additional college graduates with degrees in science, technology, engineering, and mathematics. Executive Office of the President. ERIC Digest.* Retrieved from ERIC database. (ED541511)

Osborne, J., Simon, S., & Collins, S. (2003). Attitudes towards science: A review of the literature and its implications. *International Journal of Science Education*, *25*(9), 1049-1079. doi:10.1080/0950069032000032199

Price, C. A., & Lee, H. (2013). Changes in participants' scientific attitudes and epistemological beliefs during an astronomical citizen science project. *Journal of Research in Science Teaching*, *50*(7), 773-801. doi:10.1002/tea.21090

Raddick, M. J., Bracey, G., Gay, P. L., Lintott, C. J., Murray, P., Schawinski, K., Vandenberg, J. (2010). Galaxy Zoo: Exploring the motivations of citizen science volunteers. *Astronomy Education Review*, *9*, 010103. doi:10.3847/AER2009036

Raddick, M. J., Bracey, G., Gay, P. L., Lintott, C. J., Cardamone, C., Murray, P., . . . Vandenberg, J. (2013). Galaxy Zoo: Motivations of citizen scientists. *Astronomy Education Review*, *12*(1). doi:10.3847/AER2011021

Rowe, M. P., Gillespie, B. M., Harris, K. R., Koether, S. D., Shannon, L.-J. Y., & Rose, L. A. (2015). Redesigning a general education science course to promote critical thinking. *CBE—Life Sciences Education*, *14*(3), ar30. doi:10.1187/cbe.15-02-0032

Ryan, R. M., & Deci, E. L. (2000). Intrinsic and extrinsic motivations: Classic definitions and new directions. *Contemporary Educational Psychology*, *25*(1), 54-67. doi:10.1006/ceps.1999.1020

Scientific American (2018). Citizen science. Retrieved from https://www.scientificamerican.com/citizen-science/

SciStarter (2018a). Globe at night. Retrieved from https://scistarter.com/project/169-Globe-at-Night

SciStarter (2018b). What is citizen science? Retrieved from https://scistarter.com/citizenscience.html

Schön, D. A. (1992). *The reflective practitioner: How professionals think in action*. London, United Kingdom: Routledge. doi:10.4324/9781315237473

Seymour, E. (2002). Tracking the processes of change in US undergraduate education in science, mathematics, engineering, and technology. *Science Education*, *86*(1), 79-105. doi:10.1002/sce.1044

Seymour, E., & Hewitt, N. M. (1997). *Talking about leaving: Why undergraduates leave the sciences* (Vol. 12). Boulder, CO: Westview Press.

Shulman, L. S. (1999). Taking learning seriously. *Change: The Magazine of Higher Learning, 31*(4): 10-17. doi:10.1080/00091389909602695

Simpson, R., Page, K. R., & De Roure, D. (2014). Zooniverse: Observing the World's Largest Citizen Science Platform. In *Proceedings of the 23rd International Conference on World Wide Web* (pp. 1049–1054). New York, NY, USA: ACM. doi:10.1145/2567948.2579215

Smith, A. M., Lynn, S., & Lintott, C. J. (2013). An Introduction to the Zooniverse. In *First AAAI Conference on Human Computation and Crowdsourcing*. Retrieved from https://www.aaai.org/ocs/index.php/HCOMP/HCOMP13/paper/view/7520

Steele, C. M. (2010). *Whistling Vivaldi: How stereotypes affect us and what we can do*. New York, NY: WW Norton & Co.

Sullivan, B. L., Aycrigg, J. L., Barry, J. H., Bonney, R. E., Bruns, N., Cooper, C. B., . . . Kelling, S. (2014). The eBird enterprise: An integrated approach to development and application of citizen science. *Biological Conservation*, *169*, 31-40. doi:10.1016/j.biocon.2013.11.003

Swanson, A., Arnold, T., Kosmala, M., Forester, J., & Packer, C. (2016). In the absence of a "landscape of fear": How lions, hyenas, and cheetahs coexist. *Ecology and Evolution*, *6*(23), 8534–8545. doi:10.1002/ece3.2569

The Cornell Lab of Ornithology (2018). eBird. Retrieved from https://ebird.org/home

University of Nevada (2018). Transparency in learning and teaching in higher ed. Retrieved from www.unlv.edu/provost/teachingandlearning

Villarejo, M., Barlow, A. E. L., Kogan, D., Veazey, B. D., & Sweeney, J. K. (2008). Encouraging minority undergraduates to choose science careers: Career paths survey results. *CBE—Life Sciences Education*, *7*(4), 394-409. doi:10.1187/cbe.08-04-0018

Warner, D. B., & Koeppel, K. (2009). General education requirements: A comparative analysis. *The Journal of General Education*, *58*(4), 241-258. Retrieved from http://www.jstor.org/stable/25702446

Watson, D., & Floridi, L. (2018). Crowdsourced science: Sociotechnical epistemology in the e-research paradigm. *Synthese*, *195*(2), 741-764. doi:10.1007/s11229-016-1238-2

Winkelmes, M. A. (2013). Transparency in teaching: Faculty share data and improve students' learning. *Liberal Education*, 99(2). Retrieved from https://www.aacu.org/publications-research/periodicals/transparency-teaching-faculty-share-data-and-improve-students

Winkelmes, M. A., Bernacki, M., Butler, J., Zochowski, M., Golanics, J., & Weavil, K. H. (2016). A Teaching intervention that increases underserved college students' success. *Peer Review, 18*(1/2), Retrieved from https://www.aacu.org/peerreview/2016/winter-spring/Winkelmes

Wyer, M., Schneider, J., Nassar-McMillan, S., & Oliver-Hoyo, M. (2010). Capturing stereotypes: Developing a scale to explore U.S. college students' images of science and scientists. *International Journal Gender, Science and Technology*, *2*(3). Retrieved from http://genderandset.open.ac.uk/index.php/genderandset/article/view/78

Zooinverse (n.d.). Retrieved from https://www.zooniverse.org/

SECTION IV: CITIZEN SCIENCE PROGRAMS AND STEM EDUCATION AND CAREERS

In: Enhancing STEM Motivation through Citizen Science Programs ISBN: 978-1-53616-038-3
Editors: Suzanne E. Hiller and Anastasia Kitsantas

Chapter 10

I DIDN'T KNOW WHAT REAL SCIENCE WAS OR WHAT IT COULD BE: CITIZEN SCIENCE AND INTEREST IN STEM EDUCATION AND CAREERS

***Michele Hollingsworth Koomen*[1,*], *Mary Norell Hedenstrom*[2], *Molly Moran*[3] and *Karen S. Oberhauser*[4]**
[1]Gustavus Adolphus College, MN, US
[2]University of Minnesota, MN, US
[3]Burnsville-Eagan-Savage School District, MN, US
[4]University of Wisconsin-Madison Arboretum, WI, US

ABSTRACT

Using the lens of social cultural career theory (SCCT), the purpose of the present study was to assess the role participation in informal citizen science played in youth's STEM educational and career choices. We examined fourteen student responses to semi-structured interviews and survey instruments from eighteen students for evidence of the SCCT constructs self-efficacy, outcome expectations, personal goals, interest, and learning experiences. Findings from this descriptive study showed positive self-efficacy in STEM courses and careers for youth who participated in an informal citizen science project. Through participation in citizen science, students experienced the relevance of contributing to an authentic science investigation, the rigor of citizen science protocols, and relationships with teacher mentors. Overall, the findings underscore the role of teacher mentors and authentic engagement in science in changing students' perceptions of science. Educational implications are discussed.

Keywords: citizen science, *Monarch Larva Monitoring Project,* social cognitive career theory, STEM careers

---

* Corresponding Author's Email: mkoomen@gustavus.edu

The United States is projected to have a labor shortage in the fields of Science, Technology, Engineering, and Mathematics (STEM) of at least one million professionals, more than it can supply in the next decade (Miller, Sonnert, & Sadler, 2018). The shortage of STEM professionals is predicted to seriously impact the ability of the United States to compete internationally in an economy with a backbone heavily dependent on advanced technology and science (Chen & Weko, 2009). To meet the growing demand for highly skilled STEM professionals, more students ought to be interested in pursuing STEM educational goals. Many studies have documented the importance of students' attitudes toward STEM as well as their self-efficacy (Lent & Brown, 2008). Students' interest and motivation promote STEM career choice (Hidi & Renninger, 2006). There is a developing body of research that demonstrates that outside of school experiences have the potential to cultivate interest in STEM (Bell, Lewenstein, Shouse, & Feder, 2009), including informal programs using citizen science (Hiller & Kitsantas, 2014).

Citizen science (CS) involves partnerships between volunteers and scientists where scientists design protocols to address real world questions. Participation in CS projects enhances participant engagement with scientific topics (Trumbull, Bonney, Bascom, & Cabral, 2000), promotes scientific thinking skills (Koomen, Hoffman, Schultz, Peterson, & Oberhauser, 2018), and leads to increased science knowledge (Bonney et al., 2009). However, much of this research has focused on adults in informal science education settings with little focus on the value of CS as a mechanism that might foster interest in STEM education or careers. For example, we wondered how citizen science, and projects like the *Monarch Larva Monitoring Project* (The University of Minnesota Monarch Lab, MLMP, n.d.), might inspire middle school students toward STEM. Rotman et al., (2012) used a survey to measure motivations of citizen scientists engaged in ecological studies and learned that citizens were motivated by enjoyment, but scientists were motivated by status or professional gain. Curtis (2015) also used a survey to understand the motivation of adult participants in an online CS project deciphering the 3-dimensional configuration of proteins. Participants cited the desire to engage with a scientific community and solve a problem for science as the strongest motivators. Our study builds off the work of Hiller and Kitsantas (2014) who documented the connection between a citizen program and STEM career motivation in middle school students.

Theoretical Framework

Social cognitive career theory. The theoretical framework for this study is social cognitive career theory (SCCT), a model developed by Lent, Brown, and Hackett (1994). SCCT helps researchers and career counselors understand factors that influence career choice decisions and evaluate career satisfaction. We use the SCCT lens to assess the role CS can play in STEM educational and career choices for youth (Miller, Sonnert, & Sadler, 2018). For the purpose of our study, we use the SCCT constructs of (a) self-efficacy, (b) outcome expectations, (c) personal goals, (d) interests, and (e) learning experiences (Hiller, 2012).

Self-efficacy. Described as self-efficacy, this term refers to a person's belief in their ability to successfully meet career goals which can be moderated by environmental conditions in the classroom (Lent, Brown, & Hackett, 2002). For this study, self-efficacy is described in terms of mastery experiences, vicarious experiences, social persuasion, and affective states (Hiller,

2012). Hiller and Kitsantas (2016) identify the value of developing self-efficacy scales to measure middle school students' self-efficacy in citizen science contexts since high self-efficacy has led to increased interest in science and STEM careers. *Mastery experiences* are opportunities for students to practice science skills and reflect and assess their own capability in relation to science practice (Britner & Pajares, 2006). We argue that CS provides a context for students to engage in mastery experiences. *Vicarious experiences* provide opportunities for students to observe role models and mentors engaged in skilled procedures or behaviors (Pacheco-Guffrey, 2018). Through *social persuasion*, individuals experience greater self-efficacy as a result of the self-perceived feedback from others whose opinion they care about, including peers and mentors (Hiller, 2012). *Affective or physiological states*, sometimes referred to as anxiety (Bandura, 1997), are negative factors which, when reduced, increase self-efficacy. Emotional and physiological states in the context of CS may be present as doubt about one's ability to master an observational protocol.

Outcome expectations. The term outcome expectations refers to the belief that one can successfully complete a work-related task or behavior. This expectation can help individuals meet their goals and look beyond current difficulties to future outcomes (Soldner, Rowan-Kenyon, Inkelas, Garvey, & Robbins, 2012). Outcome expectations can help students see the importance and value of their work (Lent & Brown, 2008) and, for adults, provide job satisfaction as individuals experience personal well-being as a result of job mastery recognized by self and others (Lent & Brown, 2008).

Personal goals. Individual educational or career choices are known as personal goals (Lent et al., 2000; Soldner et al., 2012). We use the term personal goals to describe students' choices for STEM courses and careers.

Interest. As a result of numerous experiences, interest leads to motivation to engage in an activity (Hiller, 2012). These interests can promote actions in pursuit of goals (Lent et al., 2000), but they are dependent on many environmental factors. In this study, we look at students' interest related to STEM courses and careers.

Learning experiences. In academic settings, learning experiences can be measured in multiple ways, including grades (Lent et al., 2000; Patrick, Care, & Ainley, 2011) or multiple-choice test scores (Hiller 2012). Because accurate scientific protocols are required to ensure the validity of data submitted to CS databases, we assessed learning experiences through students' ability to accurately complete MLMP protocols, including observations and data collection.

We chose SCCT as a theoretical lens to analyze the personal factors inherent in individuals' career selection process, and we are interested in understanding how personal factors can support or discourage choices of STEM education and careers. We used this theory to study how personal characteristics and backgrounds influence learning experiences, which influence self-efficacy and outcome expectations, which lead to interests, goals, and actions. We hypothesized that youth participation in a summer MLMP CS experience would provide opportunities for students to build their capacity to do science and foster their interest in pursuing educational and career goals in STEM fields. Our study was guided by the following research questions:

1. In what ways does engaging in a citizen science program provide opportunities for participants to build their capacity to do science?
2. In what ways does engaging in a citizen science program foster interest in pursuing educational and career goals in STEM?

Figure 1 illustrates how the various constructs for SCCT relate to our two research questions. The constructs of self-efficacy, including all four domains and learning experience, are aligned with the first research question while the constructs of outcome expectation, personal goals, and interests are tied with research question two.

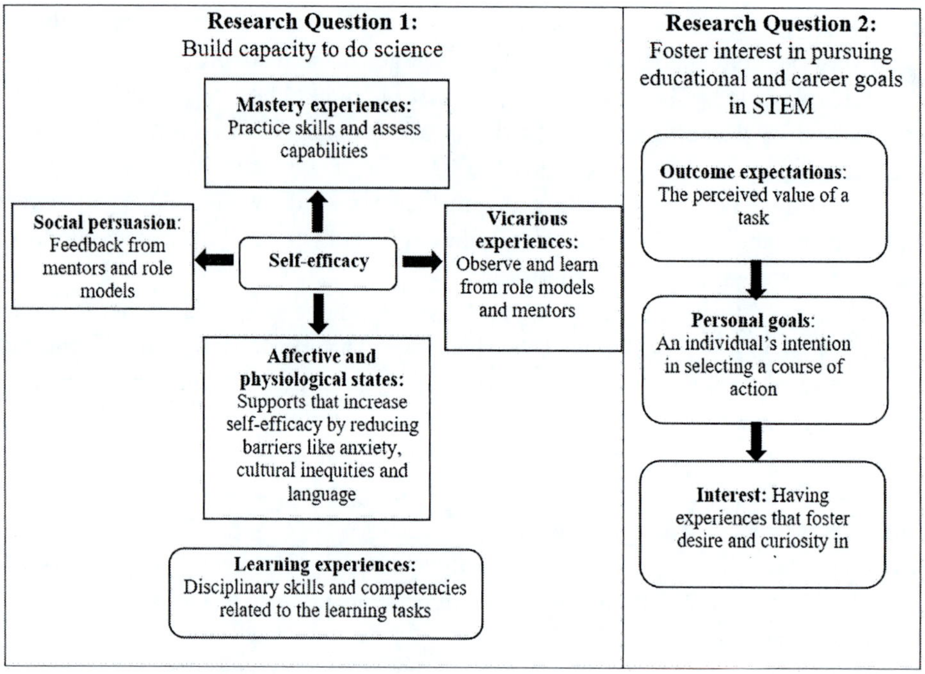

Figure 1. The various constructs of SCCT used in this study and their alignment with our two research questions.

METHODS

Context

The *Monarch Larva Monitoring Project* (MLMP) was begun by graduate student Michelle Prysby and faculty member Karen Oberhauser at the University of Minnesota in 1996. The goals are to understand the factors that affect monarch reproduction and development during the breeding season and to enhance volunteers' appreciation and concern for monarchs and their habitat. Volunteers conduct up to six data collection activities. They provide annual (a) site descriptions and (b) estimates of milkweed density, and (c) conduct weekly surveys that provide per plant densities of monarch eggs and larvae. Other activities include (d) comparing plants occupied by monarchs to random plants, (e) measuring rates of parasitism by parasitoids by collecting and rearing larvae, and (f) collecting rainfall data. Volunteers choose the activities in which they participate and also their monitoring sites, which include schoolyard and backyard gardens, abandoned fields, pastures, and restored prairies located throughout the monarch's breeding range.

In this study, teachers demonstrated the use of MLMP protocols, such as random sampling techniques to count monarchs on milkweed plants in identified monitoring sites. Student pairs followed a randomly-selected transect line and observed milkweed plants within one meter of the line. The observer identified the stage (egg, or larva to instar) of all monarchs observed, and a partner documented the observations on a recording sheet. The compiled findings from the group were submitted into the larger MLMP database. Participants needed to be able to identify various types of milkweed: *Asclepias syriaca,* (common milkweed), *Asclepias verticillata* (whorled milkweed), *Asclepias tuberosa* (butterfly weed), *Asclepias incarnata* (swamp milkweed), monarch eggs, and the five larval stages (called instars, which refer to the periods between molting the larval exoskeleton). Building participant expertise in the MLMP protocols provided opportunities to gain mastery-level observational expertise to contribute to a larger science database. Scientists use this database to understand monarch population density, disease, and habitat restoration. Recent topics addressed with MLMP data include *impacts of habitat characteristics on monarch egg densities* (Stenoien, Nail, & Oberhauser, 2015), *use of roadside habitat by monarchs* (Kasten, Stenoien, Caldwell, & Oberhauser, 2016), *species distribution models of monarch natural enemies* (McCoshum, Andreoli, Stenoien, Oberhauser, & Baum, 2016), and *impacts of genetically modified herbicide resistant crops on monarchs* (Stenoien, Nail, Zalucki, Parry, Oberhauser, & Zalucki 2016).

Annual teacher workshops (since 1997) helped teachers prepare to engage their students in natural history learning and research, with a strong focus on activities, like CS, that can be conducted in schoolyards. Two versions of these workshops, both funded through the National Science Foundation (NSF), are most relevant to the work described here. First, from 1998 to 2001, teachers and secondary students attended summer and spring or autumn week-long workshops (in Minnesota and Texas, respectively) during which they observed and documented monarch breeding and migration and conducted their own research on various aspects of monarch biology. Second, a Driven to Discover series, focused on learning in non-formal science settings and in secondary classrooms. During summer workshops from 2009 to 2018, adults who work with youth in formal and non-formal settings learned to use citizen science programs as a springboard to authentic science research (see Koomen et al., 2014; Koomen et al., 2018). Participants developed their own research projects using the *Next Generation Science Standards* (NGSS) practices of science (2013) outlined in Table 1.

Table 1. NGSS (2013) Science and Engineering Practices

1. Asking questions;
2. Developing and using models;
3. Planning and carrying out investigations;
4. Using mathematics, information and computer technology and computational thinking;
5. Analyzing and interpreting data;
6. Constructing explanations;
7. Engaging in argument from evidence; and
8. Obtaining evaluating and communicating information.

Teacher Participants

Two teachers, Thia and Leslie, with extensive experience with MLMP, facilitated an informal CS summer program for middle school students. Both teachers selected students from their formal school setting. At the beginning of each summer, teacher interaction focused on learning the MLMP protocols. As participants developed expertise in monitoring, they began learning more about monarch biology and ecology and developed research questions that led to a science fair (see Koomen et al., 2018).

Former Student Participants

Students participated in the MLMP for a range of 1-6 years. The two teachers provided the researchers with a combined list of 37 former MLMP students consisting of current high school students, college students, and college graduates. Researchers invited all 37 students to participate in an interview about their MLMP experiences; 14 students responded and were interviewed. Nine (seven females and two males) of the interviewed students attended private school and five (three females and two males) attended public school. Of the 14 students interviewed, five have professional careers and completed undergraduate and/or graduate degrees, one is a recent college graduate, four are recent high school graduates, and four are high school students (see Table 4). Eighteen of the 37 former students completed an online survey to evaluate SCCT constructs. To protect the identity of teacher and student participants, all participants and parents of minors signed consent forms in compliance with internal review board specifications. All student and teacher participant names used in the paper are pseudonyms.

Data Collection

We used a mixed method design, in which quantitative and qualitative data were collected at the same time and analyzed in a complementary manner (Tashakkori & Teddlie, 1998) to understand the impact of the MLMP monitoring on participants' capacity to do science, their interest in science, and their interest in pursuing STEM educational and career goals.

Qualitative Methods

Qualitative methods included fourteen 60-90-minute semi-structured interviews of the former student participants and one interview for each of the teachers. Our interviews began with a set of questions, with additional questions for greater clarity as needed. Teacher phone interviews, conducted by two interviewers, focused on the teacher's background, motivation for starting an informal MLMP CS program, selection of student participants, the structure and setting of the program, and limitations unique to each setting.

Student interviews were conducted by one researcher, with the exception of one which was conducted by both researchers to assess interviewer consistency. The goal of the ex post facto interviews was to understand participants' thoughts and feelings about their experiences in CS research and possible outcomes from those experiences. Interview questions focused on participants' motivation for CS involvement, their interest in science, their beliefs about their science identity, the role of their teacher in building their confidence in science, and the role of MLMP in STEM high school and college educational and career choices. All interviews were audio recorded and transcribed verbatim.

Quantitative Methods

Quantitative methods included an online survey instrument composed of three different scales. At the time of the study, participants ranged from high school students through adults, and had participated in the MLMP monitoring from 2-21 years prior to the current study. Due to the retrospective nature of the data, we used instruments with established reliability and validity to help us to understand SCCT constructs (Table 2): MLMP CS Self-Efficacy Survey based on Hiller (2012), STEM Semantics Survey (Knezek, Christensen, & Tyler-Wood, 2011) for the interest construct and the MLMP Motivation Strategies for Learning Questionnaire (MSLQ; Pintrich, Smith, Garcia, & McKeachie, 1991) for the constructs of self-efficacy (affective/physiological states), outcome expectations, personal goals, interest, and learning experiences.

Table 2. Survey Instruments Aligned with SCCT Constructs

Instruments	SCCT Constructs
MLMP CS Self-Efficacy Survey	Self-efficacy (mastery experiences, vicarious experiences, social persuasion, and affective or physiological states)
STEM Semantics Survey	Interest
MLMP Motivated Strategies for Learning Questionnaire	Self-efficacy (affective/physiological states), outcome expectations, personal goals, interest, and learning experiences

The STEM Semantics Survey (Knezek et al., 2011) and MSLQ were used in novel ways to understand SCCT constructs. The STEM Semantics Survey has been used with multiple groups to study dispositions and attitudes toward STEM but not in conjunction with SCCT constructs. Similarly, the MSLQ has been used by researchers for years, but not in tandem with SCCT constructs to understand motivation in STEM education and careers.

MLMP CS Self-Efficacy Survey. We drew from the work of Hiller (2012) to develop a 12-item self-efficacy scale (see Table 6) to assess motivation for self-regulated learning (Schunk & Zimmerman, 2008). We used a Likert-type subscale from 1 (strongly disagree) to 5 (strongly agree) for participants to rank their confidence in MLMP activities, including locating monarchs in various life cycle phases and assessing milkweed characteristics. Our adaptations of the instrument included key protocol expectations for MLMP.

Table 3. Motivation Scales of the MLMP MSLQ Defined*

Motivation Subscale	Subscale defined	SCCT Construct Comparable To
	Value Components	
Intrinsic goal orientation Items 1, 16, 22, 24	Student perception of the reasons for engaging in a learning task	Learning experiences Interest
Extrinsic goal orientation Items 7, 11, 13, 30	Degree to which student perceives importance of grades, rewards, performance evaluation by others, and competition as reasons for participation	Personal goals
Task value Items 4, 10, 17, 23, 26, 27	Student evaluation of how interesting, important, and useful the task is	Outcome expectations; Interest
	Expectancy Components	
Control of learning beliefs Items 2, 9, 18, 25	Belief that efforts will result in positive outcomes	Personal goals
Self-efficacy for Learning and Performance Items 5, 6, 12, 15, 17, 18, 29, 31	Self-appraisal of one's ability to master a task	Self-efficacy
	Affective Components	
Task anxiety Items 3, 8, 14, 19, 28	Negative thoughts that disrupt performance, and affective and physiological arousal aspects of anxiety	Self-efficacy: Affective/ physiological states

*Table modified from Pintrich, et al., 1991

MLMP Motivated Strategies for Learning Questionnaire. The Motivational Strategies for Learning Questionnaire (MSLQ; Pintrich et al., 1991) is an extensively used research instrument used to measure students' motivation to persist in an academic endeavor, and their use of learning strategies for academic study (Edzie, 2014). We drew from two recent studies that used the MSLQ, Nausheen (2016) and Nugent, Grandgnett, Barker, and Adamchuk (2010). The motivation scales of the MLMP MSLQ are composed of three components (value, expectancy, and affective) each divided into various subscales. Table 3 defines each subscale and makes comparisons to SCCT constructs that helped us to answer our research questions (Figure 1). Items in the motivation subscale can be found in Appendix 1.

STEM Semantics Survey. The STEM Semantics Survey (Knezek et al., 2011) is derived from earlier work by Zaichkowsky (1995). It uses five adjective pairs from the Teacher's Attitudes Toward Information Technology Questionnaire (TAT) as descriptors for item statements that reflect perceptions of and interest in STEM, with an additional item that measures STEM career interests. The STEM Semantics Survey is a reliable and valid instrument for a broad range of participants from Grade 6 through adult (Knezek et al., 2010), and is also short and easy to use. This survey allowed us to understand the interest construct of SCCT (see Table 8). We used the STEM Semantics Survey with no adaptations.

Data Analysis

Our study features a form of triangulation analysis that is often used in mixed methods studies such as this one (Creswell & Plano Clark, 2007) that follows in two distinct stages. Separate analyses of qualitative and quantitative data occur in Stage 1. Both the quantitative and qualitative data are merged in Stage 2 to develop a holistic picture of the impact of the citizen science program, in this case. Our qualitative analysis featured deductive coding (Miles & Huberman, 1994), which is often used to test a model or hypothesis (Marshall & Rossman, 1995), as was our intent. The first step in deductive coding is to develop a categorization matrix and to code the data according to the categories (Elo & Kyngäs, 2008). We developed an *a priori* categorization matrix in which each code was defined by the constructs of SCCT (self-efficacy, outcome expectations, personal goals, interest and learning experiences) and shown in Table 4 (Blake, 2018). The data were then reviewed line-by-line and coded for alignment with or exemplification of the identified SCCT categories (Polit & Beck, 2004). We established researcher consensus to the categories by reviewing approximately 10% of the interview data to establish consistency in the data interpretation and to diminish researcher bias (Patton, 2002). When all data were coded, we reviewed 25% of each other's coding in a cross-case analysis (Patton, 2002) which allowed us to examine themes, similarities, and differences across cases.

Our quantitative analysis featured descriptive statistics, including central tendencies of mean and standard deviation. Because of our small sample size, we cautiously report on central tendencies for each of the three survey scales and did not use additional inferential statistics.

Our final step was to merge the qualitative and quantitative findings. We began with the analysis of the quantitative data, followed by a review of the analysis of the qualitative data, looking for parallel themes a practice common in mixed method research (Creswell & Plano Clark, 2007). We validated the sum total of our findings through triangulation of survey results and interviews of teacher and students.

Table 4. Codebook Excerpt for Analysis of SCCT Construct in Participant Surveys

Interview Text	Code	Construct	Description
I think it just made me think that I was smart at science. Like oh science is fun and I'm good at it. (Tracy)	SEPPA	Self-efficacy: Personal performance accomplishments vicarious learning: or (Zeldin & Pajares, 2000) mastery experiences I can do it.	Authentic mastery of a given task can create a strong sense of efficacy from the interpreted results of one's past performance: (Lent, R. W., Brown, S. D., & Hackett, G. (2002).

Results

Table 5. Student Interview Participants

Pseudo-nym	Initial grade level	First year in MLMP	Years in MLMP	Current occupation	College major/minor	STEM in HS as a result of MLMP
Thia's students						
Brad	Eighth	1999	5	DNR Data Manager	BA Environmental Sciences and English MS Natural Resources Science Management	Biology, Zoology
Terry	Seventh	2000	2	2nd grade teacher	BA Spanish major, Biology minor MA Curriculum and Instruction	No*
Amy	Seventh	2004	2	SPED teacher	BS Chemistry MA Education	AP Biology
Ellery	Seventh		2	In college	BS Nutrition, Psychology minor	Yes*
Zelda	Eighth		1	HS Freshman	Plans child psychology	Environmental Science class
George	Eighth		1	HS Freshman	Math	AP Biology
Tracy	Eighth	2013	1	HS Senior	Plans art education and Spanish	AP Biology
Maddie	Seventh	2013	2	HS Junior	Planning genetic counselor	AP Biology
Eva	Seventh		2	HS Senior	Undecided	Joining science club not science courses
Leslie's students						
Jacob	Fifth		5	HS Senior	Planning neuroscience and biology	HS science fairs, not classes
Jess	Fifth		4	HS Sophomore	Considering nursing	Plans to take AP Biology
Greta	Fifth	2010	6	HS senior	Planning Business	No*
Evan	Sixth		4	Mechanical engineer	BS Mechanical Engineering	Yes*
Carolyn	Fifth		Un-known	Animal diagnostician	BS Biology major, Chemistry minor	Anatomy, Geology, Oceanography

*Note: SPED = special education; HS = high school. Yes or no in *STEM in HS as a result of MLMP* indicates student did not elaborate further.

We build a case for two findings: participation in the MLMP (1) built capacity to do science and (2) fostered interest in pursuing education and career goals in STEM. We begin this section with an overview of participants (Table 5), followed by a brief overview of data that reflects what the teachers did throughout the MLMP monitoring.

Overview of Former Student Participants

The fourteen interview participants represent two different communities, worked with two different teachers, and participated in MLMP different years. For example, the earliest participant began in 1999 and the two latest participants began in 2016. The length of participation was between one and six years. Due to the range of participation and small sample size, we were not able to control for program implementation changes such as site locations, number of experienced peer mentors, and seasonal differences from year to year.

Of the fourteen participants interviewed, 86% reported choosing STEM courses as a result of MLMP participation. All of the six participants enrolled in college said that participation in the MLMP influenced their college major, and three of the five participants working in a professional career said that their MLMP participation influenced their STEM career choices (Table 5).

Teacher Participants

Leslie. Leslie taught middle school science in a small, (483 students in 2018), rural public school with 93% White students and 25% of students on free and reduced lunch (MDE, 2018) located in a small town about 100 miles from a Midwestern metropolitan area. She received no remuneration for her involvement in the summer program but charged a small snack fee so she could buy treats after monitoring sessions.

Leslie chose MLMP because "it was one of the first CS programs where you could put data online. I would still say that MLMP is the easiest CS program to enter data in." Leslie used monarch curriculum in her class in order to meet state standards, but MLMP wasn't a part of her formal classroom. "MLMP didn't really work well in my classroom because the monarchs are typically gone by the time we would be able to utilize it [the CS program]." Because of this, she decided that MLMP functioned best as a summer program. Students were invited to join the program based on staff recommendations:

> I just sent some notices around to all these teachers and they gave me some names and I invited them. Then the next year, I did the same thing, but I still invited the students that had done it the prior year to continue. It was more by their interest than their academic abilities. Some had learning disabilities and they were amazing in the field. (Leslie)

Leslie used multiple locations as monitoring sites. She began with a colleague's restored prairie, followed by a school garden, a pond garden, and her own garden. Leslie served in the role of a supervisor rather than a teacher at the monarch monitoring sites:

> I would say that the older students were really responsible for making sure that data were being collected correctly and the protocol was being followed. I was the safety supervisor. These students would just run from one site to the other, and I would ride my bike along just to make sure they were safe. (Leslie)

Leslie identified many goals of her monitoring program:

> [I wanted students to] develop some strong observation and science skills and for them to do real science. [It is important] for them to see that the decisions that we make have an impact. We get so overwhelmed by the negatives. I wanted them to see that even at 10-12 years old, they can make a positive impact. (Leslie)

Leslie built science capacity by supporting students to "develop some strong observation and science skills" when they observed her or student mentors. Her participants developed research questions for science fair projects to provide them with the opportunity to complete their own science investigations.

Thia. Thia taught seventh and eighth grade science at a small K-8 (647 students in 2015-16) private school with 92% White students located in a Midwestern outer-ring suburb. She received no remuneration for the program but received a stipend for equipment from a class she had taken with MLMP program coordinators. She used that funding to buy equipment.

Thia saw the potential of student participation in the MLMP after completing the 1998 NSF-funded course with two of her students. She based her own program on activities used in this course.

> This was really fun for me to do. Those were kids that to this day are my dearest friends. The next summer I thought, well, I'll do it on my own. I liked the monitoring, I liked the whole idea of the CS project, and so I started out doing that. In middle school, it became the driving force for my seventh-grade program. (Thia)

Thia found a meadow with limited through-traffic and established this as her monitoring site. She describes her interactions with the individuals in charge of the property:

> I met the woman in charge of the restoration of the meadow, and every year I would just let her know that we were coming out and pretty soon she just expected us out there. She helped brainstorm, and she was a botanist so she would run informal sessions on plant life which taught me a lot, too. (Thia)

Thia describes how the admission process evolved: "It was very informal in the beginning, probably for the first five years," until more students started to join. Then she developed an application-based admission process, and encouraged interested students to join. Thia set specific goals for her program. She noted that,

> In the beginning we had two goals, one was to do the monitoring for MLMP and the second was to develop a question that [they] could research and develop it into [their] science fair project. As the years went on, more kids started to join after we started going to Mexico to the overwintering sites. That was when my population of students grew because the only kids that could go on those trips were the kids that had done the summer program with me. Overall, [the goal was] just to get kids excited about science. Science is something important. It's something everybody can do and learn about and you don't have to be really smart to do it and you can do it no matter what. (Thia)

Thia built capacity for science by offering her students the opportunity to learn and apply the MLMP monitoring protocols by first working closely with her until they achieved mastery. Then, like Leslie, Thia's participants began to identify research questions for a science fair project in which they were able to develop their own scientific investigation.

Outcome of Quantitative Data Sources Using Descriptive Statistics

Table 6 presents data from our MLMP self-efficacy survey (scale 1-5). Study participants reported high self-efficacy, or mastery (Hiller, 2012), in locating monarch eggs and larva, understanding the differences between larval stages, recognizing milkweed, rearing monarchs, locating a site for monitoring monarchs, and successfully collecting data and completing a scientific study that answers a question using hypotheses (see Table 2 for SCCT constructs). Participants' confidence in inputting data to the online database varied widely since that task was assigned to limited numbers of participants.

Table 6. Monarch Larva Monitoring Project (MLMP) Self-Efficacy

Item	Mean(*SD*)
I can locate monarch eggs.	4.80(.84)
I can locate monarch larvae.	4.83(.88)
I can understand the difference between different larval instar stages.	4.44(.28)
I can recognize common milkweed from other plants growing in a monarch community.	4.83(.88)
I can collect weekly monarch density information using protocols.	3.89(.55)
I can rear eggs or larvae collected from the wild indoors and record whether they survived to adulthood and, if not, possible causes of their death.	4.72(.71)
I can measure height, reproductive status, age, herbivore damage, and the presence or absence of invertebrates of plants with monarchs and a set of random plants.	3.83(.65)
I can locate a site to monitor monarch density.	4.28(.04)
I can accurately collect data that reflect what's happening at a monitoring site.	4.44(.28)
I can input monitoring data into the MLMP CS website accurately.	2.50(1.30)
I can generate wonderings from a monitoring site that might lead to an investigative field study.	4.11(.22)
I can complete an investigative study that answers a scientific research question and develops hypotheses.	4.39(.21)

*Note: N=18; Scale is 1-5; 1 is strongly disagree; 5 is strongly agree; *SD* = Standard Deviation

The revised MLMP MSLQ (Nausheen, 2016) captured aspects of constructs of SCCT within motivation (Table 2). Table 7 presents survey responses on motivation for and attitudes about MLMP monitoring on a subscale of 1-7 (see Appendix A for a complete list of items).

Students felt their performance in MLMP was contingent on their own efforts [Control of learning (personal goals): 5.31], and participation in MLMP was an end in itself [Extrinsic goal orientation (personal goals): 4.74; Pintrich, et al., 1991]. Their self-efficacy for learning and confidence in their performance was high (5.85). Students' interest in participating in MLMP were related to their own curiosity and emerging mastery [intrinsic goal orientation (learning experiences & interests): 5.82]. The MSLQ showed that students exhibited low levels of task anxiety [self-efficacy (affective & physiological state): 2.64] and felt value and interest in completing the tasks of MLMP.

In Table 8, we illustrate participant perceptions of STEM disciplines based on the STEM Semantics survey results (subscale 1-7). The participant dispositions were higher for science then for any of the other STEM disciplines with engineering having the lowest score. All of the discipline means were positive (over 50% or more participants reported interest scores $\geq$ 4).

Table 7. MLMP MLSQ Scales of Motivation

MSLQ Scales	SCCT Construct	Mean(SD) ∞
Control of Learning	Personal goals	5.31(1.47) .10
Self-efficacy for Learning & Performance	Self-efficacy	5.85(1.16) .84
Intrinsic Goal Orientation	Learning experiences & Interests	5.82(1.05) .23
Extrinsic Goal Orientation	Personal goals	4.74(1.43)-.24
Task Anxiety	Self-efficacy: affective & physiological states	2.64(1.56) .23
Task Value	Outcome expectations & Interests	5.87(.91) .82

*Note: $N = 18$, Subscale is 1 - 7. 1 = strongly disagree, 7 = strongly agree.

Table 8. STEM Semantics Survey: Perceptions of STEM Disciplines Summary

Item	Science		Technology		Math		Engineering		STEM Career	
	Mean (SD)	%	Mean (SD)	%	Mean (SD)	%	Mean (SD)	%	Mean (SD)	%
This topic is fascinating.	5.4 (.83)	77	4.8 (.94)	69	4.0 (.63)	57	4.0 (.19)	57	4.8 (.61)	69
This topic is appealing.	5.3 (1.09)	76	4.7(1.04)	67	4.1 (.51)	59	3.5 (.79)	50	4.8 (.61)	69
This topic is exciting.	5.7 (.05)	81	4.7(1.04)	67	3.7 (.97)	53	3.3 (1.03)	47	4.6 (.94)	66
This topic means a lot.	6.0 (.73)	86	5.8(.11)	83	5.7 (1.34)	81	5.3 (1.36)	76	5.8 (1.00)	83
This topic is interesting.	6.2 (1.25)	89	5.8(.05)	83	5.2 (.77)	74	4.7 (.65)	67	5.9 (1.16)	84
Combined mean/∞	5.7/ -.24		5.2/-0.31		4.5/- .25		4.2/-2.15		5.2/ -1.91	

*Note: $N = 18$; Scale is 1-7; 1 is strongly disagree; 7 is strongly agree

Findings from Qualitative Data

Our qualitative data provided rich descriptions related to the outcomes of MLMP and self-efficacy, outcome expectations, personal goals, interests, and learning experiences. We analyzed 184 pages of interview transcripts. For the sake of brevity, we include only one or two text excerpts for each construct, selected because they collectively represent the various constructs and the sentiments of the majority of our study participants

Self-Efficacy: Mastery experiences. In MLMP, study participants learned to identify monarch life cycle stages (egg, larva, pupa, and adult) and documented their frequency in a CS database, maintaining the rigor of the MLMP protocol. Students expressed how through practice and time their confidence gradually increased from having "no idea" (Eva) to mastery of the MLMP protocols.

> I mean definitely, when I started I really had no idea, like I didn't know there were like five stages of growth for the caterpillars. I definitely went from not really knowing anything to learning more about the life cycle, the monarchs, and then you started to become more confident and like oh that's a first instar. That's a third instar. And then I think by the end

think that might've been the first time in my educational career that I, that I felt like here vas something that, you know, it wasn't like I was trying out for like- I was horrible at ports in grade school. It wasn't something like that where I knew just like, I guess I'll do t because all my friends were doing it since it's like the normal thing to do, but here was something was actually catered to an interest that I hadn't been exposed to before. (Brad)

I really want to give kids a positive science experience when they're young because so many people see it as a subject that's really difficult. And like if you're bad at math, you're bad at science and they're both so hard. And really that's not true. (Amy, a teacher)

Learning experiences. For students of both teachers, an immediate outcome expectation s that the summer program and the MLMP program would provide them the opportunity to rk to complete a science fair project, required for school.

It really helped with the science fair aspect of it because all the other students in eighth grade had to come up with their own home-based science fair experiment that they had to do, like in the winter, like at their house kind of thing. But I was really able to have a really solid experiment and really more in depth just because I was able to do it over the summer and with a mentor. (Tracy)

For Evan, "going out and monitoring is really fun and all. But it's nice to have like an actual chievable milestone or something tangible to show for all that." MLMP was a way for students o take part in the NGSS practices of science (see Table 1); as students monitored, they were enacting, planning, and carrying out investigations (Practice 3) and using mathematics, information and computer technology, and computational thinking (Practice 4).

For the actual data collecting part we got in partners. I think it was just like randomized partnering, and we would have a clipboard. One of us, the one person will have a clipboard, one other person would probably go a little more to look at the leaves more closely. We would count off, how many eggs or larvae they found and then look for butterflies. Each group would get a section and cover that day. (Ellery)

MLMP offered opportunities for them to analyze and interpret data (Practice 5), construct explanations (Practice 6) and engage in argumentation from evidence (Practice 7).

It was really interesting to me to learn how we not only just took that single data point, but how we looked at . . . the data over time and really saw how the population fluctuated with, um, differences in early spring time temperatures. Just seeing how all the factors that went into these insects' lives affected the population over time. (Evan)

MLMP provided a context for students to understand how the comprehensive MLMP population data helped citizen scientists and scientists develop models (Practice 2) and explanations for broader scientific understandings like climate change (Practice 6).

of the summer I felt pretty confident going out, you know, where I didn't necessarily need to check with other- like we kind of all could divide up the work and like feel confident in what we were doing. (Eva)

As they monitored, students began to realize their capacity to do science. They saw science as something they were "good at" (Tracy), changed their image of what scientists do, and started to become "personally involved" (Evan) in doing science. "I think it just made me think that I was smart at science. Like oh science is fun and I'm good at it." (Tracy)

I would say I had never really appreciated how involved you could be in science before. Before I became involved myself with this, I would say you always kind of think of scientists in lab coats and faraway labs. You don't really think, "Oh yeah, I have what it takes to really learn and produce interesting findings." Just by becoming personally involved in the data collection and analysis you realize how approachable science really is. (Evan)

Self-Efficacy: Vicarious experiences. Students described the value of observing role models to build confidence. For example, in one of the MLMP cohort groups, several students practiced the protocols for as many as six years and were able to forge new relationships by acting as teachers and mentors to new participants in the project.

Well, Leslie, had just always pushed us to do our best and just sat down with us and helped teach us how to do each and every part of it, and make sure that we have a full understanding of it. And if other people, like your peers, understood it more, they helped explain it, so then it made it so you could help explain it to the next generation. (Jess)

I guess just seeing them do it and realizing that you don't have to be a scientist in a lab coat to record data; [it] can be anyone who's contributing. It seems less intimidating I would say. (Amy)

Self-Efficacy: Social persuasion. Participant interviews revealed the impact of hearing positive feedback and encouragement from their peers and teachers on their abilities to identify various larval stages in MLMP and be successful in science coursework.

I would just say just being with them and hearing like good job, you know, keep going, just more of the encouragement. Leslie, too, is just a great all-around person. It just boosted confidence and kept us focused. (Carolyn)

Leslie would help me with like anything. Even if I went in with a science question now, and all my classmates, like my fellow scientists they would, they would all be very helpful if I had a question about anything and it kind of helped because they kind of pushed me in the classes that I'm in because I'm like they did; they can do this class. I can do this one [too]. (Greta)

Previously-established relationships often made participants excited for the summer CS and sustained their participation. "Two of my best friends did it, which was what made me so excited to do it once a week I get to spend an hour and a half with them." (Zelda)

I said let's try it out, and it was later that I found out that all of our friends had been, offered that up. It was kind of independent choice going into it. But later on, I would say it helped having those people I knew to stick around with it. (Jacob)

Self-Efficacy. Affective and physiological states. Many of the students reported that they initially had some anxiety about their ability to monitor accurately because they knew that the data would be used by scientists in a larger study. This anxiety was buffered by mastery experiences for monitoring by teachers, vicarious experiences embodied by teacher, and social persuasion through positive comments and presence of peers and teachers.

Doing that as a seventh or eighth grader was actually pretty intimidating at first. But know this, that even at that point I knew this is kind of the stuff that I'd be doing with the rest of my life, so I'm just kinda ate it up. (Brad)

At first, I was really nervous because this data actually matters, and I didn't wanna screw anything up, and I wanted to have the most accurate data. I was a lot more confident than I was at the start because I realized that she [teacher] actually did prepare us really well. (George)

Outcome expectations. MLMP provided a context for students to understand how scientists do science in the real world—to observe, ask questions, design investigations, collect and analyze data, and report their findings, all hallmarks of the NGSS (2013). This experience demonstrated the value and usefulness of doing science, opened students' eyes to what science is, and showed them how much there is to learn.

I feel like it really just opens the door more than anything. It opened my eyes because, I mean I was a fifth grader. I didn't know what science was. I didn't know what real science was or what it could be. So, getting to do all that just sent me onto the track that I feel like I belong on it. It made me realize what's out there and what I want to do. (Jacob)

We took Biology in seventh grade. But I wasn't as interested as I would've been if I had not done the program just because it made me realize how much more there is to learn and like learning. (Zelda)

Students realized that monitoring the monarch population in their local area was relevant and contributory to understanding scientific and environmental challenges in the world around them as well as connections to global climate change.

I felt really good. In my mind, I was changing some small part of the world. We were being told that it was being used by these big-name scientists at the university and then they would be using it and collaborating with people all over the country and so that was cool knowing that I was a little piece of that puzzle of data. (Jacob)

Personal goals. MLMP teachers and scientists inspired students to think about their career choices. After traveling to Ecuador for an MLMP educational tour, Tracy was inspired to learn

Spanish and pursue an educational career. Because of MLMP, Brad was university scientists which solidified his career path to be a scientist.

Like I've been inspired by teachers including Thia and the Spanish teacher. I of great teachers like Thia, for example, that made me want to go into the edu (Tracy)

That day we spent at the university going into the monarch lab and actually see do this for a profession. I mean that just kind of solidified it even at that point grade when I knew I was still kind of the minority, it wasn't really a question a was my career path. (Brad)

Involvement in MLMP opened up some students' interest in doing scientific r

I already had an interest in science, but it definitely did change my ideas. I don't wa a physicist anymore. It made me want to go into research. (Maddie)

I felt like I was a part of a bigger research project and so, I think that made it more f more interesting. I think I definitely thought about that more as an option than I had l being a part of the project. I definitely feel like it got me interested in science and think maybe my eyes were a little more open to like looking for other ways to be inv in science. (Eva)

MLMP provided students with an opportunity to make new discoveries.

I think what the monitoring did is it really just set me up to realize how much fun all of t was. The butterflies provided me with that sense of I can learn new things. I really just lo the aspect of discovery. Like when I discover something new, it's so cool to think that I' the only person that knows this information. It's cool to know that it's like, this is, this i like my information. That idea is the best feeling I've ever come across. And that's why just love pursuing science. (Jacob)

Interest. Students acknowledged that the repeated experience of MLMP monitor enabled them to feel competent and how that competence led to an interest to do science.

By doing it at such a young age, it really makes you realize that anyone is capable of doing such things as the older you get, the more advanced you just get at it and so, it lets you know that you don't have to be the super rocket scientist, smart person to do this kind of work. You just have to . . . do what you love really. And that's what the project really taught me about science. (Jacob)

Former students helped us to understand how their interests turned into academic and career goals and influenced their actions in choosing STEM courses and careers.

> I just thought it was so cool that I knew that it was going to like a bigger thing. I knew that there are people all over monitoring like that and I think as the years went on because I would just drop in throughout the years, I could tell the monarch population started going down. (Terry)

DISCUSSION

Our findings provide evidence that participation in MLMP provided a capacity to do science (research question 1) and fostered interest in pursuit of educational and career opportunities for STEM careers (research question 2). We argue that, for the students in this study, MLMP opened a window that allowed them to see that science was in reach for them.

MLMP Develops Capacity to Do Science

Jiménez-Aleixandre, Rodriquez and Duschl (2000) argued that student discourse is more likely to reflect "doing the lesson" versus "doing science" when they engage in developing explanations in science classroom. Thus, many students were in effect not really doing science as envisioned by the NGSS (2013) but were engaged in interactions that were part of the rules or the procedures of the task or incidental talk. In MLMP, we found that as students conducted CS activities they felt that they were doing science and that science mattered. They thought of themselves as scientists accurately collecting data for a purpose bigger than themselves (Terry, Greta, Jacob, Brad and Evan). MLMP and the rigorous protocols that are part of it provided an opportunity to engage in many of the NGSS practices of science outside of regular school (Table 1). The MLMP MSLQ reveals that they were confident that they could master the skills being taught (Appendix 1: Item 29) and felt the material was useful for them to learn (Appendix 1: Item 23).

This study extends our previous work (Koomen et al., 2018) through our use of SCCT and the construct of self-efficacy to understand possible educational or career outcomes that resulted from engagement in the CS program (Figure 1). Our self-efficacy measures (Tables 6 and 7) and our interview data demonstrate positive self-efficacy for the participants. Many participants reported low confidence levels early in their engagement with the project, but over time and with gradual release towards independence by the teacher, they built their capacity and their confidence for doing CS (Maddie; research question 1). Kountoupes and Oberhauser (2008) highlighted both the scientific and social value of participation in MLMP for children but found that most adults elected to complete the difficult parts of monitoring. Our results suggest that youth are able to participate in all parts of monitoring, they just need time and mentoring in which to develop the confidence in their ability.

ENGAGING IN CITIZEN SCIENCE PROGRAMS FOSTERS INTEREST IN PURSUING EDUCATIONAL AND CAREER OPPORTUNITIES IN STEM

Maltese and Tai (2011) showed that student attitudes toward STEM influence motivation (SCCT constructs of outcome expectations and interest) to pursue STEM education and careers. Their results show that middle school students are at an important juncture in exploring career options in STEM. In our study, middle school participants monitoring monarchs portrayed strong perceptions of STEM disciplines and interest in STEM careers in the STEM Semantics survey (Table 8). Knezek, Christensen and Tyler-Wood's (2015) comparisons between middle school students and STEM education professionals using the STEM Semantics Survey provide baseline data regarding perceptions of STEM disciplines and in STEM careers. Our reported values (Table 8) for participants are in between the values of the middle school and STEM professionals, cautiously suggesting that our participants have interest in STEM fields.

Second, like the programs discussed in Knezek and colleagues (2011), MLMP and CS are contributory in nature, meaning that students felt that what they were doing was interesting and important (Jacob, Brad, Evan and Zelda). Recent research notes the importance of providing students with opportunities to learn about and experience STEM disciplines (Carrico & Matusovich, 2016), including the importance of hands on and authentic activities like CS in promoting interest in STEM (Aschbacher, Ing, & Tsai, 2013). Our results suggest that the contributory aspect of CS leads to high levels of value and outcome expectations in STEM.

Taken as a whole, the results of this mixed methods study support our hypothesis that MLMP CS positively fostered participant interest in STEM educational endeavors and possible careers (Figure 1; Table 2). Participants had high interest in the content area of the experience (Appendix 1: Item 7), preferred to learn information that aroused their curiosity (Appendix 1: Item 16), and liked the subject matter of the experience (Appendix 1: Item 26). Through MLMP, participants began to understand and were excited about how much there was to learn in science, how involved science was, how science was something that appealed to them, and opened their eyes to their own possibilities or personal goals (Evan, Zelda, Brad, Maddie Eva & Amy).

MLMP promoted a sense of self-efficacy in their capabilities of doing science, promoted their interest and curiosity in science, and provided learning experiences using science protocols that aligned with the NGSS (2013). We argue that MLMP allowed students to unlock the door to science through rigor, relevance, and relationships (McNulty & Quaglia, 2007). CS requires rigor as students work to master the protocols for monitoring. The MLMP program provides opportunities for relevant learning that is interdisciplinary, contextual, and applied to real-world predictable and unpredictable situations (McNulty & Quaglia, 2007). Through mentoring and encouragement of the nuances of CS, teachers built relationships between participants and teachers that are nurturing, accountable, and positive. We assert that building the capacity to do science through MLMP CS protocols, with mentoring and encouragement from teachers and peers, fosters interest in pursuing STEM education and career pathways.

Limitations

We would be remiss if we did not recognize limitations of this study. The small sample size eliminated the use of inferential statistics, but allowed us to develop rich, descriptive information from participants about the impacts of MLMP CS on their educational and career outcomes. Other limitations of our design are that student participation occurred across different years. We were not able to compare some specific elements of the programs, such as different collection sites or even the way that the protocols were enacted. We do not know if the same results could be found with other CS programs or other informal programs, questions we hope to address in future research. Additionally, we recognize students' motivation may have been higher as they previously volunteered for the CS and, again, volunteered for the summer MLMP program.

Like other recent research using SCCT (Blake, 2018), the framework was useful in helping us define constructs for analysis of the impact of CS on STEM educational and career goals using mixed methods (Figure 1; Table 2). We add to the sparse literature reporting on qualitative outcomes using SCCT, especially those represented by a small group of participants (Blake, 2018).

Educational Implications and Directions for Future Research

The present study has implications for science educators as we show how informal education programs like CS had positive outcomes for students' attitudes, experiences, and engagement in science. Future research might study classroom implementation as currently time and structural constraints of formal education (Carlone, Haun-Frank, & Kimmel, 2010) limit students' engagement in the practices of science. In MLMP CS informal science led to greater self-efficacy for formal science as evidenced by their enrollment in higher level science courses. It is important to leverage this phenomenon to make STEM more accessible to all learners.

The relationships between teacher and students began in a formal setting but CS was enacted in an informal setting. Though informal settings are promoted as opportunities to engage K-12 students in the practices of science (NGSS, 2013), CS experiences in these settings lack equity relative to geography, socioeconomic status, and transportation resources, which may be a barrier to access to MLMP and other CS participation. Participants in both MLMP programs were from small, predominantly White schools. Both schools are embedded in middle-class settings; one school is a private school and the other is a public school with 25% free and reduced lunch. We recommend using caution when translating these results into other settings. Students were reliant on family members for transportation to the program, a potentially limiting factor in other settings. More research is needed to promote greater accessibility and equity for all students, including underrepresented groups in CS, since it can be a pathway into STEM courses and careers.

In addition, teacher participants received no remuneration despite the time and effort required for this part of the project. Teachers also bore responsibility for any risks inherent in taking students outdoors, though efforts were made to reduce these risks. Since CS is a valuable pathway to STEM, financial resources need to be directed to support teachers and to reduce liability risks for those leading these programs.

Finally, our alignment of SCCT with the MLMP MSLQ and the STEM semantics surveys illustrated that these instruments can be used in novel ways. We used rigorous methods throughout our study; however, we had a small population of former student participants. Research with larger and more diverse populations of students would allow us to use inferential statistical measures for greater generalizability with these instruments.

APPENDIX A

MLMP MSLQ Scales of Motivation Items

Item	MSLQ #	MSLQ Scales	Mean (SD)	% Agree/ Strongly agree
If I study in appropriate ways, then I will be able to learn the material in MLMP monitoring.	2	Control of learning	6.33 (.77)	94.4%
It is my own fault if I didn't learn the material in MLMP monitoring.	9		4.72 (1.56)	22.3%
If I try hard enough, then I will understand the MLMP material.	18		6.33 (.77)	83.3%
If I don't understand the MLMP monitoring directions, it is because I didn't try hard enough.	25		4.00 (1.04)	0%
I believe I will receive excellent feedback from this experience.	5	Self-efficacy for learning and performance	5.33 (1.28)	44.4%
I'm certain I can understand the most difficult material presented in the monitoring lessons for MLMP.	6		5.22 (1.83)	55.5%
I'm confident I can learn the basic monitoring skills taught in the MLMP experience.	12		6.50 (.71)	88.9%
I'm confident I can understand the most complex material presented by the instructor in this experience.	15		5.78 (1.11)	61.1%
I'm confident I can do an excellent job on the monitoring in this experience.	17		5.72 (.96)	50%
I expect to do well in the MLMP experience.	18		6.00 (.84)	77.8%
I'm certain I can master the skills being taught in this experience.	29		6.00 (.97)	66.7%
Considering the difficulty of this experience, the teacher, and my skills, I think I did well in the MLMP monitoring.	31		6.28 (.75)	83.3%
In MLMP or experiences like this, I prefer project material that really challenges me so I can learn new things.	1	Intrinsic goal orientation	6.17 (.79)	83.3%
In an experience like this, I prefer to learn information that arouses my curiosity, even if it is difficult to learn.	16		6.22 (1.00)	83.3%
The most satisfying thing for me in the MLMP experience is trying to understand the content as thoroughly as possible.	22		5.06 (1.11)	77.2%

APPENDIX A (CONTINUED)

Item	MSLQ #	MSLQ Scales	Mean (SD)	% Agree/ Strongly agree
When I have the opportunity during MLMP monitoring, I choose experiences that I can learn from even if they don't guarantee I will be good at them.	24		5.83 (.92)	61.1%
Doing well in MLMP monitoring was the most satisfying thing for me during the summers I was doing it.	7	Extrinsic goal orientation	4.94 (1.51)	33.4%
The most important thing for me when I was monitoring was improving my ability to monitor, so my main concern in this experience was getting positive feedback from my teacher.	11		4.50 (1.50)	16.7%
If I can, I want to be better at monitoring for monarchs than most of the other students.	13		4.67 (1.33)	27.8%
I want to do well in this experience because it is important to show my ability to my family, friends, employer, or others.	30		4.83 (1.42)	27.8%
When I monitor for MLMP I think about how poorly I am doing compared with other students.	3	Task anxiety	1.78 (1.11)	0%
When I monitor for MLMP I think about items about monitoring I struggle with	8		3.78 (1.77)	11.2%
When I monitor for monarchs I think of the consequences of failing.	14		2.44 (1.04)	0%
I have an uneasy, upset feeling when I take an exam.	19		2.39 (1.38)	5.6%
I feel my heart beating fast when I participate in MLMP monitoring	28		3.13 (1.76)	11.1%
I think I will be able to use what I learn in MLMP monitoring in other experiences.	4	Task value	5.78 (1.06)	55.5%
It is important for me to learn the MLMP monitoring material in this experience.	10		5.80 (.99)	72.2%
I am very interested in the content area of this experience.	17		5.72 (1.13)	55.5%
I think the material in the MLMP experience is useful for me to learn.	23		6.06 (.87)	16.7%
I like the subject matter of this experience.	26		5.94 (.73)	72.2%
Understanding the subject matter of this MLMP monitoring is very important to me.	27		5.78 (.88)	61.1%

ACKNOWLEDGMENTS

We are grateful for the review of this manuscript by Jane Fields and for the enthusiastic assistance of undergraduate researchers Natasha Behboudi, Kari Felland, Greta Ferguson, Daniella Habib, Jenna Mortel, Nikki Neequaye, Miranda Rice, Kameryn Schafer, and Emily Webb. This work was supported by a grant from the National Science Foundation (DR K-12-1417777). Any opinions, findings, conclusions, or recommendations are those of the authors and do not necessarily reflect the position or endorsement of the funding agency. None of the authors has a conflict of interest with respect to the contents of this chapter.

References

Aschbacher, P. R., Ing, M., & Tsai, S. M. (2013). Boosting student interest in science. *Phi Delta Kappan*, *95*(2), 47-51. doi:10.1177/003172171309500210

Bandura, A. (1997). *Self-efficacy: The exercise of control.* New York, NY: W. H. Freeman and Company.

Bell, P., Lewenstein, B., Shouse A. W., & Feder, M. A. (2009). *Learning science in informal environments. People, places, and pursuits.* Washington DC: The National Academies Press.

Blake, D. (2018). Motivations and paths to becoming faculty at minority serving institutions. *Education Sciences*, *8*(1), 30. doi:10.3390/3educsci801003

Bonney R., Ballard, H., Jordan, R., McCallie, E., Phillips, T., Shirk, J., & Wilderman C. C. (2009). *Public participation in scientific research: Defining the field and assessing its potential for informal science education.* Washington, DC: Center for Advancement of Informal Science Education.

Britner, S. L., & Pajares, F. (2006). Sources of science self-efficacy beliefs of middle school students. *Journal of Research in Science Teaching*, *43*(5), 485-499. doi:10.1002/tea.20131

Carlone, H. B., Haun-Frank, J., & Kimmel, S. (2010). Tempered radicals: Elementary teachers' narratives of teaching science within and against prevailing meanings of schooling. *Cultural Studies of Science Education, 5*(4), 941-964. doi:10.1007/s11422-010-9282-6

Carrico, C., & Matusovich, H. M. (2016). A qualitative examination of rural central Appalachian high school student knowledge of the college processes needed to meet career goals. *Journal of Women and Minorities in Science and Engineering*, *22*(3). doi:10.1615/JWomenMinorScienEng.2016013308

Chen, X., & Weko, D. (2009). *Students who study science, technology, engineering, and mathematics (STEM) in post-secondary education.* Washington DC: U.S. Department of Education, NCES, 2009-161.

Creswell, J. W., & Plano Clark, V. L. (2007). *Designing and conducting mixed methods research.* Thousand Oaks, CA: SAGE Publications, Inc.

Curtis, V. (2015). Motivation to participate in an online citizen science game: A study of Foldit. *Science Communication*, *37*(6), 723-746. doi:10.1177/1075547015609322

Edzie, R. L. (2014). *Exploring the factors that influence and motivate female students to enroll and persist in collegiate STEM degree programs: A mixed methods study.* (Doctoral dissertation). Retrieved from ProQuest Dissertations & Theses Global. (UMI No. 3615058)

Elo, S., & Kyngäs, H. (2008). The qualitative content analysis process. *Journal of Advanced Nursing*, *62*(1), 107-115. doi:10111/j.1365-2648.2007.04569.x

Hidi, S., & Renninger, K. A. (2006). The four-phase model of interest development. *Educational Psychologist*, *41*(2), 111-127. doi:10.1207/s15326985ep4102_4

Hiller, S. E. (2012). *The impact of a CS program on student achievement and motivation: A social cognitive career perspective*. (Dissertation thesis). Retrieved from Proquest Information and Learning Company. (UMI Number 3547088)

Hiller, S. E., & Kitsantas, A. (2014). The effect of a horseshoe crab CS program on middle school student science performance and STEM career motivation. *School Science and Mathematics*, *114*(6), 302-311. doi:10.1111/ssm.12081

Hiller, S. E., & Kitsantas, A. (2016). The validation of the citizen science self-efficacy scale (CSSES). *The International Journal of Environmental and Science Education.* doi:10.12973/ijese.2016.405a

Jiménez -Aleixandre, M. P., Rodriguez, A. B., & Duschl, R. A. (2000). "Doing the lesson" or "doing science": Argument in high school genetics. *Science Education*, *84*(6), 757-792. doi: 10.1002/1098-237X(200011)

Kasten, K., Stenoien, C., Caldwell, W., & Oberhauser, K. S. (2016). Can roadside habitat lead monarchs on a route to recovery? *Journal of Insect Conservation*, *20*(6), 1047-1057. doi:10.1007/s10841-016-9938-y

Knezek, G., Christensen, R., & Tyler-Wood, T. (2011). Contrasts in teacher and student perceptions of STEM content and careers. *Contemporary Issues in Technology and Teacher Education*, *11*(1), 92-117.

Koomen, M. H., Blair, R., Young-Isebrand, E., & Oberhauser, K. S. (2014). Science Professional Koomen, M. H., Blair, R., Young-Isebrand, E., & Oberhauser, K. S. (2014). Science professional development with teachers: Nurturing the scientist within. *Electronic Journal of Science Education*, *18*(6), 1-28. Retrieved from http://ejse.southwestern.edu

Koomen, M. H., Rodriguez, E., Hoffman, A., Petersen, C., & Oberhauser, K. (2018). Authentic science with citizen science and student-driven science fair projects. *Science Education*, *102*(3), 593-644. doi:10.1002/sce.21335

Kountoupes, D., Oberhauser, K. S. (2008). Citizen science and youth audiences: Educational outcomes of the monarch larva monitoring project. *Journal of Community Engagement and Scholarship, 1*, 10-20.

Lent, R. W., & Brown, S. D. (2008). Social cognitive career theory and subjective well-being in the context of work. *Journal of Career Assessment*, *16*(1), 6-21. doi:10.1177/1069072707305769

Lent, R. W., Brown, S. D., & Hackett, G. (1994). Toward a unifying social cognitive theory of career and academic interest, choice, and performance. *Journal of Vocational Behavior*, *45*(1), 79-122 doi:10.1006/uvbe.1994.1027

Lent, R. W., Brown, S. D., & Hackett, G. (2000). Contextual supports and barriers to career choice: A social cognitive analysis. *Journal of Counseling Psychology*, *47*(1), 36-49. doi:10.1037/0022-0167.47.1.3

Lent, R. W., Brown, S. D., & Hackett, G. (2002). Social cognitive career theory. *Career choice and development*, *4*, 255-311.

Maltese, A. V., & Tai, R. H. (2011). Pipeline persistence: Examining the association of educational experiences with earned degrees in STEM among US students. *Science Education*, *95*(5), 877-907. doi:10.1002/sce.20441

Marshall C. & Rossman G. B. (1995). *Designing qualitative research.* London, United Kingdom: SAGE Publications, Inc.

McCoshum, S. M., Andreoli, S. L., Stenoien, C. M., Oberhauser, K. S., & Baum, K. A. (2016). Species distribution models for natural enemies of monarch butterfly (*Danaus plexippus*) larvae and pupae: Distribution patterns and implications for conservation. *Journal of Insect Conservation*, *20*(2), 223-237. doi:10.1007/s10841-016-9356-z

McNulty, R. J., & Quaglia, R. J. (2007). Rigor, relevance and relationships. *School Administrator*, *64*(8), 18-23.

Miles, M. B. & Huberman, A. M. (1994). *Qualitative data analysis: An expanded sourcebook.* Thousand Oaks, CA: SAGE Publications, Inc.

Miller, K., Sonnert, G., & Sadler, P. (2018). The influence of students' participation in STEM competitions on their interest in STEM careers. *International Journal of Science Education, Part B*, *8*(2), 95-114. doi:10.1080/21548455.2017.1397298

Minnesota Department of Education. (2018). Minnesota report card. Retrieved from https://rc.education.state.mn.us/#demographics/orgId--10345020000__p--3

Monarch Larva Monitoring Project. (n.d.). MonarchNet. Retrieved June 10, 2018 from htps://www.monarchnet.org/monarch-larva-monitoring-project.

Nausheen, M. (2016). An adaptation of the motivated strategies for learning questionnaire (MSLQ) for postgraduate students in Pakistan: Results of an exploratory factor analysis. *Bulletin of Education and Research*, *38*(1).

NGSS Lead States. (2013). *Next generation science standards: For states, by states.* Washington, DC: The National Academies Press.

Nugent, G., Barker, B., Grandgenett, N., & Adamchuk, V. I. (2010). Impact of robotics and geospatial technology interventions on youth STEM learning and attitudes. *Journal of Research on Technology in Education*, *42*(4), 391-408. doi:10.1080/15391523.2010.10782557

Pacheco-Guffrey, H. A. (2018). Oh, the places you will go! But how will you get there? In M. H. Koomen, S. Kahn, C. L. Atchison & T. Wild (Eds.), *Toward inclusion of all learners through science teacher education* (pp. 291-300). Rotterdam, Netherlands: Sense Publishers.

Patrick, L., Care, E., & Ainley, M. (2011). The relationship between vocational interests, self-efficacy, and achievement in the prediction of educational pathways. *Journal of Career Assessment*, *19*(1), 61-74. doi:10.1177/1069072710382615

Patton, M. Q. (2002). Two decades of developments in qualitative inquiry: A personal, experiential perspective. *Qualitative Social Work*, *1*(3), 261-283. doi:10.1177/1473325002001003636

Pintrich, P. R., Smith, D. F., Garcia, T., & McKeachie, W. J. (1991). A manual for the use of the Motivated Strategies for Learning Questionnaire (MSLQ).

Polit, D. F., & Beck, C. T. (2004) *Nursing research principles and methods.* Philadelphia, PA: Lippincott Williams & Wilkins.

Rotman, D., Preece, J., Hammock, J., Procita, K., Hansen, D., Parr, C., Jacobs, D. (2012, February). *Dynamic changes in motivation in collaborative citizen-science projects*. Paper presented at the annual conference of the Association for Computing Machinery, pp. 217-226, New York, NY. doi:10.1145/2145204.2145238

Soldner, M., Rowan-Kenyon, H., Inkelas, K. K., Garvey, J., & Robbins, C. (2012). Supporting students' intentions to persist in STEM disciplines: The role of living-learning programs among other social-cognitive factors. *The Journal of Higher Education*, *83*(3), 311-336. doi:10.1080/00221546.2012.11777246

Schunk, D. H., & Zimmerman, B. J. (Eds.). (2008). *Motivation and self-regulated learning: Theory, research, and applications*. New York, NY: Taylor & Francis.

Stenoien, C., Nail, K. R., & Oberhauser, K. S. (2015). Habitat productivity and temporal patterns of monarch butterfly egg densities in the Eastern United States. *Annals of the Entomological Society of America*, *108*(5), 670-679. doi:10.1093/aesa/sav054

Stenoien, C., Nail, K. R., Zalucki, J. M., Parry, H., Oberhauser, K. S., & Zalucki, M. P. (2016). Monarchs in decline: a collateral landscape-level effect of modern agriculture. *Insect Science*. doi:10.1111/17444-7917.12404

Tashakkori, A. & Teddlie, C. (1998). *Mixed methodology: Combining qualitative and quantitative approaches* (Vol. 46). Thousand Oaks, CA: SAGE Publications, Inc.

The University of Minnesota Monarch Lab. (n.d.). MonarchLab. Retrieved June 10, 2018 from hpps://monarchlab.org/

Trumbull, D. J., Bonney, R., Bascom, D., & Cabral, A. (2000). Thinking scientifically during participation in a citizen-science project. *Science Education*, *84*(2), 265-275. doi:10.1002/(SICI)1098-237X(200003)84:2<265::AID-SCE7>3.0.CO;2-5.

Zaichkowsky, J. L. (1985). Measuring the involvement construct. *Journal of Consumer Research, 12*(3), 341-352. doi:10.1086/208520

In: Enhancing STEM Motivation through Citizen Science Programs ISBN: 978-1-53616-038-3
Editors: Suzanne E. Hiller and Anastasia Kitsantas

Chapter 11

USING MOBILE LEARNING AND AUTHENTIC PRACTICE IN CITIZEN SCIENCE CONTEXTS TO FOSTER STEM INTEREST IN HIGH SCHOOL STUDENTS

Duane E. Wallace**[*] **and Alec M. Bodzin
Lehigh University, PA, US

ABSTRACT

Citizen science programs often provide outdoor learning experiences where students learn about nature and the environment while in naturalistic environments. The purpose of the present chapter is to examine the effectiveness of the mobile learning and authentic practice (MobiLAP) citizen science approach aimed to promote student interest in STEM education and careers and perceptions of mobile learning affordances. We implemented this approach with 93 secondary students using the Leafsnap app and the PictureThis app to learn about phenology across multiple phenophases. Using a qualitative design, we investigated the students' interest in STEM following participation in the MobiLAP approach. Findings showed that using mobile technologies in a citizen science project enhanced student interest in both STEM education and STEM careers.

Keywords: authentic practice, citizen science, education, mobile learning, STEM, science, technology, engineering, math

One of the major educational challenges in the United States is science, technology, engineering and mathematics (STEM) education reform. The National Science Board (2007) noted that the current educational system is failing to meet the STEM education needs for U.S.

[*] Corresponding Author's E-mail: duane@duanew.com.

students. As a result, our nation's inability to produce STEM workers will negatively impact our economic growth and national security. Likewise, The President's Council of Advisors on Science and Technology (PCAST, 2010) acknowledged that the United States is no longer a leader in STEM education and is in fact, ranked in the "middle of the pack" or lower. Further exacerbating the problem is the lack of traditionally underrepresented groups (such as women, African Americans and Hispanic Americans) in STEM educational programs and careers (Anderson, 2014). With a predicted 80% of new jobs requiring math, science and engineering (Wolfram, 2017), the challenge of adequately preparing the future workforce has been clearly set forth. Thus, promoting STEM interest for students in U.S. schools is an important priority as future generations will need to develop critical thinking skills to address STEM issues such as climate change.

Creating interest in STEM during an individual's formative years results in learners seeking out STEM education and careers (Manson, Martinez, Buchwald, Rubio, & Moss, 2015). More importantly, however, is sparking interest in STEM during high school years and before a college major is chosen (Maltese, Melki, & Wiebke, 2014). Sadler, Sonnert, Hazari, and Tai (2012) contend that career interest in STEM at the beginning of high school is the best predictor of maintaining STEM interest upon leaving high school. High school experiences are in fact, the most crucial in terms of creating STEM interest in students (Riegle-Crumb, Moore, & Ramos-Wada, 2011). STEM interest in high school students could include enhanced interest, motivation or participation in the sciences, technology, engineering and mathematics fields of study. In order to meet the deficit in STEM workers and to conquer challenges such as anthropogenic climate change, the U.S. educational system needs to find ways to generate STEM interest in high school students.

One of the ways in which educators may help facilitate STEM interest in students is through citizen science projects. Citizen science projects make use of a network of amateur scientists (often students) to make, record and submit observations on environmental occurrences in an effort to better understand our world. Citizen science projects address real-world problems, create authentic learning experiences (Meyer, Scott, Strauss, Nippolt, Oberhauser, & Blair, 2014) and help students understand the importance of STEM education as it relates to their schooling and everyday lives (Lamb, 2016). In addition to providing students opportunities to learn STEM content and becoming familiar with the scientific process (Ruiz-Mallén, Riboli-Sasco, Ribrault, Heras, Laguna, & Perié, 2016), citizen science has shown promise in creating motivation toward STEM careers (Hiller & Kitsantas, 2014). Many citizen science projects offer experiences with nature to create initial interest in STEM, while hands-on projects help to sustain that interest (VanMeter-Adams, Frankenfeld, Bases, Espina, & Liotta, 2014). This falls in line with the Next Generation Science Standards' recommendation of preparing students for college, career and citizenship by educating in an authentic manner, with real-world problems and using outdoor learning experiences (NGSS Lead States, 2013). Likewise, the *2016 Horizon Report* asserts that students should have community-connected, real-world, authentic learning experiences (Adams Becker, Freeman, Giesinger Hall, Cummins, & Yuhnke, 2016).

Mobile Learning and Authentic Practice in Citizen Science Contexts

Mobile learning has the potential to create meaningful, personalized, situated (Huang, Yang, Chiang, & Su, 2016) and authentic learning. Research has shown that mobile learning has been used effectively to create authentic learning experiences in varying contexts (see for example Baya'a & Daher, 2009; Hsu & Ching, 2012; De Pietro, 2013). Many of the tasks required in citizen science projects can be completed by using mobile technologies (Devisch & Veestraeten, 2013). For instance, a smartphone may be used to record the latitude, longitude and altitude of a species, in addition to collecting other information such as specimen photos and phenophase. The use of mobile devices seems to go hand in hand with many citizen science activities as students attain authentic experiences as they attempt to solve real-world issues and contribute to research by learning, collaborating, observing, collecting, reflecting and submitting relevant data (Oberhauser & LeBuhn, 2012).

Using mobile technologies as part of an authentic practice can act as an important bridge between the formal classroom and natural contexts (Liljeström, Enkenberg, & Pöllänen, 2013) and may extend learning potential beyond the activity. Students may continue to use mobile devices to explore and learn while outside of a formal learning environment (Baloch, Rahman, & Ihad, 2012; Mills, Knezek, & Khaddage, 2014). This blurs the line between requisite learning and interest. Moreover, interest in STEM-related fields and excitement toward science have been shown to increase in non-science majors after taking part in authentic scientific practice (Barnett, 2012).

Since many citizen science projects make use of mobile technologies, there is great potential for authentic experiences to take place. The mobile learning and authentic practice (MobiLAP) approach in citizen science contexts is reciprocal and may prove to be a valuable method in creating and helping to maintain STEM interest in participants. Mobile technologies add an additional layer in citizen science projects, which allow participants to have authentic experiences as scientists while using tools for observation, identification, communication, location, recording and reporting data. In many cases, these tools are very similar to, or the same as, the tools that real scientists may use in research. Additionally, the ubiquity, access and personalization of smartphones create even more opportunities for individuals to dive deeper into a subject and potentially shape interest. It is these technological components (i.e., authentic scientific inquiry and real-world problem solving) that may increase participant interest in STEM while citizen science may work in concert to develop interest in science-related subject matter.

The MobiLAP approach is based on the pedagogical framework of place-based education and socio-scientific issues-based instruction. Place-based education focuses on local or regional investigations, is designed around engaging students in examining local problems (Sobel, 2004), and utilizes fieldwork to gather evidence in that local setting (Semken, 2005). Socio-scientific issues, such as climate change effects, are relevant, real-world problems that are informed by science (Sadler, Barab, & Scott, 2007). Addressing these issues requires the use of observational data, evidence-based reasoning, and provides a context for understanding scientific information through an active approach to learning, while placing science content within a social context in a way that supplies both motivation to and the ownership of learning by the student (Sadler, Barab, & Scott, 2007; Zeidler & Nichols, 2009).

The approach of mobile learning and authentic practice in a citizen science context has been shown to aid in constructing STEM interest in participants (Wallace & Bodzin, 2017). This study utilized mobile learning and authentic practice in a citizen science context in order to understand participant attitudes and perceptions of citizen science identity, mobile learning and STEM interest. The treatment group showed significant increases in citizen science identity and STEM interest after using the MobiLAP approach to take part in a phenology-based citizen science project. The comparison group had a "business as usual" classroom experience and did not show any significant increase toward citizen science identity or STEM interest. This implementation had a brief, 4-day intervention and observed only one phenophase; this study highlights the positive effect that the MobiLAP approach can have on participant attitudes and perceptions toward citizen science and STEM education and careers.

While there were several salient outcomes to the aforementioned research project, the study by Wallace and Bodzin (2017) was brief and only scratched the surface of using mobile learning and authentic citizen science projects to foster STEM interest and science learning. A more in-depth qualitative study that takes place across two phenophases and offers participants multiple observation periods to recognize changes in patterns of tree phenology and climate is discussed in this chapter. Phenology can be defined as the observation of seasonal changes in plants and animals related to environmental cues such as seasonal variations in temperature, precipitation and length of days.

Research Questions

This study sought to understand how the use of the MobiLAP approach in citizen science contexts creates STEM interest and impacts perceptions of mobile learning affordances in secondary high school students. This study investigated the following research questions:

1. How did taking part in the MobiLAP approach influence high school student interest in STEM education and careers?
2. How did using the mobile technologies in the MobiLAP approach impact high school student perceptions of mobile learning affordances?

Method

Setting, Population, and Sample

This study took place at a small, suburban vocational high school in the Northeastern United States. The school had a free and reduced lunch rate of 21%. The student demographic breakdown was 58% White, 13% Asian, 11% Black, 2% Multiracial and 16% whom identify as Hispanic. The school was situated on over 80 acres with a variety of plant life and forested areas, which makes it an ideal location for observing and collecting information relevant to the study. Permission from the Institutional Review Board (IRB) was received to conduct the study. Additionally, participant identity was protected through the use of pseudonyms.

The target population for this study was ninth-grade students entering high school in the United States. Students at this age may be best suited for this application depending on curriculum alignment (study sequence of Earth Science, Environmental Science and Biology courses) and other factors such as identity formation and plasticity of educational and career choices. Research literature indicates that forming STEM interest in high school is a vital determinant of future STEM education and career interest (Duncan et al., 2007; Maltese, Melki, & Wiebke, 2014; Sadler, Sonnert, Hazari, & Tai, 2012; van Tuijl & van der Molen, 2016).

A convenience sample of 93 ninth-grade students, ages 14-16, was obtained from science classes at the high school. Intact classes were selected to participate in the project. There was a total of two groups with one group taking part in the MobiLAP approach ($n = 48$) and the other group who completed the citizen science project without mobile devices ($n = 45$).

The Citizen Science Project

This MobiLAP approach citizen science project was implemented over a 5-week period and included two different time periods of phenophase observations. Both groups had a 2-week climate change unit taught by existing teachers during the time of this project. Participants in both groups observed tree phenophase data in two distinct phenophases, which included leaves changing color and leaves falling from trees. By observing the timing of these naturally-occurring events and comparing these data to trends, participants may be able to determine the extent of climate change in their local communities.

Figure 1. Photo collage of observation area during first observation. This image displays the stages of tree phenology at the observation site during the first observation (second week of October 2017). The full-color version of this photograph displays mostly green trees with some slight yellowing of leaf coloring and a minimal amount of leaves on the ground.

Both groups observed and reported phenological changes of local tree species to a citizen science project entitled *Project BudBurst*. This national research project requests observations of the phases of phenology of various plants and trees across the United States. The purpose is to observe if changes in climate cause plant life to alter their natural patterns. For example, *Project BudBurst* data can be used to observe trends and relationships associated with warmer temperatures based on plants entering the blooming phase earlier and entering the fall phase later each year. Over time, these data may show trends regarding the impact of climate change on the patterns of plant species and the organisms that depend on them. Each group made a total of two observations over a 5-week period across October and November, 2017. The purpose of selecting this time period for observations was due to the possibility of observing two, or perhaps three, phenophases as the leaves begin to change color and fall from the trees. Figure 1 shows the stages of tree phenology at the study's observation site during the first observation period which was the second week of October 2017. Figure 2 displays the stages of tree phenology at the observation site during the second observation during the second week of November 2017.

Figure 2. Photo collage of observation area during second observation. This image displays the stages of tree phenology at the observation site during the second observation (second week of November 2017). The full-color version of this photograph shows that the leaves on the trees have mostly changed color to yellow and orange hues and most leaves have fallen to the ground.

Prior to the start of the project, the researcher delivered general information regarding the scope of the project via lecture format and facilitated instruction based on the *Project BudBurst* implementation guide, educational standards and supplemental activities for grades 9-12 (Chicago Botanic Garden, 2017). The MobiLAP group utilized a bring your own device (BYOD) approach with mobile technologies to observe and report data to *Project BudBurst*. There were 10 iPads available for use in the event participants did not have access to smartphones, but none were utilized. A series of smartphone apps enabled participants to identify tree species, locate the coordinates of the species, photograph the species, and submit informational data about the species of plant they are observing. The LeafSnap app (Columbia University, 2011), a free app for iOS, was used to identify trees from the types of leaves participants sampled. This app uses an algorithm similar to facial-recognition software which matches leaves to a tree species. Once a leaf is "snapped," a series of potential results are provided and users are able to observe additional characteristics (e.g., photos of bark, fruit, flowers, seeds and descriptive text) to properly identify each tree. Participants used scientific observation skills by matching characteristics of the species they were observing with the results that were returned from the app. Several participants read the descriptors from the app and attempted to visually or tactilely identify the species. After making careful observations, comparing data with the app and determining if the tree species was native to the location, participants selected the appropriate species. Once participants correctly identified trees, they completed an observation report where they identified the date, tree species, location and the tree's current stage of phenology. Figure 3 shows a collection of screenshots from processes in the LeafSnap app (Columbia University, 2011).

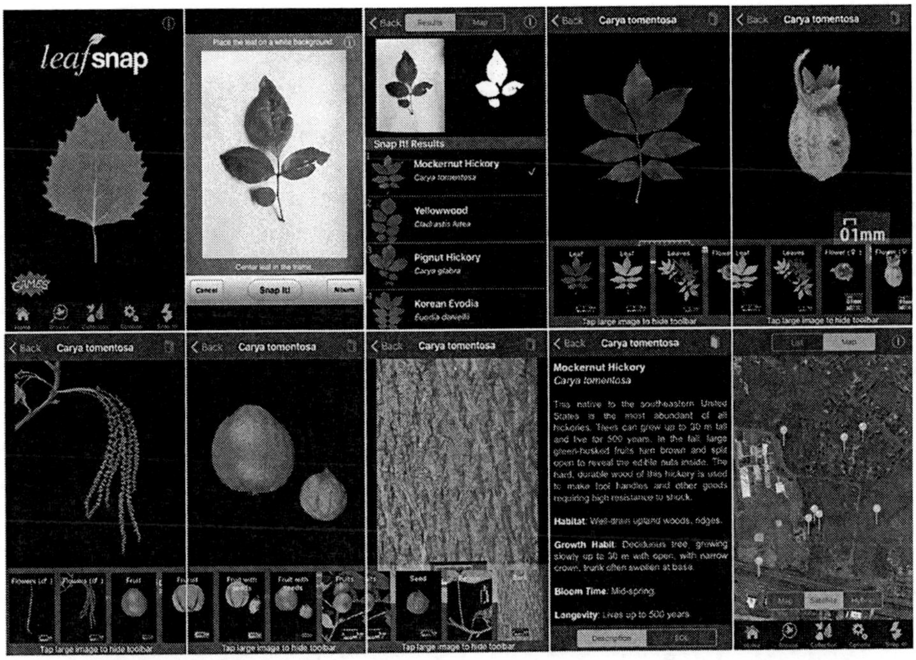

Figure 3. Screenshots from the LeafSnap app (Columbia University, 2011). This image shows a series of screenshots and processes a user might experience when using the LeafSnap app for tree identification. Leafsnap app (2019). Retrieved from https://itunes.apple.com/us/ app/leafsnap/ id4306 49829?mt=8

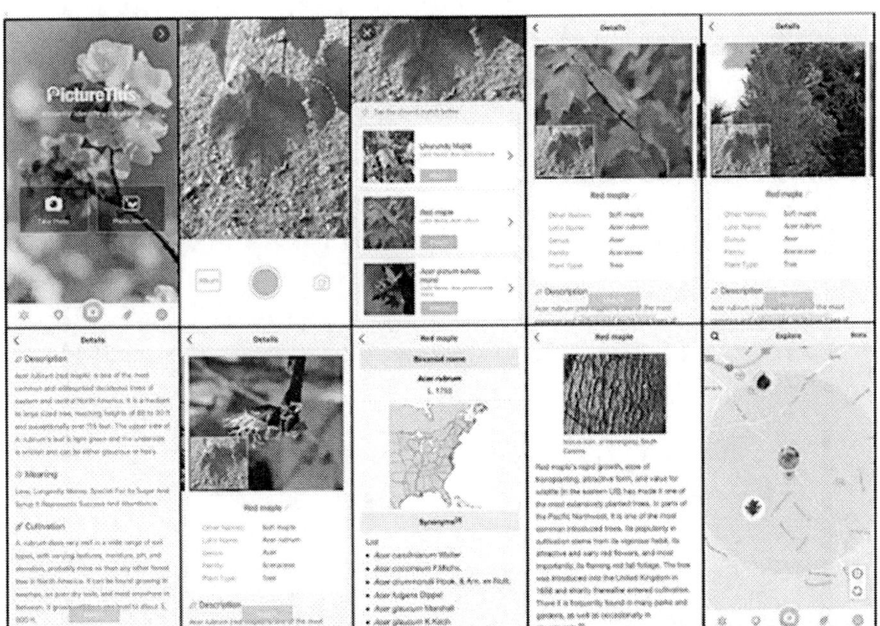

Figure 4. Screenshots from the PictureThis app. This image shows a series of screenshots and processes a user might experience when using the PictureThis app for tree identification. PictureThis – Plant identifier (2019). Retrieved from https://itunes.apple.com/us/app/picturethis-plant-identifier/ id1252497129?mt=8.

Participants in the MobiLAP group with Android devices also made use of the PictureThis app (Dana Technology Inc., 2017), a free app for iOS and Android, which functions similarly to the Leafsnap app (Columbia University, 2011). Figure 4 shows screenshots and processes from the PictureThis app. Once the observations were made and data were collected, participants returned to the classroom.

While in the classroom, the MobiLAP group submitted their observations to *Project BudBurst* via the project website. After the data were submitted, the researcher led a recap, reflection and discussion about the project. MobiLAP participants then completed the 10-item open-ended questionnaire (Appendix A). In addition to this project's activities, the MobiLAP group had a standard 2-week climate change curriculum based on global climate change concepts.

The non-mobile technology (NMT) group was provided information about the project in a lecture format by the researcher. The researcher facilitated instruction based on the *Project BudBurst* implementation guide, educational standards and supplemental activities for grades 9-12 (Chicago Botanic Garden, 2017). This group identified trees using a pocket-sized field guide with dichotomous key (Watts, 1963). Figure 5 shows images from the pocket-sized field guide.

Figure 5. Images of the pocket-sized field guide. This image shows several examples of the field guide the non-mobile technology group used to identify trees. "Tree Finder - A Manual for the Identification of Trees by Their Leaves" by May Theilgaard Watts. Permission to reprint from KEEN Communications, LLC.

Procedures

After each tree was identified, participants observed and recorded the stages of phenology for each tree species. A paper form, clipboard and pen were used to record their observations while out in the field. Once back in the classroom, they submitted their findings through the *Project BudBurst* website. A recap, reflection and group discussion took place to discuss the project. In addition to project activities, the NMT group took part in a 2-week climate change curriculum similar to the MobiLAP group. Following the intervention, participants completed an open-ended survey with 10 items (Appendix B). Qualitative analysis was employed for context and insight into participants' attitudes, perceptions and beliefs. Survey responses were collected through Qualtrics (2019) web-based service. In regard to privacy, each participant was assigned an identifier to ensure anonymity throughout the study.

Data Instruments

A group-specific, open-ended questionnaire was created by the authors and distributed to both the MobiLAP and NMT groups following the intervention. The instruments were created with the purpose of gleaning pertinent information related to high school-aged individuals and their perceptions related to developing STEM interest. Several experts in the areas of environmental science and education provided feedback on the questions and refinement of the instruments. In addition, students in the appropriate age group pilot-tested the instruments and offered further feedback in an effort to use language and formatting appropriate for learners of that age and ability.

The open-ended questionnaire (MobiLAP group) has 10 items and was developed to gain a greater understanding of students' perceptions of using mobile technologies to increase STEM interest, climate change awareness and citizen science. The items related to the topics of citizen science identity, mobile learning, STEM interest, climate change and environmental stewardship.

The open-ended questionnaire (NMT group) consists of 10 items and was created in order to obtain information regarding the perceptions of high school students related to STEM interest, climate change and citizen science. The NMT questionnaire differs from the MobiLAP group questionnaire as it has several items related to completing the project without the use of mobile technologies.

While the MobiLAP group questionnaire (see Appendix A) and the NMT group questionnaire (see Appendix B) are similar, there are a few minor differences related to mobile learning. Select items from the MobiLAP group questionnaire include, "How might using mobile technologies to gather environmental data change your future career path, course of study or interest in STEM?" "How has using a smartphone for this project to identify trees, plot location, conduct research and take photos impacted your thoughts on learning with your smartphone/mobile device?" and "Did using your smartphone for this project make you more interested in any STEM fields? If so, how?" Select items from the NMT group questionnaire include, "How important is it to consider the environment when creating STEM innovations?" "Would using your smartphone for this project make you more interested in any STEM fields?

If so, how?" and "How might your feelings about this project have changed if you were able to use your smartphone to identify trees, plot location, conduct research and take photos?"

Data Analysis

An initial round of a priori coding occurred (Table 1) based on the results from the literature review. Following the initial round of coding, additional coding methods were utilized to revise codes for emerging categories and themes. Due to the epistemological nature of this study's research questions, Saldaña (2015) suggests several coding methods, which will discover the most important concepts from participant responses. Of the coding methods recommended, a simultaneous, line-by-line and split-coding technique was used for first cycle coding, and then researchers employed several coding methods such as *In Vivo* and thematic analysis in an effort to identify patterns in the dataset.

Table 1. Initial Round of a Priori Coding

Code	Description
STEM Interest	Participants responded with affirmations of interest in STEM or science.
STEM Career	Participants responded with positive feelings related to careers in STEM.
No STEM Interest Gained	Participants reported that they did not have any increased interest in STEM.
Mobile Tech Affordances	Participants recorded opportunities for learning and working with mobile technologies.

Participant responses were coded to nodes in NVivo V 11 software and further analyzed as outlined below. While each group was coded separately, many of the codes were similar despite the group of origin. Codes shared by at least one quarter of respondents warranted consideration for further analysis (Harding, 2013) with many becoming major themes. Codes were aggregated into categories and further refined into emerging themes. Analytical memos were recorded after each analytical phase.

The second cycle coding utilized a pattern coding technique to understand relationships and further develop themes. Emergent themes were examined for insight into participants' beliefs, attitudes and perceptions of interest in STEM. These themes led to theoretical assumptions and assertions by the researchers described in the validity section. An example of the coding process is seen in Figure 6.

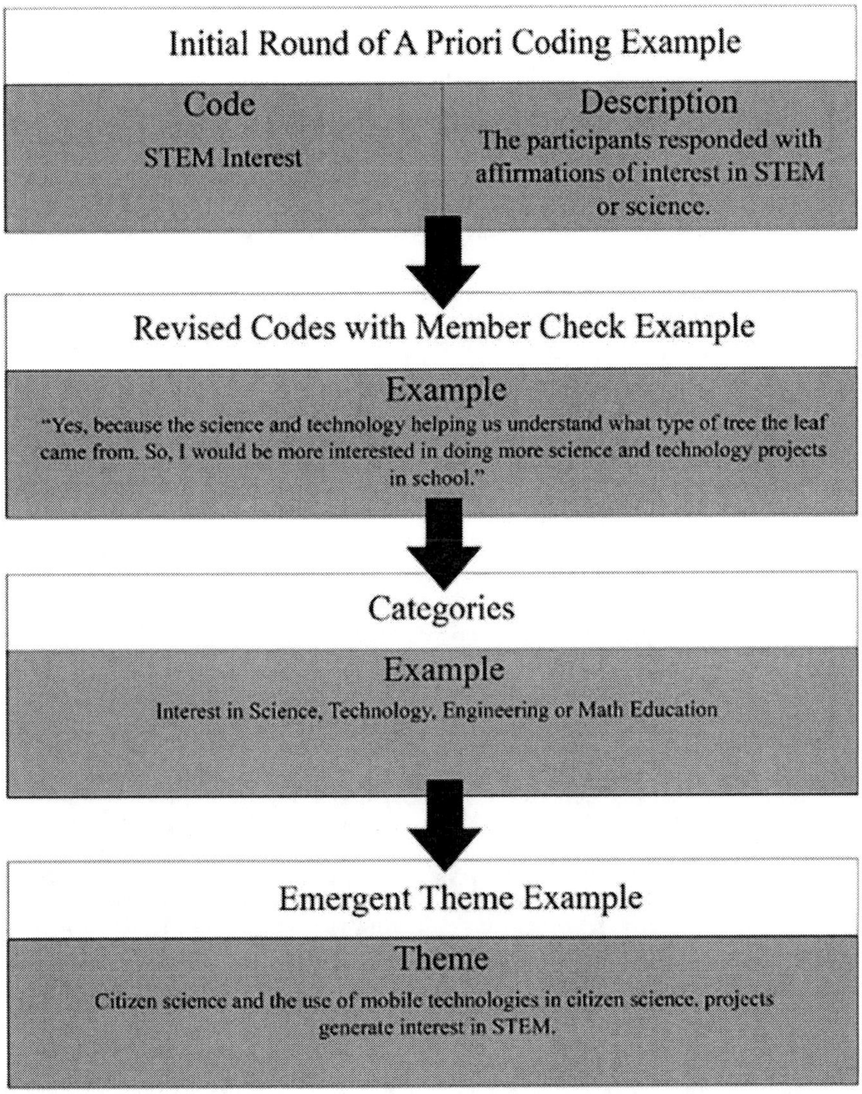

Figure 6. Example of coding process.

Validity

Validity in qualitative studies refers to the appropriateness of the tools, processes, and data (Creswell & Creswell, 2018). Our choice of methodology must enable detection of findings in the appropriate context with due regard to our curriculum implementation context that includes learners making naturalistic observations and using mobile technology to verify and document their observations. Our research assumptions were guided by our personal experiences with secondary-age learners. From our prior work with secondary students, we assumed that they had past experiences with making observations and were quite capable of using mobile technologies for recording observations. For data gathering and analysis, our methods to enhance validity included a well-documented trail of materials and coding processes (Carcary, 2009) that included respondent verification. A member check with 10-12 participants in each group took place to verify open-ended responses. This process consisted of the researcher reading several open-ended responses to each group of participants and asking if the responses represented their thoughts on each item. In addition, a peer-debriefing took place in order to ensure reliability and validity. A listing of revised codes with member check can be found in Table 2.

Table 2. Listing of Revised Codes with Member Check

Code	Description	Example
STEM Interest	Participants responded with affirmations of interest in STEM or science.	"Yes, because the science and technology helping us understand what type of tree the leaf came from. So I would be more interested in doing more science and technology projects in school."
STEM Career	Participants responded with positive feelings related to careers in STEM.	"Citizen science has made me consider different careers. It has increased my interest in STEM and how it could help the environment."
No STEM Interest Gained	Participants reported that they did not have any increased interest in STEM.	"I don't believe it will change my career path or course of study but it could be a hobby for me"
CS Affordances	Participants commented on the attributes related to taking part in citizen science.	"I feel very happy to have helped scientists and the environment by collecting information that may help slow down climate change."
Mobile Learning Affordances	Participants recorded opportunities for learning and working with mobile technologies.	"It would have been a lot easier. I feel that smartphones and technology have become efficient. They could show me easily which species, the location and everything which would have made my study easier."
Mobile Generates STEM Interest	Participants pointed to the use of mobile technologies as a contributing factor in generating STEM interest.	"Using technology to gather data may make you have a stronger connection to the subject. Using technology like phones may make you more interested in STEM since its a daily item you use and can relate it to the subject."

RESULTS

The results of this study addressed questions related to how using the MobiLAP approach in citizen science contexts may create STEM interest in secondary high school students. Several coding methods such as *In Vivo* and thematic analysis were employed across two coding cycles to code the data via NVivo V 11 software.

Research Question One

"How did taking part in the MobiLAP approach influence student interest in STEM education and careers?" resulted in an emerging theme. The theme, "Using mobile technologies in citizen science projects generates interest in STEM," emerged from a large swath of data, which highlights learners' propensity to become more interested in STEM education and careers after taking part in the MobiLAP approach.

Research Question Two

"How did using the mobile technologies in the MobiLAP approach impact high school student perceptions of science learning?" resulted in an emergent theme related to how students perceive their own science learning. This theme, "Mobile technologies provide a personalized, accessible, engaging and efficient way to learn science," was developed from learner feedback about their own perceptions of using mobile technology to observe, understand and take part in scientific study.

Using Mobile Technologies in Citizen Science Projects Generates Interest in STEM

The qualitative data resulted in the emergence of the following theme: Citizen science and the use of mobile technologies in citizen science projects generate interest in STEM. A list of codes, code definitions and examples related to this theme can be found in Table 3. The open-ended survey questions which yielded data supporting this theme and addressing the first research question were: (a) How might using mobile technologies to gather environmental data change your future career path, course of study or interest in STEM? (b) How has this project influenced how you think about citizen science? (c) Did using your smartphone for this project make you more interested in any STEM fields? If so, how? (d) Did using your smartphone for this project make you feel like a citizen scientist? If so, how? (e) Would using your smartphone for this project make you more interested in any STEM fields? If so, how? (f) Would using your smartphone for this project make you feel like a citizen scientist? If so, how? Student responses were generally limited to one or two sentences but provided insight into their feelings toward STEM after taking part in this project.

Participants responded with statements that conveyed an increased general interest in STEM. Often, they pointed to citizen science or "the project" as having a direct correlation to their enhanced interest due to increased understandings of climate change. Jack stated, "Taking part in this citizen science project advanced my interest in STEM, since it helped me to understand the natural world and the effect of climate change on it." Additionally, students who took part in the study may not have previously had local outdoor learning experiences where they interacted with and observed nature for the purpose of solving real-world issues. Taking part in this citizen science project provided them with opportunities that they may have thought were unattainable in high school. Hugo responded, "This definitely changed my interest in STEM because this project helped me interact with citizen science and environmental problems that I thought were beyond me in school."

Several participants, such as Claire, noted an enhanced interest in a specific STEM subject area (science education). "I feel that taking part in this citizen science project will persuade me to take it one step further and maybe take a science course or something similar to learn more about climate change and nature." John, another participant, expressed a desire to further his learning in science as well as potentially looking into a science-focused career: "This project gave me more insight into what scientists do. I may pursue more advanced scientific jobs or studies in my future."

Participants reported having increased interest in STEM related careers after taking part in this citizen science project. Penny stated, "This citizen science project is making me more likely to take a STEM-related career as I now better see how making better tools and innovation can lead to scientific observation." Often, they associated outdoor activities and interaction with plant life as key reasons for an increased interest in STEM careers such as in the following example from Ricardo:

> Being outside and studying the leaves has definitely increased my interest in STEM, especially the science aspect of it. I did not know that research is so specific, like knowing the rate at which the leaves unfold. I think this could somewhat change my future career plan. (Ricardo)

Table 3. Codes, Definitions, and Examples for the Theme: Using Mobile Technologies in Citizen Science Projects Generates Interest in STEM

Code and Definition	Examples
STEM Interest Participants responded with affirmations of interest in STEM or science.	1. "Citizen science has made me consider different careers. Also, it has made me consider a different course of study. It has increased my interest in STEM and how it could help the environment." 2. "Taking part in this citizen science project advanced my interest in STEM, since it helped me to understand the natural world and the effect of climate change on it." 3. "This definitely changed my interest in STEM because this project helped me interact with citizen science and environmental problems that I thought were beyond me in school."
STEM Career Participants responded with positive feelings related to careers in STEM.	1. "This citizen science project is making me more likely to take a STEM related career as I now better see how making better tools and innovations can lead to scientific observation." 2. "Being outside and studying the leaves has definitely increased my interest in STEM, especially the science aspect of it. I did not know that research is so specific, like knowing the rate at which the leaves unfold. I think this could somewhat change my future career plan." 3. "It gave me another career path to look into. STEM is a very flexible science meaning as there is a lot you can do in going in STEM."
Science Interest Participants specifically pointed to having an increased interest in science.	1. "This project gave me more insight into what scientists do. I may pursue more advanced scientific jobs or studies in my future." 2. "I feel that taking part in this citizen science project will persuade me to take it one step further and maybe take a science course or something similar to learn more about climate change and nature." 3. "Taking part in this can help making science more interesting."
Mobile Generates STEM Interest Participants pointed to the use of mobile technologies as a contributing factor in generating STEM interest.	1. "Using technology to gather data may make you have a stronger connection to the subject. Using technology like phones may make you more interested in STEM since it's a daily item you use and can relate it to the subject." 2. "Naturally using my smartphone would have made the project easier so it's more likely that with the positive opinion towards citizen science projects using a smartphone would have generated, I would have been more likely to be interested in STEM fields." 3. "Using my smartphone would have made me more interested in any STEM fields because I would have been using technology. I may have become more interested in the way that it works while using it."

The primary difference between the MobiLAP group and the NMT group was that the MobiLAP group was permitted to use smartphones for the project and the NMT group was not. As such, the questionnaires were identical except for the mobile technology-related questions. Both groups reported that using smartphones for the project had (or would have) generated interest in STEM. For example, Charlie, a participant in the MobiLAP group stated, "Mobile technologies are going to make STEM more interesting because it's a hands-on thing for kids and adults to do." Similarly, the NMT group perceived that the use of smartphones would have resulted in an increased interest in STEM. Libby stated, "Using my smartphone would have made me more interested in any STEM fields because I would have been using technology. I may have become more interested in the way that it works while using it." Table 3 shows the codes, definitions, and select examples for the theme: "Using mobile technologies in citizen science project generates interest in STEM."

The qualitative findings revealed that participants in the both the NMT and MobiLAP groups generated interest in STEM education and careers following the intervention. Participants responded to open-ended questions with affirmations of interest in general STEM or science. Codes related to general STEM interest totaled 32 in the MobiLAP group and 28 in the NMT group. Many participants made statements with positive feelings toward STEM careers. There were 19 instances of data coded in the "STEM career" category for the MobiLAP group and 16 instances in the NMT group. Some participants responded with specific interest toward science. The "science interest" code had 12 occurrences in the MobiLAP group and 15 in the NMT responses. The final code related to STEM interest is "Mobile generates STEM interest." This code consisted of 13 instances (in each treatment group) where participants pointed to the use of mobile technologies as a contributing factor in creating STEM interest.

Mobile Technologies Provide a Personalized, Accessible, Engaging and Efficient Way to Learn Science

The qualitative data analysis resulted in the following emerging theme which relates to research question 2: Mobile technologies, such as smartphones afford learners a personalized, accessible, engaging and efficient way to learn science and scientific principles. A list of codes, code definitions and examples related to this theme can be found in Table 4. The open-ended survey items related to this research question include: (a) How might using mobile technologies to gather environmental data change your future career path, course of study or interest in STEM? (b) How has this project influenced how you think about citizen science? (c) How has using a smartphone for this project to identify trees, plot location, conduct research and take photos impacted your thoughts on learning with your smartphone/mobile device? (d) Did using your smartphone for this project make you feel like a citizen scientist? If so, how? (e) Did using your smartphone for this project make you more interested in any STEM fields? If so, how? (f) Would using your smartphone for this project make you feel like a citizen scientist? If so, how? (g) Would using your smartphone for this project make you more interested in any STEM fields? If so, how? Most participants responded to the questions with at least one sentence.

Many of the participants in both treatment groups were able to cite benefits of mobile technologies regarding their learning and ability to do school-related tasks. The majority of respondents noted that using smartphones in conjunction with the citizen science project would make identifying tree species faster and easier when compared to paper-based methods. A few responses regarding this perception included Paulo's, Ilana's and Pierre's respective comments: "I feel that smartphones and technology become efficient. They could show me easily which species, the location and everything, which would have made my study a lot easier," and "Using a smartphone is making learning easier and more fun. If we didn't use a smart phone, then we would have had a very hard time getting information," and "It made the discovery of different tree types easier and quicker." More specifically, several participants like Eloise pointed to the accessibility of information that comes with having access to the internet:

> The project would be much, much, MUCH easier with smartphone technology as there would be real pictures to compare to and an unlimited amount of information to base findings off of . . . If we didn't use a smart phone, then we would have had a very hard time getting information . . . Using a phone to identify and research makes me realize how dependent we are on technology for the right reasons. Phones can help us in a million ways with finding out new information. (Eloise)

Table 4. Codes, Definitions, and Examples for the Theme Mobile Technologies

Code	Definition	Examples
Mobile Learning Affordances	Participants recorded opportunities for learning and working with mobile technologies.	1. "It would have been a lot easier. I feel that smartphones and technology have become efficient. They could show me easily which species, the location and everything which would have made my study easier." 2. "I really liked working with my phone to take pictures and look at databases full of different types of trees and I will probably keep the app I downloaded to look at more trees sometimes later." 3. "Using a phone to identify and research makes me realize how dependent we are on technology for the right reasons. Phones can help us in a million ways with finding out new information."
Mobile Generates STEM Interest	Participants pointed to the use of mobile technologies as a contributing factor in generating STEM interest.	1. "Mobile technology can change the way I look at STEM. It can make it seem more fun and might make me want to be more involved." 2. "Using technology to gather data may make you have a stronger connection to the subject. Using technology like phones may make you more interested in STEM since it's a daily item you use and can relate it to the subject." 3. "I'm more interest in STEM because it made it seem more real when we're using something we always have with us."

Students such as Rose and Ethan also identified the "just in time" nature and accessibility of mobile learning with comments, such as "(Mobile technology) can help us learn what we need to know faster than reading an entire book about it," and "It was very easy and faster to use a smartphone rather than looking through a book or something or having to remember".

It should be noted that smartphones were used by students to verify their own naturalistic observations. Students used the smartphone app to identify a tree with a digital image. While students did not go through a detailed series of observations with a taxonomic key to identify a tree species (leaf shape, veins, fruit, etc.), the use of the smartphone app likely decreased the number of identification errors that typically occur when high school students use paper-based images that often do not reflect the actual leaf appearance in nature. Therefore, there is a trade-off between making multiple observations with a taxonomic key versus using a smartphone to verify an authentic image-based observation.

Participants in the MobiLAP group touched on the personalized aspect of using their own smartphones and appreciated the familiarity by making statements like Tom's: "It made it seem more real when we're using something we always have with us," and Boone's: "It's easy to learn on something you are so used to using and know how to work with." Several participants even mentioned that they may extend the learning beyond the classroom as a result of having the apps available on their own smartphones. A couple of examples of this sentiment included Sayid's comments: "I really liked working with my phone to take pictures and look at data bases full of different types of trees, and I will probably keep the app I downloaded to look at more trees sometimes later," and Elizabeth's: "It's interesting so I might explore trees in my own area."

Participants in both groups reported that mobile technologies aided their learning. Several examples of this metacognition from the MobiLAP group include responses similar to Cindy's:

> For me, learning with my phone is easy and fun, so this project helped with that. It made this project more entertaining and interactive, which helped me learn better. . .The smartphone is really helpful in learning new things, especially to do with science and climate change . . . I think learning with phones helps us relate and understand more than if we didn't."

Participants in the NMT group also shared thoughts related to understanding their own learning with mobile technologies, such as Vincent's and Emma's responses: "Yes, because I would have quick access to a very large amount of information about STEM fields. In addition, I feel that using technology would help enhance my learning experience," and "Using my smartphone would definitely have been easier, and it would have allowed for a deeper understanding."

Several NMT respondents touched on a crucial component of citizen science projects (i.e., obtaining accurate data). These participants remarked that using mobile technologies for tree identification would result in more accurate results. A few of these comments such as Scott's, Steve's and Martin's were: "Using a smartphone would make me feel like a citizen scientist because I could be accurate in my results and record them more efficiently than on a worksheet," "Using a smartphone/mobile device would have been a lot faster and accurate," and "It could be more accurate and it could take less time." Participants in the MobiLAP group did not record any issues with accuracy or lack of confidence in their tree identifications. They recorded statements which conveyed confidence in tree identification, such as "We could determine the type of tree or plant in a matter of seconds," "We were deep into learning about the trees and finding more information about them," and "It made the process of figuring out what were the names of the plants and leaves simple." The MobiLAP group pointed to the use of their smartphones as a method to explore nature. Several examples like Goodwin's include, "It really shocked me that my phone can take a picture of a leaf and the app shows the leaf's information. Exploring outside was better when you have the information right there," and "Using cell phones and technology throughout this project made seeing nature more in detail and more interesting." Goodwin also noted that using mobile technology in a natural setting helped him to be engaged in the activities and learning. Select examples of this sentiment included responses like Zoe's and Roger's: "Using my smartphone made me more engaged with looking at the trees," and "It made this project more entertaining and interactive."

Participants in both treatment groups reported that the use of mobile technologies created (or would create) STEM interest. The NMT group responded with perceptions of how using mobile technologies in the project would have impacted their interest in STEM. A few examples included "Using my smartphone would make me more interested in STEM fields because I could see the connection between technology and science," "Naturally using my smartphone would have made the project easier, and I would have been more likely to be interested in STEM fields," and "Using my smartphone would have made me more interested in any STEM fields because I would have been using technology. I may have become more interested in the way that it works while using it." Several examples of a gain in STEM interest from the MobiLAP group included Omar's: "Yes, because the science and technology helping us understand what type of tree the leaf came from. So, I would be more interested in doing more science and technology projects in school." Stuart stated, "It (the use of a smartphone) did make me more interested in STEM fields," and Danny said, "Using technology to gather data may make you have a stronger connection to the subject. Using technology like phones may make you more interested in STEM since it's a daily item you use and can relate to the subject."

The qualitative findings highlighted participants' understandings of learning with mobile devices. Many statements from both groups indicated that participants view mobile technologies as tools that make learning faster and easier. These and other examples comprised the "mobile learning affordances" code, which had 38 MobiLAP and 33 NMT instances where participants identified opportunities for learning and working with mobile devices. Another aspect of learning with mobile devices was apparent as 13 responses from each group were coded to the "Mobile generates STEM interest" code. Responses in this code point to the use of mobile technologies as a contributing factor in generating STEM interest.

Discussion

Participants that took part in this citizen science project had an increased interest in STEM education and careers. While both treatment groups had similar outcomes in these areas, the MobiLAP group's responses showed that the use of mobile technologies benefitted their interest in STEM, while the NMT group seemed to recognize the need for technology in order to enhance the project. The use of mobile technologies was expected to have a greater impact in terms of participants' understanding of the affordances of mobile learning. The emergent themes for both groups revealed that participants felt that smartphones made learning faster or easier. The participants understood the value their smartphones can bring to their learning, but they did not connect that idea with this project. Since smartphones are such an integral part of teen life, it is possible that teens do not grasp the metacognition associated with learning via mobile technologies. The ubiquity and accessibility of mobile devices may cause individuals to take their smartphones for granted in terms of using them as a learning tool. According to Terras and Ramsay (2012), learners need to develop specific metacognitive skills to understand the benefits and challenges of mobile learning.

In using any technology to aid the learning process, there are challenges with the user's proficiency and comfortability with the technology. Compounding this issue is the introduction of new course material and building scientific observational skills. For learners to properly connect outdoor observations with deeper scientific learning, they require a scaffolding of knowledge, which stems from access to content, a natural example and supportive technology tools to capture data (Tabak, 2004). In addition, learners' scientific knowledge construction will greatly benefit from socio-technical guidance from those that have a firmer grasp on scientific subject matter, observation skills and technical ability (Land & Zimmerman, 2015)

Motivation is another contributing factor, which helps to determine a learner's level of metacognition in mobile learning (Sha, Looi, Chen, Seow, & Wong, 2012). Terras and Ramsay (2012) contend that metacognition is essential to mobile learning and that learners must have the psychological infrastructure to learn in this manner. While teens are motivated to use their smartphones for a variety of personal reasons, they may not be motivated to use them as a formal learning tool or understand how mobile device impact their learning. In fact, research by Common Sense Media (2016) revealed that 59% of parents believe that their teens are addicted to smartphones while 50% of teens feel as though they are themselves addicted to smartphone use. However, a study by Chen, Liu, and Hwang (2016) found that the type of mobile learning used can improve student motivation to learn. The Chen et al. study examined two groups of learners using two different forms of mobile learning. While both groups gained in terms of learning achievement, only the mobile learning group that used a gaming strategy significantly improved motivation to learn. The MobiLAP study did not measure motivation; however, participants did cite being engaged in the learning via use of their mobile devices.

Creating STEM Interest

Several studies have determined that an individual's high school years are a crucial time period to create and maintain career interest in STEM (Maltese et al., 2014; Sadler et al., 2012). Other research has shown that having STEM experiences between the ages of 8-16 years is a vital determinant of STEM education and career decisions (van Tuijl & van der Molen, 2016). Participants in both the MobiLAP group and the NMT group displayed an interest in STEM as a result of their participation in this citizen science project. This study also corroborates studies that contend that citizen science can be a motivating factor in choosing STEM careers (Hiller & Kitsantas, 2014) and building and sustaining STEM interest (Toomey & Domroese, 2013; VanMeter-Adams et al., 2014).

This study expands on the work of Wallace and Bodzin (2017) in that using mobile learning and the authentic practice of a citizen science project created STEM interest in participants. The MobiLAP approach, when used in citizen science contexts, leads to interest in both STEM education as well as STEM careers. Results from this study reaffirm findings from the first iteration of the MobiLAP approach. Students in both the MobiLAP and MNT groups reported much interest in STEM at the end of the project implementation. The findings from the qualitative data suggested that the MobiLAP group's use of mobile technologies did have an impact in terms of creating STEM interest. Participants in this group pointed specifically to the use of smartphones as a cause for increased interest in STEM fields.

Affordances of Mobile Learning in Citizen Science

Prior research has shown that mobile learning can help to create STEM interest in participants (Gilliam, Bouris, Hill, & Jagoda, 2016; Metcalf, Milrad, Cheek, Raasch, & Hamilton, 2008; Terkowsky, Haertel, Bielski, & May, 2013). Using mobile technologies instills learner confidence in the subject matter (Al Mosawi & Wali, 2016) and, when used to learn about climate change, can provide learners with additional tools to scaffold knowledge and understand content (Pettit, & Kukulska-Hulme, 2011). The MobiLAP group used apps on their smartphones to contribute to a citizen science project related to climate change. Students used scientific processes to make observations, identify tree species, determine phenophases, compare data and consider the impact that climate change has on their local environment. Interestingly, several participants in the NMT group seemed to lack confidence in their observations and commented that using mobile devices to identify tree species would be faster and more accurate than their paper-based method. While measuring accuracy in tree identification was not a goal of this study, the MobiLAP group did not provide statements, which exhibited a lack of confidence in their ability to identify tree species with the use of their apps. Research has shown that learners who receive positive external feedback are more confident and self-sufficient when using mobile technologies to learn. Likewise, if learners have consistent negative experiences with mobile learning, they will lack confidence in regard to mobile learning (Razzaq, Samiha, & Anshari, 2018). The apps used in this study instantly returned results for the user to further examine and verify their naturalistic observations. Qualitative data from both groups noted that students had a better understanding of climate change after taking part in the study.

Even though school-issued iPads were available, all participants owned and used a smartphone, which highlights the availability of mobile devices and their accessibility within education (Willemse & Bozalek, 2015). Participants in this study were able to access tree information on-demand, which is a central benefit of mobile technology (Hummel & Hlavacs, 2003). Tools such as this afford learners with an ever-present knowledge construction that they can take advantage of during formal and informal learning opportunities (Peng, Su, Chou, & Tsai, 2009). Several participants from the MobiLAP group expressed interest in continuing learning with mobile technologies outside of regular class time.

The results of this study indicated that the mobile technologies were especially effective in aiding the contextualization and visualization of knowledge (Baloian & Zurita, 2012). MobiLAP participants benefitted by comparing tactile samples and observations with vivid photos and descriptions of tree species in their own community. Allowing participants to observe tree phenology in the context of their own community, seems to have greatly shaped their understandings of climate change. Many participants made statements related to personally experiencing climate change, which turned an abstract concept into a reality. If "seeing is believing" then students who observed the delay in tree phenophase due to unseasonably warm temperatures were afforded an appropriate context and visualization of knowledge to believe climate change is real and happening in their communities. Experiential learning, such as this, is similar to findings from MacCallum, Day, Skelton, and Verhaart (2017).

In addition to visualization of knowledge and contextualization, mobile learning also provides opportunities for collaborative learning and connectivism (MacCallum et al., 2017).

using authentic data, concurrently instilling important STEM-related skills while providing experiences that develop students' professional capacity and potential for self-identification as members of a STEM community. Future studies may seek to understand how students develop skills that are essential in the STEM workforce and add new knowledge on how the MobiLAP approach may increase students' motivation to pursue STEM career trajectories. There are many opportunities for place-based mobile citizen science curriculum to be designed to promote understandings of science phenomena and environmental issues on a variety of topics related to the NGSS including trees and ecological services, urban heat island effects, and land use change.

CONCLUSION

The need for qualified STEM workers and the urgency in solving major issues such as climate change could potentially be solved by promoting citizen science in schools. The approach of mobile learning and the authentic practice of a citizen science project is a mechanism which builds STEM interest in participants. Utilizing citizen science in schools may lead individuals to care more deeply about the environment, have an increased interest in STEM and have a more positive attitude toward science and technology. The individuals in this study may continue to strengthen their respective interests, which may lead to positive change in the areas of continuing STEM education and choosing STEM careers.

This study also demonstrated the practical application of the MobiLAP approach, which adheres to many key components of the NGSS. The conceptual framework of NGSS lends itself to mobile learning and the authentic practice of taking part in a citizen science project. Participants who took part in this project benefitted from outdoor learning activities, addressing real-world problems, engaging in community and educating in an authentic manner. The aforementioned aspects of this project directly connect to NGSS standards, which may aid teachers who are required to adopt the standards.

In an era where teachers feel pressure to align content with standards and meet learning objectives, utilizing the MobiLAP approach may be a method by which students can be engaged in their learning while at the same time meeting important learning goals. The access and availability of smartphones to use in learning environments, both inside and outside of school, creates enhanced experiential opportunities in which educational institutions, teachers and students may improve teaching and learning. The MobiLAP approach may ultimately help to shape student interest and prepare learners with important skills to meet the needs of the 21st Century.

APPENDIX A

Open-Ended Questionnaire (Mobile Technology Group)

1. How might using mobile technologies to gather environmental data change your future career path, course of study or interest in STEM?
2. How important is it to consider the environment when creating STEM innovations?

Appendix A (Continued)

3. How have your feelings toward climate change been affected by taking part in this citizen science project?
4. How has this project influenced how you think about citizen science?
5. How has using a smartphone for this project to identify trees, plot location, conduct research and take photos impacted your thoughts on learning with your smartphone/mobile device?
6. How has this citizen science project changed the way you think about humankind's impact on climate change?
7. How has taking part in this project impacted your thoughts on caring for the environment?
8. Did using your smartphone for this project make you feel like a citizen scientist? If so, how?
9. Did using your smartphone for this project make you more interested in any STEM fields? If so, how?
10. Would you like to provide any additional feedback about this project? If so, please respond.

Appendix B

Open-Ended Questionnaire (Non-Mobile Technology Group)

1. How might taking part in this citizen science project change your future career path, course of study or interest in STEM?
2. How important is it to consider the environment when creating STEM innovations?
3. How have your feelings toward climate change been affected by taking part in this citizen science project?
4. How has this project influenced how you think about citizen science?
5. How might your feelings about this project have changed if you were able to use your smartphone to identify trees, plot location, conduct research and take photos?
6. How has this project changed the way you think about humankind's impact on climate change?
7. How has taking part in this project impacted your thoughts on caring for the environment?
8. Would using your smartphone for this project make you feel like a citizen scientist? If so, how?
9. Would using your smartphone for this project make you more interested in any STEM fields? If so, how?
10. Would you like to provide any additional feedback about this project? If so, please respond.

REFERENCES

Adams Becker, S., Freeman, A., Giesinger Hall, C., Cummins, M. & Yuhnke, B. (2016). *NMC/CoSN horizon report: 2016 K-12 edition*. Austin, TX: The New Media Consortium.

Anderson, M. (2014, September 15). *The race gap in science knowledge*. Retrieved from http://www.pewresearch.org/fact-tank/2015/09/15/the-race-gap-in-science-knowledge/

Al Mosawi, A. & Wali, E. A. (2016). Exploring the potential of mobile applications to support learning and engagement in elementary classes. In M. Khosrow-Pour (Ed.), *Human-computer interaction: Concepts, methodologies, tools, and applications*, (pp. 1538-1550). Hershey, PA: IGI Global.

Baloch, H. Z., Rahman, A. A. & Ihad, N. A. (2012). Mobile collaborative informal learning design: Study of collaborative effectiveness using activity theory. *International Journal of Interactive Mobile Technologies (iJIM)*, *6*(3), 34-41. doi:10.3991/ijim.v6i3.2090

Baloian, N. & Zurita, G. (2012). Ubiquitous mobile knowledge construction in collaborative learning environments. *Sensors*, *12*(6), 6995-7014. doi: 10.3390/s120606995

Barnett, R. (2012). Learning for an unknown future. *Higher Education Research & Development*, *31*(1), 65-77. doi:10.1080/07294360.2012.642841

Baya'a, N. F. & Daher, W. M. (2009). Learning mathematics in an authentic mobile environment: The perceptions of students. *International Journal of Interactive Mobile Technologies*, *3*(1), 6-14. doi:10.3991/ijim.v3s1.813

Carcary, M. (2009). The research trail audit – enhancing trustworthiness in qualitative inquiry. *The Electronic Journal of Business Research Methods*, *7*(1), 11-24.

Chen, C., Liu, G. & Hwang, G. (2016). Interaction between gaming and multistage guiding strategies on students' field trip mobile learning performance and motivation: Gaming and guiding strategies for mobile learning. *British Journal of Educational Technology*, *47*(6), 1032-1050. doi:10.1111/bjet.12270

Chicago Botanic Garden. (2017). *Project BudBurst for grades 9-12 educators*. Retrieved from http://budburst.org/education_9-12

Columbia University. (2011). *Leafsnap*. Retrieved from http://leafsnap.com/

Common Sense Media. (2016). *Dealing with devices: The parent-teen dynamic. Are we addicted?* Retrieved from https://www.commonsensemedia.org/technology-addiction-concern-controversy-and-finding-balance-infographic#

Creswell, J. W. & Creswell, J. D. (2018). *Research design. Qualitative, quantitative, and mixed methods approaches* (5th ed.). Los Angeles, CA: SAGE Publications, Inc.

Dana Technology Inc. (2017). *PictureThis*. Retrieved from https://www.picturethisai.com

De Pietro, O. (2013). Authentic and situated learning with the use of an adaptive search engine and a QR-code in mobile mode. *International Journal of Digital Literacy and Digital Competence*, *4*(3), 19-30. doi:10.4018/ijdldc.2013070103

Devisch, O. & Veestraeten, D. (2013). From sharing to experimenting: How mobile technologies are helping ordinary citizens regain their positions as scientists. *Journal of Urban Technology*, *20*(2), 63-76. doi:10.1080/10630732.2013.769313

Duncan, G. J., Dowsett, C. J., Claessens, A., Magnuson, K., Huston, A. C., Klebanou, P. & Japel, C. (2007). School readiness and later achievement. *Developmental Psychology*, *43*(6), 1428-1446. doi:10.1037/0012-1649.43.6.1428

Gilliam, M., Bouris, A., Hill, B. & Jagoda, P. (2016). The source: An alternate reality game to spark STEM interest and learning among underrepresented youth. *Journal of STEM Education: Innovations and Research*, *17*(2), 14-20. doi: 10.1007/s10956-016-9679-4

Harding, J. (2013). *Qualitative data analysis from start to finish*. Thousand Oaks, CA: SAGE Publications Inc.

Hiller, S. E. & Kitsantas, A. (2014). The effect of a horseshoe crab citizen science program on middle school student science performance and STEM career motivation. *School Science and Mathematics*, *114*(6), 302-311. doi:10.1111/ssm.12081

Hsu, Y. C. & Ching, Y. H. (2012). Mobile microblogging: Using Twitter and mobile devices in an online course to promote learning in authentic contexts. *The International Review of Research in Open and Distributed Learning*, *13*(4), 211-227. doi:10.19173/irrodl. v13i4.1222

Huang, C. S., Yang, S. J., Chiang, T. H. & Su, A. (2016). Effects of situated mobile learning approach on kearning motivation and performance of EFL students. *Journal of Educational Technology & Society*, *19*(1), 263-276.

Hummel, K. A. & Hlavacs, H. (2003, January). Anytime, anywhere learning behavior using a web-based platform for a university lecture. Session presented at the Proceedings of the SSGRR 2003 Winter Conference, L'Aquila, Italy.

Lamb, A. (2016). Citizen science part 1: Place-based STEM projects for school libraries. *Teacher Librarian*, *43*(4), 64-69.

Land, S. & Zimmerman, H. (2015). Socio-technical dimensions of an outdoor mobile learning environment: A three-phase design-based research investigation. *Educational Technology Research & Development*, *63*(2), 229–255. doi:10.1007/s11423-015-9369-6

Liljeström, A., Enkenberg, J. & Pöllänen, S. (2013). Making learning whole: An instructional approach for mediating the practices of authentic science inquiries. *Cultural Studies of Science Education*, *8*(1), 51-86. doi:10.1007/s11422-012-9416-0

Liu, M., Horton, L., Kang, J., Kimmons, R. & Lee, J. (2013). Using a ludic simulation to make learning of middle school space science fun. *International Journal of Gaming and Computer-Mediated Simulations (IJGCMS)*, *5*(1), 66-86. doi:10.4018/jgcms. 2013010105

MacCallum, K., Day, S., Skelton, D. & Verhaart, M. (2017). Mobile affordances and learning theories in supporting and enhancing learning. *International Journal of Mobile and Blended Learning (IJMBL)*, *9*(2), 61-73. doi: 10.4018/ijmbl.2017040104

Maltese, A. V., Melki, C. S. & Wiebke, H. L. (2014). The nature of experiences responsible for the generation and maintenance of interest in STEM. *Science Education*, *98*(6), 937-962. doi:10.1002/sce.21132

Manson, S. M., Martinez, D. F., Buchwald, D. S., Rubio, D. M. & Moss, M. (2015). Vision, identity, and career in the clinical and translational sciences: Building upon the formative years. *Clinical and Translational Science*, *8*(5), 568-572. doi:10.1111/cts.12316

Metcalf, D., Milrad, M., Cheek, D., Raasch, S. & Hamilton, A. (2008). My sports pulse: Increasing student interest in STEM disciplines through sports themes, games and mobile technologies. *5th IEEE International Conference on Wireless, Mobile, and Ubiquitous Technology in Education (wmute 2008)*. doi:10.1109/wmute.2008.38

Meyer, N. J., Scott, S., Strauss, A. L., Nippolt, P. L., Oberhauser, K. S. & Blair, R. B. (2014). Citizen science as a REAL environment for authentic scientific inquiry. *Journal of Extension*, *52*(4), 1-6.

Mills, L. A., Knezek, G. & Khaddage, F. (2014). Information seeking, information sharing, and going mobile: Three bridges to informal learning. *Computers in Human Behavior*, *32*(1) 324-334. doi:10.1016/j.chb.2013.08.008

National Science Board (US). (2007). *A national action plan for addressing the critical needs of the US science, technology, engineering, and mathematics education system*. National Science Foundation. Retrieved from https://www.nsf.gov/pubs/2007/nsb07114/nsb07 114.pdf

NGSS Lead States. (2013). *Next generation science standards: For states, by states*. Washington, DC: The National Academies Press. Retrieved from https://www.nap.edu/ read/18290/chapter/1

Oberhauser, K. & LeBuhn, G. (2012). Insects and plants: Engaging undergraduates in authentic research through citizen science. *Frontiers in Ecology and the Environment*, *10*(6), 318-320. doi:10.1890/110274

Palermo, E., Laut, J., Nov, O., Cappa, P. & Porfiri, M. (2017). A natural user interface to integrate citizen science and physical exercise. *PloS One*, *12*(2), e0172587. doi: 10.1371/journal.pone.0172587

Peng, H., Su, Y., Chou, C. & Tsai, C. (2009). Ubiquitous knowledge construction: Mobile learning redefined and a conceptual framework. *Innovations in Education and Teaching International*, *46*(2), 171-183. doi:10.1080/14703290902843828

Pettit, J. & Kukulska-Hulme, A. (2011). Mobile 2.0: crossing the border into formal learning? In M.J. Lee & C. McLoughlin (Eds.), *Web 2.0-based E-Learning: Applying social informatics for tertiary teaching*, (pp. 192–208). Hershey PA: Information Science Reference.

President's Council of Advisors on Science and Technology. (2010). *Prepare and inspire: K-12 education in science, technology, engineering, and math (STEM) for America's future*. Retrieved from https://www.whitehouse.gov/sites/default/files/microsites/ostp/pcast-stem-ed-final.pdf

Qualtrics. (2019). Retrieved from https://www.qualtrics.com

Ranieri, M. & Pachler, N. (2014). Inventing and re-inventing identity: Exploring the potential of mobile learning in adult education. *Prospects*, *44*(1), 61-79. doi:10.1007/s11125-014-9294-1

Razzaq, A., Samiha, Y. T. & Anshari, M. (2018). Smartphone habits and behaviors in supporting students' self-efficacy. *International Journal of Emerging Technologies in Learning*, *13*(2), 94–109. doi:10.3991/ijet.v13i02.7685

Riegle-Crumb, C., Moore, C. & Ramos-Wada, A. (2011). Who wants to have a career in science or math? Exploring adolescents' future aspirations by gender and race/ethnicity. *Science Education*, *95*(3), 458-476. doi:10.1002/sce.20431

Robinson, C. F. & Kakela, P. J. (2006). Creating a space to learn: A classroom of fun, interaction, and trust. *College Teaching*, *54*(1), 202-206. doi:10.3200/CTCH.54.1.202-207. doi:10.3200/ctch.54.1.202-207

Ruiz-Mallén, I., Riboli-Sasco, L., Ribrault, C., Heras, M., Laguna, D. & Perié, L. (2016). Citizen science: Toward transformative learning. *Science Communication*, *38*(4), 523-534. doi:10.1177/1075547016642241

Sadler, T. D., Barab, S. A. & Scott, B. (2007). What do students gain by engaging in socio-scientific inquiry? *Research in Science Education*, *37*, 371–391. doi:10.1007/s11165-006-9030-9

Sadler, P. M., Sonnert, G., Hazari, Z. & Tai, R. (2012). Stability and volatility of STEM career interest in high school: A gender study. *Science Education*, *96*(3), 411-427. doi:10.1002/sce.21007

Saldaña, J. (2015). *The coding manual for qualitative researchers*. Thousand Oaks, CA: SAGE Publications, Inc.

Sha, L., Looi, C. K., Chen, W., Seow, P. & Wong, L. H. (2012). Recognizing and measuring self-regulated learning in a mobile learning environment. *Computers in Human Behavior*, *28*(2), 718-728. doi:10.1016/j.chb.2011.11.019

Semken, S. (2005). Sense of place and place-based introductory geoscience teaching for American Indian and Alaska Native undergraduates. *Journal of Geoscience Education, 53,* 149-157. doi:5408/1089-9995-53.2.149

Sobel, D. (2004). *Place-based education: Connecting classrooms and communities*. Great Barrington, MA: Orion Society.

Tabak, I. (2004). Synergy: A complement to emerging patterns of distributed scaffolding. *The Journal of the Learning Sciences*, *13*(3), 305-335. doi:10.1207/s15327809jls1303_3

Terkowsky, C., Haertel, T., Bielski, E. & May, D. (2013). Creativity @ school: Mobile learning environments involving remote labs and E-Portfolios. A conceptual framework to foster the inquiring mind in secondary STEM education. In J. G. Zubía and O. Dziabenko (Eds.), *IT innovative practices in secondary schools: Remote experiments*, (pp. 255-280). Spain: University of Deusto Bilbao.

Terras, M. M. & Ramsay, J. (2012). The five central psychological challenges facing effective mobile learning. *British Journal of Educational Technology*, *43*(5), 820-832. doi:10.1111/j.1467-8535.2012.01362.x

Toomey, A. H. & Domroese, M. C. (2013). Can citizen science lead to positive conservation attitudes and behaviors? *Human Ecology Review*, *20*(1), 50.

VanMeter-Adams, A., Frankenfeld, C. L., Bases, J., Espina, V. & Liotta, L. A. (2014). Students who demonstrate strong talent and interest in STEM are initially attracted to STEM through extracurricular experiences. *CBE-Life Sciences Education*, *13*(4), 687-697. doi:10.1187/cbe.13-11-0213

van Tuijl, C. & van der Molen, J. H. W. (2016). Study choice and career development in STEM fields: An overview and integration of the research. *International Journal of Technology and Design Education*, *26*(2), 159-183. doi:10.1007/s10798-015-9308-1

Wallace, D. E. & Bodzin, A. M. (2017). Developing scientific citizenship identity using mobile learning and authentic practice. *Electronic Journal of Science Education*, *21*(6), 46-71.

Watts, M. T. (1963). *Tree Finder: A manual for the identification of trees by their leaves*. Rochester, NY: Nature Study Guild Publishers.

Willemse, J. J. & Bozalek, V. (2015). Exploration of the affordances of mobile devices in integrating theory and clinical practice in an undergraduate nursing program. *Curationis*, *38*(2), 1-10. doi:10.4102/curationis.v38i2

Wolfram. (2017). *STEM – Science, technology, engineering, and mathematics*. Retrieved from https://www.wolfram.com/solutions/stem/

Zeidler, D. L. & Nichols, B. H. (2009). Socioscientific issues: Theory and practice. *Journal of Elementary Science Education*, *21*(2), 49-58. doi:10.1007/bf03173684

In: Enhancing STEM Motivation through Citizen Science Programs ISBN: 978-1-53616-038-3
Editors: Suzanne E. Hiller and Anastasia Kitsantas

Chapter 12

CITIZEN SCIENCE IN HIGHER EDUCATION: EVALUATING COLLABORATIVE PRACTICES

Lisa Lobel[1,*], Elisabeth Cianciola[2] and Jason Erichsen[3]
[1]Wheelock College, MA, US
[2]Charles River Watershed Association, MA, US
[3]Simmons University, MA, US

ABSTRACT

Citizen science is a rapidly growing method for large-scale data collection that can serve as a valuable tool for connecting scientists, communities, and public policy makers. However, research into purported benefits of citizen science participation including increased engagement and improved learning outcomes in higher education have produced mixed results. For several years, the Charles River Watershed Association has partnered with higher education institutions in the Boston area to monitor water quality in the Charles River watershed. Non-STEM-major college students and adult volunteers were assessed to evaluate changes in motivation for doing science and taking environmental action, as well as knowledge gains due to participation in the project. Motivation for science and environmental action did not change for either group, but student motivation was significantly lower than volunteer motivation. After participating in the study, knowledge gains in water quality science knowledge gains were significant for students but not for the volunteers. Neither group showed significant changes in awareness of watershed issues but both groups indicated an increased sense of self-efficacy. Anecdotal evidence suggests that participation in citizen science can inspire STEM students to continue to pursue STEM careers. Participating organizations also gain data and opportunities to expand activism networks.

Keywords: civic engagement, community-based organization, partnership, self-efficacy, volunteer monitoring, water quality

---

* Corresponding Author's Email: llobel@bu.edu

INTRODUCTION

Citizen science is a rapidly growing practice described as public participation in scientific research. Differing models of citizen science are based upon the desired scientific outcomes or the level of input participants have in the scientific process (Bonney & Dickenson, 2012). Citizen science projects are described as variations on three basic models or project types such as contributory projects, where citizen scientists collect and contribute data, but scientists ask the scientific questions, design the study and conduct the analysis. While scientists also design collaborative projects, citizen scientists might contribute to analysis, refining the project, or dissemination of findings, in addition to data collection. Projects where scientists and the general public collaborate for the majority of the scientific process are categorized as co-created projects (Bonney et al., 2009). Within these models, citizen science projects are also diverse in their methodological approaches, ranging from mass participation, where it is easy for anyone to participate anywhere, to projects with systematic monitoring, where volunteers receive extensive training and support in repeated monitoring of specific locations (Pocock, Tweddle, Savage, Robinson, & Roy, 2017).

Research into the benefits of citizen science has moved beyond the basic utility and efficacy of using volunteers as data collectors to understanding the greater impacts of public engagement in science in generating new scientific knowledge, improved educational practices and promoting civic engagement (Turrini, Dorler, Richter, Heigl, & Bonn, 2018). Citizen science is at the nexus between science, science education and civic engagement.

Technological advances have given contributory citizen science projects the potential to generate large amounts of data (Theobald et al., 2015). The potential for large numbers of volunteers collecting reliable and verifiable data over large spatial and temporal scales can have tremendous impacts in fields such as ecology and conservation (Dickenson et al., 2012). Large-scale data on the distribution of organisms as well as geospatial data on abiotic factors influencing these distributions has had significant impact on studies of climate change, landscape ecology, macro-ecology and urban ecology (Dickenson et al., 2012). Broader impacts of using citizen science in basic ecological research include using the generated scientific knowledge to inform policy and to connect the research with formal and informal science education. Both factors encourage citizen scientists to become increasingly engaged in action surrounding their projects (McKinley et al., 2017; Toomey & Domroese, 2013).

Citizen science projects can be viewed as an important tool in formal and informal education, thought to increase student performance and problem-solving abilities by engaging citizen scientists in active, inquiry-based and contextualized learning strategies including problem, place or service-based citizen science projects (Freeman et al. 2014; Gormally, Brickman, Brittan, & Armstrong, 2009; Markus, Howard & King, 1993; Tawfick, Trueman, & Lorz, 2014). However, assessing whether citizen science as a pedagogical practice results in scientific literacy gains or changes in attitudes and behavior through increased interest in science, self-efficacy, knowledge and attitudes, and/or professional skill sets/identity formation has produced variable results (Lynch, Dauer, Babchuk, Heng-Moss, & Golick, 2018). In some cases, participation in citizen science projects has been associated with increases in science and environmental literacy for adult and student participants (Bonney et al., 2009; Bonney, Phillips, Ballard, & Enck, 2016) while other studies have not detected changes in knowledge or attitudes toward science (Brossard, Lewenstein, & Bonney, 2005; Overdevest, Orr, & Stepenuck, 2004).

Citizen science has also been suggested as a mechanism to provide connections to the environment. Citizen science projects might provide nature-based learning experiences for urban dwellers who have been isolated or have few experiences with the natural world, and therefore lack personal connections, interest or knowledge of their local environment (Schuttler, Sorenson, Jordan, Cooper, & Shwartz, 2018). Cultivating these emotional, cultural or material connections with the natural world and with a specific "place" such as a park, watershed or forest are important in generating interest in conservation and civic action (Newman et al., 2017).

Increased knowledge gains, environmental awareness and change in behaviors can result in programmatic and community-level outcomes that have the potential for broad-scale impacts on ecosystems, human society and their interactions (Crain, Cooper, & Dickenson, 2014; Jordan, Ballard, & Phillips, 2012). As place-based monitoring programs generate ecological data for local resources, the data are shared with the community, potentially spurring local conservation action and connections with higher governmental authorities to support regulatory action. Citizen science as a convergence between science and society is an effective mechanism for civic engagement (Dillon, Stevenson, & Wals, 2016). These programs have the capacity to increase social networks or social capacity, as citizens are involved in contributing environmental data to local communities and ultimately to government agencies promoting meaningful socioecological outcomes (Dillon et al., 2016; Overdevest et al., 2004). Citizen science projects can potentially increase resilience by providing data for an adaptive management approach and by exposing non-scientists to factors that affect resource management and may require responses in the future. In the dynamic and complex interactions between human and ecological systems, resources are best managed with flexible management objectives that can be modified based on continual biological, ecological and social feedback (Folke, 2006; Overdevest et al., 2004).

Water monitoring programs in North America include more than 8,600 volunteers from 30 states or territories monitoring wells, lakes, streams, estuaries and beaches (CREES, 2005). These programs, often called participatory or community-based monitoring programs, are invaluable for their contributions to understanding ecological conditions and the application of these data for management and conservation actions (Overdevest et al., 2004). Water monitoring programs often focus on chemical, abiotic or bacterial determinates of water quality, which are of interest to public health. More advanced programs also include biological water monitoring, including projects using volunteer macroinvertebrate monitoring (VMM) that focus on collecting and identifying macroinvertebrates from streams and rivers as indicators of water quality and the waterbody's ability to support a living ecosystem. Using macroinvertebrates (or other assemblages of organisms) as indicator organisms is based on the differential tolerance of macroinvertebrates to degraded water quality – certain species are only found in "pristine" stream areas while other species are usually found in degraded areas, often urban stream areas (Barbour, Gerritsen, Snyder & Stribling, 1999). Therefore, the studies are based upon basic and applied research questions focused on understanding ecological patterns that inform both research and management. While many of the citizen scientists in these water monitoring or VMM programs are not undergraduate students, features of VMM make it easy to engage undergraduates in data collection, data analysis and in additional research.

Citizen Science Program

One of the country's oldest watershed organizations, the Charles River Watershed Association (CRWA) was formed in 1965 in response to public concern about declining conditions in the Charles River. The Charles has a long history of human impacts and flows from regions with relatively low population density through highly urbanized areas, ultimately discharging into Massachusetts Bay in Boston Harbor. CRWA continues to be a prominent figure in major clean-up and watershed protection efforts, working with government officials and citizen groups from 35 Massachusetts towns. Initiatives over the last four decades have dramatically improved the quality of water in the watershed and fundamentally changed approaches to water resource management. As a result of their work, the International River Foundation awarded the organization the Theiss International River Prize in 2011 in recognition and support of the development and implementation of an outstanding, visionary and sustainable river management program.

CRWA's mission includes developing science-based understanding of watershed interactions, providing long-term solutions to watershed problems, promoting sustainable watershed management practices and advocating for the protection and revitalization of the Charles River. While CRWA's mission of "protecting, preserving and enhancing the Charles River and its watershed through science, advocacy and the law" is locally focused, the implications of their work reaches far beyond the watershed's boundaries. CRWA's unique strengths and effectiveness lie in the Association's use of science-based knowledge to develop innovative, workable solutions to watershed problems. CRWA advocates persuasively for beneficial outcomes but also partners with government agencies and environmental groups to protect the natural environment and promote sustainable ecosystem policies and practices.

The goals of the citizen science program at CRWA are to engage local residents, train volunteers to collect water and biological samples, and educate individuals of all ages about watershed management challenges and solutions. It addresses large gaps in state agencies' capacities to collect environmental data. Started in 1995, the CRWA's water quality monitoring program is one of the oldest volunteer water quality monitoring programs in the United States. However, in its first 15 years, the program was limited to testing chemical water quality parameters. In 2011 and 2013, CRWA partnered with university faculty and graduate students.

Here we describe how undergraduates were involved in a VMM project through the partnership between an academic institution and a community-based organization. Students from introductory non-majors and advanced majors environmental science courses participated in an established citizen science VMM project created by the CRWA in Boston, MA. The focus of the partnership was not only to increase participation in the program but also to increase understanding of environmental impacts in urban watersheds, increase general knowledge of watershed ecology and management, and to create positive changes in environmental attitudes and engagement. Benefits and challenges faced in creating and maintaining the collaboration are discussed. Specific measurable results participants experienced included increased skills and knowledge in sampling benthic macroinvertebrates (BMIs), identifying microhabitats and BMIs, assessing habitat quality, and calculating water quality indices. We hypothesized that these experiences would increase interest in and understanding of the health and function of participants' local watershed. We assessed whether participation in this citizen science project resulted in increased knowledge and engagement in terms of motivation for learning and doing science, and motivation for continued environmental action. Responses of non-major student participants were compared to volunteer citizen scientists' responses, who were expected to have relatively high self-motivation for science and environmental action. Therefore, non-major students working with local scientists to understand impacts in their local watershed were expected to have a greater change in motivation for doing and learning science and environmental action as compared to volunteers after participating in this study.

METHOD

CRWA uses biological monitoring to assess whether improvements or degradation in Charles River water quality affect aquatic communities over time. The program aims to identify areas in need of land conservation, landscape restoration, stormwater management, or other interventions to inform the organization's future projects. In order to maintain the program from year to year, CRWA recruits and trains volunteers annually to do the necessary fieldwork and identify the organisms collected. Volunteers were recruited through web and email announcements. Students enrolled in an introductory environmental science course were also trained as biological monitors and participated in the sampling program as part of their course work.

Participants and Setting

To assess if changes in motivation for science and environmental action occurred as a result of participating in the project, 13 students enrolled in a non-majors, general education, introductory Environmental Science course at a small liberal arts college and 14 adult volunteers were given pre- and post- participation questionnaires. Demographic information was not collected for either group. The study received approval from the College Institutional Review Board (IRB). Students were required to participate in the citizen science project as part of the course but were informed that their participation in the pre- and post-participation surveys was voluntary and anonymous.

Data Collection

Students completed paper questionnaires in class during the first week of class before training in the citizen science project began. The post-participation paper questionnaires were also administered in class by one of the project leaders who was not the course instructor after the project had been completed, the class assignments had been returned, and after the midterm exam containing material on the project. Volunteers were given a paper questionnaire during their first training session before formal training was initiated. Most volunteers completed paper questionnaires when they returned their identified samples to CRWA, but volunteers who did not return to CRWA after completion of the project received an electronic copy of the questionnaire by email from a staff member who was not affiliated with the project. Students and volunteers were asked to label their questionnaires with a unique code and record the code in their notebooks to use for the post questionnaire so that the pre and post-participation individual responses could be linked.

Intervention

Training for both groups of participants began with an introductory lecture given by a CRWA scientist about the history of the Charles River, the mission of the CRWA and the biological monitoring project. Participants were also given background information on habitat types and factors (urbanization, historic and current sources of pollution, abiotic conditions) that influence both habitat quality and the types of organisms inhabiting an area. Training also included safe sampling techniques to collect and preserve the benthic macroinvertebrates. CRWA holds a second, shorter training in the fall focused on macroinvertebrate identification and calculating a water quality score using the U.S. Environmental Protection Agency's Water Quality Index. CRWA includes the program results in its annual water quality reports and shares them with local, state and federal agencies as well as other interested parties. Training occurred in class for students and at CRWA for volunteers.

Volunteers were given the half-day sampling training in June including the introductory presentation on project concepts and safety protocols as well as a hands-on practice in assessment and sampling at a local stream. CRWA staff divided volunteers into teams, assigned them to the priority monitoring locations for that year, and provided the equipment for the fieldwork including waders, nets, collection jars, forceps and preservative. Training concluded with practice in habitat assessment on a section of river using the Environmental Protection Agency (EPA) Rapid Bioassessment protocol as well as practice in collecting benthic macroinvertebrates using kick nets. Successful participation in the project consisted of volunteers completing a visual habitat assessment and sample collection at their team's convenience over the summer. One team member was assigned to "pick" or prepare the sample for identification by removing rocks, sticks, and leaves. The separated benthic macroinvertebrates (BMIs) were preserved in labeled containers for later identification.

To incorporate student-collected samples into the field season, CRWA staff conducted student training in habitat assessment and sampling during the first class lab periods of the semester and sampling was completed by September 30. Additional class periods were used to pick the samples, provide training in identifications of BMIs and calculate the water quality scores.

Participants in both groups assessed water quality by conducting a habitat assessment and by sampling, picking, then identifying and quantifying the BMIs in order to calculate a water quality index. On their designated 100-meter stream reach, citizen scientists conducted a visual characterization of the different habitats using the Rapid Bioassessment Protocol Habitat Assessment (Barbour et al., 1999) of the reach beginning at the downstream end. This provided a quantitative, but subjective assessment of habitat conditions and impact status. There were 10 habitat parameters (e.g., substrate cover, channel flow, sediment deposition) assessed and each parameter was ranked according to the scale of "poor" (0-5), "marginal" (6-10), "sub-optimal" (11-15), or "optimal" (16-20). These rankings were summed for all parameters and final scores were characterized as "poor" (0-59), "marginal" (60-109), "sub-optimal" (110-159), or "optimal" (160-200) (Barbour et al., 1999).

BMI sampling was conducted using a 20-jab multi-habitat approach (Barbour et al., 1999). Participants first estimated the percentage of each of seven microhabitats (run, riffle, pool, submerged vegetation, overhanging vegetation, roots and large wood) found within their stream reach. Twenty sub-samples were collected from each 100-meter reach using a D-frame kick net, dividing the locations of the sub-samples proportionately across each of the microhabitats represented. After each sub-sample was collected, nets were emptied and rinsed over a bucket with river water. Once the sampling was complete, large debris was removed and the complete sample was sieved to remove excess water, transferred to a container and preserved with 90% ethanol. Each container was labeled with the location, date, and sampler's initials and stored at room temperature until the sample was picked. Samples were picked to remove BMIs from sediment and debris within 72 hours. Picking of the samples was accomplished by removing BMIs with forceps and placing in 70% ethanol for later identification. Identification was completed by first sorting BMIs into morphospecies and then to genus or species using a dichotomous key developed by CRWA.

A water quality index was calculated based on the number of individuals of each BMI taxon collected. Each BMI taxon was classified as rare (1-9), common (10-99), or dominant (100+) in the sample. The number of BMI taxa in each of these three groups was summed and weighted using a multiplier to calculate the final water quality score. This score categorized the river water quality condition as "excellent" (59-78), "good" (40-78), "fair" (20-39), or "poor" (0-19) (US EPA, 1997).

Measures

Motivation for Learning and Doing Science (Cornell Lab of Ornithology, 2013). The motivation for doing science survey instrument was developed and tested by the Cornell Laboratory of Ornithology (2013) in the context of informal science learning settings and primarily with participants of citizen science projects. The scale includes 16 items, half of which assessed internal motivation and the other half of which addressed external motivation for engaging in the science activities in question. The instrument used a Likert 5-point scale with internal reliability of .81/.85 (Phillips & Bonney, 2016). The customizable version of the questionnaire was used with the term "water quality" substituted as the subject for each question. Example items from the survey include the general question of "In general, why do you choose to learn and understand water quality topics?" followed by ranked responses to statements including, "Because I like learning about water quality topics" (internal motivation) and "Because I am required to understand water quality topics" (external motivation). Internal reliability for this measure from this study was .83/.85 (pre/post-survey).

Motivation for Environmental Action (Cornell Lab of Ornithology, 2013). Similar to the Motivation for Learning and Doing Science instrument (Cornell Lab of Ornithology, 2013), the motivation for environmental action instrument was developed and tested by the Cornell Laboratory of Ornithology. The scale assessed internal and external motivations for engaging in environmental action activities ***using*** a Likert 5-point scale with internal reliability of 0.84/0.75 (Phillips & Bonney, 2016). The customizable version of the questionnaire was used with the term "water quality" substituted as the subject for each question. Participants were asked to "Think about some of the things you do to protect water quality or help solve water quality problems. Why do you do these things?" Participants were asked to respond to 14 questions with seven questions addressing internal motivation and seven questions addressing external motivation. Example items included, "Because I think it's a good idea to do something for water quality" (internal motivation) and "Because other people will be disappointed in me if I don't" (external motivation). Internal reliability for this measure was .81/.89 (pre/post) in this study.

Water Quality Concepts and BMI and Water Quality Knowledge and Skills (Lobel, & Cianciola, 2017). To assess changes in skills and knowledge for this project, questionnaires were developed that assessed self-efficacy in knowledge and skills gained during participation surrounding water quality, habitat assessment and benthic macroinvertebrate sampling. The instrument contained two sections; the first addressed knowledge of water quality concepts, where participants indicated whether or not they recognized 23 water quality concepts. Example water quality concepts included watershed, riffle, and sediment deposition. The measure used a dichotomous scale (yes or no) with internal reliability of .91/.74 (pre/post, this study).

The second section of this instrument contained 14 items, addressing BMI and Water Quality Knowledge and Skills. These questions focused on four constructs meant to assess self-efficacy in skills and knowledge gained through participation as well as questions that addressed their interest and engagement with watershed ecology. The instrument used a Likert 5-point scale with internal reliability of .88/.68 (pre/post, this study). Example items included "I know how to assess habitat quality" (skills), and "I pay attention to issues influencing my local watershed" (engagement). The post-survey included an additional category of 6 for "I thought I knew before, but now I really know."

All three questionnaires were combined into one instrument given in paper format, with 16 Motivation for Science, 14 Motivation for Environmental Action, 23 Water Quality Concepts and 14 BMI Knowledge items, for a total of 67 items.

Data Analysis

Response data were entered into a Microsoft Excel spreadsheet and cleaned by removing responses from individuals that did not provide answers to two or more questions. For the Motivation for Learning and Doing Science and for the Motivation for Environmental Action scales (Cornell Lab of Ornithology, 2013), scores were calculated for each participant by taking the average of the set of scores from both the internal external motivation questions for each participant. The averaged external motivation score was subtracted from the averaged internal motivation score to create the overall motivation score for each participant (internal motivation - external motivation = motivation). Positive scores indicate that participants were internally motivated while negative scores indicate that participants were internally motivated, while negative scores indicate that external motivation drives participants. These motivation scores for students and volunteers were used to test for difference between median pre- and post-survey results for the two groups.

Scores for the Water Quality Concepts (Lobel & Cianciola, 2017) measure were calculated by scoring a "yes" answer as one, while a "no" was scored as zero. The number of "yes" responses divided by the number of concepts (#yes/23) was used as the final score for each participant. Scores for the BMI and Water Quality Knowledge and Skills (Lobel & Cianciola, 2017) measure were taken as the average of all responses to the 14 questions on the 5- point Likert scale for the pre-survey and the 6-point Likert scale for the post-survey.

Surveys were not consistently given an identifying code by student or volunteer participants and were treated as aggregate pre- and post-participation samples. Sample sizes were small, data were not normally distributed, and variances were heterogeneous so the Kruskal-Wallis rank sum test for multiple independent samples with the Dunn post-hoc pairwise multiple comparison method was used to test for differences between the student and volunteer pre- and post-survey scores (SPSS version 25). The Dunn method was used for pairwise comparison due to ties in the ranks, with the p-value adjusted using the Bonferroni correction for multiple comparisons.

RESULTS

A Kruskal-Wallis test provided strong evidence of a difference (p = .001) between the median ranks of pairs of groups for the Motivation for Learning and Doing Science measure (Cornell Lab of Ornithology, 2013). Dunn's pairwise tests were carried out for the four pairs of groups. There was no evidence of a difference between the student and volunteer pre- versus post-survey scores in motivation for doing science. There was very strong evidence (p < .001, adjusted using the Bonferroni correction) that student pre- versus post-survey scores were significantly lower than the volunteer pre- and post-survey scores. The median score for the student pre-survey was .44 compared to 1.32 for the volunteers (p = .015), while the median scores post-survey were .40 for the students and 2.25 for the volunteers (p = .002). These results indicate that there was no change in motivation for learning and doing science as a result of participation in the project, but that, overall, student motivation was lower than that of volunteers' for learning and doing science.

For the Motivation for Environmental Action measure (Cornell Lab of Ornithology, 2013), there was very strong evidence of a difference (p < .001) between the median ranks of pairs of groups. However, Dunn's pairwise tests showed no evidence for differences between the student and volunteer pre- versus post-survey scores. There was very strong evidence (p < .001, adjusted using the Bonferroni correction) that the student pre- and post-survey scores were significantly lower than the volunteer pre- and post-survey scores. The median score for the student pre-survey was .43 compared to 1.50 for the volunteers (p = .044; Table 1), while the median scores post-survey were .14 for the students and 1.57 for the volunteers (p = .001). The decline in student motivation for environmental action was not significant. These results indicate that participation in the project did not result in a change in motivation for environmental action and that student motivation for environmental action is lower than that for volunteers' motivation for environmental action.

Results from the Kruskal-Wallis test provided very strong evidence of a difference (p < .001) between the median ranks of pairs of groups for recognition of Water Quality Concepts (This study). There was very strong evidence (p < .001, adjusted using the Bonferroni correction) of a difference between the student pre- and post-survey scores. The median score for the student pre-survey was .43 compared to .96 in the post-survey (p < .001). While the scores increased for volunteers, the increase was not significant. The post-survey median scores of .96 for students and .96 for volunteers were not significantly different (Table 1). Even though the pre-survey median student score (.43) suggests that students made greater knowledge gains, it was not significantly different from the volunteer pre-survey score (.70). These results suggest that after participation in the project, students' ability to recognize water quality concepts increased, while volunteers' abilities to recognize water quality concepts did not change. Students increased their ability to the same level as volunteers.

For the BMI and Water Quality Knowledge and Skills measure (this study), there was strong evidence of a difference ($p < .001$) between the median ranks of pairs of groups. Dunn's pairwise tests were carried out for the four pairs of groups. There was strong evidence ($p < .001$) of a difference between both the student and volunteer pre- and post-survey scores. The median score for the student pre-survey was 2.7, compared to 4.0 in the post-survey ($p < .001$) while the pre-survey score for volunteers was 3.1 and increased significantly to 4.1 in the post-survey ($p = .002$). The pre- (2.7, 3.1) and post-survey (4.0, 4.1) median scores of students and volunteers were not significantly different (Table 1). These results indicate that both groups improved in their recognition of water quality concepts.

Table 1. Pre- and Post-Survey Median Scores for Students and Volunteers Participating in the 2017 Benthic Macroinvertebrate (BMI) Field Training and Sampling Program

Survey		Students		Volunteers	
		Pre (N= 13)	Post (N= 10)	Pre (N= 14)	Post (N= 9)
Motivation for Learning and Doing Science (16 questions)	Median	.44^a	.40^a	1.32^b	2.25^b
	SD	.46	.50	.78	.88
Motivation for Environmental Action (14 questions)	Median	.43^a	.14^a	1.50^b	1.57^b
	SD	.74	.32	1.71	.86
Water Quality Concepts (23 concepts; Y = 1, N = 0)	Median	.43^a	.96^b	.70^{ab}	.96^b
	SD	.24	.06	.15	.10
BMI & Water Quality Knowledge (14 questions; Scale of 1-5)	Median	2.71^a	4.00^b	3.07^a	4.07^b
	SD	.69	.40	.91	.45

Note: The superscript letters are used as symbols to indicate statistical differences between treatments (columns) in each intervention. For scores with the same letter, or scores across columns sharing an "a" or "b," the difference was not statistically significant. Scores with both letter indicators "ab" do not differ from scores in columns with "a" or "b." For scores in a column with a different letter, such as "a" versus "b," the difference was statistically significant at $p < .05$.

Both students and volunteers showed improvements in self-efficacy regarding skills gained during participation. Scores on the post-survey indicated that students' and volunteers' responses changed from disagreement to agreement with statements that they knew how to identify and collect BMIs, identify and assess habitats and calculate a water quality index (Table 2). There was a significant increase in scores from a median of 2.0-4.0 for students ($p < .001$) and a significant increase from a median of 2.1-3.8 for volunteers ($p = .004$). Both groups also showed improvements in knowledge of BMIs and their connections to water quality, changing their scores from "neutral" in the pre-survey to "agree" in the post-survey. The increase in scores from a median of 2.8-4.0 ($p = .002$) for students and 3.5-4.0 ($p = .04$) for volunteers were statistically significant. Two questions addressed interest in how actions influence watershed ecology and knowledge of the watershed in which the participant's community was located in. The increase in student scores from "disagree"/"neutral" (2.5) "to agree" (4.0) was significant (p = .005) while the increase in volunteer scores from close to "agree" (4.0) to "agree"/"strongly agree" (4.5) was not statistically significant. The single question that addressed engagement showed that students' responses changed from "disagree" (2.0) to between "neutral" and "agree" (4.0) with respect to paying attention to issues influencing their local watershed. However, this change was not statistically significant. Volunteer pre- and post-survey median scores did not change for this question (Table 2). These results indicate that both student and volunteer groups felt increased self-efficacy around skills, knowledge and engagement after participation in the project. While student interest increased, it did not change for volunteers.

Table 2. Comparison of Pre- and Post-Survey Questionnaire Results Related to Skills, Knowledge, Interest and Engagement between Students and Volunteers

		Student		Volunteer	
Construct	Question	Pre	Post*	Pre	Post*
Skills	I know how to collect BMIs.	1.8	4.4	1.8	3.3
	I know how to identify microhabitats.	2.4	4.0	2.7	3.8
	I know how to assess habitat quality.	2.3	4.2	2.4	3.6
	I know how to identify BMIs.	2.2	4.2	2.1	4.2
	I know how BMIs are used to calculate a water quality index.	2.2	4.1	2.1	4.6
	Median	2.0^{a}	4.0^{b}	2.1^{a}	3.8^{b}
	SD	.9	.5	.9	.9
Knowledge	I know what BMIs are.	2.5	4.6	3.2	4.8
	BMIs are good indicators of water quality.	3.1	4.2	3.6	4.6
	Some BMIs are sensitive to changes in water quality.	2.8	4.2	3.8	4.6
	BMIs are found in every habitat.	2.6	4.2	2.8	3.7
	Habitat assessment tells us how much habitat has been altered.	3.5	3.7	3.8	3.4
	Habitat quality influences the type of BMIs found.	3.3	4.6	4.0	4.7
	Median	2.8^{a}	4.0^{b}	3.5^{a}	4.0^{b}
	SD	1.0	.4	1.0	.6
Interest	I know how my actions influence my local watershed.	2.9	4.2	4.1	4.2
	I know which watershed my community is in.	2.0	4.2	3.4	4.7
	Median	2.5^{a}	4.0^{b}	4.0^{ab}	4.5^{b}
	SD	.9	.7	1.4	.6
Engagement	I pay attention to issues influencing my local watershed.	2.0	4.0	4.0	4.0
		1.1	.7	1.6	.3
Overall Median Score		2.7^{a}	4.0^{b}	3.1^{a}	4.1^{b}

Note: The superscript letters are used as symbols to indicate statistical differences between treatments (columns) in each intervention. For scores with the same letter, or scores across columns sharing an "a" or "b," the difference was not statistically significant. Scores with both letter indicators "ab" do not differ from scores in columns with "a" or "b." For scores in a column with a different letter, such as "a" versus "b," the difference was statistically significant at $p < .05$.

Overall results suggested that participation in this citizen science project did not change motivation for learning or doing science, or motivation for environmental action in either students or volunteers. Student motivation was significantly lower than volunteers before and after participation. Students demonstrated knowledge gains in recognizing water quality concepts as compared to volunteers. Both students and volunteers showed improved self-efficacy in knowledge and skills relating to sampling and using BMIs for water quality assessment. Students also showed increased interest in their local watershed.

DISCUSSION

A citizen science VMM program was successfully incorporated into an introductory undergraduate, non-majors Environmental Science course over four fall semesters. Pre- and post-participation surveys were used during the fourth offering to assess student self-efficacy in identifying water quality concepts, skills and knowledge gains, as well as motivation for science and for environmental action as compared to adult volunteers. Positive scores for both groups reflected intrinsic motivation. Student and volunteer participants showed no change in motivation for learning and doing science or motivation for environmental action after participating in the project. Additionally, for both constructs, students scored significantly lower than volunteers and showed a non-significant decrease in motivation toward environmental action. This may be a function of this particularly small cohort of non-major students who were required to do the project as part of their course. However, an additional factor to assess would be to determine if students become overwhelmed when confronted with complex environmental issues. These introductory non-majors courses may be students' first exposure to environmental science. As students learn how environmental processes operate over large temporal and geographic scales, they may feel as though individual actions may not be able to significantly improve their environment.

Knowledge gains were documented for both groups including increased recognition of water quality concepts and gains in skills and knowledge regarding sampling BMIs and using these skills to assess water quality. While the motivation for science and motivation for environmental action questionnaires did not document changes in student motivation, there was an increase in the student scores for the two questions related to watershed interest. However, these questions might also be interpreted as knowledge questions. While the gains in knowledge and in skills were not coupled with changes in motivation, these gains lead to increased feelings of self-efficacy, or a student's perception of his or her capabilities, that is essential to student motivation and is a predictor of academic success (Zajacova et al., 2005; Zimmerman, 2000). In their comments on the survey, students mentioned their appreciation for the opportunity to actually do the work, as well as their feelings of accomplishment. Better control of coding to properly pair pre- and post-surveys would have allowed for quantitative analysis of changes in motivation for individuals. Anecdotally, some individuals were inspired by participation in a project collecting data with real implications for management as opposed to traditional labs with known "endpoints." Students also gained an opportunity to meet and work with aquatic scientists as a method to envision future career opportunities. Some students applied for internships with CRWA, others continued with the project through subsequent years, and yet other motivated students contributed to the dissemination of the monitoring results through presentations at regional and national meetings.

Partnership Benefits

The goals of the partnership between a college from the Northeast region of the United States and CRWA were to integrate classroom work with citizen science field research in urban stream health to create connections between science, scientists and students; to increase knowledge of issues impacting local watersheds; and to create advocates for protecting and preserving these urban waterways. Having students participate in the CRWA citizen science program engages them as residents of their watershed, trains them as volunteers to collect biological samples, and challenges them to engage in understanding watershed management issues and solutions. Participating in the biological monitoring program theoretically allows students to link their experiences in fieldwork, data analysis and data interpretation for management solutions directly to their course topics. Students engaged in classroom learning, translated that learning into field research, completed the analysis and contributed their results to the program.

If you take a minute to look around, you can find a community group with their boots on the ground advocating for just about any cause. Non-governmental organizations (NGOs) supporting environmental causes are particularly common and often eager to work with community partners because environmental problems often require collaborative solutions. Partnerships between community groups and academic institutions can be especially effective for several reasons. Partnerships with academic institutions give community groups the opportunity to expand their programs and provide an avenue to expose an impressionable audience to their work. This exposure invites students to share their experience with relatives and peers, explore volunteer and internship offerings with the NGO and can even lead to lifelong commitment to community engagement or a professional career with the NGO or supporting its mission.

The *Watershed Watch* of the University of Rhode Island (University of Rhode Island Watershed Watch, 2019) and the *Watershed Alliance Program* (UVM Extension Watershed Alliance, 2019) at the University of Vermont are two regional examples of well-established university community-based monitoring partnerships. Partnering with universities can extend significant benefits to the program. For example, faculty might serve as expert advisors, encourage their students to participate as volunteers, or train volunteers, analyze samples or interpret results (Savan, Morgan, & Gore, 2003). The initial partnership between CRWA and the University of Massachusetts, Boston (UMB) to create the biological monitoring program included real-world project management experience for a graduate student who assisted CRWA in developing the sampling methodology and training materials, selecting monitoring locations, and evaluating volunteer data quality. The significant benefit to CRWA was the creation of a sustainable water quality monitoring program. A continuing partnership between CRWA and the university still benefits students through invitations to attend volunteer trainings, to collect data with their peers or citizen scientist volunteers according to the established protocols for independent research. Additionally, CRWA pays UMB students to use their expertise to do a blind quality control check on citizen scientist BMIs.

Graduate students at UMB have also had the opportunity to integrate the results of CRWA's citizen science program with other information such as mapping and modeling tools into their theses, enabling them to be co-creators of scientific studies with CRWA staff. In years when a UMB student elects to do an independent study with CRWA, the number of sites monitored typically doubles.

At small liberal arts colleges without graduate training opportunities, partnerships to increase participation in citizen science projects can work well as part of general education courses. The Fenway neighborhood of Boston, Massachusetts, named because of its proximity to the Muddy River and surrounding "fens" or wetlands, is home to several academic institutions, making the Muddy River an ideal living laboratory for these neighboring institutions. The Muddy River is a tributary to the Charles River, and a significant source of pollution, making it an area of interest for further study for CRWA. As faculty members are continually looking for place, problem, and inquiry-based investigations, or just authentic research experiences to include in their curricula (Oberhauser & LeBuhn, 2012), a faculty member attended CRWA's citizen scientist volunteer training. Faculty participation in the citizen science program lead to the suggestion to have an Environmental Science course commit to sampling the Muddy River site as one of CRWA's annual monitoring locations each year. The partnership was to integrate classroom work with citizen science field research in urban stream health having students assess BMI communities as a measure of water quality. Classroom instruction includes an overview of the program and organization by a CRWA staff scientist, but the course instructor provided the instruction and training on watershed ecology, habitat identification, sampling, identification and safety using the volunteer training materials developed by CRWA (Cardamone & Lobel, 2016). The citizen science volunteer training in fieldwork and sample identification protocols is usually provided by a recent graduate in environmental science or a related field who is the recipient of a CRWA fellowship and overseen by a CRWA staff scientist. Having additional trainings are great opportunities for the fellowship recipients to gain experience instructing others. The aim of the partner institutions was to create connections between CRWA scientists and students in science and general education programs in order to increase knowledge of issues impacting local watersheds and to create advocates for protecting and preserving these urban waterways.

Academic institutions are valuable partners because they are relatively reliable and consistent. This leads to a mutually beneficial relationship in which the NGO needs participants to collect data and to increase its audience, while faculty want a place-based, experiential learning opportunity that is an authentic research experience linked to course topics (Oberhauser & LeBuhn, 2012; Mitchell et al., 2017). If a partnership between an NGO and an academic institution is built into the curriculum for a particular course or internship experience, it can guarantee some level of participation every year. However, as students are required to participate in the partnership in order to receive academic credit, they are not truly "volunteers," which should be considered when designing and conducting assessments. Partnering with an academic institution can also help an NGO improve or expand its program. The academic institution's commitment to sample the same site every year allows volunteers to be used to sample additional sites. While the NGO's primary objective for the project is likely to complete a particular task, partnering with an academic institution adds a new project objective: for the students participating in the project to learn skills and learn to appreciate values relevant to course learning goals and civic engagement. The academic institution's need to evaluate students' learning also creates an opportunity for the partnering NGO to evaluate the training protocols and outcomes of its program at the same time. Program participants from the community at large often are not motivated to provide feedback to an NGO about its programs, especially if they are one-time participants, but students and teachers often reflect upon and share their learning experiences. The NGO can use this feedback to deliver its program more efficiently and effectively in the future.

Lastly, the NGO expands its network of influence by partnering with academic institutions. Academic institutions may have sister schools and corporate connections through alumni. Partnerships that are sustained over multiple years can open the NGO to partnering with these secondary groups on the same or other projects.

Citizen Science Project Growth

Challenges in incorporating the program in additional formal education settings occurred due to the nature of the biological monitoring program. Specifically, the sampling season encompassed the summer months and ended on September 30. This posed a challenge for instructors who needed to incorporate the training and sampling at the very beginning of the semester. The time commitment to complete the entire sampling and identification protocol required multiple class or lab sessions. The solution has been to modify the protocol to allow different audiences to experience the educational aspects of the program, but CRWA does not gain "usable" data for the project.

Attempts to have faculty adopt the program at neighboring institutions were difficult as courses did not run due to under enrollment, courses were not being offered every year, or faculty were unwilling to commit the needed course/lab time to the project. However, the citizen science partnership program was expanded to include students in an Ecology class at a neighboring institution whose campus also borders the Muddy River. The Ecology class was only offered every other year, so additional coverage of sites was limited to this timeline.

Expanding the program to other potential audiences, such as local middle and high schools, had begun as a few schools approached CRWA to explore ways to integrate benthic macroinvertebrate sampling into their science programs. The science director at a local middle school decided to develop a Charles River Day field trip. While CRWA does not have funds dedicated to supporting local school programs, the middle school was able to provide some financial support so that CRWA staff could work with the science teachers to plan the citizen science experience. Because the trip was conceived as a one-day activity for the entire seventh grade, the multi-day model we had used for a single class of 10-20 college students would not be adequate to have students participate in the sampling, identification and analysis required for the program. The one-day field trip also included a learn-to-row component. The school was only able to accommodate the field trip in late October, which is outside of the accepted timeframe for collecting samples under CRWA's protocols. While these students' findings could not be used as part of the watershed-wide monitoring effort, schools participating in September or June would be able to obtain usable results. Therefore, for some schools, benthic macroinvertebrate sampling would only be used as an outreach and learning experience. Both the middle school science teachers and CRWA scientists collected the samples in advance, which allowed the students to identify the benthic macroinvertebrates and participate in a learn-to-row experience during the allotted time for the field trip. In this case, offering both activities were valuable to help students of all abilities build a connection with the river.

In addition to working with the first middle school, a second middle school asked CRWA to help lead a benthic macroinvertebrate sampling program for several of its classes. A stream that feeds into the Charles River, runs behind the school property, making it possible for classes to walk to their fieldwork site. Because benthic macroinvertebrate sampling was the only activity planned for this program, students were able to help collect the samples. However, this program was also held in October and, consequently, outside of CRWA's accepted sample collection period, meaning that data that could contribute to the watershed-wide dataset were not collected. The school expressed interest in running the program again, but was not able to schedule a field trip date the next year.

CRWA also began exploring ways to integrate the biological water quality monitoring program into more informal settings. A CRWA staff member brought a collection of preserved benthic macroinvertebrates, a microscope and slides, and photos to the Museum of Science in Boston on a field trip day for high school students. This provided organizational exposure both to high school students and other museum guests. As one of the high schools that visited the Museum that day was planning a benthic macroinvertebrate sampling activity for its students, CRWA was able to connect with a teacher, giving the students a preview of the work they would be doing and its practical application in their community. CRWA also led a benthic macroinvertebrate identification activity during its Free Fun Friday event hosted with the Highland Street Foundation in July of 2018. This family-oriented activity was be similar to the approach taken for the middle school field trips, where scientists collected samples in advance of the event, and the primary objective was to inspire learners to participate in science and take environmental action as opposed to generating high-quality scientific data. If successful, this model can also be applied to summer camps and other community events.

The biological monitoring project including college students was expanded to include assessment of chemical contaminants in sediments. As part of the pedagogy surrounding habitat assessment and impacts of urbanization on streams, students were introduced to the concepts of point and non-point source pollutants and discussed potential sources of pollutants relevant to their study site. During the project, student observations of pollution in the river (e.g., oil residues, garbage) resulted in questions surrounding how pollution would impact the habitat and the organisms they were sampling. To answer these questions, the scope of the project was expanded with a goal of including students interested in other fields of inquiry in the project. As the environmental science students were focused on measuring biotic factors to assess stream health, they were partnered with chemistry students in an introductory Analytical Chemistry course at a nearby college who would quantify chemical contaminants in the river. The integration of the chemistry students into the program is continuing with the goal of CRWA scientists providing presentations on the organization's history and goals, the biological monitoring project, and discussion of how the abiotic factors that the chemistry students measured would influence the biotic factors measured by students in the Environmental Science course. The chemistry students successfully assessed metal and polycyclic aromatic hydrocarbon contamination in sediments collected from the same biological monitoring sites, but the courses and projects were still fairly independent. Student volunteers from both courses met to combine their biological and chemical results into a poster presented back to their respective courses and at a regional sustainability symposium. While these students used the combination of abiotic factors and biotic factors in their assessment of the river health, understanding of the interdisciplinary connections between the biological and chemical assessments has not been assessed. To generate data on patterns and relationships between biological indicators and contaminants that would be beneficial to the community partner, sediments should be analyzed from additional biological monitoring sites. However, having these interdisciplinary courses collaborate enables students to observe and participate in a realistic model of interdisciplinary scientific investigation.

Environmental Implications

Environmental monitoring conducted by government and non-governmental organizations such as CRWA provides essential information to regulators, legislators and the public about the quality of our environment, particularly in a complex and dynamic system like an urban watershed. Information collected by community-based monitoring groups such as CRWA is critical to improve regulatory programs and to ensure that environmental policy is updated to meet changing conditions as a part of an adaptive management scheme (Biber, 2011; Overdevest et al., 2004). Today, due to significant budget cuts, many government agencies have eliminated or severely reduced their monitoring efforts, leaving CRWA's program as the only robust, year-round sampling program spanning the full extent of the Charles River. In this economy, with local, state and federal funds being cut further every year, non-profit organizations like CRWA are often the only resource available to collect basic water quality data for rivers and streams (Conrad & Hilchey, 2011).

The data collected via environmental monitoring are essential to understanding the problems the river faces, for protecting human health, as well as for identifying trends and directing informed solutions to the areas where they are needed. Through this program, CRWA engages local residents as citizen scientists, training volunteers to collect water and biological samples, increasing the organization's capacity to conduct monitoring while educating individuals of all ages about watershed management challenges and solutions. The ultimate goal is to create citizen advocates that become further involved in the management of their watershed.

Thus, the goal of this partnership between CRWA and faculty at a small liberal arts college was to engage students with a place-based experiential opportunity quantifying environmental changes in their local watershed, allowing students to deepen connections to their course topics. Students worked on a real-world issue during class knowing that the work they were doing and the data they were collecting was to be used toward assessing and managing their local river. Students learned about potential research options, pathways for civic engagement, as well as an introduction to future job options. Students participating in these citizen science research experiences created important connections between self, society and environment in addition to providing support for long-term community partnerships. Students' experiences as citizen scientists, regardless of their major, has the potential to provide the interest and tools to create a lifetime of civically engaged citizens.

Educational Implications

Increasing and retaining graduates in STEM fields is a national priority, but many students with the potential to excel in science and technology fields are choosing other careers (Bettinger, 2010; Chen, 2013; Zumeta & Raveling, 2002). Post-secondary education plays an important role in producing STEM majors, but more than half of STEM majors change fields before graduation (Chen, 2013).

Pressures on undergraduate STEM students are intensifying as they are expected to work and contribute to increasingly quantitative, interdisciplinary fields (Brewer & Smith, 2011). Specifically, biology in the 21st century requires all undergraduates to learn to "integrate concepts across levels of organization and complexity and to synthesize and analyze information that connects conceptual domains" (National Research Council, 2009). This poses challenges for faculty, as the current curricula in high school and college laboratory experiences are not designed to integrate learning of science content with science process (National Research Council, 2006). Specifically, students do not frame their own research questions and design their own experiments and, therefore, are not engaged in higher-level learning such as analysis, synthesis and evaluation. Instead, instruction is focused on lower cognitive skills such as knowledge and comprehension, where students conduct experiments designed by instructors that reinforce learned concepts (Cambell & Bohn, 2008; Momsen et al., 2010).

Calls for changes in teaching and learning practices to incorporate student-centered, active inquiry-based pedagogies have been occurring for decades and perhaps for more than a century (American Association for the Advancement of Science, 1993; National Research Council, 2003). As early as 1910, Dewey suggested that inquiry should be used in K-12 classrooms and that students will increase their scientific knowledge if they are engaged as active learners in scientific inquiry relating to their personal experiences (Barrow, 2006). Inquiry-based pedagogies are still viewed as the best method for students to gain scientific literacy, providing opportunities for students to discuss and debate the scientific process and ideas. Some surveys suggest that inquiry-based practices have been widely adopted in undergraduate science classrooms and laboratories (Sundberg, Armstrong, & Wischusen, 2005). However, further investigations have determined that many of the inquiry-based investigations in current use are modifications of standard cookbook investigations where instructors guide students through the inquiry process (Beck, Butler, & Burke da Silva, 2014; Gormally et al., 2009). These guided inquiry investigations still have benefits, particularly to alleviate stress and frustration in students who are inexperienced and not yet prepared to attempt inquiry-based problems on their own. To further increase student problem-solving skills, problem-based, project-based or service learning components have been shown to support active student-centered learning while also providing context to inquiry-based investigations (Tawfik et al., 2014). Specifically, these methods provide support for students to strengthen weak skills as they work collaboratively with other students and faculty to ask questions and test solutions applied to complex problems relevant to themselves and the community (Tawfik et al., 2014).

Participation in citizen science projects supports these teaching and learning practices but also supports increased learning regarding the importance of civic engagement that is needed to understand and resolve complex socioecological problems (Wals Brody, Dillon, & Stevenson, 2014). Depending on the type of project, citizen science provides opportunities ranging from data collection to guided inquiry, to more in-depth, inquiry-based learning that can also focus on local civic issues. The hands-on practice used by citizen science projects has been suggested as critical to making science relevant to students and as an important step to support continued interest in science (Jenkins, 2011).

Citizen science has broad applications in science, education and civic engagement. The inclusion of undergraduates as citizen scientists through partnerships with NGOs, community-based monitoring programs or through faculty-led research projects can result in significant benefits at the intersection of these three aspects. Students benefit from educational best practices, data are generated for scientific knowledge, and students are also introduced to ways to engage in the civic process. The challenge is to clearly define projects and/or form collaborations that promote full participation and generate the most civic, scientific and educational benefits.

Here we describe an exemplary citizen science partnership formed initially between a large, research-based academic institution (University of Massachusetts, Boston) and a non-profit watershed association (CRWA) and expanded to include small liberal arts colleges. Partnering with a large, research-based institution benefitted this community-based organization by providing access to faculty and graduate student experts, who assisted in the program design, sample analysis and continued quality control analyses. Benefits of partnerships with small liberal arts institutions include guaranteed participation, increased educational outreach, and, ideally, increased civic participation in the focal issues.

Student participants in this citizen science project benefitted from the place, problem-based, and experiential learning experience surrounding their local watershed. Student knowledge and self-efficacy increased regarding understanding and assessing watershed ecosystem function. Students connected course topics and laboratory work to real problems, collecting data that had real-world application. Students had increased interest in the health and function of their local watershed, increased self-efficacy, feelings of achievement, STEM career motivation as well as ideas regarding future career paths.

Faculty who want to improve their educational practices can benefit from participation in citizen science projects that provide students with the place, problem, service learning and/or inquiry-based learning that has been shown to deepen learning and increase interest. Since personal interest is a major factor in future career choice and STEM major retention (Hall, Dickerson, Batts, Kauffman, & Bosse, 2011), it would seem that these educational changes are necessary to train and retain future STEM leaders. Citizen science projects also provide opportunities for skill development within the core competencies and disciplinary practices suggested for the "New Biology" (Brewer & Smith, 2011). These skills are the

- ability to apply the process of science,
- ability to use quantitative reasoning,
- ability to use modeling and simulation,
- ability to tap into the interdisciplinary nature of science,
- ability to communicate and collaborate with other disciplines, and
- ability to understand the relationship between science and society.

These competencies, when aligned with core concepts, also reinforced in citizen science projects, can provide a framework for course design in preparing students to practice biology and address complex biological issues in current society (Brewer & Smith, 2011). STEM students often feel overwhelmed by the skills and interdisciplinary knowledge they feel that they need to acquire before graduation. Participation in citizen science projects can provide the experience and scaffolding to bolster their self-efficacy and persistence for additional training or a transition to a STEM career.

The use of citizen science as a way to engage students in university education can support faculty in overcoming some of the barriers they face in improving their teaching methods. These barriers may include lack of training, incentives, and time, which could be minimized by working with a partner organization. The inconsistent use and difficulty in administering survey instruments for this project reflect the lack of training, incentives and time scientists have for conducting educational assessments. While the culture of science often values research above teaching, engaging in citizen science projects both improves teaching methods and can result in research opportunities (Brownell & Tanner, 2012).

Partnerships between academia and NGOs or community groups are beneficial, but careful consideration should go into planning the project type (contributory, collaborative, co-created), potential for growth, and desired outcomes needed (data, education, engagement). Specifically, this project required many hours of lab work to complete, making it difficult to recruit faculty from other higher education institutions or have high schools or middle schools adopt the activity. The partners must determine the correct balance between the need for data collection and educational/environmental engagement. In this example, modifications to original biological sampling goals resulted in a mixture of the two. The activity was modified to introduce the project to high and middle school audiences, providing an opportunity for outreach and education, but research-quality data were not collected. However, providing education and outreach experiences related to local watersheds and their impacts might increase institutional participation in the future as well as individual interest in the subject, resulting in increased civic engagement. Studies have found mixed results in terms of increased knowledge gains or changes in motivation and engagement through citizen science as compared to traditional teaching methods, suggesting that assessment of these impacts on student learning and engagement should be consistent and have metrics or instruments that are comparable with other studies (Bonney et al., 2016; Crall, Jordan, & Hollfelder, 2012).

This project has impacted teaching and learning in the undergraduate classroom by providing a place-based, experiential learning opportunity for students who also contribute important data to a community-based monitoring, citizen science project. Since this citizen science project was contributory, the students did not participate in the higher-level aspects of the science process. The students did not ask the questions or design the study. Participation did not increase student or volunteer motivation toward environmental action or learning and doing science. However, results from pre and post-surveys suggested that student and volunteers participating in the citizen science project with "real-life" and "hands-on" methods increased their skills and knowledge in assessing water quality likely to influence participant self-efficacy. Adding inquiry-based modules to create more of a collaborative model would engage students in asking new questions and designing experiments that would benefit student learning, potentially generating new scientific data for scientific and management purposes, and potentially producing the additional benefit of increasing student interest in civic engagement and environmental action.

ACKNOWLEDGMENTS

The project was partially funded by the National Center for Science and Civic Engagement, Partnership Champions, SENCER-ISE and professional development through Mentoring to Enhance Learning Environments. Thank you to the CRWA scientists and Rita Barron Fellows for assistance and support (Julie Wood, Katie Friedman, Logan Bailey, Bryan Dore). Thank you to Michael Berger for providing chemistry support and the students who conducted sediment analyses. Students and CRWA volunteers provided all of the data, insight and enthusiasm to the project.

REFERENCES

American Association for the Advancement of Science (1993). *Benchmarks for science literacy*. New York, NY: Oxford University Press.

Barbour, M. T., Gerritsen, J., Snyder, B. D., & Stribling, J. B. (1999). *Rapid bioassessment protocols for use in streams and wadeable rivers: periphyton, benthic macroinvertebrates and fish* (Vol. 339). Washington, DC: U.S. Environmental Protection Agency, Office of Water.

Barrow, L. H. (2006). A brief history of inquiry: From Dewey to standards. *Journal of Science Teacher Education, 17* (3), 265-278. doi:10.1007/s10972-006-9008-5

Beck, C., Butler, A., & Burke da Silva, K. (2014). Promoting inquiry-based teaching in laboratory courses: Are we meeting the grade? *CBE-Life Sciences, 13*, 444-452. doi:10.1187/cbe.13-12-0245

Bettinger, E. (2010). To be or not to be: Major choices in budding scientists. In C. T. Clotfelter (Ed.), *American Universities in a Global Market* (pp. 68-98). Chicago, IL: University of Chicago Press.

Biber, E. (2011). The problem of environmental monitoring. *University of Colorado Law Review 83*,1-82.

Bonney, R., Ballard, H., Jordan, R., McCallie, E., Phillips, T., Shirk, J., & Wilderman, C. C. (2009). *Public participation in scientific research: defining the field and assessing its potential for informal science education. A CASE Inquiry Group Report.* Washington, DC: Center for Advancement of Informal Science Education (CAISE). Retrieved from http://www.birds.cornell.edu/citscitoolkit/publications/CAISE-PPSR-report-2009.pdf

Bonney, R. E., & Dickenson, J. L. (2012). Overview of citizen science. In J. L. Dickenson & R. E. Bonney (Eds.), *Citizen science: Public participation in environmental research* (pp. 1-14). Ithaca, NY: Cornell University Press.

Bonney, R., Phillips, T. B., Ballard, H. L., & Enck, J. W. (2016). Can citizen science enhance public understanding of science? *Public Understanding of Science*, 2-16. doi:10.1177/0963662515607406

Brewer, C. A., & Smith, D. (2011). *Vision and change in undergraduate biology education: A call to action.* Washington, DC: American Association for the Advancement of Science.

Brossard, D., Lewenstein, B., & Bonney, R. (2005) Scientific knowledge and attitude change: The impact of a citizen science project. *International Journal of Science Education, 27*, 1099–1121. doi:10.1080/09500690500069483

Brownell, S. E., & Tanner, K. D. (2012). Barriers to faculty pedagogical change: lack of training, time, incentives, and . . . tensions with professional identity? *CBE-Life Science Education, 11*, 339-346. doi:10.1187/cbe.12-09-0163

Campbell, T., & Bohn, C. (2008). Science laboratory experiences of high school students across one state in the US: Descriptive research from the classroom. *Science Educator, 17*(1), 36-48.

Cardamone, C, & Lobel, L. (2016). Using citizen science to engage introductory students: From streams to the solar system. *Journal of Microbiology and Biology Education, 17*, 117-119. doi:10.1128/jmbe.v17i1.1082

Chen, X. (2013). STEM attrition: College students' paths into and out of STEM fields. Statistical Analysis Report. NCES 2014-001. Washington, DC: National Center for Education Statistics.

Conrad, C. C., & Hilchey, K. G. (2011). A review of citizen science and community based environmental monitoring: issues and opportunities. *Environmental Monitoring and Assessment, 176*, 273-291. doi:10.1007/s10661-010-1582-5

Cornell Lab of Ornithology (2013). Evaluation instruments. Retrieved December 11, 2018, from http://www.birds.cornell.edu/citscitoolkit/evaluation/instruments

Crain, R. L., Cooper, C., & Dickinson, J. L. (2014). Citizen science: a tool for integrating studies of human and natural systems. *Annual Review of Environment and Resources, 39*, 641-665. doi:10.1146/annurev-environ-030713-154609

Crall, A. W., Jordan, R., Holfelder, K., Newman, G. J., Graham, J., & Waller, D. M. (2013). The impacts of an invasive species citizen science training program on participant attitudes, behavior, and science literacy. *Public Understanding of Science*, *22*(6), 745-764. doi:10.1177/0963662511434894

Dickenson J. L., Shirk J., Bonter D., Bonney R., Crain R. L., Martin J., . . . Purcell K. (2012). The current state of citizen science as a tool for ecological research and public engagement. *Frontiers in Ecology and the Environment, 10*(6), 291–297. doi:10.1890/110236

Dillon, J., Stevenson, R. B., & Wals, A. E. J. (2016). Introduction to the special section moving from citizen to civic science to address wicked conservation problems. *Conservation Biology, 19*, 589-594. doi:10.1111/cobi.12689

Folke, C. (2006). Resilience: The emergence of a perspective for social–ecological systems analyses. *Global Environmental Change, 16*, 253-67. doi:10.1016/j.gloenvcha.2006.04.002

Freeman, S., Eddy, S. L., McDonough, M., Smith, M. K., Okoroafor, N., Jordt, H., & Wenderoth, M. P. (2014). Active learning increases student performance in science, engineering, and mathematics. *PNAS, 111*(23), 8410-8415. doi:10.1073/pnas.1319030111

Gormally, C., Brickman, P., Brittan, H., & Armstrong, N. (2009). Effects of inquiry-based learning on students' science literacy skills and confidence. *International Journal for the Scholarship of Teaching and Learning*, 3(2), Article 16. doi:10.20429/ijsotl.2009.030216.

Hall, C., Dickerson, J., Batts, D., Kauffmann, P., & Bosse, M. (2011). Are we missing opportunities to encourage interest in STEM fields? *Journal of Technology Education*, *23*(1), 32-46. doi:10.21061/jte.v23i1.a.4

Jenkins, L. L. (2011). Using citizen science beyond teaching science content: A strategy for making science relevant to students' lives. *Cultural Studies of Science Education*, *6*(2), 501-508. doi:10.1007/s11422-010-9304-4

Jordan, R. C., Ballard, H. L., & Phillips, T. B. (2012). Key issues and new approaches for evaluating citizen-science learning outcomes. *Frontiers in Ecology and the Environment, 10*(6), 307-309. doi:10.1890/110280

Lobel, L., & Cianciola, E. (2017). Water quality concepts and benthic macroinvertebrate knowledge evaluation instrument. Unpublished manuscript.

Lynch, L. I., Dauer J. M., Babchuk W. A., Heng-Moss, T., & Golick, D. (2018). In their own words: The significance of participant perceptions in assessing entomology citizen science learning outcomes using a mixed methods approach. *Insects, 9*(16), 1-15. doi:10.3390/insects9010016

Markus, G. B., Howard, J. P., & King, D. C. (1993). Integrating community service and classroom instruction enhances learning: Results from an experiment. *Educational evaluation and policy analysis, 15*(4), 410-419. doi:10.3102/01623737015004410

McKinley, D. C., Miller-Rushing, A. B., Ballard, H. I., Bonney, R., Brown, H., Cook-Patton, S. C., Soukup, M. A. (2017). Citizen science can improve conservation science, natural resource management, and environmental protection. *Biological Conservation, 208*, 15–28. doi:10.1016/j.biocon.2016.05.015

Mitchell, N., Trska, M., Liberatore, A., Ashcroft, L., Weatheril, R., & Longnecker, N. (2017). Benefits and challenges of incorporating citizen science into university education. *PLoS ONE, 12* (11): e0186285. doi:10.1371/journal.pone.0186285

Momsen, J. L., Long, T. M., Wyse, S. A., & Ebert-May, D. (2010). Just the facts? Introductory undergraduate biology courses focus on low level cognitive skills. *CBE-Life Science Education 9*, 435-440. doi:10.1187/cbe.10-01-0001

National Research Council. (2003). BIO 2010: *Transforming undergraduate education for future research biologists*. Washington, DC: National Academies Press.

National Research Council. (2006). America's Lab Report: Investigations in high school science. Washington, DC: The National Academies Press. doi:10.17226/11311

National Research Council (NRC). (2009). *A new biology for the 21st century: Ensuring the United States leads the coming biology revolution*. Washington, DC: National Academies Press.

Newman, G., Chandler, M., Clyde, M., McGreavy, B., Haklay, M., Ballard, H., Gallo, J., (2017). Leveraging the power of place in citizen science for effective conservation decision making. *Biological Conservation, 208*, 55-64. doi:10.1016/j.biocon.2016.07.019

Oberhauser, K., & LeBuhn, G. (2012). Insects and plants: Engaging undergraduates in authentic research through citizen science. *Frontiers in Ecology and the Environment, 10*(6), 318-320. doi:10.1890/110274

Overdevest, C., Orr, C. H., & Stepenuck, K. (2004). Volunteer stream monitoring and local participation in natural resource issues. *Human Ecology Review*, 177-185.

Phillips, T., & Bonney, R. (2016). Developing, validating, and implementing situated evaluation instruments (DEVISE). Paper presented to Advancing Informal STEM Learning (AISL) PI Meeting held in Bethesda, MD, 29th February-2nd March. Retrieved from http://www.informalscience.org/posterdevisedeveloping-validating-and-implementing-situated-evaluations.

Pocock, M. J. O., Tweddle, J. C., Savage, J., Robinson, L. D., & Roy, H. E. (2017). The diversity and evolution of ecological and environmental citizen science. *PLoS ONE, 12*(4), e0172579. doi:10.1371/journal.pone.0172579

Schuttler, S. G., Sorenson, A. E., Jordan, R. C., Cooper, C., & Shwartz, A. (2018). Bridging the nature gap: can citizen science reverse the extinction of experience? *Frontiers in Ecology and the Environment, 16*(7), 405-411. doi:10.1002/fee.1826

Sundberg, M. D., Armstrong J. E., & Wischusen E. W. (2005). Reappraisal of the status of introductory biology laboratory education in U.S. Colleges & Universities. *The American Biology Teacher, 67*, 525-529. doi:10.2307/4451904

Tawfik, A., Trueman, R. J., & Lorz, M. M. (2014). Engaging non-scientists in STEM through problem-based learning and service learning. *Interdisciplinary Journal of Problem-Based Learning, 8*(2). doi:10.7771/1541-5015.1417

Theobald, E. J., Ettinger, A. K., Burgess, H. K., DeBey, L. B., Schmidt, N. R., Froehlich, H. E., Parrish, J. K. (2015). Global change and local solutions: tapping the unrealized potential of citizen science for biodiversity research. *Biological Conservation, 181*, 236-244. doi:10.1016/j.biocon.2014.10.021

Toomey, A. H., & Domroese, M. C. (2013). Can citizen science lead to positive conservation attitudes and behaviors? *Human Ecology Review*, 50-62.

Turrini, T., Dorler D., Richter, A., Heigl, F., & Bonn, A. (2018). The threefold potential of environmental citizen science – Generating knowledge, creating learning opportunities and enabling civic participation. *Biological Conservation, 225*,176-186. doi:10.1016/j.biocon.2018.03.024

University of Rhode Island Watershed Watch (2019). Retrieved from https://web.uri.edu/watershedwatch/

University of Vermont Extension Watershed Alliance (2019). Retrieved from https://www.uvm.edu/watershed/about

United States Environmental Protection Agency (1997). Volunteer Stream Monitoring: A Methods Manual. EPA 841-B-97-003. *US Environmental Protection Agency, Office of Wetlands, Oceans, and Watersheds,* Washington, DC. https://archive.epa.gov/water/archive/web/html/stream_index.html

Wals, A. E. J., Brody, M., Dillon, J., & Stevenson R. B. (2014). Convergence between science and environmental education. *Science, 344*, 583-584. doi:10.1126/science.1250515

Zajacova. A., Lynch, S. M., & Espenshade, T. J. (2005). Self-efficacy, stress and academic success in college. *Research in Higher Education, 46*(6), 677-706. doi:10.1007/s11162-004-4139-z

Zimmerman, B. J. (2000). Self-efficacy: An essential motive to learn. *Contemporary Educational Psychology, 25*, 82-91. doi:10.1006/ceps.1999.1016

Zumeta, W., & Raveling, J. (2002). Attracting the best and the brightest *Issues in Science Technology, 19*(2), 36-40.

SECTION V: ASSESSMENT AND EVALUATION OF CITIZEN SCIENCE PROGRAMS

In: Enhancing STEM Motivation through Citizen Science Programs ISBN: 978-1-53616-038-3
Editors: Suzanne E. Hiller and Anastasia Kitsantas

Chapter 13

AN IMPACT-DRIVEN APPROACH TO EVALUATING CITIZEN SCIENCE PROGRAMS FOR YOUTH

Stephanie Downey* and Emily Skidmore
Randi Korn and Associates, Inc., DC, US

ABSTRACT

This chapter details the "who," "what," "how," and "why" of evaluating citizen science programs for youth (ages 10-25 years) using an *impact-driven, outcomes-based* approach. Impact is *the difference you make in the quality of people's lives*, and this approach to evaluation places impact at the center of everything you do—program planning, implementation, evaluation, and reflection. The "who" is about how to narrowly define youth audiences; the "what" deals with the kind of impact you intend to have on your audiences and how to define the different elements of impact, including outcomes (intended results) and indicators (evidence of those results), which are used as guideposts for evaluation; the "how" guides you to think about the methods, processes, and limitations that need to be considered in order to evaluate a citizen science program in the context of intended impact; and, finally, the "why" emphasizes the main value of evaluation—reflecting on and learning from findings to strengthen impact. In this chapter, inquiry is used as the primary method to push readers to consider impact throughout every step of the planning and evaluation process.

Keywords: citizen science, evaluation, impact planning, intentional practice

This chapter is about the "who," "what," "how," and "why" of evaluating citizen science programs. The intent is not for you to read this chapter and then evaluate a citizen science program entirely on your own; instead, we suggest that you work with or utilize the resources of those with expertise in evaluation. That being said, we think it is vitally important to understand certain things about evaluation in order to benefit from it. Many people are asked

* Corresponding Author's Email: downey@rka-learnwithus.com.

by a funder to evaluate a program. While understandable, this is not the true value of evaluation. This entire chapter is predicated on the notion that evaluation is a *learning tool, not a judgment tool.* We learn from evaluation how to make citizen science programs better for the audiences who participate in them. By doing so, we enhance the likelihood that these programs have an impact on those audiences.

Impact is *the difference you make in the quality of people's lives*, and, in the case of this book, "people" are youth (10-25 years) participating in citizen science programs. This chapter advocates for an *impact-driven, outcomes-based* approach to evaluation (Korn, 2018). This approach places impact at the center of everything—program planning, implementation, evaluation, and reflection. The process is cyclical and never-ending because impact is difficult to achieve even under the best of circumstances. To understand the benefits and limitations of this approach and how it works, we will delve into four common aspects of evaluation—the who, what, how, and why. Our greatest hope is that, by the end of this chapter, you, the reader, can be a more effective collaborator in the evaluation process of your citizen science program.

The "Who"

The key to starting any evaluation is defining the target audience. The target audience is the "who"—who are you designing the citizen science program for? The focus of this book is on youth in formal education settings, defined as those ages 10-25 years of age. While it may be tempting to say that your program is for "everyone" in this age range, this is not wise. Remember that this chapter is advocating for an impact-driven evaluation approach and impact is difficult to achieve. In other words, your program cannot be everything to everyone if you expect to have an impact. Therefore, we suggest that you define your audience as narrowly as possible and consider several factors when doing so.

1. Are there audiences for whom your program is more relevant?
2. Which grade-level curriculum best aligns with your program focus? Similarly, which grade levels are covering the knowledge and/or skills that you are trying to enhance through your program?
3. Does your program require your audience to have any prior knowledge and/or skills?
4. Are there audiences that have a greater need for the program you are offering? For example, female students and students of color are underrepresented in STEM (National Science Foundation, 2017; Noonan, 2017). There may also be specific schools in your area that have a greater need for access to STEM opportunities. Consider filling these gaps.

When we say "define your audience as narrowly as possible," we mean, for example, sixth grade students and teachers or even females in sixth grade. We also mean "how many?" How broadly are you trying to reach students within your local area? Again, you cannot serve everyone. So, how many students are you trying to serve? To answer this, you will need to consider your available resources—financial resources and staff resources. At the most basic level, you need to know the following, specifically in relation to maintaining quality and preserving your ability to achieve impact:

1. How often will you be able to offer the program during the school year?
2. How many students can *realistically* participate in a single program activity?
3. Will the same students repeat or will it be new students each time?

When you consider your available financial and staff resources, you will know how many students you can realistically serve while still maintaining quality and preserving your ability to achieve impact. One final note of caution that bears repeating—err on the side of fewer, more narrowly defined audiences (Korn, 2010, 2018). Non-profit organizations are full of individuals who are passionate about what they do but the flip side of this is that these individuals often try to accomplish too much. Start small and narrow and then expand your reach when program resources allow.

THE "WHAT"

The "what" of an evaluation encompasses a few different things. First, what is the purpose of the evaluation of your program? As we mentioned at the beginning of this chapter, we view evaluation as a learning tool, not a judgment tool. It's possible that a funder is asking you to do evaluation as part of a grant you received and, that left to your own devices, you would choose not to use your limited resources to evaluate the program. This is not uncommon. However, try to view evaluation as a learning tool because at the end of the day, you are the one who will benefit the most from the evaluation findings. So, decide right now what about the evaluation is important to you. To help define the purpose of the evaluation, consider the following:

1. Who are the stakeholders of the program (your staff, teachers, students, funders, board members, etc.)? What do these stakeholders hope to learn from an evaluation?
2. What stage of development is your program in? Do you need to understand more about your audience to inform program development? Are you piloting the program for the first time and need to understand what is working well (and not so well) about its implementation (so you can make improvements)? Or, have you been offering this program for a long time, feel confident in the program design, and are ready to measure its impact?
3. Are you trying to answer a research or evaluation question that will inform the citizen science field at-large? If so, is it a new question or are you building on prior knowledge?

You may have more than one purpose but choose a primary purpose in order to focus your available resources for evaluation. Again, you will not be able to explore or measure everything so it helps to know what you most want to learn from the evaluation. In a later section in this chapter (The "How"), knowing the purpose of the evaluation will help you decide which phase of evaluation—front-end, formative, or summative—you want to conduct (Davidson, 2005). Knowing the purpose of the evaluation up front also will help you focus your reflections on the data at the end of the evaluation—what did you learn about what you wanted to know? How can you do better, or what are your next steps?

Creating a List of Outcomes

Another important "what" to consider are the outcomes for the evaluation. Outcomes are the *intended results of program participation on the audiences served.* Outcomes are the "So what?" of a program—as in, "So what does it matter that you are doing this program?" While there are many ways to approach evaluation, this chapter advocates for an outcomes-based model of evaluation where you, as the individual or organization offering the program, articulate intended outcomes for your program and align everything else with these outcomes. In this section, our intent is to help you articulate these outcomes. To do so, you need to consider the following question: *What difference do you intend to make in the quality of your audience's life?* For example, what difference do you intend to make in the quality of sixth graders' lives?

When answering this question, your first step can simply be to brainstorm a list of possible outcomes alone or in a small group. However, the ultimate goal is for you to prioritize your list of outcomes so that you end up with no more than three to five outcomes (erring on the side of three if possible)! So, after this first step, there are several things for you to consider, in no particular order, to help you prioritize. We expand on some of these ideas in the next few sections because there are some existing strategies and knowledge that can serve as a baseline to inform your discussions (see Figure 1).

Questions to Consider

1. Who, within and outside your organization, is going to contribute to answering this question?
2. What does the citizen science field at-large already know about particular outcomes? How might you build on this knowledge or fill in gaps in that knowledge?
3. Which outcomes are most relevant for your audience given their grade level and what they are learning in school (i.e., curricular alignment)?
4. Which outcomes are most realistic for your audience given their developmental level?
5. Will your program focus on breadth or depth?
6. Which outcomes are most realistic given that focus?
7. Which outcomes are most important to you and other stakeholders, such as where do your priorities lie?

Figure 1. Questions to consider.

Who is going to contribute to answering the question? First, a note about the process. Consider who, within and outside your organization, is going to contribute to answering the question that helps you define outcomes: *What difference do you intend to make in the quality of your audience's life?* It is important not to operate in a silo when it comes to planning, implementing, evaluating, and reflecting on your citizen science program. Therefore, consider soliciting feedback from individuals across your organization (i.e., in different departments) as well as up and down your organization (i.e., entry level positions to leadership positions). You do not have to involve everyone but think strategically about who might have important insights about your audiences, program goals, program structure, etc., remembering that these insights do not only come from leadership and management. Also, consider soliciting feedback from those outside your organization, including teachers (especially if your primary audience is their students), staff from other organizations who are conducting citizen science programs, and other relevant community members. The process of soliciting feedback from many different stakeholders challenges you to clarify (and likely revise) your thinking and also has the added benefit of creating buy-in for your program within and outside the organization (Korn, 2018).

What does the citizen science field already know (and where are the knowledge gaps)? Some weight should be given to the citizen science field at-large and what it already knows about certain outcomes. We all fall into the trap of "reinventing the wheel" and thinking, perhaps, that our approach or what we want to know is unique. However, this is rarely the case. So, what does the field already know about outcomes in citizen science programs? Numerous articles and studies have described outcomes of citizen science programs. For instance, studies have looked at individuals' motivations for participating in citizen science programs, which include, among other things, the opportunity to contribute to scientific research and science learning, though few studies have focused on youth (Lee, Miller, Oesterlund, & Crowston, 2017; Measham & Barnett, 2008; Raddick et al., 2013). Studies have also explored individuals' learning in citizen science programs, including science content, scientific processing skills, and creative problem-solving (Ballard, Dixon, & Harris, 2017; Bonney et al., 2016; Gommerman & Monroe, 2012; Jennett et al., 2016; Masters et al., 2016; Prather et al., 2017). And, studies have also explored how citizen science programs affect individuals' interest in science and STEM, including STEM careers (Brossard, Lewenstein, & Bonney, 2005; Hiller & Kitsantas, 2014).

The present book also has a particular focus when it comes to the types of outcomes that are important to explore in the context of citizen science programs. These outcomes fall into three main categories and include student motivational beliefs (e.g., self-efficacy, interest, etc.), science achievement, and STEM career motivation. Please refer to Section 2 of this book for more information about students' motivations, engagement, and science achievement in citizen science programs and Section 4 of this book for more information about students' STEM career motivation in citizen science programs.

Given what the field already knows and may want to know, ask yourself: Does your list of outcomes build on existing knowledge and/or fill a knowledge gap? Do your outcomes align with the categories that are a focus of this book? If not, how might you revise or prioritize your list of outcomes to better align with the citizen science field at-large?

Which outcomes are most appropriate and relevant for your audience? If you are a practitioner, you may already have the experience and knowledge to determine which outcomes are developmentally appropriate for the target audience. However, if you are involving internal and external stakeholders, such as other practitioners, teachers, and community members, in your discussion of outcomes, they can also help you determine which outcomes are developmentally appropriate. If you will be working with students in a specific grade, collaborate with teachers and administrators to determine which outcomes they perceive as most important for their students based on students' interest and the curriculum they will be focusing on at certain times of the year (when your program is taking place), etc. Soliciting this type of feedback will help place the outcomes of the citizen science program firmly within the reality in which your audience(s) operate(s) and can help cement a true collaboration between your organization and the audience(s) being served (Ibarra & Hansen, 2011; Korn, 2018).

What is the primary focus or emphasis of your program—breadth or depth? Choosing to focus on breadth or depth is a tricky thing because most individuals try to do both simultaneously when implementing programs, including citizen science programs. That is, they try to be everything to everyone but that is not realistic. Breadth and depth can refer to many aspects of a citizen science program, including how many individuals you serve as well as what those individuals do as part of the program. Both are important when considering outcomes; however, the information here refers more to what audiences do as part of the program. An audiences' level of participation in citizen science programs can vary widely, and sometimes this level of participation is within your control and sometimes it is not. For instance, a teacher might have his or her class participate in a single data collection activity (that is part of a larger research project), or he or she might have students participate in several aspects of the research process. The former scenario reflects a program experience that has more breadth than depth and the latter scenario reflects a program experience that, on some level, assumes repeat participation, and, thus, has the potential for more depth. Common sense dictates (and program evaluations, whether of citizen science programs or not, reveal) that some outcomes are not realistic in a one-time experience—changes in behavior, for example (Crall et al., 2012; Heimlich & Ardoin, 2008; Jordan, Gray, Howe, Brooks, & Ehrenfeld, 2011). So, ask yourself, which outcomes are most realistic to achieve given the structure of your citizen science program?

Finally, which outcomes are most important to you and other stakeholders? All of the questions up to this point were meant to help you prioritize your list of outcomes so that you have no more than three to five. However, you may still have more than that, and you are not alone. This is an all too common problem. But, considering that even one outcome can be hard to achieve, how can you prioritize your list of outcomes further? Go back to your internal and external stakeholders and have them vote on the final list of draft outcomes. This may seem overly simplistic, but this strategy can be highly efficient and effective at narrowing the list of outcomes for many organizations. One way to do this is to give everyone a list of the outcomes and pose the question, "Which outcomes are most important to achieve over the next few years of this program?" Each person can select three outcomes. You can simply choose the three (to five) outcomes that receive the most votes or you may want to dig a little deeper. For instance, you may want to understand *why* people made the selections they did in order to make the final determinations (especially if two outcomes received the same number of votes). Even if the voting eliminates a certain outcome, there may be a compelling reason to keep it, for example, advancing the knowledge of the field (Korn, 2018).

Crafting Indicators (or Evidence) of the Outcomes

Another important "what" is the indicators for the evaluation. Indicators are *the evidence that an outcome has been achieved.* Indicators are extremely important to an evaluation because they help evaluators *measure* results. You will be very happy that you did the hard work to prioritize both audiences and outcomes in the previous sections because each audience has a set of outcomes and each outcome has a set of indicators. And, an evaluation quickly becomes unwieldy if you have multiple audiences, multiple outcomes and, therefore, hundreds of indicators to measure. Very few people have the time and resources to do that kind of evaluation so, prioritize (again). As with outcomes, best practice is to define a limited number of indicators per outcome—again, three to five, erring on the side of less.

To define indicators for your outcomes, consider this question: *What will your audience do or say to show that they have achieved an outcome?* By considering what your audience will do, you are starting to think about behaviors that an evaluator can capture during an open-ended or standardized observation of your citizen science program. By considering what your audience will say, you are starting to think about data an evaluator can capture during an interview or focus group, on a standardized questionnaire, or even during an observation (if the conditions allow). For instance, if a student demonstrates their science process skills then they might pose their own questions or classify specimens during your citizen science program. Likewise, if a student increases their interest in science then they might participate in another citizen science program or STEM activity after program participation.

When crafting indicators, it is important to be as specific and concrete as possible, leaving very little room for ambiguity. So, ask yourself some important questions to push for clarity and consensus:

1. What do we mean by X?
2. How is X different from Y?

Also, much of the considerations and questions we described when discussing how to prioritize outcomes can come in useful when defining and prioritizing indicators. For instance:

1. Is X or Y realistic given the depth or breadth of the program's structure (in other words, will audiences actually have the opportunity to do something that leads to X or Y)?
2. Is X or Y realistic given the developmental level of the audience? Is there a range or continuum (e.g., beginner to advanced) of behaviors or language that might result from this audience?
3. Are there existing, vetted indicators (or evidence) that you can use from the citizen science field (or another STEM or even arts program)?

Once you have defined up to three to five indicators per outcome, you should take the additional step to have select stakeholders review the list to ensure they are as specific and concrete as possible. The more you are pushed to explain what you mean, the better the indicators will be and the more likely it is that everyone will be on the same page about what is being measured.

THE "HOW"

Defining the "what" is the hardest part of conducting an evaluation because it influences everything else. If you do not take the time to clearly define the "what," it will result in a bad evaluation and resources wasted. If you have clearly defined the "what," it's now time to consider the "how" of the evaluation. That is, all the mechanics of conducting the evaluation so you end up with data that you can actually use to improve your program or demonstrate impact. These mechanics include:

1. choosing the type of evaluation (front-end, formative, summative, or some combination),
2. determining appropriate methodologies and data collection strategies,
3. ensuring proper procedures for the ethical treatment of human subjects,
4. considering how to effectively communicate findings, and
5. defining the limitations of the evaluation and adjusting the evaluation accordingly.

Before moving into more detail about the mechanics, it is important to mention that evaluation is a field that requires training and expertise; it also takes a lot of time and resources. Ideally, if you are a practitioner, you will not be conducting the evaluation entirely on your own but using the expertise or resources of someone with evaluation experience (Patton, 2008; Phillips, Bonney, & Shirk, 2012; Sheppard, 2010). As such, this section does not include such details as how to develop a data collection instrument (though we cite examples and resources that may help inform this process) or how to use statistics to analyze quantitative data (though, again, we may provide resources that help you understand what can and cannot be elicited from this type of data). Instead, the information in this section is meant to inform so that you can collaborate more effectively with others to assess your citizen science program.

Choosing the Evaluation Type

There are typically three types of evaluation that coincide with the implementation stage of your citizen science program: front-end evaluation (which coincides with program planning), formative evaluation (which coincides with program development), and summative evaluation (which coincides with measuring the program's results or impact). We like to think of evaluation as a cyclical, learning process not a linear, judgmental process. Thus, depending on the context and circumstances of your citizen science program, you may not need to conduct all three phases of evaluation, and you may not need to start with front-end evaluation. This section provides you with some baseline information to help you understand which type(s) of evaluation might be appropriate. When defining the "what" of the evaluation, you collaborated with others to define its primary purpose. That purpose can now help you determine which type(s) of evaluation you need.

When might you need front-end evaluation? Front-end evaluation is conducted when *planning* a public program like citizen science in order to understand more about your audience vis a vis the program being designed. Some common questions that would require front-end evaluation are:

1. What are teachers' motivations for having their students participate in citizen science?
2. What are teachers' goals or expectations for the citizen science program?
3. What are teachers' and/or students' perceptions or understandings of themes and concepts being introduced in the program?
4. What are effective communication strategies for recruiting teachers and students?
5. What are barriers to participation in the citizen science program?

Understanding, for example, how teachers and students think about a particular theme or concept, such as birds or water quality, can help when designing the program. If teachers and students are coming in with very basic knowledge of themes and concepts, then it makes sense to start where they are to build a somewhat more complex (but not too complex) understanding. Likewise, if scheduling or transportation emerge as significant barriers to participation, the program structure may need to be adjusted to accommodate school schedules and/or program resources may need to be diverted to partially subsidize transportation costs.

When might you need formative evaluation? Formative evaluation is conducted when *implementing (or piloting)* a program to iterate (or modify) design and implementation. Because of its iterative nature, formative evaluation can happen multiple times (or in multiple rounds) until the program design is "perfected" (though, in reality, no program will ever be perfect). Some common questions that would require formative evaluation include:

1. How is the program recruitment/registration process working for schools/teachers?
2. Which aspects of the program structure (themes/concepts, facilitation strategies, scheduling, etc.) work well to support your intended outcomes? Why?
3. Which, if any, aspects of the program structure (themes/concepts, facilitation strategies, scheduling, etc.) are barriers to engagement/participation/outcome achievement? Why?

You can also begin to explore whether the program is having its intended impact though we suggest doing so from the perspective of program implementation. For example, if students are not engaged in program activities and/or not taking away key understandings or skills, you might try to understand what about the program structure might be contributing to or acting as a barrier to impact. Or, you can wait to explore impact since it typically takes time to emerge.

When might you need summative evaluation? Summative evaluation is conducted when a program is well established and/or its structure and implementation have been well honed through formative evaluation. The intent of summative evaluation is to measure intended results, which is where the outcomes and indicators (or evidence) you have defined as the "what" are critical (Davidson, 2005; Korn, 1994; W. K. Kellogg Foundation, 2017). Some common questions that would require summative evaluation include:

1. To what extent did the program affect outcomes related to students' motivational beliefs (e.g., self-efficacy, interest, etc.)?
2. To what extent did the program affect outcomes related to students' science achievement?
3. To what extent did the program affect outcomes related to students' STEM career motivation?
4. To what extent did the program have other, unintended outcomes?

Achieving impact is difficult. Thus, you may encounter certain barriers to the program's impact, and some of these barriers may be out of your control. For instance, it is possible that not enough time will have passed for any meaningful impact to emerge (which is common because it takes time and, in some cases, repeat participation to see results); program constraints, such as staff capacity, resources, and logistical barriers inherent to citizen science programs (e.g., bad weather conditions, transportation to data collection sites), may prevent achievement of impact (and, perhaps, the program's intended results need to be revised in order to be more realistic or you need additional support/resources to achieve intended results); and/or audiences' previous knowledge, interests, skills, etc. may be affecting the achievement of impact (which is where data from a front-end evaluation might be useful since it forms a baseline for your audience). An example of how audiences' prior experiences can affect the achievement of impact is that often audiences self-select into citizen science programs due to a strong previous interest in science and/or individuals have some prior knowledge of themes and topics being presented; in such circumstances, you may not see as significant of an impact because the level of interest and/or knowledge is already high (Bell, Lewenstein, Shouse, & Feder, 2009; Brossard et al., 2005). However, this may not apply as often for students whose school or teacher made the choice to participate in the program.

Choosing Evaluation Methods

Consulting evaluation resources (Brody, Bangert, & Dillon, 2007; Cornell Lab of Ornithology, 2018; Davidson, 2005; Phillips et al., 2014) or working with those who have evaluation expertise can be helpful in selecting the overall methodology for the evaluation; however, outlined here are some common data collection methods and when they are typically employed. Some methods will result in qualitative data, and some methods will result in quantitative data. Both are useful depending on the intent of the evaluation and, if resources allow, it is often helpful to employ a mixed methodological approach that yields both qualitative and quantitative data, as doing so provides a holistic picture of participants' experiences in citizen science.

When might you use methods that result in qualitative data? Common methods that result in qualitative data are interviews, focus groups, naturalistic observations, and case studies, which often employ a combination of qualitative methods such as observations and interviews. These methods are particularly useful when you need to understand the nuances of your audience vis a vis a citizen science program. For instance, to harken back to the "what" section of this chapter where you defined indicators (or evidence) of impact, interviews and focus groups can be useful for capturing what participants *say* before, during, or after a program experience, and are useful when you need to understand the language audiences use to think about or discuss a particular topic or idea. This language can be useful data on its own, or it can be used to develop a more standardized data collection instrument that includes words and phrasing more reflective of (or recognized by) your audience. Likewise, naturalistic observations can be useful for capturing what participants *do* during or after a program. A description of some of these methods and when they are commonly used follows (Merriam, 2009; Patton, 2015).

Interviews. Interviews can take many forms but typically consist of a series of open-ended questions, often evolving from simplistic to complex as the interview progresses. Interviews can be short and to-the-point if the nature of the encounter is such that data must be collected quickly and/or what is being explored is fairly straightforward. Or, interviews can be in-depth and involve many probing, "why" questions to understand audiences more deeply. For example, front-end evaluations, where the intent is to understand audiences' prior knowledge, interests, etc. to inform program development, often require in-depth interviews. Interviews can also be conducted using a variety of data collection strategies, including on the telephone, online, or in person, and interview data can be captured using handwritten notes or through transcribed audio-recordings.

Focus groups. Focus groups are, in essence, group interviews, where a moderator engages 10-12 participants in a discussion using a series of open-ended questions. Focus groups are useful when the intent is to have participants brainstorm ideas in response to a relevant theme or topic. For instance, teachers might be able to brainstorm program ideas for engaging their students in citizen science if given some initial parameters or ideas for the program. This type of brainstorming is more challenging for an individual to accomplish during a one-on-one interview. On the other hand, focus groups are less useful when the subject matter is controversial or sensitive in nature; these are conditions better suited to a one-on-one confidential interview (but also conditions that are less likely in a citizen science program evaluation where the topics tend not to be personal in nature). Focus groups can be expensive depending on the necessity of a rental facility, audio- or video-recording, recruitment of and stipends for participants, and refreshments, all things that can be part of conducting focus groups.

Naturalist observations. Naturalistic observations (sometimes also referred to as thick description or ethnography) are a detailed account of a particular context and the behaviors of participants in that context at a certain point in time. Naturalistic observations of a program are usually of a "point in time," as opposed to an account of everything that has or will ever happen in the program, simply because the latter is not realistic given typical resources for evaluation. Naturalistic observations are useful for providing as objective an account as possible of what facilitators and participants are doing (their behaviors) in a citizen science program. In contrast, interviews and focus groups are participants' accounts of what they did in a program, which can be less objective. Data from naturalistic observations usually take the form of detailed notes describing exactly what facilitators and participants do (and, on occasion, say) in a program, *not an interpretation of what they do or say.*

Case studies. Case studies tell the story of a program at the micro level by examining the interplay of variables to provide as complete an understanding of one event or situation as possible. Case studies often comprise mixed methods, such as naturalistic observations and interviews, to understand individual contexts or experiences in depth. The information gleaned from case studies is not generalizable, but they are particularly useful for understanding best-case scenarios of what might be possible to achieve given a certain set of circumstances. For instance, one might have 10 case studies that explore the experiences of 10 students in a citizen science program. The case study data might include observations of each student at different points in the program, interviews with the students about their program experiences, as well as interviews with those who can speak to students' experiences in the program, such as teachers, parents, mentors, peers, etc. In the end, all these different accounts can provide a holistic account of that student's experiences in the program and what contributed to those experiences.

The selection of case studies is often purposeful to represent a range of program experiences, and case studies often include an element of time (to examine change over time), though they can also examine one point in time.

When might you use methods that result in quantitative data? Common methods that result in quantitative data are standardized questionnaires and observations. There are also methods of analysis that take qualitative data and transform it into quantitative data such as rubric-scored interviews. These methods are particularly useful when you need to collect large quantities of data from a more representative group of people. Like methods that produce qualitative data, these methods can capture what participants *do* or *say* before, during, or after a program experience; yet, unlike most qualitative data, which tends to be collected in smaller quantities, quantitative data can be useful for generalizing trends and demonstrating statistically significant relationships among a myriad of different factors. Thus, it can be useful to include methods that produce quantitative data if your intent is to produce results that can be applied more broadly to the citizen science field at-large. A description of some of these methods and when they are commonly used follows (Bitgood, 2011; Diamond et al., 2016; Taylor-Powell, 1998).

Standardized questionnaires. Standardized questionnaires are a method with which many are familiar. Questionnaires can assess attitudes, understandings, and behaviors from a broad sample of participants and typically employ closed-ended question formats (i.e., multiple choice, ranking, and rating scale questions). Qualitative data from interviews can be helpful in shaping the language used in questionnaires, making what might otherwise be a generic data collection instrument more authentic using participants' own words to describe relevant ideas, concepts, and experiences. Questionnaire data can be collected online or in person and can be self-administered (completed on one's own) or administered by a data collector (which are often referred to as "interview-style"). Having data collectors administer a questionnaire strengthens reliability and completeness but it also takes additional time and resources which may not be realistic.

Standardized observations. Standardized observations are useful in many of the same ways as naturalistic observations except one can typically collect more data from more people in a shorter period of time, and the results can be applied more generally due to the larger sample sizes and applied statistics. In other informal settings, such as museums, standardized observations can take the form of timing and tracking, where data collectors record where visitors stop in an exhibition and what they do there (behaviors). For citizen science programs, a more appropriate method might be scans, where a data collector records times and the occurrence of certain behaviors at designated intervals (for example, every 5 minutes). Scans take into account the non-linear environment of citizen science programs and systematically capture the behaviors of researchers and participants performing a variety of tasks and activities (Martin & Bateson, 2007). In either case, behaviors must be very clearly defined ahead of time so that data collectors are not interpreting and recording a behavior in different ways. This is where the indicators (or evidence) you have defined in the "what" section of this chapter come in useful; specific and concrete indicators ensure everyone is on the same page and that there is no room for interpretation.

Rubric-scored interviews (or observations). Rubric-scoring is actually a method of analysis, but its usefulness lies in the fact that you can capture the authenticity of participants' program experiences by collecting qualitative, descriptive data from an interview (or naturalistic observation) and turn that data into quantifiable measures (or scores) that can be analyzed using statistics. For instance, there might be four levels on a rubric—beginning, emerging, developing, and accomplished (or similar). What participants do or say is scored at these levels to show the range of knowledge, interests, behaviors, etc. occurring. The downside of rubric-scoring is that it is very resource-intensive. Not only does the method require the time and resources to collect qualitative data but the analysis of that data requires multiple steps—analyzing a portion of the qualitative data to develop the rubric, scoring the data using the finalized rubric, and applying statistics to the scored data. A large expenditure of resources probably means an evaluator is conducting summative evaluation of your program to demonstrate or measure results, which may mean using a control and treatment group and collecting a larger sample. Still, if resources allow and the purpose of the evaluation is to demonstrate or measure impact, rubric-scoring is an authentic means of doing so.

IRB: Ensuring Ethical Treatment of Human Subjects

There is a long history that supports the necessity of having an independent body of experts, referred to as an Institutional Review Board (or IRB), review and approve an evaluation plan for the ethical treatment of human subjects as espoused by The Belmont Report (Ryan et al., 1979). In the case of citizen science programs conducted in partnership with formal education institutions, it is vitally important to secure IRB approval because one of your typical audiences includes minors (students) who deserve special protections when it comes to collecting data. Working in a formal education setting also adds an extra layer of complexity because you may also have to seek approval from the school district using any formal review processes they have in place. Unfortunately, there is no universal process or centralized system for IRB or school district review. However, there are common elements and requirements so we have attempted to address questions and concerns that typically arise when undergoing this process.

Do you have to get IRB approval? The short answer is yes; the evaluator should submit the evaluation for IRB approval. There are certain circumstances where the evaluation will receive an IRB exemption because it *does not* include: working with a vulnerable population (e.g., minors), exploring sensitive topics or questions, or collecting identifiable information (i.e., your subject's identity is truly anonymous). Informal education programs, like citizen science programs, may often qualify for exemptions, except in the case where the evaluation includes minors (for the purposes of this book, 10-17 years) or if data collectors are audio- or video-recording (and true anonymity is unlikely).

Which IRB should you use? If you or one of your partners is affiliated with a college or university, they typically have their own IRB, as do some other types of institutions that conduct a lot of research and evaluation like the Smithsonian Institution. If you or a partner are not affiliated with an institution that has its own IRB, there are companies that offer IRB approval for a fee.

What is typically involved in securing IRB approval? IRB approval typically includes providing a description of the evaluation plan, including subjects (age, gender, ethnicity, etc.), proposed methods, and sample sizes. An IRB also wants to know how you are going to protect study participants' information (where are you going to store data, when are you going to destroy data, how are you going to maintain participants' anonymity or protect their identity if known to evaluators). An IRB also wants to know how you are going to obtain consent; if there are minors involved, a guardian's consent will be required. In addition to the paperwork required, all evaluators and hired data collectors (and sometimes, you, as the program developer) have to receive training in protecting human subjects. In the past, common trainings have been offered online through the National Institutes of Health (2019) and the Collaborative Institutional Training Initiative (CITI, 2019).

Do you have to get IRB approval and approval from the school district? Again, the short answer is yes. However, not all school districts have a formal review process. Public school districts in major cities almost always have a formal review process (e.g., New York City, Chicago, District of Columbia, San Francisco, Los Angeles, Miami, etc.), but many smaller districts do not. Going to the official website for a school district often reveals whether they have a formal review process. If it is not obvious, call and ask the central office. If there is a formal review process, it almost always requires IRB approval as the first step, unless as is the case in New York City, the school district has its own IRB. School district review may duplicate elements of the IRB as well as require permission from individual principals or administrators and/or background checks to enter school property.

Communicating Evaluation Findings

Evaluation's greatest value is as a learning tool; therefore, effectively communicating the findings is essential to your own and others' learning, whether that communication is internal (staff, leadership, etc.) or external (field at-large, audiences, other stakeholders, etc.). Whether or not a full evaluation report is prepared, there are two important strategies for communicating findings in a way that will help facilitate learning (as opposed to letting the report sit on a shelf). One suggested strategy is to isolate the actionable information from the rest of the findings—have it brought front and center in some kind of executive summary or discussion of the findings. Evaluations can yield a lot of interesting information, but only some of that information will be actionable; that is, only some information will help you improve the program or demonstrate to others which aspects of the program structure are most likely to result in intended outcomes so maybe their program can follow a similar structure. The "what" of the evaluation, its purpose and intended results, will act as an effective road map for pulling out the actionable information. For instance, if an intended result (or outcome) of the program is to increase students' ability to ask questions, and evaluation findings show that there are certain program activities where students are generating more questions, it would be helpful to present what about those activities is contributing to that result. Hopefully, the evaluation was designed to capture data related to program structure and facilitation so that findings can be considered in that context.

The second suggested strategy for effectively communicating findings is to present them in a shareable format, such as a PowerPoint presentation with salient elaboration in the notes section. A full, detailed report of findings serves a purpose. For example, if someone wants more specific information about the evaluation methods or needs to know if a specific statistical test was run (and if it yielded anything of note). However, for most people, knowing the topline, actionable information is most important, and something like a PowerPoint presentation is more accessible and easily digestible while allowing data visualizations that can be more efficient at communicating findings than words and numbers alone. Again, the intent is to share the information so that everyone can use the findings to create more engaging and effective citizen science programs.

Acknowledging Limitations

Throughout this chapter, we have mentioned and/or hinted at limitations to evaluation. Since no one operates in a vacuum with an endless amount of resources, it is important to acknowledge these limitations and consider how they will affect the evaluation. There are, in essence, two types of limitations you probably need to consider when choosing to evaluate a citizen science program. The first has to do with the overall context of operating these kinds of programs in a formal education setting. The second has to do with staff and financial resources, which would apply to evaluating any type of program.

What limitations will you encounter when evaluating citizen science programs? Since the focus of this book is youth, 10-25 years of age, any evaluation of a citizen science program likely will encounter limitations related to audience access. Evaluation or research involving minors (17 years and younger) requires guardian consent. School districts also tend to limit access to their students except in circumstances where the evaluation or research is aligned with district-wide goals. The good news is that most districts have goals related to STEM programming; thus, citizen science programs likely will be a natural fit. However, access to a youth audience, especially a large sample of students, will always pose some limitations for evaluation of citizen science programs in formal education settings. There may be greater access to teachers as part of an evaluation since it is less complicated to secure their consent for an evaluation. Though, school districts are still mindful of teachers' time when it comes to obligations outside of the classroom, which would include participating in an evaluation. Another aspect of audience access that may prove limiting is the primary language that the majority of the audience speaks. More and more school districts are requiring that consent forms and data collection instruments be translated into the primary languages spoken in their schools (other than English, Spanish is the most common). This may impact which students are able to participate in the evaluation.

Another common limitation of evaluating citizen science programs relates to how these programs are commonly facilitated. While programs typically have a structure (including training) that strives for consistency, there are still natural variations in how programs are facilitated. In the case of citizen science programs, scientists are often involved in part of the facilitation. Some scientists are natural facilitators, while others may need a great deal of support. For instance, they may be a subject-matter expert and well versed in the scientific method; however, they may be less comfortable engaging youth in their research as a teacher or facilitator. Lack of consistency in program facilitation complicates an evaluation since it becomes difficult to identify which aspects of the program structure are truly barriers to impact and which might be due to one or two inconsistent or unengaging facilitators. Of course, formative evaluation can be helpful for identifying challenges related to program implementation; so, ideally, you can use it to identify inconsistencies in program facilitation before moving to measuring impact.

A third limitation for evaluating citizen science programs is how they are typically structured as isolated data collection activities rather than holistic involvement in all aspects of a research project. It makes sense that most audiences are involved in the data collection aspects of citizen science since these activities presumably follow standard protocols set out by the researcher. These activities also tend to be repetitive so that, over the course of a few hours, participants likely have the opportunity to get the hang of and complete a task from start to finish. However, involving participants primarily in the data collection aspects of citizen science has its limitations. Namely, the outcomes, or intended results, need to be realistic and consider the depth of participants' experiences. For instance, it may not be realistic to expect students to make great gains in scientific achievement from a single data collection activity. What may be more realistic is that the activity piqued students' interest in science (Ballard et. al, 2017; Randi Korn & Associates, Inc., 2016). In cases where citizen science programs involve participants in more aspects of the research process (e.g., generating research questions, designing the research strategy or approach, analyzing and reporting results), there may be greater potential for the achievement of more complex outcomes since participants have opportunities for repeat participation.

What limitations will you encounter due to resources and timing? The initial brainstorming of outcomes for a citizen science program is best done without placing limitations on potential ideas due to available resources. However, once ideas related to outcomes are generated, available resources must become part of the equation. There are typically two options if an outcome is not realistic given available resources. One, the outcome can be eliminated or revised to reflect available resources. Or, two, you can determine ways to secure the resources needed, perhaps through partnerships or collaborations with other organizations or through additional funding. When it comes to outcomes-based evaluation of a program, available resources are always a factor. There are several aspects of an evaluation that can influence timing and cost, including but not limited to the

1. number of audiences,
2. sample sizes needed for those audiences,
3. phases of evaluation needed,
4. number and types of methods being used,
5. complexity of the IRB and school district review process, and
6. logistical barriers that impede or extend the length of data collection (e.g., poor weather conditions, transportation to/from data collection sites).

It may be obvious why most of the items in this list add time or cost. When you increase the amount of data that needs to be collected, either by increasing sample sizes or methods, naturally the time and resources needed to complete the evaluation increase. IRB and school district review are not obvious time- and cost-factors unless you are familiar with the process; however, a typical IRB process takes six to eight weeks and most school districts warn that the review process can take up to 90 days. These two processes alone can add up to five months to program planning (because, remember, data collection of program activities cannot take place until the evaluation is approved). And, if you are not affiliated with an organization that has an IRB, the cost of securing IRB approval also becomes a potential limitation (Silberman & Kahn, 2011).

THE "WHY"

Hopefully, at this point, the value of doing evaluation has been communicated. But it bears repeating. Evaluation is a learning tool not a judgment tool. Its value lies in revealing successes and failures to learn from them and do better. If every citizen science program was presented as perfect, no one would ever learn anything. And, the most effective tool in your arsenal for facilitating learning is reflection (or reflective practice). The process of program improvement is cyclical. You are not done when the evaluation is done. Instead, when the evaluation is complete, ideally, everyone asks, "What did we learn?" and "How can we improve?" Considering those questions leads right back into program planning and improvement which then leads to implementation and probably more evaluation (Korn, 2018). The process is not meant to be daunting; instead, in our experience, reflecting on the evaluation findings is exciting and invigorating because it deals with the most important part of the citizen science program and the evaluation, the "why." Why are you doing the program? Why are you evaluating the program? The initial reason for doing the evaluation may have been because a funder requested the evaluation, and that is okay. But, take advantage of the opportunity to learn from the evaluation by reflecting on the findings with your colleagues. We encourage everyone to structure the reflection instead of doing it haphazardly in a staff meeting one day. Inquiry is the best way to structure a reflection, and here we pose some questions that you might consider when reflecting on evaluation findings (though, we also encourage you to come up with questions that are specific to your program and evaluation).

1. What did you learn from the evaluation findings?
2. To what extent did you achieve outcomes? Did this achievement vary among audiences? How so (or in what ways)?
3. What about the program may have led to achievement of outcomes? What makes you say that (e.g., evidence from the evaluation, information from prior studies)?
4. What barriers or challenges may have prevented achievement of certain outcomes? What are ways you might address these barriers (e.g., funding, collaboration)?
5. What can you improve about program implementation (e.g., recruitment, activity structure, facilitation, etc.) to better achieve outcomes?
6. What did you learn that has broader implications for the citizen science field?
7. What, if any, new questions emerged that should be explored in future research or evaluations of citizen science programs?

Consider doing the reflection with the original group of stakeholders who contributed to helping you define the "what" of your program and evaluation. Document the process so you can turn the reflection into action steps when planning the next phase of the program.

Educational Implications

Potential consumers of impact-driven evaluation of citizen science programs can be educators—either those facilitating the programs or those bringing their students to participate: funders—those supporting the programs; and other evaluators and researchers—those who will design research and evaluation projects that are intentionally designed to build on, not unnecessarily repeat or recycle, current findings and knowledge. For educators who provide citizen science programs, impact-driven evaluation findings have a myriad of uses in every phase of their work. For instance, front-end evaluation can help educators plan their programs to achieve impact, formative evaluation can help educators hone best practices toward achieving impact, and summative evaluation can help educators measure whether these practices are achieving intended results or impact. For educators who are bringing their students to participate in citizen science programs, impact-driven evaluation findings are useful for understanding the value of citizen science for achieving their teaching and learning goals if care is taken to align those goals with those of the program from the beginning. Thus, as stated in previous sections, soliciting feedback from teachers about intended program outcomes is important if there is to be a true partnership and collaboration between the informal and formal entities involved.

Funders may, understandably, want to use impact-driven evaluation findings to determine which citizen science programs are worthy of continued support. However, a word of caution that stems from the idea of using evaluation as a learning tool not a judgement tool; demonstrating impact takes significant time and barriers to impact are plentiful. Thus, allowing grantees adequate time to learn from evaluation findings is of the utmost importance if they are to use funding resources effectively to achieve impact. Funders can also be a vital resource for grantees looking for advice on best practices in planning, implementing, and evaluating citizen science programs. Directing grantees to be intentional every step of the way by requiring an impact-driven approach to planning and evaluation can go a long way toward achieving a more consistent, systemic change in how citizen science programs are delivered and the impact they have on audiences (Clewell & Fortenberry, 2009).

Evaluators and researchers are examining citizen science programs to ideally reveal best practices when it comes to teaching and learning. While researchers are typically exploring open-ended questions or testing hypotheses and evaluators are measuring intended results, they can both use evidence-based findings to help educators or practitioners solve problems and create exceptional citizen science programs. One continued struggle for citizen science programs is how to engage participants more holistically in the research process (Randi Korn & Associates, 2016). Whether holistic engagement is a worthwhile pursuit, or whether citizen science programs should operate within more narrow parameters, such as single data collection activities, is an interesting question about the inherent value of citizen science programs that still has not been put to rest. In other words, at their core, what are citizen science programs trying to accomplish? In some sense, the answer to this question has to do with whether the field prioritizes breadth—reaching more people to spark an interest in science—or depth—attempting to change STEM-related knowledge, attitudes, and behaviors at a deeper level. And, how can a network of sustained partnerships between informal and formal education institutions contribute to this ultimate goal, whatever it may be? These questions are all worthy of exploration by evaluators and researchers of citizen science programs to help move the field forward in an intentional way.

CONCLUSION

This chapter was about evaluating the "who," "what," "how," and "why" of citizen science programs using an impact-driven, outcomes-based approach that treats evaluation as a learning tool, not a judgment tool. If you are a practitioner, the intent was not to turn you into an evaluator, because, after all, that is probably not where your passions lie. However, there was a clear purpose—to help encourage more practitioners to plan citizen science programs with evaluation in mind and to encourage practitioners and evaluators to collaborate intentionally toward impact. Evaluation and planning are inextricably linked. Planning with an evaluation in mind requires more thought and discipline and ultimately results in a more intentional citizen science program. Since impacting audiences through citizen science can be difficult to achieve, intentionally directing resources toward that impact is vital if we actually want to make a difference, large or small, in people's lives.

APPENDIX A

Evaluation Planning Checklist

This checklist is intended to support those wishing to follow an impact-driven approach to evaluation, as described in this chapter. Thus, this checklist assumes that steps have been taken to define outcomes (intended results) and indicators (evidence of those results) as part of planning a citizen science program. It can be used by staff internally and/or in collaboration with external evaluators. Not every field or category in this checklist will apply to every evaluation—so you should only consider and complete those fields that are applicable to your program and evaluation. For questions regarding terminology used in this checklist, please refer to the chapter.

Audience: The "Who"	
Who is the target audience for this program evaluation?	
☐ Grade level(s)/age(s): ____________	☐ First-time/repeat program participant (circle)
☐ Language(s) spoken: ____________	☐ Other (e.g., Title 1, gender): ____________
☐ Geographic area(s): ____________	
Purpose: The "What"	
What is the purpose of evaluation for this audience? Prioritize 1 or 2 reasons from the list below, which will suggest the study type.	
☐ To inform planning or development of programs or materials	
☐ To assess program implementation/facilitation toward intended results	
☐ To measure results	
☐ Other: ____________	
Outcomes and indicators: The "What"	
Which <u>2-3 outcomes</u> (and associated indicators) are a priority to examine in this evaluation? This assumes you have already brainstormed outcomes and indicators for the program as described in the chapter. *You don't have to evaluate everything at once!*	
1. ____________ 2. ____________ 3. ____________	

APPENDIX A (CONTINUED)

Other questions and considerations: The "What"

What other information about the program structure do you need to collect to compare the data or provide context?

- ☐ # facilitators, scientists or other staff: ___________
- ☐ # students: ___________
- ☐ # teachers: ___________
- ☐ Demographics (gender, age): ___________
- ☐ Weekday/weekend: ___________
- ☐ Time of day: ___________
- ☐ Weather: ___________
- ☐ Other (e.g., site context, transportation): ___________

What additional questions do you have (not covered by the outcomes/indicators)?

- ☐ Program experience (e.g., what the audience liked/did not like, what resources were most helpful, etc.)
- ☐ Concerns about the program (e.g., a specific aspect of a program that seems ineffective or problematic?)
- ☐ Other audience characteristics (e.g., years in the program, barriers to participation, etc.)

Study type: The "How"

Refer to the purpose to select the study type; there is overlap among these categories.

- ☐ Front-end (planning for…)
- ☐ Formative (pilot testing, implementation)
- ☐ Summative (measuring results)
- ☐ Other: ___________

Evaluation methods: The "How"

What are the best method(s) given your outcomes, indicators, and additional questions? Refer to the chapter for detailed descriptions.

Qualitative methods

- ☐ Interviews
  - ☐ Short-answer interviews
  - ☐ In-depth interviews
- ☐ Focus groups
- ☐ Case studies
- ☐ Naturalistic observations

Quantitative methods

- ☐ Standardized questionnaires
  - ☐ Self-administered
  - ☐ Interview-style
- ☐ Standardized observations
- ☐ Rubric-scored interviews (analysis)
- ☐ Rubric-scored observations (analysis)

Ethical treatment of human subjects (IRB): The "How"

If any of the following apply, the evaluation probably requires full/expedited IRB review.

- ☐ Research or evaluation contributing to generalizable knowledge
- ☐ Collecting data from human subjects (especially under 18 years)
- ☐ Data can be linked to human subjects (not anonymous)
- ☐ Data are about personal thoughts and opinions

If any of the following apply, the evaluation probably requires school district review.

- ☐ School district is in a major city (NYC, Chicago, San Francisco, Los Angeles, Washington D.C., Houston, Dallas, etc.)

Communicating evaluation findings: The "How"

Refer to the chapter for more detailed information about reporting strategies.

- ☐ Isolate actionable findings using outcomes/indicators
- ☐ Succinct communication (one-page executive summary, PowerPoint)

Acknowledging limitations: The "How"

Refer to the chapter for more detailed information about how these limitations can impact the evaluation study.

- ☐ IRB and consenting procedures (especially for minors)
- ☐ Costs associated with IRB and consent procedures
- ☐ Primary language spoken (translation required)
- ☐ Inconsistent program facilitation (mitigate w/ formative evaluation)
- ☐ Depth of program participation (single vs. repeat participation)
- ☐ Cost of evaluation (given sample size, methods, etc.)

Reflecting on and learning from evaluation findings: The "Why"

Refer to the chapter for specific reflection questions that can help you answer these larger questions:

- ☐ What did we learn from the evaluation findings?
- ☐ How can we improve the program to strengthen impact?

References

Ballard, H. L., Dixon, C. G. H., & Harris, E. M. (2017). Youth-focused citizen science: Examining the role of environmental science learning and agency for conservation. *Biological Conservation*, *208*, 65-75. doi:10.1890/110280

Bitgood, S. (2011). Sampling for a visitor survey: Finding your way through the maze. *Social Design in Museums: The Psychology of Visitor Studies Collected Essays*, *1*, 208-219. doi:10.1080/10645578909443580

Bell, P., Lewenstein, B., Shouse, A. W., & Feder, M. A. (2009). *Learning science in informal environments: People, places, and pursuits*. Washington, DC: National Academies Press.

Bonney, R., Phillips, T. B., Ballard, H. L., & Enck, J. W. (2016). Can citizen science enhance public understanding of science? *Public Understanding of Science*, *25*(1), 2-16. doi:10.1177/0963662515607406

Brody, M., Bangert, A., & Dillon, J. (2007). *Assessing learning in informal science contexts*. Commissioned paper by the National Research Council for Science Learning in Informal Environments Committee. Retrieved from https://sites.nationalacademies.org/cs/groups/dbassesite/documents/webpage/dbasse_080082.pdf

Brossard, D., Lewenstein, B.V., & Bonney, R. (2005). Scientific knowledge and attitude change: The impact of a citizen science project. *International Journal of Science Education*, *27*(9), 1099-1121. doi:10.1080/09500690500069483

Clewell, B. & Fortenberry, N. (Eds.) (2009). *Framework for evaluating impacts of broadening participation projects*. The National Science Foundation, The Division of Research on Learning in Informal and Formal Settings (DRL). Retrieved from https:// www.nsf.gov/ od/ broadeningparticipation/ framework- evaluating- impacts- broadening- participation-projects_1101.pdf

Collaborative Institutional Training Initiative (2019). Retrieved from www.citiprogram.org

Cornell Lab of Ornithology (2018). Citizen science central: Evaluation instruments. Retrieved from http://www.birds.cornell.edu/citscitoolkit/evaluation/instruments

Crall, A. W., Jordan, R., Holfelder, K., Newman, C. J., Graham, J., & Waller, D. M. (2012). The impacts of an invasive species citizen science training program on participant attitudes, behavior, and science literacy. *Public Understanding of Science*, *22*(6), 745-764. doi:10.1177/0963662511434894

Davidson, J. E. (2005). *Evaluation methodology basics: The nuts and bolts of sound evaluation*. Thousand Oaks, CA: SAGE Publications, Inc.

Diamond, J., Horn, M., & Uttal, D. H. (2016). *Practical evaluation guide: Tools for museums and other informal educational settings* (3rd ed.). Lanham, MD: Rowman & Littlefield.

Gommerman, L., & Monroe, M. C. (2012). *Lessons learned from evaluations of citizen science programs*. School of Forest Resources and Conservation Department, UF/IFAS Extension, FOR291.

Heimlich, J. E., & Ardoin, N. M. (2008).Understanding behavior to understand behavior change: A literature review. *Environmental Education Research*, *14*(3), 215-237. doi:10.1080/13504620802148881

Hiller, S. E., & Kitsantas, A. (2014). The effect of a horseshoe crab citizen science program on middle school student science performance and STEM career motivation. *School Science and Mathematics*, 114(6), 302-311. doi:10.1111/ssm.12081

Ibarra, H., & Hansen, M. T. (2011). Are you a collaborative leader? *Harvard Business Review*, July-August, 69-74. Retrieved from https://hbr.org/2011/07/are-you-a-collaborative-leader

Jennett, C., Kloetzer, L., Schneider, D., Iacovides, I., Cox, A., Gold, M., Talsi, Y. (2016). Motivations, learning, and creativity in online citizen science. *Journal of Science Communication*, *15*(3), 1-23. doi:10.22323/2.15030205

Jordan, R. C., Gray, S. A., Howe, D. V., Brooks, W., & Ehrengeld, J. G. (2011). Knowledge gain and behavioral change in citizen-science programs. *Conservation Practice and Policy*, *25*(6), 1148-1154. doi:10.1111/j.1523-1739.2011.01745.x

Korn, R. (1994). Studying your visitors: Where to begin. *History News*, *49*(2), 23-26.

Korn, R. (2010). Less is More. *Museum*, *89*(5), 25-27.

Korn, R. (2018). *Intentional practice for museums: A guide for maximizing impact.* Lanham, MD: Rowman & Littlefield.

Lee, T. K., Crowston, K. G., Oesterlund, C., & Miller, G. (2017). Recruiting messages matter: Message strategies to attract citizen scientists. In *CSCW 2017 - Companion of the 2017 ACM Conference on Computer Supported Cooperative Work and Social Computing* (pp. 227-230). Association for Computing Machinery, Inc. doi:10.1145/3022198.3026335

Martin, P. & Bateson, P. (2007). *Measuring behavior: An introductory guide* (3rd ed.). New York, NY: Cambridge University Press.

Masters, K., Oh, E. Y., Cox, J., Simmons, B., Lintott, C., Graham, G., Holmes, K. (2016). Science learning via participation in online citizen science. *Journal of Science Communication*, *15*(3), 1-33. doi:10.22323/2.15030207

Measham, T. G., & Barnett, G. B. (2008). Environmental volunteering: Motivations, modes, and outcomes. *Australian Geographer*, *39*(4), 537-552. doi:10.1080/00049180802419237

Merriam, S. B. (2009). *Qualitative research: A guide to design and implementation*. San Francisco, CA: Jossey-Bass.

National Institutes of Health (2019). Retrieved from https://phrp.nihtraining.com

National Science Foundation (2017). *Women, minorities, and persons with disabilities in science and engineering*. Retrieved from www.nsf.gov/statistics/wpmd/

Noonan, R. (2017). *Women in STEM: 2017 update*. (ESA Issue Brief #06-17). Retrieved from the Office of the Chief Economist, Economics and Statistics Administration, U.S. Department of Commerce website: https://www.esa.gov/reports/women-stem-2017-update

Patton, M. Q. (2008). *Utilization-focused evaluation* (4th ed.). Beverly Hills, CA: SAGE Publications Inc.

Patton, M. Q. (2015). *Qualitative research & evaluation methods* (4th ed.). Thousand Oaks, CA: SAGE Publications Inc.

Phillips, T., Bonney, R., & Shirk, J. L. (2012). What is our impact? Toward a unified framework for evaluating outcomes of citizen science participation. In J. L. Dickinson & R. Bonney (Eds.), *Citizen Science: Public participation* (pp. 82-96). Ithaca, NY: Cornell University Press.

Phillips, T. B., Ferguson, M., Minarchek, M., Porticella, N., & Bonney, R. (2014). *User's guide for evaluating learning outcomes in citizen science.* Ithaca, NY: Cornell Lab of Ornithology. Retrieved from http://www. citizen science. org/evaluation

Prather, E. E., Cormier, S., Wallace, C. S., Lintott, C. Raddick, M. J., & Smith, A. (2013). Measuring the conceptual understandings of citizen scientists participating in Zooniverse projects: A first approach. *Astronomy Education Review, 12*(1). doi:10.3847/AER2013002

Raddick, M. J., Bracey, G., Gay, P. L., Lintott, C. J., Cardamone, C., Murray, P., Schawinsky, K., Vandenberg, J. (2013). Galaxy zoo: Motivations of citizen scientists. *Astronomy Education Review, 12*(1). doi:10.3847/AER2011021

Randi Korn & Associates, Inc. (2016). *Citizen science project: Summative evaluation.* Unpublished manuscript. Manati, PR: Conservation Trust of Puerto Rico. http://informalscience.org/sites/default/files/RKA_2016_CTPR_Citizen%20Science%20Program_Summ_Rpt_April2016_distribution.docx.pdf

Ryan, K. J., Brady, J. V., Cooke, R. E., Height, D. I., Jonsen, A. R., King, P., Turtle, R. H. (1979). *The belmont report.* United States Department of Health, Education, and Welfare: The National Commission for the Protection of Human Subjects of Biomedical and Behavioral Research. Retrieved from https://www.hhs.gov/ohrp/regulations-and-policy/belmont-report/read-the-belmont-report/index.html

Sheppard, B. (2010). The gift of the ISE frameworks: A better language for museum learning. *Curator, 53*(3), 261-269. doi: 10.1111/j.2151-6952.2010.00027.x

Silberman, G., & Kahn, K. L. (2011). Burdens on research imposed by institutional review boards: The state of the evidence and its implications for regulatory reform. *The Milbank Quarterly, 89*(4), 599-627. doi:10.1111/j.1468-0009.2011.00644.x

Taylor-Powell, E. (1998). *Questionnaire design: Asking questions with a purpose.* University of Wisconsin-Extension, Program Development and Evaluation. Retrieved from https://learningstore.uwex.edu/Assets/pdfs/G3658-02.pdf

W. K. Kellogg Foundation (2017). *The step-by-step guide to evaluation: How to become savvy evaluation consumers.* Retrieved from http:// ww2.wkkf.org/ digital/ evaluationguide/ view.html#p=1

In: Enhancing STEM Motivation through Citizen Science Programs ISBN: 978-1-53616-038-3
Editors: Suzanne E. Hiller and Anastasia Kitsantas

Chapter 14

EVALUATING THE IMPACTS ON YOUNG PEOPLE AND THEIR TEACHERS WHO PARTICIPATE IN GENUINE SCIENTIFIC RESEARCH WHILST AT SCHOOL

Elizabeth A. C. Rushton[1,*] and Becky Parker[2]
[1]University College London, Institute of Education,
The Institute for Research in Schools, London, UK

ABSTRACT

The evaluation methods and approaches of the Institute for Research in Schools (IRIS) provide templates for those working in citizen science, especially those projects involving young people. Qualitative and quantitative approaches are used to better understand the outcomes for students, teachers and technicians who participated in this work. Evaluation demonstrates that participation in IRIS research projects is linked to increased participation in STEM subjects for students aged 16-18 years and in undergraduate courses, especially female students. Five distinctive features of IRIS' evaluation include evaluation that has a 360° and longitudinal approach, evaluation that recognizes and values the individual experience, evaluation that is co-created and a celebration point, and identifying evaluation as an opportunity to develop knowledge and understanding.

Keywords: co-created citizen science, co-created evaluation, multifaceted and longitudinal evaluation, student researchers, teacher scientists

This chapter begins by considering the different approaches to citizen science and the ways in which these can be evaluated. Evaluation of citizen science has yet to be fully developed and mainly focuses on simple measures for example, the number of participants and the nature of their engagement, including their levels of accuracy (Barr, Haas, & Kalderon, 2017; 2018; Bonney & Dickinson, 2012; Bonter & Cooper, 2012; Havens, Vitt, & Masi, 2012). There has been some evaluation of what participants have gained from their engagement in citizen science

* Corresponding Author's Email: erushton@researchinschools.org

projects, including skills development (Becker-Klein, Peterman, & Stylinski, 2016; Cronje, Rohlinger, Crall, & Newman, 2011; Shah & Martinez, 2016), love of science (Barr et al., 2017; 2018; Wallace & Bodzin, 2017), and improved curriculum-based attainment (Weiman & Gilbert, 2015). Underpinning increases in participants' achievement and motivation is increased self-efficacy and self-confidence (Shah & Martinez, 2016). IRIS' approach to citizen science is that of co-creation [IRIS' vision and aims are outlined in Rushton & Parker, (2019)], and IRIS' evaluation framework also draws on a partnership between evaluators, researchers and school students. A variety of quantitative and qualitative methods used to evaluate the impact of IRIS projects on students and teachers are described with specific examples from both well-established and emerging IRIS research projects. The purpose of this section of the chapter is to provide other researchers and practitioners in this field examples of how they can develop projects and evaluation frameworks for contributory, collaborative and co-created citizen science projects. Within this framework, stakeholders perhaps consider how young people can make a genuine contribution to knowledge that benefits not only the students and their teachers but is of value to the wider scientific community.

Citizen Science Approaches

In the context of a rapid expansion of citizen science projects over the last fifteen years (Curtis, 2015; Gura, 2013), the range of approaches taken when involving the public in science focused research activities has also increased (Bonney et al., 2009; Wiggins & Crowston, 2011). This growth has led to classifications or typologies of citizen science projects (Bonney, Phillips, Ballard, & Enck, 2016; Bonney et al., 2009; Cooper, Dickinson, Phillips, & Bonney, 2007; Shirk et al., 2012; Wiggins & Crowston, 2011; Wilderman, 2007). Not only is it important to understand the nature of the citizen science project to accurately set aims and objectives, it is crucial that this understanding informs the evaluation strategy so that the most appropriate methods can be implemented at the outset, reviewed during the life of the project, and the evaluation be concluded in a way that informs best practice for future versions of the project or for others designing similar or related citizen science initiatives.

The most frequent approach to grouping citizen science projects is focused on the varying levels that public participants have in directing or leading the scientific process, creating three categories; *contributory* projects, *collaborative* projects and *co-created* projects (Bonney et al., 2009). *Contributory* projects provide participants with the opportunity to collect data and/or analyse data from both the physical environment and virtual sources (e.g., images) to support research questions and methodologies that are established by scientists. Here the scientific goals are frequently established first and then learning goals develop from the processes and activities that best support meeting the scientific goals. The approach is "top-down," with participants usually only active in data collection and do not contribute to forming research questions, analysing data or communicating the research findings. It is perhaps because of this "top-down" approach and due to the focus on the scientific aims, that *contributory* citizen science projects produce the greatest number of peer reviewed publications, particularly from ecological research (Bonney et al., 2009).

Collaborative projects involve participants in more than simply the co-production of data; they can be active in the analysis of the data or may have a role in refining research aims or disseminating findings. Depending on the nature of the collaboration, this type of citizen science project may have a more dialogic and participatory focus, with participants able to shape the questions that are asked, the methods that are chosen and the format and delivery of research findings. *Co-created* projects are at the other end of the citizen science spectrum compared to contributory projects, with participants fully engaged in a dialogue with scientists and other expert partners in each stage of the research process. Due to the complex nature of the dialogue, often with multiple stakeholders (e.g., university-based scientists, museums, charities, funding bodies, schools, community groups), these projects may involve fewer non-expert participants and may evolve and develop over a longer and less well-defined period than both contributory or collaborative projects. *Co-created* projects that are successful have stakeholders or partners that are responsive, willing to listen and have a longer-term commitment to the project (Tweddle, Robinson, Pocock, & Roy, 2012). Bonney et al. (2016) further developed their typologies of citizen science projects into four groups that are defined by the nature of the activities in which participants engage. This classification is consistent with the approach from Bonney et al. (2009) but has resulted in four categories – "data collection," "data processing," "curriculum-based" and "community science." Bonney et al. (2016) found some evidence that citizen science projects extend participants' scientific knowledge and knowledge about scientific process and that some projects can provide participants a voice in local decision making and the opportunity to shape research questions. When participants contribute in this way it can have a positive impact on their social well-being (Bonney et al., 2016).

Wiggins and Crowston (2011) developed five types of citizen science projects, classified according to organisational and macrostructural features, including *Action*, *Conservation*, *Investigation*, *Virtual* and *Education*. Action features grassroots organisations, such as participatory research approaches to encourage intervention in a local or civic concern. Conservation has a grassroots participatory approach but also features stewardship and education goals. In contrast, Investigation involves large scale data collection, mainly from the physical environment, which has a more implicit educational approach. As with projects in the category Investigation, Virtual projects involve large scale data sets; however, in this group, the public participates in research almost entirely through information and communications technology (ICT). Projects in the fifth category, Education, have an explicit focus on education in both formal and informal settings. Wiggins and Crowston (2011) suggested that this typology should be used alongside the traditional three categories that are defined by participation level, and that their typology enables the greater incorporation of technology mediated citizen science projects, that had previously been largely overlooked.

IRIS research projects are founded on a dialogic and participatory approach where students and their teachers co-create knowledge with scientists and other specialist partners. The learning and scientific goals are distinct but are established at the same time and are of equal importance. The learning goals are appropriate and achievable and must be rooted in genuine scientific research that contributes knowledge, and is recognized by the wider scientific community as valid, reliable, accurate and useful. This means that IRIS projects are mostly collaborative or co-created citizen science projects, rather than contributory. Of the categories identified by Bonney et al. (2016), IRIS research projects have aspects of curriculum-based and community science projects. IRIS research projects also use a range of organisational typologies recognized by Wiggins and Crowston (2011) as part of this co-creative approach. Evaluation of IRIS projects is in alignment with this approach, and participants are involved in shaping the nature of the evaluation questions, the methods used, and are involved in both the co-creation of understanding derived from the project evaluations as well as communicating that knowledge to others.

Evaluation and Citizen Science Approaches

Evaluation has had only general consideration in the wider literature around citizen science and has mainly focused on identifying opportunities for quantitative measures of outputs and outcomes, with limited discussion of appropriate methods (Bonney et al., 2016; Curtis, 2018; Jordan, Ballard, & Phillips, 2012; Phillips, Bonney, & Shirk, 2012). Jordan et al. suggested that appropriately planned and implemented evaluation of citizen science programs can generate information at three different levels: individual learning outcomes, programmatic outcomes and community-level outcomes. Impact on a community level provides an opportunity for impact that is on a larger temporal and spatial scale, suggestive that these programs increase social capital (Adger, 2003; Schwartz, 2006), science capital (Tweddle et al., 2012), networks and positive relationships between scientists, decision makers and the public (Fernandez-Gimenez, Ballard, & Sturtevant, 2008), and community capacity (Donoghue & Sturtevant, 2007).

In their survey of 48 citizen science projects, Reeves, Tinati, Zerr, Van Kleek, and Simperl (2017), identified common success criteria that are all quantitatively measurable and can be collected as part of the project framework or platform, including engagement in terms of the number of people reached, the number of contributions received, and the accuracy and quality of results gathered as part of the project. Bonney et al. (2009) included evaluation in terms of the scientific contribution made and measures of scientific literacy. Scientific contributions can be measured through the number of peer-reviewed papers published and the number of citations of those papers, the number and size of grants to support the project and the size and quality of databases produced during the lifetime of the project (Bonney et al., 2009). Bonney et al. (2016) recognized that measuring scientific literacy is more challenging and requires the use of research methods from social sciences to accurately evaluate outcomes that include skill development, attitudes towards science, self-efficacy, self-confidence and behaviour change.

Becker-Klein et al. (2016) suggested that citizen science projects that have embedded assessments (EAs) have an opportunity to measure the skill development of participants as part of their regular contribution to the citizen science programme. There is a range of citizen projects that use online tools and quizzes to assess the validity of participant contributions (Barr et al., 2017, 2018; Bonney & Dickinson, 2012; Bonter & Cooper, 2012; Havens et al., 2012). This approach allows evaluation data to be collected and analysed without the need to implement further measures (e.g., surveys, interviews, focus groups) and ensures the contributed data is of publishable quality and that through this recognition and feedback, participants develop their identity as contributors to science (Becker-Klein et al., 2016). Masters et al. (2016) used a survey to measure the motivations of participants contributing to one of five citizen science projects hosted on the largest online platform, *Zooniverse* (n.d.). In addition to the online survey, Masters et al. used an image quiz to test general politics, general science knowledge and project specific science knowledge and observed a statistically significant link between increased participation and an increase in project specific knowledge. No increase was observed in general science knowledge which suggests that participation in

Zooniverse projects does not increase broad scientific literacy. Interviews with participants may have generated further insight as to how participants viewed the development of their scientific knowledge and understanding.

Barr et al. (2017, 2018) evaluated the ability of citizen scientists to detect unusual features in ATLAS (A Torodial LHC ApparatuS) images of collisions at the Large Hadron Collider in a project called *Higgs Hunters*. This citizen science project is global in scale, involving 32,000 participants from 179 countries who classified 1.2 million features on 39,000 images (Barr et al., 2017; 2018). Participants were able to identify the decays of long-lived particles at least as well as the ATLAS algorithms and demonstrated a high-level of scientific engagement (Barr et al., 2017, 2018). IRIS' collaboration with the *Higgs Hunters* project gave students aged 16 and above access to the data set created by the citizen scientists involved in Higgs Hunters so that they could develop their own investigation and research questions. Some students looked at the "weird" clicks that seemed to make little sense, looking for patterns that might explain why citizen scientists had clicked on certain images. Other students compared the accuracy of citizen scientists depending on the way the data was presented, comparing head-on and side views. In order to try and answer these questions students had to write code to analyse large data sets. This research culminated in students preparing research posters that they presented to academics at the Physics Department at Oxford University, hosted by Barr and Merton College (IRIS, 2018). This is an example of a citizen science project that evolved from a contributory project to one that was wholly co-created. *Higgs Hunters* began with members of the public analysing data with research questions set by academics leading the project and the first phase culminated in a genuine dialogue between school students and leading academics in the field.

Curtis (2013) used both quantitative and qualitative data to explore the motivations and expectations of those attending astronomy outreach sessions at the University of Cambridge. Surveys were used to collect both quantitative demographic data and qualitative data to explore motivations and attitudes. Curtis considered the contrasting views of surveys as a useful methodological tool, recognising that the self-selecting bias was seen as both a limitation (Creswell, 2009) and a desirable feature (Jensen & Holliman, 2009).

Price and Lee (2013) combined quantitative survey data from over three hundred participants with qualitative data from nine interviews to explore whether participation in an astronomical citizen science project shaped attitudes towards science and epistemological beliefs and how levels of participation related to any observed change. Price and Lee found statistically significant attitudinal changes towards science and increases in scientific literacy linked to their level of social participation, distinct from the amount of data individuals had contributed. This is an important example of where the combination of quantitative and qualitative data provides a more nuanced understanding of outcomes and identifies which feature(s) of a citizen science programme are linked to different outcomes. Whilst participants showed more positive attitudes towards science and a greater understanding of the nature of science, self-confidence and self-efficacy decreased. Price and Lee suggested that this reduction in self-confidence and self-efficacy could be due to an increased awareness of how much participants still had to learn and understand.

The explicit consideration of the methods and analysis used in the evaluation by Curtis (2013) and Price and Lee (2013) is not frequently seen in citizen science literature, although there is a large and growing body of work that explores the motivations of participants, and this has developed models of motivation (Krebs, 2010; Nov, Arazy, & Anderson, 2011; Rotman et al., 2012). Motivation to participate in citizen science projects can be linked to an individual's interest in the scientific area (Barr et al., 2017; 2018; Nov et al., 2011), a sense of satisfaction derived from being part of community, virtual or otherwise (Clery, 2011), and shapes self-motivational beliefs including self-efficacy, interest, outcome expectations and goals (Bonney et al., 2016; Shah & Martinez, 2016). Tools used in collecting, analysing and reporting information to better understand participant motivation can also be used more broadly to explore the outcomes and impacts upon participants. These evaluation tools and frameworks are frequently used in more general research concerning a wide variety of public engagement. Reed, Tinati, Zerr, Van Kleek, and Simperl (2017) recognized that the dearth of published evaluation findings has limited the capacity of the public engagement field to bring about measurable improvements to practice. Reed et al. address this issue through the publication of common conceptual and methodological frameworks for evaluating public engagement activities. The methodological toolkit provides those working in the field of citizen science to select the approach and tools that best meet the aims and objects of their type of project.

As citizen science has developed as a discipline, different categories and types have emerged. These contrasting approaches require appropriate evaluation frameworks and methods. The following section sets out how an evaluation framework was formed to understand the outcomes for students and teachers who participate in IRIS activities. The methods used are discussed in terms of the opportunities and challenges associated with them, although this is not an exhaustive consideration of either the different methods or their limitations as this has already been thoroughly considered by other researchers (Reed et al., 2017).

IRIS' Evaluation Framework

The work of IRIS is centred in schools, with an initial focus on physics research projects (Furnell, Shenoy, Fox, & Hatfield, 2018; Hatfield, 2010; Hatfield et al., 2019; Parker, Rushton, Thomas & Hatfield, 2019; Whyntie & Harrison, 2014; 2015) moving into subjects including biology, chemistry, maths, engineering, geography and psychology. The co-created approach and educational context is distinctive amongst other citizen science work as there are few examples of science outreach and engagement that involves young people in each phase of the research process. The nature of schools is important to consider at the outset. Schools are dynamic and, currently in the United Kingdom, places of considerable competing pressures, including those of teacher recruitment and retention (which is particularly acute amongst science teaching staff) (Allen & Simms, 2017) curriculum evolution and regulatory focus (Adams, 2018). This is the context in which IRIS is working with schools and sometimes individual teachers to implement science research projects that can extend across a year group, key stages and in some cases an entire school. As is the case with IRIS research projects, evaluation requires a dynamic, opportunistic and pragmatic approach that has its starting point with the views and perspectives of the students and teachers.

IRIS' evaluation framework has four phases: pre-baseline, baseline, formative and summative. The baseline, formative and summative phases are broadly representative of evaluation undertaken in other citizen science and public engagement projects (Reed et al., 2017; Tweddle et al., 2012) and, these phases of activity have porous boundaries so that findings from earlier phases can inform and shape future ones, enabling best uses of resources and increasing the likelihood of positive and long-lasting outcomes. During the pre-baseline phase of IRIS projects, data, ideas and perspectives are gathered from a range of sources, including summative evaluations of previous projects, interviews with former participants and ideas sharing meetings with IRIS' core team. The focus of the pre-baseline phase is to establish evaluation aims and objectives that are specific, measurable, achievable, relevant, and timely (SMART) and are co-created by the former, current and/or future participants. This ensures that evaluation of IRIS projects provides opportunities for students and teachers to make a genuine contribution to both understanding around the impact of IRIS projects but also ensures their role in the co-creation of future projects.

IRIS' evaluation is broadly focused in the key areas of engagement, considering the questions, who is participating? And, what is the nature of their participation? The second area is around understanding the outcomes for both students and their teachers who participate in IRIS projects. These two areas are considered in more detail alongside a discussion of the methods used to evaluate both engagement and outcomes.

Measuring engagement in IRIS–Who is participating? Since launching in March 2016, IRIS has developed a network of schools that are predominantly U.K.-based. This network has been built up through those networks already established by IRIS staff in their roles as teachers, university staff and through interaction with learned societies, university research and outreach departments and teaching networks. Schools have been recruited through students, teachers and technicians attending conferences; accessing the website; and teachers engaging with IRIS through subject-specific curriculum support groups. IRIS' approach to science education has been informally shared amongst the teaching community and word-of-mouth is an important part of how schools have signed up to participate with IRIS.

Fundamental to understanding the impact and outcomes of a project is having a clear and accurate grasp of who the participants are and how have they participated. The "who" aspect in the context of IRIS projects concerns the demographic information of schools, students and their teachers. Data collected includes the location of the school; the type of school; the age, school year group and gender of students; the primary subject taught by the teacher; and any research experience, qualifications, and interests they may have. More detailed socio-economic information at a school level is available using the school postal address to identify the Lower Super Output Area (LSOA) of the school. The Indices of Multiple Deprivation (Index of Multiple Deprivation 2015, 2017) rank the 34 thousand LSOAs according to levels of deprivation and divide the LSOAs into deciles, ranging from the 10% most deprived to the 10% least deprived areas in England. Using the post code of the school address it is possible to place the school within one of these deciles. A further socio-economic measure is to identify the school's local authority and, using the Social Mobility Commission report (2017) to ascertain whether the school is in a local authority that is a cold spot or hotspot of social mobility. Other measures publicly available to compare the socio-economic status of school populations includes the percentage of students who are eligible or have been eligible for Free School Meals (FSM) in the past five years and the percentage of students who are eligible for Pupil Premium funding. These socio-economic measures enable an evaluation to compare the outcomes for students from economically disadvantaged backgrounds who have consistently made less academic progress than their peers (Raffe, Croxford, Iannelli, Shapira, & Howieson, 2006). This socio-economic data can be gathered by a researcher remotely, using the school postcode, through publicly available electronic records. IRIS has stored this information using the Customer Relationship Management platform *Salesforce*, which enables searchable participation records to be collated over time and ensures that data in collected and stored in compliance with U.K. regulations, including the General Data Protection Regulations (GDPR), which came into effect in May 2018. Other demographic data is recorded by the teacher as part of a register of sessions and a list of participants.

Measuring engagement in IRIS–The nature of participation. IRIS measures this aspect of engagement using both qualitative and quantitative methods. Teachers and students engage with evaluation through informal conversations with IRIS staff as part of school visits, and this type of interaction enables IRIS to identify the varying ways students have participated in each school context. Formative and summative presentations and reports from individuals and groups provide an understanding of how students have engaged with the research topic and the different roles that they have assumed as part of that process (i.e., organisational roles, communicating findings, leading practical work, mentoring other students, engaging the support of non-school partners). The information gathered from these informal conversations helps identify other measures of participation. For example, school students often use their experiences working on IRIS research projects as the basis for British Science Association CREST Awards, Extended Project Qualifications (EPQs) or for entries to competitions such as Big Bang. School students also frequently cite their experiences of research in personal statements for university admission, and teachers, students and university admissions' tutors suggest that this is highly beneficial, providing students both academic extension and practical skills development. These aspects of participation would not necessarily be included in evaluations of outputs or outcomes as the awarding bodies are separate from IRIS; however, it is a useful way to understand how individual students have participated in research projects and how that participation has been used to achieve other related learning goals.

Developing a quantitative measure of the nature of individual participation has been a challenging aspect of research that is based in multiple schools with different groups of students working on the same or similar projects concurrently. IRIS' *Genome Decoders* project is a good example of where measuring activity and accuracy levels of a large number of participants has been achieved. *Genome Decoders* is a collaboration between IRIS and the Parasite Genomics team at the Wellcome Sanger Institute, the Wormbase team at EMBL-EBI (European Molecular Biology Laboratory-European Bioinformatics Institute) and the Wellcome Genome Campus Public Engagement team. All the scientific partners involved in this project are based at the Wellcome Genome Campus, Cambridge, United Kingdom (Rushton & Reiss, 2019).

Genome Decoders is a research project that enables school students to annotate the estimated 15,000 genes of the human whipworm (*Trichuris trichuria*) genome. The whipworm is a parasite that causes *Trichuriasis*, a currently incurable neglected tropical disease (NTD) that impacts over 500 million people globally, mainly children living in the poorest parts of the tropics, including Africa, Asia and South America (Barda, Keiser, & Albonico, 2015). High parasite burden has been linked to malnutrition, as well as physical and cognitive developmental problems, which can have chronic socio-economic consequences (Adegnika, Lötsch, Mba, & Ramharter, 2015). Annotating the genome enables the identification of the genes that code for proteins important during infection and disease which can lead to identifying a treatment and or cure.

Students access the information using *Apollo* software, a tool that is normally used by post-graduate and post-doctoral researchers (Dunn et al., 2019). This contrasts with other citizen science projects, which use platforms that are specifically created for citizen science projects (e.g., *Zooniverse*). The use of *Apollo* software in *Genome Decoders* gives school students access to a research tool that they otherwise would not encounter until at least post-graduate studies. Each student using the software has a unique avatar allowing their participation and accuracy to be measured both at an individual and school level. This ability to electronically track and measure participation in this way has been a common part of previous contributory citizen science projects (Becker-Klein et al., 2016; Bonney et al., 2009); however, the use of this software solely with school-age children is unique. Combining this quantitative data that is gathered remotely with qualitative information gathered through focus groups enables an understanding of both the levels and nature of participation and the factors and contexts that shape that participation. A school group may appear to be underperforming according to the electronically-generated participation data; however, a focus group may reveal that the school teacher has chosen to deliver a series of taught introductory sessions that enable students with a great range of abilities to access the academic content. Those students' participation may be much greater than previously thought.

The purpose of IRIS is to provide students and their teachers with opportunities to participate in authentic STEM research and to make valuable, recognized contributions to the scientific community. These projects are across the range of STEM disciplines and each have specific outcomes and outputs. Some of these are discussed in the section that follows.

Measuring Outcomes of IRIS Projects –What are the Outcomes for Students and Teachers?

Outcomes for students and teachers are measured using both quantitative and qualitative approaches. The traditional outcome measure of the number of peer-review papers published and citations produced as part of citizen science projects is part of the evaluation framework of IRIS projects. An important distinction made in IRIS' approach is that these peer-reviewed papers include students as authors and co-authors (Furnell et al., 2018; Hatfield, 2010; Hatfield et al., 2019; Parker, Fox, & Rushton, 2018; Parker et al., 2019; Whyntie & Harrison, 2014, 2015). These are outcomes that are commonly understood and widely recognized but that are not necessarily achievable for all young people who participate in IRIS research projects. Therefore, this section considers measurements of outcomes that capture a range of student experiences and outputs. All IRIS evaluations were conducted in accordance with the ethical guidelines set out by the British Educational Research Association (BERA, 2018). When schools registered with IRIS, they gave their written consent for students and teachers to participate in programme evaluations. In addition, consent was required from all participants (students and teachers) at each data collection point for their contributions to be analyzed, and they were informed that they could withdraw consent at any point. Students under the age of 18 who gave contributions that were video recorded also provided a parental consent and waiver form. All focus groups and interviews with school students (aged 18 and under) were held at the school site and in the presence of an adult employed by the school (e.g., teacher, technician). This was to ensure best safeguarding practice; however, this required careful management so that students felt comfortable sharing their experiences with both the evaluator and school staff member. The study of teacher experiences (Rushton & Reiss, 2019) received Institutional Review Board (IRB) approval in November 2017.

One of the key evaluation challenges that IRIS has encountered is the number of school partners and the range of projects that students participate in. This diversity has meant that, to-date, evaluation of projects has taken a pragmatic approach, balancing both the essential and desirable aspects of evaluation alongside the varying willingness and capacity of schools to support the evaluation work. However, there is broad commonality of output which includes measures of student attainment, social and emotional skill development, career progression and aspiration, attitudes towards science, and measures of teacher professional identity, career and skill development. Examples of these measures and associated methods are now considered.

Attainment – formal measures. Raising attainment is a frequent aim of education programs, especially those based in schools, and there are many different approaches to measurement. One measure of attainment is to explore the relative progress of students who participate in the project compared to students in the same school and year group who do not. This requires schools to be willing to share attainment data, and understandably, this is a sensitive area even when data is collected and analyzed anonymously.

IRIS has used this type of attainment data as an evaluation measure as part of the evaluation of a Wellcome Trust funded project called *Authentic Biology*. Although IRIS has only been formally established since early 2016, seven schools have been working on bioscience research since 2013 as part of *Authentic Biology*. This program involved students aged 16-18 years and linked schools with university departments working on a variety of bioscience research projects (Colthurst & Tuite, 2018). Over the course of the funding period of this project, the *Authentic Biology* schools partnered with IRIS, giving them support to continue to develop their school-based research. Some of these schools continued this by participating in the *Genome Decoder* project.

Using attainment data from one of the participating schools, IRIS has been able to explore the attainment outcomes of students who participated in the *Authentic Biology* project on four cohorts of A2 Biology students (aged 17-18 years) from 2013-2014 to 2016-2017. During the four years of this intervention, 54 A level Biology students participated and 198 did not. Student participation was completely voluntary and was open to any student as there were no academic pre-requisites.

The L3VA scores stand for "Level Three Value Added," and this is a measure of progress made between the end of Key Stage (KS) 4 (aged 14-16 years) and KS5 (aged 16-18 years) as it compares the predicted progress at the end of KS4 with actual progress made at the end of KS5. A value of 1.0 indicates the student has made progress equivalent to a whole grade at A2 level. The use of this measure enables the evaluation of progress rather than simply the attainment of students, and this methodology has been used to assess the gap in science attainment in a review of science education commissioned by the Education Endowment Foundation and the Royal Society (Nunes et al., 2017).

An independent-samples t-test was conducted to compare progress scores for Biology students who had participated in the *Authentic Biology* project (2013-2017) compared with students from the same period and school who did not. There was a significant difference in the progress scores for the *Authentic Biology* students ($M = .62$, $SD = .87$) and non-*Authentic Biology* students [($M = .12$, $SD = .11$); $t(105) = 2.1$, $p = .001$]. This shows differences in average progress scores between these groups of students is significant. As these are students from the same school, taught the same curriculum by the same teachers, it is justifiable to attribute this increase in progress to the student development through participation in the *Authentic Biology* project. This measure is a useful part of evaluation and the ability to run statistical tests to demonstrate reliability and accuracy is appealing as it provides a measure that is seen by many to be rigorous and independent. However, it can be challenging to attribute that observed differences in attainment or progress as an outcome of the research project solely through quantitative means.

Attainment – informal measures. IRIS projects such as *Genome Decoders* involve elements of embedded assessments (EA), with students electronically submitting their annotations for validation by scientists and receiving feedback so that they can correct any inaccuracies before re-submitting their work. This enables the work of each individual and school to be evaluated for the level of accuracy of the contribution as well as the amount of activity. Over time, comparisons can be made between different year and school groups and trends observed for example, levels of engagement across a school year. For trends to be best understood, this informal, quantitative measure of attainment should be coupled with other qualitative measures (e.g., interviews and focus groups) so that an evaluation can provide explanations that can potentially improve delivery of the project in subsequent phases. For example, reductions or increases in productivity may be linked to changes in teaching staff, senior management or access to the necessary technology and none of this information is visible from the quantitative attainment and participation data. The assessment and feedback procedure incorporated into *Genome Decoders* substantiates rigorous protocol and the validation from the scientists provided through their feedback develops students' self-efficacy.

During informal conversations with teachers participating in the *Genome Decoders* project, it emerged that they had observed that A level Biology students who had participated in the *Genome Decoders* project were better able to grasp aspects of the A level Biology curricula. This suggestion was initially consistent with the findings of Masters et al. (2016), who found that participants in five citizen science projects hosted as part of the *Zooniverse* (n.d.) platform did not develop their general scientific knowledge through their participation, but they did develop project specific knowledge. To test whether participation in the *Genome Decoders* did support subject specific knowledge IRIS worked with two A level Biology teachers and put together a short assessment that included multiple choice and free-form questions that was sent to Year 12 students via survey software. All the students who completed the assessment had been taught the same exam curriculum at a similar point in the academic year, although students were taught by one of two teachers and were from two different schools. Of the 66 students who took the assessment, 33 were participants in the *Genome Decoders* project and 33 were not. There was a significant difference observed in the scores for the *Genome Decoder* students (M= 12.6, SD= 3.54) and non-*Genome Decoder* students [(M = 10.1, SD = 4.1); t (73) = 2.01, p = .02)]. This provides a quantitative measure that suggests that students who participate in *Genome Decoders* had a better understanding of the wider topic of genetics and that the knowledge gained, and experience of the *Genome Decoders* project had supported their wider lesson-based learning.

The use of these informal measures of student attainment developed as part of IRIS projects is in its infancy, particularly the curriculum content assessments. However, the pilot example from the *Genome Decoders* project suggests that this type of informal assessment could provide a useful formative measure of attainment that could both quantify the impact of a project but also provide insight into the progress of individual students and school groups that could support a teacher's ability to respond to varied needs at different stages of learning.

Student Destination

The subject choices and university destinations of students have been used as a measure of outcomes. Women continue to be under-represented in STEM careers and, although more girls are choosing to study Physics at A level (1.9% in 2016, up from 1.6% in 2011), there are still more than three times more boys choosing to study Physics at A level (6.5% in 2016) (Institute of Physics; IoP, 2018). IRIS programmes do not have explicit messages to promote the participation of women in STEM university courses or careers. IRIS provides an environment where both male and female students can work with research active scientists to further knowledge. Not only does this provide students with opportunities to further their academic understanding, it also gives them an insight into the careers and lives of scientists enabling them to make informed choices about their future careers.

Impact on A level choices. As the informal *Genome Decoders* assessment evolved from the observations of teachers, it was the anecdotal evidence from teachers that linked an increase in student numbers studying STEM subjects at both A level and at university with their participation in IRIS projects. The uptake of STEM subjects at A level (17-18 years) was analyzed using national data from the IoP and data collected from four schools that have been involved with early IRIS research projects since at least 2013.

Table 1. Impact of IRIS upon Uptake of STEM Subjects at A Level

Comparison of A2 student numbers: % increase from 2009-20011 to 2014-2017		
Subject	Schools with IRIS projects	National average
Biology	61	6
Chemistry	120	14
Physics	132	15
STEM subjects	60	2

This data clearly shows the substantial increase in uptake of all science subjects at A level, particularly Physics courses in schools with IRIS projects, compared to the national average.

Impact on university choices. As with attainment data, accessing student destination data was difficult because of the potential that individual students could be identified from sharing this type of data, even once it was anonymised. To explore the destination outcomes of IRIS students without compromising their identities, IRIS worked with media service, EXACT, at the Universities and Colleges Admissions Service (UCAS). This enabled IRIS to access anonymised destination data from groups of English schools. Each of the four groups had up to 20 schools and the groups were compiled so that they were representative of school type, location and socio-economic profile. Group A contained schools that were actively involved in IRIS projects for at least six months. Group B contained schools that were not part of IRIS, Group C contained schools that had initially registered with IRIS but were not currently active, and Group D contained active IRIS schools that required students to pay fees for admission. Using university destination data gathered during 2012 (before IRIS was formally established and research activities were limited to less than ten English schools) and 2016 (IRIS' first formal phase), it was possible to compare numbers of universities places accepted to study STEM subjects.

Table 2. The Impact of IRIS Projects on Uptake of Universities Places to Study STEM Subjects

School Group	% change between 2012 and 2016
Group A (Active IRIS schools)	24
Group B (Non-IRIS schools)	-3
Group C (Inactive IRIS schools)	8
Group D (Active IRIS independent schools)	13

The number of accepted offers for Physics university courses, divided by gender, were compared between 2012 to2013 and 2015 to 2016.

Table 3. Accepted Offers for Physics Course at University

School Group	% change between 2012 and 2016 – male students	% change between 2012 and 2016 – female students
Group A (Active IRIS schools)	3.5	27.5
Group B (Non-IRIS schools)	-17	0
Group C (Inactive IRIS schools)	0	25
Group D (Active IRIS independent schools)	-6	-33

These data suggest that greater numbers of students choose to study physics at university who come from schools working on IRIS research projects. This increase is even greater amongst female students.

These measures of student destination are a quantitative way of demonstrating impact, and with the other quantitative measures considered here including both formal and informal measures of student attainment. Other citizen science studies have used questionnaires with Likert scale measurements to establish a quantitative metric of skill development and attitude change (Curtis, 2013). This use of questionnaires to produce quantitative data has formed part of IRIS' evaluation framework; however, because there a number of other, independent measures of academic attainment that are appropriate to this context that have been outlined above, they are not described in this section. In addition to the quantitative data, a range of methodologies from the social sciences have been used to generate a detailed and contrasting data set to better understand the outcomes of IRIS programmes upon students and teachers.

Qualitative Approaches to Citizen Science Evaluation

As with quantitative methods, there are a range of qualitative methods that can be used to explore the impact of citizen science programmes on participants. Three examples from evaluating the work of IRIS are outlined as they provide contrasting case studies, each using a different methodology to answer a different evaluation question. Comprehensive accounts of the challenges and opportunities of using social science methods for evaluations are provided by Krathwohl (1993) and Neuman (2013).

Does participation in IRIS projects shape the aspirations of students? Part of IRIS' work is to provide students with the opportunity to publish their research by supporting them to publish in peer-reviewed journals, and speak at conferences and more informal events. Some of the students who were involved in early forerunners of IRIS physics projects are now graduate, doctoral and post-doctoral scientists, continuing to work in fields that are sometimes directly related to their research begun whilst a school student, and who continue to support the work of IRIS by mentoring current students. Anecdotal conversations with three former students revealed that they linked their participation in research projects whilst at school with the development of aspirations for a career in science (Rushton & Parker, 2019). To further understand this, case studies of these three students were developed through a series of interviews, written responses to questions, and then the individual's responses to drafts of the case study. Through an iterative process, the case study captured individual experiences (Rushton & Parker, 2019).

The case study participants were positive about their experiences of participating in research projects whilst at school, and all participants stated that these experiences had an important role in shaping their aspirations to continue with research at the undergraduate level and beyond. Participants suggested that the experience had provided them with opportunities to develop important skills, including analysis, research design and written and oral communication. The two female participants suggested that their experience of research gave them both the self-efficacy to participate in science research beyond school, partly because they encountered examples of women working in this way during their time in school research because the research projects showed them that there was much more work to be done in scientific research, and that there was a place for them to contribute. Opportunities to visit science research facilities at university and other locations, including CERN, impressed upon the participants that science was a dynamic subject and that they were part of a larger collaboration with scientists, all working to further knowledge and understanding of concepts that were not yet known. These three participants all credited their experience of school-based science research projects with providing them with the opportunity to develop skills and experiences that not only raised their aspirations in science but gave them a pathway to reach them.

These case studies were published on the IRIS website so that they could provide a source of inspiration for current and future generations of IRIS students. The case study developed into a peer-reviewed paper with a student co-author, that further explored the impact of participating in science research projects on school students and their teachers (Parker et al., 2018). This provided an additional opportunity for a school student to make a significant contribution to both the science project and resulting evaluation. Case studies can be viewed as a limited method of evaluation because they do not provide evidence that can be generalised across a wider population; the idea that the experiences of one person is not representative of the many. This is to misunderstand the purpose of a case study entirely. Case studies give the researcher detailed insight into complex phenomena in their context. This can also provide detailed information that can inform the construction of questionnaires that generate quantitative data and develop other qualitative measures including focus group question schedules.

Does participation in IRIS Projects develop students' soft skills? One of the themes that emerged from the case studies was that IRIS gave students the opportunity to develop soft skills including confidence, teamwork, communication, enjoyment of science and leadership. This discussion of soft skill development was also a frequent part of informal conversations with teachers as part of school visits. These discussions informed the design of a questionnaire survey with responses that were constructed as per a Likert scale, with five possible responses ranging from "strongly disagree" to "strongly agree." This was given to students at the outset of a project and then after participating for at least four months. Survey questions included:

- I have lots of positive ideas and hopes about my future when I leave school.
- I enjoy a challenge and like to continue with projects and hobbies even when I find them difficult.
- I believe that I can achieve what I set out to do, and I can overcome challenges.
- I have lots of new ideas and work to make a change in the world around me, at home, school or with my friends.
- I work well with my teachers and friends in small groups and large teams. I can listen and share my own ideas.
- I love science and enjoy thinking and learning about science in school and in my own time.

Analyses of responses indicated that students self-reported an increase in soft skill development during the project, but this evaluation method could not further understandings as to what aspect of the IRIS experience supported the development of certain soft skills. A qualitative approach was more suitable to explore this, and focus groups were a chosen method. This approach enabled more students to participate than interviews and provided students with a more comfortable, relaxed environment as they were contributing to a discussion with their peers rather than responding to questions from a previously unknown interviewer. The students' experience of presenting research findings at IRIS conferences was the focus of the discussion.

Ten focus groups were held with a total of 46 students (1:1 gender balance) from five schools. Students were aged 16-18 and had all taken part in a conference in the last year that was at a venue external to their school and which had an audience that included people from outside their school communities. The focus groups were recorded, transcribed and anonymised. Interview questions included:

- Can you tell me about the conference you presented at?
- What was your role in preparing for the conference?
- What was your experience of the conference?
- Can you tell me about any of the challenges you encountered? How did you overcome them?

- What would you do to prepare for a future conference?
- Did the conference change how you saw your research/saw yourself?

An A level psychology student aged 18 was recruited to jointly analyse anonymized focus group transcripts in a shared process using the six phases of thematic analysis outlined by Braun and Clarke (2006; 2012). After undergoing training and support to develop their qualitative research skills, the student worked with the first author, and both met twice a month over a 6-month period and began the analysis after reviewing relevant literature and exploring appropriate analytical processes (Rushton, Charters & Reiss, 2019). Each of the six phases of thematic analysis were discussed before the two of them began the formal analysis. Each worked independently between each of the twelve meetings, sharing ideas via email.

Analysis began from the premise that what participants say about their experience is a reflection of their reality and lived experience (Braun & Clarke, 2006; 2012). As such, the analysis that was used in this study employed an inductive, semantic approach that allowed theory to be identified from the data. A semantic, rather than a latent, approach was chosen to code the data, as this enables those undertaking the analysis to focus on experiential meanings, rather than possible social constructions of meaning that participants could and/or should have intended. In reality, is not possible totally to separate analysis from the research context (Clarke, Braun & Hayfield, 2015). In Phase 5, the identification and naming of themes, both descriptive (data used to illustrate themes) and interpretative (data used to understand latent meaning) analysis were used. Braun and Clarke demonstrated that these two approaches can be successfully combined.

This student co-authored a paper for submission to a peer-reviewed journal (Rushton, Charters & Reiss, 2019). This research suggests that participating in student conferences supports the development of soft skills, including communication (both written and oral) and teamwork, as well as confidence, motivation and enjoyment of science.

Students valued the opportunity to develop their ability to communicate their research and experiences to audiences of experts as well as to the general public, and their school communities. The students recognized that speaking at conferences enabled them to receive feedback that would improve their work as well as receiving recognition for their work and described how they needed to be able to talk about the scientific context of their research in different ways to different audiences, moving between technical and lay language. Students also recognized that their knowledge and understanding was more fixed in their minds once they had explained their research to other audiences, especially younger students.

Teamwork was a soft skill students identified as critical for successful and enjoyable research experiences. Some students, particularly female students, described how important friendship groups were as a part of the research team experience. For some students these were pre-existing friendship groups and for others these were friendships developed whilst working as part of a research team:

> We have ended up making friends through research groups because I ended up being friends with someone who before I didn't really speak to, and we are friends now because we have similar interests.

High-quality teamwork included identifying the different roles students had within the team and using the skills of each student to good effect, setting shared goals and aims and having a sense of fun and social interaction during research activities that were more informal and relaxed than the traditional classroom setting. Linked to teamwork is the confidence that students gained from participating in research projects. Students felt more confident working with other students and described how their confidence increased during the project, often resulting in achievements that they would not have thought possible at the outset.

> We were school students, and we were able to present alongside university students, and we were able to talk to academics and engage them in what we were doing . . .We had been doing either a similar thing or more advanced or at the same level. . .and we were just like, 'Wow!' We could really do something with this.

Students described how they were able to take confidence developed as part of research projects into other aspects of their academic life; they were more resilient when they encountered challenges because they were better able to problem solve, ask for support and find out the answers themselves.

Motivation and enjoyment of science were closely linked. Students identified that they were intrinsically motivated to participate in research projects, giving up their free time and taking on additional work in the form of presenting at conferences and external qualifications such as CREST awards. This sense of motivation came from an increased enjoyment of science and of research. Students appreciated the autonomy and freedom research gave them and recognized the opportunity that they had to develop themselves and be part of something that extended beyond the boundaries of their normal school experience:

> I feel I am able to do this. I am able to think about the science in a way that other scientists can. I am able to think in the same way because I am working in the same way. I have the same tools.

The role of the teacher in providing a supportive and engaging environment was implicitly clear. Students had the freedom to explore their own interests but were given the encouragement and direction needed when they faced challenges and unexpected difficulties

Focus groups enabled the researcher to understand the detail of the individual and group experience and to observe how students interact with their peers and with their teacher. Once they are relaxed and comfortable, students are more than willing to say exactly what they think, and this also provides them with the chance to reflect on their experiences and identify what they have learnt. Students value the opportunity to discuss their experiences and to share their thoughts and ideas with both their teachers and the evaluator. By involving students in the authorship of a peer-reviewed paper that communicates the experiences of the students and their peers to a wider audience, this further demonstrates to students the importance of their contribution to both science and to the evaluation of citizen science.

It is important to consider the tone and atmosphere of focus groups run with young people so that this is a wholly positive experience, even if they are talking about experiences that are challenging. Every student must have the opportunity to speak and to respond to questions at a later point in the focus group if they need some time to provide a response. To safeguard young people, members of school staff should be present as part of a focus group with an independent evaluator. Sometimes, it is appropriate to have a member of school staff present who is there to observe. At other times, teachers who have worked with students can form part of the focus group. In both cases, it is important to understand that the responses of students may be shaped by the presence of teachers.

Does participation in IRIS Projects shape the professional identity of teachers? Anecdotal conversations with teachers during school visits revealed that those teachers involved with IRIS research projects described a renewed sense of professional worth and increased enjoyment of their subject (Parker et al., 2018). To better understand this, science teachers (biology, chemistry, physics) were recruited to take part in a wider study which used thematic analysis of semi-structured interviews with each participant (Rushton & Reiss, 2019). These interviews were conducted with 18 teachers, 17 secondary teachers and one primary school teacher. Two thirds of the participants were women and teachers worked in different types of schools (including Multi-Academy Trusts, Comprehensives, Independents and selective-Grammars) from different geographic and socio-economic areas. All participants had an undergraduate degree in science (apart from one participant whose undergraduate degree was in mathematics), and some had experience of research at doctoral and post-doctoral level. Teaching experience ranged from newly qualified teachers (NQTs) to those with over 30 years of experience. The interviews were transcribed and analyzed using thematic analysis (Braun & Clarke, 2006; Braun, Clark, & Terry, 2012), the same approach used in the analysis of the student focus groups. A detailed description of the methods, analysis and findings of this research can be found in Rushton and Reiss (2019); however, a brief outline is given below.

Five key themes emerged from this analysis, the first was *Freedom to teach*. In their experience of research, both the teacher and the student described a flexible approach to teaching and learning, and a freedom from curriculum constraints and the pressures of external exams. The second theme was that of *(Re)connection with science/research*, where teachers identified that collaborating in research with their students enabled them to (re)connect with their subject and with research. Some teachers also described how they were able to (re)connect with their identity as a scientist, as opposed to a teacher of science. *Collaboration* was the third theme identified by teachers. There was a widely held view that the opportunity to participate in a research collaboration with their students, other scientists, and industry partners gave them a sense of satisfaction developed from being a contributing part of the science community and establishing a wider professional network. *Professional development* was the fourth theme and this focused on the use of research to develop pedagogy. For example, using research software in *Genome Decoders* to more effectively teach the abstract concept of DNA. Teachers also described the professional development they had experienced as part of their involvement in research, gaining recognition from their colleagues, and sharing their experiences of research with other teachers in their department. Some participants discussed that their experience of research projects had given them a pathway to career development that did not necessarily involve management. They suggested this would enable school communities to draw on the skills and attributes of a wider range of staff, including those who had not experienced research beyond their undergraduate studies and were in part-time roles. The final theme explored *Student/societal development through research, both* the contribution that research made to the development of students and the contribution that the teachers and their students made to the science community and the wider world through their research. The capacity of students to contribute to the lives of their peers in the tropics who might benefit from the *Genome Decoders* project was frequently mentioned as a key incentive and continued motivation for student participation.

At the outset of IRIS' work, students were the focal point of the vision which sought to enable all young people to make a genuine contribution to knowledge through research that is valued by the science community. An important outcome of the study of teachers' perspectives is the substantial contribution that participating in research projects has on the professional identity, career development and job satisfaction of the teacher. There is a recruitment and retention crisis in science teaching in the United Kingdom, as research suggests that science teachers are 26% more likely to leave their school within 5 years (Allen & Sims, 2017). Newly-qualified science teachers are 35% more likely to leave their first school within 5 years than newly qualified teachers in other subjects and if those NQTs have a physics or engineering degree they are 87% more likely to leave (Allen & Simms, 2017). A model of science teaching, where the teacher is given the time and freedom to undertake research with their students as part of a collaboration with other scientists alongside their classroom-based teaching is a model to consider when trying to improve the recruitment and retention of science teachers in the United Kingdom and beyond.

Evaluation of citizen science projects should seek where possible to capture the perspectives of as many people involved in the design and delivery of the project, especially when they are not perceived as a "beneficiary" at the outset of the project. This approach to evaluation is longitudinal and draws on perspectives from 360°. This is a key feature of IRIS' approach to evaluation and is described in more detail, along with a further four features of the evaluation of co-created citizen science projects. These features have educational implications and are important areas for future research in school-based citizen science and are useful for all citizen science approaches.

Educational Implications and Directions for Future Research: Five Features of Evaluations of Co-Created Citizen Science Projects

Through this review of the evaluation of IRIS co-created citizen science projects, five distinctive features have emerged that provide useful directions for future research concerning the evaluation of citizen science projects, as well as broad public engagement activities.

360° and Longitudinal Approach to Evaluation

The evaluation framework for IRIS involves a wider variety of participants and associated partners including current students, teachers, school management staff, student peers, teaching colleagues, scientists and other industry partners. This provides the greatest range of perspectives on the different impacts of the projects. This approach can also include formal and informal conversations with students and teachers who decide not to participate to better understand the barriers to participation. Interviews with, and case studies from former students give the evaluation a longitudinal perspective that includes student reflections on the impacts of their experiences over an extended period, in some cases more than five years. A longitudinal approach also gives information regarding the evolution of IRIS' approach. This approach to evaluation necessarily takes much longer than general summative approaches which may have data collection phases 3-6 months after the end of the project or engagement activities and, as such, is not often appropriate for evaluation linked to a funding report or period. However, this form of evaluation can be incorporated as part of regular communication with former participants or alumni and so can be gathered in an opportunistic way until sufficient information can be accumulated.

Evaluation That Recognizes and Values the Individual Experience

Quantitative approaches to evaluation are important as they can provide data that can be robustly and independently statistically measured and so are seen as accurate, reliable and valid evaluation outputs. By their nature, qualitative approaches gather information from individuals, and researchers use this information to explore individual experiences that could be shared by wider groups of students and teachers. These qualitative approaches can be viewed as less reliable and valid compared to quantitative approaches; however, the latter does not provide the rich, detailed information that recognizes and values the individual experience. Quantitative methods can be viewed as inextricably linked with the researchers' perspectives and views and are less likely to enable the voices or experiences of the participants to emerge. One of the challenges facing researchers who wish to evaluate using qualitative methods is that they may lack experience in social science research. This can often be overcome if evaluation is seen as a collaborative approach that draws on the skills of social scientists who may be gathered from university departments as well as school teachers and students. Future citizen science projects should seek to build research collaborations that, from the outset, include social scientists and/or evaluators, so that a blended approach to evaluation can include both quantitative and qualitative measures that includes the individual experience.

A Co-Creative Approach to Evaluation

The starting point for many of the evaluation questions has been the experiences and ideas of students and teachers that have been shared anecdotally. This includes the study of the professional identity of teachers, the experiences of students presenting at conferences, and the role of research projects in shaping the aspirations and developing the soft skills of the students who participate. IRIS' approach to citizen science is co-creative and this extends to the evaluation outputs as students have been involved in writing a peer-reviewed paper that explores their experiences of student conferences. This ensures that the ideas and aims of students are incorporated into every phase of the research process and can support students in developing subsequent projects. This approach to evaluation is challenging as it requires significant amounts of time from students, teachers and the evaluation team. This can be especially difficult in the context of changes in teaching staff and school management. However, where it is possible, this method of evaluation can be a further opportunity to provide students with experiences that support skill development and shape the project outputs, rather than as a necessary burden that is linked to successful funding applications.

Evaluation as a Celebration Point

Linked to the idea that evaluation is a co-creative process is the idea that evaluation can provide a celebration point for students who participate in research and that this celebration point can be an opportunity to inspire the next generation of participants. This celebration point does not have to be separate to the evaluation process; a summative focus group can provide students with the opportunity to share their experiences so that these can be heard by those who have the capacity to both recognize and validate their ideas and implement these in the future. An example of this is a focus group held in a school with students discussing their project with an evaluator, their teacher, observed by members of the school management team. This needs to be carefully managed so that students feel comfortable sharing their ideas and opinions in the presence of adults who are in leadership roles in their schools. However, in most cases, students value the opportunity to share what they have learnt and done with staff who have the capacity to provide tangible support for their work and can enable more students to participate in similar activities. Simple elements, such as providing refreshments and asking students to write postcards to future project members with advice and encouragement can give the focus group a celebratory atmosphere that enable students to share the opportunities and challenges they have encountered and provide them with the sense that they have been a recognized and valued part of a research project.

Evaluation as an Opportunity to Develop Knowledge and Understanding

As with the previous four aspects of IRIS' evaluation, this approach views evaluation as an opportunity to develop knowledge and understanding in areas that go beyond establishing the efficacy of an individual project or even broader evaluative approach. An example of this is the qualitative evaluation of the impact of participating in IRIS projects on teachers. This evaluation generated a series of findings that have developed a better understanding of the professional identity of teachers and how this can be supported through developing a multi-faceted professional identity that includes the roles of teacher, subject specialist, researcher and mentor. This evaluation of the experiences of teachers has further linked understanding of how to create pathways of career development for those already working in the teaching profession and frameworks for recruiting science teachers who wish to develop their scientific knowledge and practical skills through research collaborations with their students, university, and other partners. This model of a "teacher scientist," who is given time to work as an active researcher in school could increase the attractiveness of teaching science as a profession, enhance the career development of teachers and provide students, teachers and technicians with the opportunity to make a genuine contribution to knowledge that is valued by the science community (Rushton & Reiss, 2019).

Evaluation is a central part of most public engagement activities (Reed et al., 2017) and citizen science projects are increasingly incorporating evaluation into the delivery of their programs (Barr et al., 2017; 2018; Becker-Klein et al., 2016; Bonney et al., 2009; 2016; Curtis, 2013; Price & Lee, 2013). This focus on evaluation has begun to facilitate and approach to evaluation that goes beyond consideration of the efficacy of using participants drawn from the public and their levels of accuracy compared to computer generated algorithms (Barr et al., 2017; 2018). IRIS' experience of co-created research projects with school students has demonstrated through evaluation, the range of attainment, career development and socio-economic skills gained by students and teachers who participate (Parker et al., 2018; Rushton, Charters & Reiss, 2019; Rushton & Reiss, 2019). A unique aspect of IRIS' approach to evaluation is to actively include participants in every stage of the evaluation process using approaches that range from anecdotal conversations to co-authorship of peer-reviewed publications (Parker et al., 2018; Rushton, Charters & Reiss, 2019; Rushton & Reiss, 2019). This approach enables the perspectives of the participants to remain at the forefront of research, allowing students and teachers to continue to identify different pathways to making a genuine contribution to knowledge that is valued by the wider scientific community.

Authors' Note

The authors would like to thank the school students, teachers and technicians and academic and industry partners who have worked with IRIS over the past three years. IRIS' work has been supported by grants from Wellcome, the Garfield Weston Foundation, the UK Space Agency and the Royal Commission for the Exhibition of 1851.

References

Adams, R. (2018, April 1). Vast majority of teachers considered quitting in past year – poll. *The Guardian*. Retrieved from https://www.theguardian.com/education/2018/apr/01/vast-majority-of-teachers-considered-quitting-in-past-year-poll

Adegnika, A. A., Lötsch, F., Mba, R. M. O., & Ramharter, M. (2015). Update on treatment and resistance of human trichuriasis. *Current Tropical Medicine Reports*, *2*(4), 218-223. doi:10.1007/s40475-015-0061-z

Adger W. N. (2003). Social capital, collective action, and adaptation to climate change. *Economic Geography, 79*, 387-404. doi: 10.1111/j.1944-8287.2003.tb00220.x

Allen, R. & Sims, S. (2017). *Do National STEM Learning Network professional development courses keep science teachers in the classroom?* Education Datalab. Retrieved from https://wellcome.ac.uk/sites/default/files/science-teacher-retention.pdf

British Educational Research Association (BERA) (2018) *Ethical Guidelines for Educational Research*, fourth edition, London. Retrieved from https://www.bera.ac.uk/researchers-resources/publications/ethical-guidelines-for-educational-research-2018

Barda, B. D., Keiser, J., & Albonico, M. (2015). Human trichuriasis: Diagnostics update. *Current Tropical Medicine Reports*, *2*(4), 201-208. doi:10.1007/s40475-015-0063-x

Barr, A. J., Haas, A. & Kalderon, C. W. (2017). 'That looks weird' – evaluating citizen scientists' ability to detect unusual features in ATLAS images of collisions at the Large Hadron Collider. *arXiv preprint arXiv:1610.02214*

Barr, A. J., Haas, A. & Kalderon, C. W. (2018). Citizen Science community engagement with the HiggsHunters project at the Large Hadron Collider. *Research for All, 2*(2), 359-373. doi:10.18546/RFA.02.2.13

Becker-Klein, R., Peterman, K., & Stylinski, C. (2016). Embedded assessment as an essential method for understanding public engagement in citizen science. *Citizen Science: Theory and Practice*, *1*(1). doi:10.1177/0963662515607406

Bonney, R., & Dickinson, J. L. (2012). Overview of citizen science. *Citizen science: Public participation in environmental research,* (pp. 19-26). Ithaca, NY: Cornell University Press.

Bonney, R., Cooper, C. B., Dickinson, J., Kelling, S., Phillips, T., Rosenberg, K. V., & Shirk, J. (2009). Citizen science: A developing tool for expanding science knowledge and scientific literacy. *BioScience*, *59*(11), 977-984. doi:10.1525/bio.2009.59.11.9

Bonney, R., Phillips, T. B., Ballard, H. L., & Enck, J. W. (2016). Can citizen science enhance public understanding of science? *Public Understanding of Science*, *25*(1), 2-16. doi:10.1177/1075547016656191

Bonter, D. N., & Cooper, C. B. (2012). Data validation in citizen science: A case study from Project FeederWatch. *Frontiers in Ecology and the Environment*, *10*(6), 305-307. doi:10.1890/110273

Braun, V., & Clarke, V. (2006). Using thematic analysis in psychology. *Qualitative Research in Psychology*, *3*(2), 77-101. doi:10.1191/14780088706qp063oa

Braun, V., Clark, V., & Hayfield, N. (2015). *Thematic analysis*. In J.A. Smith (Ed.), *Qualitative psychology: a practical guide to research methods* (pp. 222-248) London. SAGE Publications, Inc.

Braun, V., Clarke, V., & Terry, G. (2012). Thematic analysis. *APA handbook of research methods in psychology*, *2*, 57-71. doi:10.1037/13620-004

Clery, D. (2011). Galaxy Zoo volunteers share the pain and glory of research. *Science, 333*(6039), 173-175. doi:10.1126/science.333.6039.173

Colthurst, D. R., & Tuite, M. F. (2018). Authentic Biology: Student-led research and discovery in schools. *Research for All*, *2*(1), 6-15. doi:10.18546/RFA.02.1.02

Cooper, C., Dickinson, J., Phillips, T., & Bonney, R. (2007). Citizen science as a tool for conservation in residential ecosystems. *Ecology and Society*, *12*(2). doi:10.5751/es-02197-120211

Creswell, J. W. (2009). *Research Design: Qualitative, quantitative and mixed methods approaches*. London, United Kingdom: SAGE Publications, Inc.

Cronje, R., Rohlinger, S., Crall, A., & Newman, G. (2011). Does participation in citizen science improve scientific literacy? A study to compare assessment methods. *Applied Environmental Education & Communication*, *10*(3), 135-145. doi:10.1080/1533015X.2013.876302

Curtis, V. (2013). Evaluating the motivations and expectations of those attending a public astronomy event. *Communicating Astronomy to the Public*, *13*, 14-18. Retrieved from http://www.capjournal.org/issues/13/13_14.php

Curtis, V. (2015). Motivation to participate in an online citizen science game: A study of Foldit. *Science Communication*, *37*(6), 723-746. doi:10.1177/1075547015609322

Curtis, V. (2018). *Online citizen science and the widening of academia* (pp. 167-187). Cham, Switzerland: Palgrave Macmillan.

Donoghue, E. M. & Sturtevant, V. E. (2007). Social science constructs in ecosystem assessments: Revisiting community capacity and community resiliency. *Society & Natural Resources 20*: 899–912. doi: 10.1080/08941920701561114

Dunn N. A., Unni D. R., Diesh C., Munoz-Torres, M, Harris, N. L., Yao, E., . . . Lewis, S. E. (2019) Apollo: Democratizing genome annotation. PLOS Computational Biology 15(2): e1006790. doi:10.1371/journal.pcbi.1006790

Fernandez-Gimenez, M. E., Ballard, H. L., & Sturtevant, V. E. (2008). Adaptive management and social learning in collaborative and community-based monitoring: A study of five community-based forestry organizations in the western USA. *Ecology and Society*, *13*(2). Retrieved from http://www.ecologyandsociety.org/vol13/iss2/art4/

Furnell, W., Shenoy, A., Fox, E. & Hatfield, P. (2018) First results from the LUCID-Timepix spacecraft payload onboard the TechDemoSat-1 satellite in LEO. *Advances in Space Research*. Submitted for publication.

Gura, T. (2013). Citizen science: amateur experts. *Nature*, *496*(7444), 259-261. doi:10.1038/nj7444-259a

Hatfield, P., (2010). Using line intensity ratios to determine the geometry of plasma in stars via their apparent areas. *High Energy Density Physics*, *6*(3), pp.301-304. doi:10.1016/j.hedp.2009.10.001

Hatfield, P., Furnell, W., Shenoy, A., Fox, E., Parker, B., Thomas, L. & Rushton, E.A.C. (2019). IRIS opens pupils' eyes to real space research. *Astronomy and Geophysics 60*(1) 1.22-1.24. doi: 10.1093/astrogeo/atz046

Havens, K., Vitt, P., & Masi, S. (2012). Citizen science on a local scale: the plants of concern program. *Frontiers in Ecology and the Environment*, *10*(6), 321-323. doi: 10.1890/110258.

Index of Multiple Deprivation 2015 (2017). Retrieved from https://www.arcgis.com/home/item.html?id=14b9617e617c4ae09c0a5b0cab06044b

Institute for Research in Schools (2018). Higgs – an exciting discovery. Retrieved from http://researchinschools.org/higgs_hunters/home.html

Institute of Physics (2018, May). *Why not physics? A snapshot of girls' uptake at A-level.* Institute of Physics Report. Retrieved from http://www.iop.org/publications/iop/2018/file_71495.pdf

Jensen, E., & Holliman, R. (2009). *Investigating science communication to inform science outreach and public engagement* (pp. 55-71). Oxford: Oxford University Press.

Jordan, R. C., Ballard, H. L., & Phillips, T. B. (2012). Key issues and new approaches for evaluating citizen-science learning outcomes. *Frontiers in Ecology and the Environment*, *10*(6), 307-309. doi: 10.1890/110280

Krathwohl, D. R. (1993). *Methods of educational and social science research: An integrated approach*. New York, NY: Longman/Addison Wesley Longman.

Krebs, V. (2010). Motivations of cybervolunteers in an applied distributed computing environment: MalariaControl.net as an example. *First Monday, 15*(2). doi:10.5210/fm.v15i2.2783

Masters, K., Oh, E. Y., Cox, J., Simmons, B., Lintott, C., Graham, G., . . . Holmes, K. (2016). Science learning via participation in online citizen science. *arXiv preprint arXiv:1601.05973*.

Neuman, W. L. (2013). *Social research methods: Qualitative and quantitative approaches*. Whitewater, WI: Pearson education.

Nov, O., Arazy, O., & Anderson, D. (2011, February). Dusting for science: Motivation and participation of digital citizen science volunteers. Paper presentation of the annual conference of the *Association for Computing Machinery*, (pp. 68-74), Seattle, WA.

Nunes, T., Bryant, P., Strand, S., Hillier, J., Barros, R., & Miller-Friedmann, J. (2017) *Review of SES and Science Learning in Formal Education Settings.* A Report prepared for the Education Endowment Foundation and The Royal Society. Retrieved from https://educationendowmentfoundation.org.uk/public/files/Review_of_SES_and_Science_Learning_in_Formal_Educational_Settings.pdf

Parker, B., Fox, E. & Rushton, E. A. C. (2018). IRIS – promoting young people's participation and attainment in STEM and reigniting teacher's passion for science education. *Impact, 2.* Retrieved from https://impact.chartered.college/article/parker-iris-stem-students-teachers-participation-research/

Parker, B., Rushton, E. A. C., Thomas, L., & Hatfield, P. (2019). Transforming Education with the Timepix detector – ten years of CERN@school. *Radiation Measurements.* doi:10.1016/j.radmeas.2019.03.008

Phillips, T., Bonney, R., & Shirk, J. (2012). What is our impact? Towards a united framework for evaluating the impacts of citizen science. In Dickinson, J. L. & Bonney, R. (Eds) *Citizen science: Public collaboration in environmental research*, (pp. 82-96). Ithaca, NY: Cornell University Press.

Price, C. A., & Lee, H. S. (2013). Changes in participants' scientific attitudes and epistemological beliefs during an astronomical citizen science project. *Journal of Research in Science Teaching, 50*(7), 773-801. doi:10.1002/tea.21090

Raffe, D., Croxford, L., Iannelli, C., Shapira, M., & Howieson, C. (2006). *Social-class inequalities in education in England and Scotland* (CES special briefing no. 40). Retrieved from the Centre for Educational Sociology, University of Edinburgh, Edinburgh, website: <http://www.ces.ed.ac.uk/PDF%20Files/Brief040.pdf>

Reed M. S., Duncan S., Manners P., Pound D., Armitage L., Frewer L., Thorley C., & Frost B. (2017). A common standard for the evaluation of public engagement with research. *Research for All, 2*(1): 143–162. doi:10.18546/RFA.01.1.13

Reeves, N., Tinati, R., Zerr, S., Van Kleek, M. G., & Simperl, E. (2017). From crowd to community: A survey of online community features in citizen science projects. Paper presentation of the annual conference of the *Association for Computing Machinery on Computer Supported Cooperative Work and Social Computing*, (pp. 2137-2152), Portland, OR.

Rotman, D., Preece, J., Hammock, J., Procita, K., Hansen, D., Parr, C., . . . Jacobs, D. (2012, February). Dynamic changes in motivation in collaborative citizen-science projects. Paper presentation of the annual conference of the *Association for Computing Machinery*, (pp. 217-226), Seattle, WA.

Rushton, E. A. C., Charters, L., & Reiss, M. J. (2019) The experiences of active participation in academic conferences for high school science students. Manuscript submitted for publication.

Rushton, E. A. C. & Parker, B. (2019). Empowering young people through co-created scientific research. In S. E. Hiller, & A. Kitsantas (Eds.), *Enhancing STEM Motivation in Citizen Science Programs* (pp. 97-127). Hauppauge, NY: Nova Science Publishers, Inc.

Rushton, E.A.C. & Reiss, M.J. (2019) From science teacher to 'teacher scientist': Exploring the experiences of research-active science teachers in the UK. *International Journal of Science Education.* doi.org/10.1080/09500693.2019.1615656

Schwartz, M. W. (2006). How conservation scientists can help develop social capital for biodiversity. *Conservation Biology*, *20*(5), 1550-1552. doi:10.1111/j.1523-1739.2006.00421.x

Shah, H. R., & Martinez, L. R. (2016). Current approaches in implementing citizen science in the classroom. *Journal of Microbiology & Biology Education*, *17*(1), 17. doi:10.1128/jmbe.v17i1.1032

Shirk, J. L., Ballard, H. L., Wilderman, C. C., Phillips, T., Wiggins, A., Jordan, R., . . . Bonney, R. (2012). Public participation in scientific research: A framework for deliberate design. *Ecology and Society*, *17*(2). doi:10.5751/ES-04705-170229

Social Mobility Commission (2017). *State of the nation 2017: Social mobility in Great Britain*. Retrieved from www.gov.uk/government/publications

Tweddle, J. C., Robinson, L. D., Pocock, M. J. O., & Roy, H. E. (2012). *Guide to citizen science: Developing, implementing and evaluating citizen science to study biodiversity and the environment in the UK*. NERC/Centre for Ecology & Hydrology. Retrieved from http://nora.nerc.ac.uk/id/eprint/20678/

Wallace, D., & Bodzin, A. (2017). Developing scientific citizenship identity using mobile learning and authentic practice. *Electronic Journal of Science Education*, *21*(6), 46-71. doi:17688-69716-1-PB

Whyntie, T., & Harrison, M. A., (2014). Simulation and analysis of the LUCID experiment in the Low Earth Orbit radiation environment. In *Journal of Physics: Conference Series, 513*(2), 022038. doi:10.1088/1742-6596/513/2/022038

Whyntie, T., & Harrison, M. A., (2015). Full simulation of the LUCID experiment in the Low Earth Orbit radiation environment. *Journal of Instrumentation*, *10*(03), C03043. doi:10.1088/1748-0221/10/03/c03043

Wiggins, A., & Crowston, K. (2011, January). From conservation to crowdsourcing: A typology of citizen science. Paper presentation at the annual conference of *System Sciences (HICSS)*, (pp. 1-10), Honolulu, HI.

Wilderman, C. C. (2007, June). *Models of community science: Design lessons from the field*. Conference presented at the Citizen Science Toolkit Conference, Cornell Lab of Ornithology, Ithaca, NY. Retrieved from https:// www.google.com/ url? sa=t& rct=j& q=& esrc=s& source=web& cd=1& ved= 2ahUKEwjruPnAkfbdAhUSneAKHRiQDEIQFjAAegQICBAC&url= https% 3A% 2F% 2Fthrivingearthexchange. org% 2Fwp-content% 2Fuploads% 2F2015% 2F01% 2FModels- of- Community- Science- Lessons- from- the-Field.pdf&usg=AOvVaw37xBym-YkKAfG4jfh9bR06

Zooinverse (n.d.). Retrieved from https://www.zooniverse.org/

ABOUT THE EDITORS

Suzanne E. Hiller, Ph.D. has been involved in the field of education for nearly three decades as a practitioner, professor, researcher, and writer. She has taught in K-12 and higher education settings. In addition, she has served as the Teacher Naturalist with the Smithsonian Institution's National Museum of Natural History during which time she developed curricula with natural specimens as well as overseeing a specimen loan program for instructional purposes. Her research interests center on STEM career motivation, particularly in citizen science contexts and self-regulated learning. She has authored books, book chapters, and her research has appeared in various peer reviewed journals.

Anastasia Kitsantas, Ph.D. is professor of Educational Psychology in the College of Education and Human Development at George Mason University. Her research interests focus on the role of self-regulation and student motivational beliefs on learning and performance across diverse areas of functioning. She is the editor, co-author or author of two books and over 100 scholarly publications, many of which are directed toward the training of self-regulation. She has been an active participant in many international student and scholarly groups. She is the recipient of a George Mason University Teaching Excellence Award and the 2019 Barry J. Zimmerman Award for Outstanding Contributions to the fields of studying and self-regulated learning research by the Studying and Self-Regulated Learning SIG of the American Educational Research Association (AERA). She is also a Fellow of the American Psychological Association's (APA): Division 15, Educational Psychology. Presently, she is serving as the Editor-in-Chief of the Journal of Experimental Education - http://www.tandfonline.com/toc/vjxe20/current.

INDEX

A

B

C

D

E

K

L

M

N

O

P

Q

R

S

T

U

V

W

Y

ANIMAL SCIENCE, ISSUES AND PROFESSIONS

TURTLES

ANATOMY, ECOLOGY AND CONSERVATION

Animal Science, Issues and Professions

Additional books in this series can be found on Nova's website under the Series tab.

Additional E-books in this series can be found on Nova's website under the E-books tab.

ANIMAL SCIENCE, ISSUES AND PROFESSIONS

TURTLES

ANATOMY, ECOLOGY AND CONSERVATION

MELANIE J. COSGROVE
AND
SOPHIE A. ROE
EDITORS

Nova Science Publishers, Inc.
New York

For permission to use material from this book please contact us:
Telephone 631-231-7269; Fax 631-231-8175
Web Site: http://www.novapublishers.com

Library of Congress Cataloging-in-Publication Data

Turtles : anatomy, ecology, and conservation / editors, Melanie J. Cosgrove and Sophie A. Roe.
p. cm.
Includes bibliographical references and index.
ISBN 978-1-61470-554-3 (hardcover : alk. paper) 1. Turtles--Anatomy. 2. Turtles--Ecology. 3. Turtles--Conservation. I. Cosgrove, Melanie J. II. Roe, Sophie A.
QL666.C5T862 2011
597.92--dc23

2011024675

Published by Nova Science Publishers, Inc. † New York

CONTENTS

PREFACE

This book presents current research in the study of the anatomy, ecology and conservation of turtles. Topics discussed in this compilation include the accumulation and tissue distribution of metals and other elements in sea turtles across the globe as bioindicators of marine pollution; the pathological findings and causes of mortality of sea turtles in the Canary Islands; the painted turtle as a model of natural anoxia tolerance; the evolutionary history of turtles and responses of freshwater turtles to drought.

Chapter 1- All the sea turtle species have been classified as vulnerable, endangered or critically endangered by IUCN Red List 2010 (except for *Natator depressus*: data deficient). Among others, the greatest threats to the survival of marine turtles in the world include loss of nesting habitats, accidental capture by some types of fishing and marine pollution. For example, unnatural quantities of inorganic pollutants (metals and other toxic elements) have been released and continue to be released into the sea by contaminant human activities altering the natural biological equilibrium. Many of these inorganic elements are biologically essential, but they can be toxic to biota above certain exposure thresholds. The risk of toxicity is higher on those long-living species at the top of the marine food chains, such as sea turtles, which have the potential to accumulate them. Because of that, the scientific community has increased the interest in the assessment of metal exposure and accumulation in different species of marine turtles, mainly in the last decade. Stranded individuals of the seven existing species of sea turtles (*Caretta caretta*, *Chelonia mydas*, *Dermochelys coriacea*, *Eretmochelys imbricata*, *Lepidochelys kempii*, *Lepidochelys olivacea* and *Natator depressus*) have been analyzed in several studies for inorganic pollution monitoring in different areas around the world (America, Europe, Asia and Oceania). Concentrations of several metals and other elements of toxicological interest (Cd, Pb, Hg, As,

Cu, Zn, Fe, Mn, Ni, Se, Cr, etc.) have been measured in most of their organs and tissues (liver, kidney, muscle, lung, bone, fat, blood, carapace and eggs, among others). Elements related to biomagnification phenomena in marine environments, such as Hg and Cd, showed especially high levels in samples of liver and kidney, respectively, from areas with a great human presence (Japan, Mexico or Mediterranean Sea). The majority load of other elements was found in muscle (As) and hard tissues (Pb, Zn).

In spite of the difficulty to study these animals (due to their solitary nature or long pelagic phases), the information about metal levels in sea turtles has increased considerable in the last years. In this chapter, authors review the available scientific literature in order to provide a global view of the matter. This global view allow us to get a deeper knowledge of the presence of inorganic elements in the different species of sea turtles, to estimate the risk of metal exposure for them in different areas of the world, and to assess their role as a potential long-term bioindicators of inorganic marine pollution.

Chapter 2- Because all species of sea turtles are included on the Red List of the World Conservation Union, the efforts to conserve sea turtles, the advances in their medical management, and the studies on diseases and causes of mortality and/or stranding of sea turtles must to be increased. A multidisciplinary approach to this focus carried out by veterinarians, biologists, and scientists, is necessary in order to compile data that prove the main threats for these endangered reptiles, and to design adequate strategies of conservation. This chapter lists the pathological findings and causes of mortality of 49 sea turtles (46 *Caretta caretta*, 2 *Chelonia mydas*, and 1 *Dermochelyscoriacea*) stranded on the coasts of the Canary Islands, Spain, between 2002 and 2009.Of these, 12 turtles (24.49%) had died of spontaneous diseases including different types of hepatitis, pneumonia, and septicemic processes. However, 37 turtles (75.51%) died from lesions associated with human activities: ingestion of hooks and monofilament lines (34.69%), entanglement in fishing nets (24.49%), and boat-strike injuries (16.33%). In addition, polychlorinated biphenyls (PCBs 28, 31, 52, 101, 138, 153, 180, and 209) and DDT and its metabolites (*o,p'*-DDT, *o,p'*-DDE, and *o,p'*-DDD) were measured in liver and fat from 32 sea turtles. Tissues from these turtles contained higher levels of PCBs and *o,p'*-DDT than those reported in turtles from other geographical regions. Statistically, a positive correlation was detected between ∑PCBs concentration and cachexia, and between ∑DDT concentrations and cachexia. No histological lesions exclusively attributed to the acute effects of PCBs and DDTs were described. However, chronic effects of organochlorines can not be discarded.

Chapter 3- Painted turtles range throughout the USA and Canada. The most northern member of this species is the western painted turtle *Chrysemys picta bellii;* which is also the most anoxia-tolerant vertebrate identified, surviving for up to 6 months buried in the mud of ice covered pond at 3°C. *C. picta* provides a working model of an anoxia-tolerant vertebrate, and of particular interest - a brain that functions without oxygen. Authors have recently been exploring the role of GABA, the primary vertebrate inhibitory neurotransmitter, in anoxia tolerance. Increases in endogenous GABAergic signalling appear to mediate neuronal electrical depression in *C. picta* and in a number of other anoxia-tolerant species. Unlike most other vertebrate species that undergo severe neural injury within minutes of an anoxic episode, *C. picta* neurons reduce action potential frequency and suppress metabolism rapidly with the onset of anoxia, and avoid injury. Anoxia-sensitive mammals lack a hypoxia/anoxia-mediated increase in GABA concentrations and suffer electrical hyper-excitability and excitotoxic cell death following the onset of an anoxic/ischemic episode. Experimental interventions aimed at increasing inhibitory GABAergic signalling prevent over-excitation and rescue mammalian neurons from anoxic/ischemic cell death. Drawing upon evidence from a number of studies on a variety of naturally anoxia-tolerant species authors will explore in detail the role of GABA in neuroprotection. The anoxic response could be interpreted as a natural anaesthetic mechanism and is achieved by a reduction in excitatory neurotransmitter receptor activity (glutamate) and an increase in inhibitory receptor activity (GABA). An understanding of the complex regulation of GABAergic signaling in facultative anaerobes may provide insight into interventions that could prove efficacious against low oxygen situations in mammals, such as ischemic insult due to stroke.

Chapter 4- The history of turtle lineage began around 225 Mya. Since then, this group of reptiles developed diverse ecological strategies and colonized a wide range of environments, from marine to fully terrestrial habitats. In spite of their great ecological diversity, the bauplan of turtles is peculiar and little variable, with a body encased in a rigid shell consisting of a dorsal carapace and a ventral plastron. This fact has made the morphological comparison of turtles with other vertebrates very complicate. Thus, the understanding of the origins and phylogenetic relationships of turtles within the Amniota and the evolution of turtle biology are stimulating challenges for researchers. Several attempts in order to resolve these mysteries have been carried out from a multidisciplinary approach, but lacks consensus. The knowledge of the evolutionary history of turtle species and their inclusive groups is relevant for their conservation management, especially taking into account that many species are among moderate and high risk of extinction. In this chapter, authors update and review the state of the art of the systematics of turtles from the point of view of several often conflicting disciplines, embryology, morphology, paleontology and molecular systematics. In addition, authors sound out the amount of threatened phylogenetic diversity in the turtle tree of life based on fossil evidence, recent phylogenetic hypotheses and data from the red list of the International Union for the Conservation of Nature.

Chapter 5- The study of the hematological characteristics of loggerhead turtles can be very helpful in the rescue and rehabilitation of this endangered species. Normal hemocellular and biochemical values are the basis for the evaluation of a subject's health status. The identification of the blood cells through cytochemical or immunocytochemical methods is the first step to obtain a correct classification of the cells. The following blood cell types were found: lymphocytes, monocytes, heterophils (the counterparts of mammalian neutrophils), eosinophils, basophils (rarely observed), erythrocytes along with some of their precursors, and thrombocytes. Lymphocytes and monocytes are easily distinguishable, as they are similar to the mammalian counterparts, whereas heterophils (which are the equivalent of mammalian neutrophils) and eosinophils both contain eosinophilic granules in their cytoplasm, making their identification difficult. The thrombocytes are nucleated cells, whose shape is variable in dry film smears of circulating blood, either round or oval; they take part in the hemostatic process, as suggested by their tendency to form aggregates. The erythrocytes are also nucleated cells, showing an ellipsoidal and flattened configuration. In all specimens of this marine reptile, both healthy and unhealthy, the mature erythrocytes contain one characteristic inclusion in their cytoplasm, which was recognized as a Heinz body, formed by the precipitation of hemoglobin. The loggerhead's hemoglobin was studied in depth and revealed some specific characteristics, one of which is that under physiologic conditions it precipitates due to its instability. While no data are available about the erythropoiesis in the primary organs, circulating erythropoiesis was revealed to occur in the loggerhead's blood, just as it also occurs at this site in other non-mammalian vertebrates. Loggerhead blood infestation by spirorchids was reported by Wolke *et al.* (1982), while Di Santi *et al.* (2010) found two specimens of loggerheads hemoparasitized by *Theileria*.

Chapter 6- Australia was not always arid, and the freshwater turtles that have survived to this day are either relicts residing in river systems that have themselves had a history or continuity through the drying of the continent, or they are species that have adapted in some way to meet the challenges of drought. In this chapter, authors report on how Australian freshwater turtles cope with periodic loss of habitat through drought at a range of spatial and temporal scales, and consider how they might fare under changing climates predicted to occur through global warming. Global warming is not occurring in isolation of other environmental changes at a landscape scale, and authors look at what interactions there might be between climate change and the increasing demands of agriculture, industry and our cities for water, in presenting challenges for our unique freshwater turtle fauna.

In: Turtles: Anatomy, Ecology and Conservation ISBN: 978-1-61470-554-3
Eds: M. J. Cosgrove and S. A. Roe

Chapter 1

Accumulation and Tissue Distribution of Metals and other Elements in Sea Turtles from all over the World

***Miguel Motas Guzmán*[1] *and Silvia Jerez Rodríguez*[2]**
Toxicology Area, Department of Socio-Sanitary Sciences,
Faculty of Veterinary Medicine, University of Murcia,
Campus of Espinardo 30100, Murcia, Spain

Abstract

All the sea turtle species have been classified as vulnerable, endangered or critically endangered by IUCN Red List 2010 (except for *Natator depressus*: data deficient). Among others, the greatest threats to the survival of marine turtles in the world include loss of nesting habitats, accidental capture by some types of fishing and marine pollution. For example, unnatural quantities of inorganic pollutants (metals and other toxic elements) have been released and continue to be released into the sea by contaminant human activities altering the natural biological equilibrium. Many of these inorganic elements are biologically essential, but they can be toxic to biota above certain exposure thresholds. The risk of toxicity is higher on those long-living species at the top of the marine food chains, such as sea turtles, which have the potential to accumulate them. Because of that, the scientific community has increased the interest

[1] Correspondence address by e-mail: motas@um.es
[2] Correspondence address by e-mail: silviajerez@um.es

in the assessment of metal exposure and accumulation in different species of marine turtles, mainly in the last decade.

Stranded individuals of the seven existing species of sea turtles (*Caretta caretta*, *Chelonia mydas*, *Dermochelys coriacea*, *Eretmochelys imbricata*, *Lepidochelys kempii*, *Lepidochelys olivacea* and *Natator depressus*) have been analyzed in several studies for inorganic pollution monitoring in different areas around the world (America, Europe, Asia and Oceania). Concentrations of several metals and other elements of toxicological interest (Cd, Pb, Hg, As, Cu, Zn, Fe, Mn, Ni, Se, Cr, etc.) have been measured in most of their organs and tissues (liver, kidney, muscle, lung, bone, fat, blood, carapace and eggs, among others). Elements related to biomagnification phenomena in marine environments, such as Hg and Cd, showed especially high levels in samples of liver and kidney, respectively, from areas with a great human presence (Japan, Mexico or Mediterranean Sea). The majority load of other elements was found in muscle (As) and hard tissues (Pb, Zn).

In spite of the difficulty to study these animals (due to their solitary nature or long pelagic phases), the information about metal levels in sea turtles has increased considerable in the last years. In this chapter, we review the available scientific literature in order to provide a global view of the matter. This global view allow us to get a deeper knowledge of the presence of inorganic elements in the different species of sea turtles, to estimate the risk of metal exposure for them in different areas of the world, and to assess their role as a potential long-term bioindicators of inorganic marine pollution.

INTRODUCTION

Six of the seven existing species of sea turtles have been included in the IUCN Red List 2010 [1]: *Caretta caretta* (endangered), *Chelonia mydas* (endangered and decreasing population trend), *Dermochelys coriacea* (critically endangered and decreasing population trend), *Eretmochelys imbricata* (critically endangered and decreasing population trend), *Lepidochelys kempii* (critically endangered) and *Lepidochelys olivacea* (vulnerable and decreasing population trend). About the other one, *Natator depressus*, the available data are deficient. Among others (fisheries impacts, direct take, coastal development or global warming), marine pollution is one of the five major hazards to sea turtles conservation [2]. Plastics, discarded fishing gear or light pollution pose a threat to sea turtles. In addition, chemical pollution related to petroleum by-products, agricultural runoff or sewage could affect these animals weakening, for example, their immune system [2].

Several anthropogenic activities have been pointed out as sources of inorganic pollutants in marine environments, mainly related to growing population and industrialization: industrial plants (metallurgy, paints, tannery, etc.), chemical plants, mining, harbour activities, ship traffic, intensive farming, aquatic breeding, solid waste and sewage (urban and others), etc. [3, 4, 5]. It is quite clear that inorganic elements are natural components of rocks and soils, which find their way into the marine environment as a consequence of weathering and erosion [6]. However, unnatural quantities of several inorganic elements coming from human activities (previously cited) have been released, and continue to be released into the sea altering the natural biological equilibrium [7]. Many of these elements are biologically essential but they can be toxic to biota above certain exposure thresholds [8]. Mainly, some toxic elements may cause adverse effects in organisms located at the top of the marine food-chains [9] and long-living species such as sea turtles, which have the potential to accumulate them [10, 11], although the concentration of metals in the water column is often low due to their poor solubility [12].

With a view towards the preservation of these species, the interest in monitoring the presence of inorganic pollutants (metals and others) in sea turtles has increased in the last years [13]. Nowadays, a large number of studies on the matter have been carried out throughout the world: European Mediterranean area [14, 15, 16, 17, 18, 19, 20, 21, 22, 23, 24, 25], European Atlantic area [10, 26, 27], American Atlantic area [28, 29, 30, 31, 32, 33, 34, 35, 36, 37], American Pacific area [38, 39, 40, 41, 42, 43, 44, 45, 46], Asia-Pacific area and Indian Ocean [11, 13, 47, 48, 49, 50, 51, 52, 53, 54, 55, 56, 57, 58, 59, 60, 61, 62].

Nevertheless, the information is scarce and fragmented in comparison with other big marine organisms (marine mammals or seabirds), probably due to the solitary nature of the sea turtles, the length of their pelagic phase and long lasting apnoeas, so they are among the most difficult marine animals to study [63]. The purpose of this chapter is to put in order and unify all this information to provide a global view of the presence and distribution of eleven elements (key elements in the toxicological field, namely: cadmium, lead, mercury, arsenic, copper, zinc, iron, manganese, nickel, selenium and chromium) in several biological samples of marine sea turtles (liver, kidney, muscle, blood, fat, egg and stomach content). The majority of the data reported as wet weight in the original studies have been converted to dry weight, in order to avoid errors related to different water contents of the samples (conversion ratios: 3.13 for liver, 4.17 for kidney, 4.76 for muscle, 5.10 for blood, 1.18 for fat tissue, 5.21 for egg content, 37.04 for albumen, 2.67 for yolk and 2.44 for eggshell) [22, 23, 35, 46].

This global view allows us to recognize areas, species or elements that need to be studied further. With this chapter we get a deeper knowledge of the presence of inorganic elements in the different species of sea turtles, in order to estimate the risk of metal exposure for them in different areas of the world. Besides, we assess the marine turtle role as potential long-term bioindicator of the inorganic marine pollution.

LOGGERHEAD SEA TURTLE (*CARETTA CARETTA*)

Referring to the information about levels of inorganic pollutants in marine turtles, *Caretta caretta* is the most studied species. The levels of heavy metals and essential elements (Cd, Pb, Hg, As, Cu, Zn, Fe, Mn, Ni, Se and Cr) that have been measured in this species (samples of liver, kidney, muscle, blood, fat and egg) are shown in Table 1. No data about levels of these elements in stomach contents of loggerhead sea turtles have been found. The metals most frequently analyzed are mercury, cadmium and lead (in that order). On one hand, probably mercury and cadmium are largely studied due to the greater relevance that these metals present in the marine environment (mercury and cadmium are often related to biomagnification and bioaccumulation phenomena in aquatic ecosystems). On the other hand, lead is largely studied because of this metal is related with several human contaminant activities, and it is one of the most suitable metals for anthropogenic pollution monitoring [64]. The tissues of this species mainly analyzed are liver, kidney and muscle (see Table 1). Generally, the highest mercury concentrations were found in liver [13, 14, 15, 18, 20, 21, 23, 31, 43, 51], whereas the highest cadmium levels were found in kidney [10, 13, 14, 18, 20, 23, 24, 25, 27, 42, 48, 51]. These patterns of metal accumulation are similar to those described for other marine vertebrates [65, 66]: mercury levels tend to be highest in hepatic tissue and cadmium levels tend to be highest in renal tissue. In the case of lead, soft tissues showed similar levels (see Table 1), although the studies that analyzed calcified tissues found the highest lead levels in bone and carapace [25, 51]. Arsenic showed the highest levels in muscle tissue [14, 16, 25, 49] and chromium in kidney [14].

Table 1. Heavy metal and essential element levels (µg g^{-1} dry weight) in tissues and other samples of *Caretta caretta* from different locations

Tissue - Location	Cadmium		Lead		Mercury		Arsenic		References
	Mean ± SD	Min - Max	Mean ± SD	Min - Max	Mean ± SD	Min - Max	Mean ± SD	Min - Max	
Liver									
Mediterranean Sea: Italy	7,60 ± 6,05	3,06 - 20,23	1,23 ± 1,01	BDL - 3,38	1,68 ± 1,04	0,35 - 3,72	21,67 ± 17,22	0,83 - 56,55	[14]
Mediterranean Sea: Italy					2,19 ± 1	1,16 - 3,44			[15]
Mediterranean Sea: Cyprus	8,64	5,14 - 12,97	BDL	BDL - 4,90	2,41	0,82 - 7,50			[20]
Mediterranean Sea: Italy							20,94 ± 14,03	8,38 - 40,84	[16]
Mediterranean Sea: Turkey	1,26 ± 0,43	(Em)	2,48 ± 0,46	(Em)	0,51 ± 0,05	(Em)			[21]
Mediterranean Sea: Italy	2,84 ± 0,72								[22]
Mediterranean Sea: Italy	19,30 ± 34,20	1,60 - 114,00			1,10 ± 1,70	0,42 - 8,76			[23]
Mediterranean Sea: Italy	10,50 ± 6,06	3,44 - 20,47	0,50 ± 0,16	BDL - 0,91	1,34 ± 0,91	0,41 - 3,94			[18]
Mediterranean Sea: Italy	2,40 ± 0,40		0,10 ± 0,08						[24]
Mediterranean Sea: Spain	0,81 ± 0,48	0,05 - 1,88	0,20 ± 0,11	0,08 - 0,51	0,39 ± 0,41	0,08 - 1,57	12,73 ± 13,03	3,50 - 43,43	[25]
Atlantic Ocean: France	8,06 ± 12,88	0,94 - 36,88							[10]
Atlantic Ocean: Canary Is.	7,91 ± 1,41	0,13 - 68,69	9,19 ± 1,84	0,16 - 103,41	0,13 ± 0,03	0,003 - 1,47	53,34 ± 9,25	0,25 - 412,06	[27]
Atlantic Ocean: US					1,84 ± 0,50	1,09 - 4,19			[31]
Pacific Ocean: Mexico	1,75	BDL - 30,62	BDL						[42]
Pacific Ocean: Mexico					0,15	0,12 - 0,18			[43]
Pacific Ocean: Japan	29,03 ± 10,31	17,69 - 45,63			4,72 ± 9,16	0,78 - 25,47			[13]

Table 1 (Continued)

Tissue - Location	Cadmium		Lead		Mercury		Arsenic		References
	Mean ± SD	Min - Max	Mean ± SD	Min - Max	Mean ± SD	Min - Max	Mean ± SD	Min - Max	
Pacific Ocean: Australia**	51,25 ± 10,31	22,81 - 109,69			0,05 ± 0,02	0,00 - 0,09	1,44 ± 0,75	0,00 - 4,88	[48]
Pacific Ocean: Japan							6,32 ± 1,56	4,24 - 9,43	[49]
Pacific Ocean: Japan	30,44 ± 10,53	(F)	0,25 ± 0,09	(F)	1,25 ± 0,50	(F)			[51]
	20,63	(M)	0,66	(M)	25,47	(M)			[51]
Pacific Ocean: Japan							11,20 ± 3,00		[54]
Kidney									
Mediterranean Sea: Italy	24,23 ± 21,40	0,39 - 64,00	0,70 ± 0,35	BDL - 1,35	0,65 ± 0,34	0,30 - 1,53	29,91 ± 39,48	6,09 - 139,60	[14]
Mediterranean Sea: Cyprus	30,50	18,80 - 42,20	2,45	BDL - 4,90	0,47	0,13 - 0,80			[20]
Mediterranean Sea: Italy	57,20 ± 34,60	10,90 - 158,00			0,90 ± 0,70	0,37 - 3,41			[23]
Mediterranean Sea: Italy	34,79 ± 20,13	5,25 - 68,34	0,50 ± 0,29	BDL - 0,88	0,67 ± 0,29	0,25 - 1,29			[18]
Mediterranean Sea: Italy	5,80 ± 1,10		0,10 ± 0,07						[24]
Mediterranean Sea: Spain	6,88 ± 6,26	0,02 - 16,16	1,22 ± 0,86	0,08 - 2,26	0,44 ± 0,51	0,07 - 1,32	34,72 ± 31,79	6,84 - 97,75	[25]
Atlantic Ocean: France	55,42 ± 56,67	7,00 - 148,76							[10]
Atlantic Ocean: Canary Is.	22,88 ± 4,25	0,04 - 254,52	10,17 ± 2,00	0,08 - 72,05	0,17 ± 0,04	0,04 - 1,38	57,50 ± 10,00	4,83 - 508,87	[27]
Atlantic Ocean: US					0,88 ± 0,21	0,54 - 1,83			[31]

Pacific Ocean: Mexico	73,11	13,72 - 140,00	0,03	BDL - 69,89					[42]
Pacific Ocean: Mexico					0,10	0,06 - 0,14			[43]
Pacific Ocean: Japan	164,18 ± 67,51	75,42 - 235,44			1,04 ± 0,54	0,17 - 1,83			[13]
Pacific Ocean: Australia**	117,93 ± 23,75	47,50 - 164,18			0,21 ± 0,04	0,13 - 0,29	2,96 ± 1,08	1,00 - 4,79	[48]
Pacific Ocean: Japan							9,47 ± 5,37	4,01 - 20,20	[49]
Pacific Ocean: Japan	159,60 ± 72,92	(F)	0,67 ± 0,21	(F)	1,00 ± 0,58	(F)			[51]
	189,60	(M)	BDL	(M)	1,25	(M)			[51]
Muscle									
Mediterranean Sea: Italy	0,55 ± 0,63	0,09 - 2,21	0,54 ± 0,17	BDL - 0,74	0,69 ± 0,46	0,17 - 1,81	68,94 ± 45,80	11,21 - 139,60	[14]
Mediterranean Sea: Italy					1,00 ± 0,62	0,33 - 2,05			[15]
Mediterranean Sea: Cyprus	0,57	0,30 - 1,43	2,46	BDL - 5,53	0,48	BDL - 1,78			[20]
Mediterranean Sea: Italy							73,67 ± 56,72	12,57 - 154,24	[16]
Mediterranean Sea: Italy	0,36 ± 0,11								[22]
Mediterranean Sea: Italy	0,20 ± 0,20	0,06 - 0,78			0,40 ± 0,30	0,14 - 1,92			[23]
Mediterranena Sea: Italy	0,33 ± 0,14	BDL - 0,62	0,19 ± 0,14	BDL - 0,43	0,86 ± 1,00	0,14 - 3,14			[18]
Mediterranean Sea: Italy	0,81 ± 0,04		BDL						[24]
Mediterranean Sea: Spain	0,08 ± 0,06	0,02 - 0,24	0,20 ± 0,25	0,01 - 0,62	0,14 ± 0,22	0,01 - 0,82	40,95 ± 37,32	6,36 - 133,48	[25]
Atlantic Ocean: France	0,38 ± 0,24	0,02 - 0,86							[10]
Atlantic Ocean: Canary Is.	5,43 ± 1,33	0,71 - 59,43	10,76 ± 2,43	1,05 - 100,34	BDL		35,00 ± 6,52	7,38 - 320,10	[27]
Atlantic Ocean: US					0,76 ± 0,33	0,24 - 2,38			[31]

Table 1 (Continued)

Tissue - Location	Cadmium		Lead		Mercury		Arsenic		References
	Mean ± SD	Min - Max	Mean ± SD	Min - Max	Mean ± SD	Min - Max	Mean ± SD	Min - Max	
Pacific Ocean: Mexico	0,01	BDL - 1,45	0,01	BDL - 1,57					[42]
Pacific Ocean: Mexico					0,03	0,02 - 0,04			[43]
Pacific Ocean: Japan	0,29 ± 0,14	0,19 - 0,57			0,52 ± 0,24	0,24 - 0,90			[13]
Pacific Ocean: Japan							20,60 ± 13,10	5,19 - 45,50	[49]
Pacific Ocean: Japan	0,29 ± 0,14	(F)	0,10 ± 0,14	(F)	0,43 ± 0,19	(F)			[51]
	0,24	(M)	BDL	(M)	0,90	(M)			[51]
Blood									
Mediterranean Sea: Spain	0,12 ± 0,21	0,01 - 0,5	0,31 ± 0,31	0,02 - 0,64	0,02 ± 0,01	0,01 - 0,03	6,99 ± 9,28	2,07 - 23,53	[25]
Atlantic Ocean: US					Dead: 0,51 ± 0,20	0,20 - 1,58			[31]
					Alive: 0,15 ± 0,05	0,03 - 0,97			[31]
Atlantic Ocean: US					0,15 ± 0,01	0,03 - 0,41			[32]
Atlantic Ocean: US*					Red blood cell: 0,07 ± 0,02				[33]
Altantic Ocean: US					0,03 ± 0,04				[33]
Fat									
Mediterranean Sea: Italy	2,33 ± 0,52								[22]

Mediterranean Sea: Italy	0,09 ± 0,07	BDL - 0,21	0,11 ± 0,04	0,07 - 0,17	0,05 ± 0,04	BDL - 0,11			[18]
Pacific Ocean: Mexico	0,50	0,20 - 1,37	BDL						[42]
Pacific Ocean: Mexico					0,01	BDL - 0,03			[43]
Pacific Ocean: Japan	0,08 ± 0,05	(F)	BDL	(F)	0,007 ± 0,004	(F)			[51]
	0,11	(M)	BDL	(M)	0,05	(M)			[51]
Egg									
Mediterranean Sea: Turkey	Yolk: 0,36 ± 0,14		Yolk: 1,31 ± 0,23		Yolk: BDL				[21]
	Shell: 0,65 ± 0,13		Shell: 0,63 ± 0,16		Shell: BDL				[21]
Atlantic Ocean: US	Yolk: 0,10 ± 0,05		Yolk: 1,58 ± 1,06		Yolk: 0,95 ± 0,04				[28]
Atlantic Ocean: US			Egg content: 0,69	BDL - 2,74	Egg content: 0,12	BDL - 0,22	Egg content: 1,70	BDL - 14,01	[29]
Pacific Ocean: Japan*	Egg: 0,01 ± 0,004	0,008 - 0,02			Egg: 0,006 ± 0,002	0,004 - 0,007			[13]
Pacific Ocean: Japan	Yolk: 0,08 ± 0,03	0,05 - 0,11			Yolk: 0,03 ± 0,01	0,02 - 0,04			[13]
	Albumen: BDL				Albumen: 0,02 ± 0,01	0,004 - 0,03			[13]
	Shell: BDL				Shell: 0,01 ± 0,002	0,005 - 0,01			[13]
Pacific Ocean: Japan*			Egg: BDL						[51]
Pacific Ocean: Japan			Yolk: BDL						[51]
			Albumen: BDL						[51]
			Shell: BDL						[51]

Table 1 (Continued)

Tissue - Location	Cadmium		Lead		Mercury		Arsenic		References
	Mean ± SD	Min - Max	Mean ± SD	Min - Max	Mean ± SD	Min - Max	Mean ± SD	Min - Max	
Liver									
Mediterranean Sea: Turkey	21,21 ± 2,62	(Em)	23,84 ± 3,11	(Em)	35,83 ± 9,98	(Em)			[21]
Mediterranean Sea: Italy	7,40 ± 3,90		27,90 ± 6,50		377,40 ± 211,20		6,23 ± 2,80		[22]
Mediterranean Sea: Italy	37,30 ± 8,70	9,40 - 41,80	66,00 ± 42,70	23,80 - 178,00					[23]
Mediterranean Sea: Italy	24,03 ± 14,47	4,47 - 55,63	91,56 ± 24,09	58,75 - 145,31	1425,00 ± 686,56	656,25 - 2465,63			[18]
Mediterranean Sea: Italy	17,50 ± 2,44		103,00 ± 14,00		1232,00 ± 404,00		7,48 ± 1,04		[24]
Mediterranean Sea: Spain			30,23 ± 12,24	17,03 - 51,96					[25]
Atlantic Ocean: France	25,78 ± 20,59	7,25 - 65,31	78,13 ± 26,69	45,31 - 120,00					[10]
Atlantic Ocean: Canary Is.	46,94 ± 6,47	0,03 - 204,91	42,13 ± 5,31	0,28 - 285,56	1071,03 ± 133,75	1,09 - 6813,63			[27]
Atlantic Ocean: US									[31]
Pacific Ocean: Mexico	33,94	16,60 - 58,98	69,14	42,45 - 91,87	301,00	71,88 - 2042,00	1,29	0,11 - 8,60	[42]
Pacific Ocean: Mexico									[43]
Pacific Ocean: Japan	55,94 ± 25,53	20,22 - 105,94	87,19 ± 13,59	72,50 - 109,69	2028,13 ± 1203,13	706,25 - 3937,50	6,47 ± 1,44	4,50 - 9,19	[13]
Pacific Ocean: Australia**			71,25 ± 9,38	42,81 - 101,88					[48]
Pacific Ocean: Japan									[49]

Pacific Ocean: Japan	55,31 ± 27,91	(F)	87,81 ± 14,78	(F)	1887,50 ± 1253,13	(F)	6,81 ± 1,25	(F)	[51]
	60,63	(M)	83,13	(M)	2865,63	(M)	4,50	(M)	[51]
Kidney									
Mediterranean Sea: Italy	2,60 ± 0,70	1,70 - 4,70	97,00 ± 31,70	62,40 - 206,00					[23]
Mediterranean Sea: Italy	5,04 ± 2,25	1,50 - 8,83	96,26 ± 18,88	69,17 - 116,26	200,85 ± 157,51	72,51 - 508,37			[18]
Mediterranean Sea: Italy	5,56 ± 0,96		119,00 ± 8,00		764,00 ± 160,00		7,01 ± 1,32		[24]
Mediterranean Sea: Spain			32,64 ± 12,25	19,73 - 55,16					[25]
Atlantic Ocean: France	9,21 ± 1,92	7,33 - 11,79	98,34 ± 28,75	68,76 - 140,84					[10]
Atlantic Ocean: Canary Is.	19,17 ± 4,04	0,54 - 204,43	37,88 ± 4,46	0,29 - 160,55	158,18 ± 23,92	1,21 - 1138,17			[27]
Pacific Ocean: Mexico	4,35	1,39 - 8,23	32,47	2,68 - 130,00	237,00	94,59 - 660,00	6,00	2,37 - 9,97	[42]
Pacific Ocean: Japan	5,42 ± 0,83	4,13 - 6,50	107,51 ± 17,38	80,01 - 126,68	149,60 ± 146,26	47,50 - 458,37	6,54 ± 2,04	3,38 - 8,21	[13]
Pacific Ocean: Australia**			76,67 ± 3,75	69,59 - 88,76					[48]
Pacific Ocean: Japan	5,42 ± 0,92	(F)	105,84 ± 18,29	(F)	125,01 ± 152,51	(F)	6,25 ± 2,13	(F)	[51]
	5,38	(M)	118,34	(M)	47,50	(M)	8,21	(M)	[51]
Muscle									
Mediterranean Sea: Italy	1,50 ± 0,40		30,90 ± 8,00		60,90 ± 38,30		2,70 ± 1,20		[22]
Mediterranean Sea: Italy	2,70 ± 1,40	0,80 - 7,00	107,00 ± 26,10	76,40 - 177,00					[23]
Mediterranena Sea: Italy	2,81 ± 1,95	0,90 - 6,43	132,86 ± 23,10	94,29 - 167,15	151,91 ± 50,00	102,86 - 246,67			[18]

Table 1 (Continued)

Tissue - Location	Cadmium		Lead		Mercury		Arsenic		References
	Mean ± SD	Min - Max	Mean ± SD	Min - Max	Mean ± SD	Min - Max	Mean ± SD	Min - Max	
Mediterranean Sea: Italy	2,40 ± 0,24		105,00 ± 14,00		85,40 ± 9,90		1,35 ± 0,24		[24]
Mediterranean Sea: Spain			113,29 ± 267,33	25,42 - 1002,40					[25]
Atlantic Ocean: France	3,48 ± 2,14	1,62 - 10,62	93,34 ± 27,14	58,10 - 172,86					[10]
Atlantic Ocean: Canary Is.	13,57 ± 2,48	0,05 - 129,76	31,91 ± 4,57	0,24 - 154,15	41,19 ± 6,67	0,76 - 468,44			[27]
Pacific Ocean: Mexico	0,41	BDL - 3,44	31,11	0,63 - 100,00	77,44	52,48 - 97,12	0,84	BDL - 5,40	[42]
Pacific Ocean: Japan	3,95 ± 1,24	2,52 - 6,10	115,24 ± 18,10	92,86 - 147,62	95,72 ± 38,10	53,81 - 167,62	1,43 ± 0,57	0,62 - 2,14	[13]
Pacific Ocean: Japan	3,86 ± 1,33	(F)	119,05 ± 16,62	(F)	94,29 ± 41,48	(F)	1,33 ± 0,52	(F)	[51]
	4,52	(M)	92,86	(M)	105,72	(M)	2,05	(M)	[51]
Blood									
Mediterranean Sea: Spain			7,07 ± 2,85	3,60 - 11,95					[25]
Fat									
Mediterranean Sea: Italy	3,40 ± 1,70		68,20 ± 34,70		132,30 ± 121,00		8,40 ± 2,50		[22]
Mediterranean Sea: Italy	0,22 ± 0,17	0,12 - 0,60	76,48 ± 26,56	44,44 - 112,17	13,47 ± 5,80	8,87 - 26,36			[18]
Pacific Ocean: Mexico	0,69	0,53 - 1,15	12,66	0,53 - 44,76	1,33	BDL - 13,61	1,82	0,80 - 3,20	[42]

Pacific Ocean: Japan	0,13 ± 0,04 0,28	(F) (M)	113,59 ± 22,22 101,18	(F) (M)	11,73 ± 6,63 22,34	(F) (M)	0,14 ± 0,11 0,21	(F) (M)	[51] [51]
Egg									
Mediterranean Sea: Turkey	Yolk: 0,93 ± 0,10		Yolk: 57,21 ± 2,23		Yolk: 15,79 ± 3,62				[21]
	Shell: 5,29 ± 0,48		Shell: 5,00 ± 2,20		Shell: 17,75 ± 7,00				[21]
Atlantic Ocean: US	Yolk: 5,75 ± 1,08		Yolk: 77,41 ± 6,30		Yolk: 72,65 ± 12,38				[28]
Atlantic Ocean: US	Egg content: 5,75	2,36 - 51,10	Egg content : 92,83	60,62 - 139,30	Egg content: 77,66	44,56 - 210,80	Egg content: 1,96	0,82 - 17,62	[29]
Pacific Ocean: Japan*	Egg: 1,05 ± 0,20	0,77 - 1,31	Egg: 14,70 ± 1,44	13,20 - 16,50	Egg: 11,50 ± 1,29	10,50 - 13,70	Egg: 0,52 ± 0,26	0,30 - 0,90	[13]
Pacific Ocean: Japan	Yolk: 4,19 ± 0,19	3,95 - 4,49	Yolk: 91,85 ± 8,49	81,44 - 101,46	Yolk: 67,02 ± 5,82	61,14 - 75,56	Yolk: 2,43 ± 1,12	1,39 - 3,74	[13]
	Albumen: 4,82 ± 2,96	1,11 - 8,89	Album en: 21,85 ± 21,48	2,22 - 57,04	Albumen: 32,22 ± 14,08	18,52 - 48,15	Albumen: 6,30 ± 11,11	BDL - 26,30	[13]
	Shell: 13,59 ± 1,88	10,37 - 15,08	Shell: 5,29 ± 1,44	4,05 - 7,00	Shell: 25,86 ± 5,37	17,64 - 32,45	Shell: 1,66 ± 1,17	BDL - 2,93	[13]

Table 1 (Continued)

Tissue - Location	Nickel		Selenium		Chromium		References
	Mean ± SD	Min - Max	Mean ± SD	Min - Max	Mean ± SD	Min - Max	
Liver							
Mediterranean Sea: Italy			15,88 ± 7,40	2,12 - 27,44	1,05 ± 0,58	0,20 - 2,07	[14]
Mediterranean Sea: Italy			15,19 ± 2,66	12,50 - 19,09			[15]
Mediterranean Sea: Italy	4,38 ± 1,43						[22]
Mediterranean Sea: Italy			9,80 ± 5,30	1,00 - 24,90			[23]
Mediterranean Sea: Italy			11,06 ± 4,66	3,16 - 17,13			[18]
Mediterranean Sea: Spain			3,10 ± 1,45	1,46 - 6,59			[25]
Atlantic Ocean: Canary Is.	9,00 ± 1,09	0,03 - 43,16					[27]
Pacific Ocean: Mexico	0,35	BDL - 3,26					[42]
Pacific Ocean: Australia**			6,91 ± 0,63	4,44 - 8,44			[48]
Pacific Ocean: Japan	BDL	(F)					[51]
	BDL	(M)					
Kidney							
Mediterranean Sea: Italy			10,33 ± 3,25	5,73 - 15,57	1,57 ± 2,05	0,20 - 6,80	[14]
Mediterranean Sea: Italy			15,50 ± 9,10	4,50 - 41,80			[23]
Mediterranean Sea: Italy			9,17 ± 2,75	2,75 - 11,58			[18]
Mediterranean Sea: Spain			5,66 ± 5,77	0,56 - 17,72			[25]
Atlantic Ocean: Canary Is.	24,21 ± 4,58	0,17 - 200,56					[27]
Pacific Ocean: Mexico	0,04	BDL - 3,38					[42]
Pacific Ocean: Australia**			6,33 ± 0,58	5,33 - 7,42			[48]
Pacific Ocean: Japan	0,92 ± 0,38	(F)					[51]
	0,21	(M)					
Muscle							
Mediterranean Sea: Italy			10,81 ± 2,89	6,51 - 15,45	1,43 ± 0,87	0,30 - 2,89	[14]
Mediterranean Sea: Italy			11,10 ± 3,19	5,67 - 15,43			[15]
Mediterranean Sea: Italy	2,76 ± 0,60						[22]
Mediterranean Sea: Italy			11,20 ± 4,90	4,00 - 24,10			[23]

Mediterranena Sea: Italy			7,86 ± 2,76	3,24 - 10,95			[18]
Mediterranean Sea: Spain			3,27 ± 1,65	0,87 - 6,67			[25]
Atlantic Ocean: Canary Is.	8,29 ± 1,38	0,14 - 62,62					[27]
Pacific Ocean: Mexico	0,01	BDL - 0,65					[42]
Pacific Ocean: Japan	0,38 ± 0,14	(F)					[51]
	0,29	(M)					[51]
Blood							
Mediterranean Sea: Spain			2,80 ± 1,29	0,92 - 4,48			[25]
Fat							
Mediterranean Sea: Italy	18,77 ± 4,40						[22]
Mediterranean Sea: Italy			2,17 ± 0,74	1,42 - 3,58			[18]
Pacific Ocean: Mexico	0,17	BDL - 1,63					[42]
Pacific Ocean: Japan	BDL	(F)					[51]
	BDL	(M)					[51]
Egg							
Atlantic Ocean: US	Yolk: 0,63 ± 0,24				Yolk: 1,28 ± 0,19		[28]
Atlantic Ocean: US	Egg content: 0,56	BDL - 1,07	Egg content: 5,46	1,10 - 15,84	Egg content: 1,20	BDL - 11,53	[29]
Pacific Ocean: Japan*	Egg: BDL						[51]
Pacific Ocean: Japan	Yolk: BDL						[51]
	Albumen: BDL						[51]
	Shell: BDL						[51]

BDL = bellow detection limit; * = $\mu g\ g^{-1}$ wet weight; ** = Mean ± SEM; F = female; M = male; Egg = whole egg.

Concerning essential metals, a large variety of these have been measured in *Caretta caretta*. The most frequently analyzed are zinc, copper and iron (in that order). The tissues mainly analyzed are liver, kidney and muscle (see Table 1). The highest copper and iron levels were found in liver [10, 13, 18, 21, 22, 23, 24, 27, 42, 48, 49, 51]. Muscle and fat tissues contained considerable zinc levels, although zinc also showed a high affinity for calcified tissues such as bone and carapace [18, 22, 25, 27, 42, 43, 51]. Regarding fat tissue, Sakai et al. [51] suggested that the accumulation of zinc in this tissue could be associated with pigment proteins, which probably play an important role in zinc storage in the body. Among the analyzed metals in the turtle eggs, zinc and iron exhibited the highest levels in whole egg, while mercury, cadmium, lead, arsenic, copper, manganese, nickel, selenium and chromium showed the lowest ones [13, 21, 28, 29, 51]. Concerning the excretion of metals via eggs, Sakai et al. [13] reported that essential metals can be freely transferred from mother to eggs, whereas toxic metals, such as mercury and cadmium, are poorly transferred.

Metals in eggs were mainly present in the yolk: higher levels of cadmium, mercury, zinc and iron were found in yolk than in albumen [13], whereas eggshell contained the highest levels of iron and copper [13, 21].

GREEN SEA TURTLE (*CHELONIA MYDAS*)

Chelonia mydas is the second most studied marine turtle with respect to levels of heavy metals and essential elements. These levels (Cd, Pb, Hg, As, Cu, Zn, Fe, Mn, Ni, Se and Cr) measured in samples of this species (liver, kidney, muscle, blood, fat, stomach content and egg) are shown in Table 2.

A large variety of metals have been measured in green turtles, but the most frequently analyzed are again cadmium, lead and mercury (in that order), probably due to the importance of these heavy metals in the marine ecosystems because of the features previously described (see section "Loggerhead Sea Turtle"). The tissues mainly analyzed are liver, kidney and muscle (see Table 2). Generally, kidney showed the highest concentrations of cadmium [11, 19, 24, 38, 42, 43, 44, 48, 49, 50, 51, 56, 58, 59, 61] and lead [11, 20, 42, 44, 50, 56, 61]. But when calcified tissues were included in the studies, normally the highest lead levels were found in bones and carapaces (2.35 and 2.70 $\mu g\ g^{-1}$ wet weight, respectively) [51]. The highest mercury levels were found in liver [11, 20, 50, 51, 56, 61] or kidney [43, 48], while arsenic was mainly retained in muscle [49, 56, 58, 59, 61] and blood [61], and chromium in kidney [11] and muscle [56].

Table 2. Heavy metal and essential element levels (µg g^{-1} dry weight) in tissues and other samples of *Chelonia mydas* from different locations

Tissue - Location	Cadmium		Lead		Mercury		Arsenic		References
	Mean ± SD	Min - Max	Mean ± SD	Min - Max	Mean ± SD	Min - Max	Mean ± SD	Min - Max	
Liver									
Mediterranean Sea: Cyprus	5,89	2,53 - 10,73	BDL	BDL - 1,84	0,55	0,27 - 1,37			[20]
Mediterranean Sea: Italy	13,31 ± 9,44	6,94 - 28,84							[19]
Atlantic Ocean: Costa Rica	10,60 ± 1,10		0,07 ± 0,01						[24]
Pacific Ocean: Hawaiian Is.	27,06 ± 26,69	1,22 - 81,25						BDL - 20,00	[38]
Pacific Ocean: Mexico	3,30	BDL - 102,00	BDL						[42]
Pacific Ocean: Mexico					0,09	0,03 - 0,17			[43]
Pacific Ocean: Mexico	16,92	BDL - 72,57	0,00	BDL - 0,07					[44]
Pacific Ocean: Australia**	39,06 ± 6,25	7,81 - 177,81			0,06 ± 0,01	0,00 - 0,16	0,81 ± 0,13	0,13 - 2,31	[48]
Pacific Ocean: Japan							1,76 ± 0,95	0,44 - 5,34	[49]
Pacific Ocean: Japan	17,44 ± 12,66	0,94 - 58,13	BDL		0,91 ± 0,50	0,16 - 2,00			[50]
Pacific Ocean: Japan	25,00			BDL - 0,38	0,59				[51]
Pacific Ocean: Japan	18,20 ± 9,70	3,58 - 38,30	0,51 ± 0,41	0,06 - 1,55	0,42 ± 019	0,23 - 0,88			[11]
Pacific Ocean: Japan	28,00	10,03 - 67,50	0,44	0,09 - 1,13					[52]
Pacific Ocean: Japan							3,65 ± 2,03		[54]

Table 2 (Continued)

Tissue - Location	Cadmium		Lead		Mercury		Arsenic		References
	Mean ± SD	Min – Max	Mean ± SD	Min - Max	Mean ± SD	Min - Max	Mean ± SD	Min - Max	
Pacific Ocean: Japan							3,75 ± 1,88	2,50 - 7,19	[55]
Pacific Ocean: China	1,45 ± 0,61	(A)	0,83 ± 0,09	(A)	0,13 ± 0,08	(A)	19,57 ± 1,98	(A)	[56]
	1,10 ± 0,10	(J)	0,15 ± 0,04	(J)	0,78 ± 0,19	(J)	4,65 ± 3,96	(J)	[56]
Pacific Ocean: Japan							5,30	0,90 - 9,70	[58]
Pacific Ocean: Japan							4,90 ± 3,30		[59]
Pacific Ocean: Australia	42,31 ± 7,50	13,53 - 101,44	0,28 ± 0,06	BDL - 0,88	0,59 ± 0,13	BDL - 1,69	9,97 ± 2,47	1,13 - 30,34	[61]
Kidney									
Mediterranean Sea: Cyprus	3,46		1,81		BDL				[20]
Mediterranean Sea: Italy	21,09 ± 9,29	9,00 - 31,21							[19]
Atlantic Ocean: Costa Rica	39,20 ± 3,10		0,04 ± 0,002						[24]
Pacific Ocean: Hawaiian Is.	108,32 ± 87,89	19,67 - 292,52						BDL - 28,34	[38]
Pacific Ocean: Mexico	121,00	6,09 - 653,00	0,01	BDL - 0,36					[42]
Pacific Ocean: Mexico					0,09	0,003 - 0,31			[43]
Pacific Ocean: Mexico	110,00	65,08 - 653,00	0,05	BDL - 1,74					[44]
Pacific Ocean: Australia**	63,76 ± 10,42	7,08 - 316,28			0,08 ± 0,02	0,00 - 0,21	0,79 ± 0,21	0,00 - 2,88	[48]
Pacific Ocean: Japan							5,72 ± 2,99	0,15 - 9,99	[49]

Pacific Ocean: Japan	160,43 ± 88,76	30,46 - 336,28	0,75 ± 0,29	0,21 - 1,17	0,54 ± 0,33	0,13 - 1,04			[50]
Pacific Ocean: Japan	171,89			BDL - 0,58	0,21				[51]
Pacific Ocean: Japan	142,00 ± 64,00	20,30 - 285,00	0,81 ± 0,56	0,16 - 2,38	0,30 ± 0,14	0,11 - 0,55			[11]
Pacific Ocean: China	2,49 ± 1,75	(J)	0,31 ± 0,19	(J)	0,34 ± 0,04	(J)	6,97 ± 0,05	(J)	[56]
Pacific Ocean: Japan							16,50	4,60 - 44,30	[58]
Pacific Ocean: Japan							17,00 ± 14,00		[59]
Pacific Ocean: Australia	191,60 ± 37,75	52,92 - 421,91	0,38 ± 0,04	BDL - 0,50	0,25 ± 0,08	BDL - 0,83	10,29 ± 3,71	0,50 - 38,63	[61]
Muscle									
Mediterranean Sea: Cyprus	0,37	0,12 - 0,78	BDL	BDL - 2,45	0,09	BDL - 0,37			[20]
Pacific Ocean: Mexico	0,01	BDL - 39,24	0,01	BDL - 1,23					[42]
Pacific Ocean: Mexico					0,02	0,003 - 0,06			[43]
Pacific Ocean: Japan							24,1 ± 13,1	2,58 - 74,9	[49]
Pacific Ocean: Japan	0,24 ± 0,38	0,05 - 2,57	BDL		0,10 ± 0,14	0,005 - 0,57			[50]
Pacific Ocean: Japan	0,10		BDL		0,02				[51]
Pacific Ocean: Japan	0,24 ± 0,17	0,07 - 0,64	0,09 ± 0,05	0,03 - 0,21	0,04 ± 0,07	BDL - 0,23			[11]
Pacific Ocean: China	0,17 ± 0,06	(A)	0,26 ± 0,11	(A)	0,05 ± 0,04	(A)	14,61 ± 7,47	(A)	[56]
		BDL - 0,10 (J)	0,08 ± 0,10	(J)	0,43 ± 0,22	(J)	14,45 ± 4,88	(J)	[56]
Pacific Ocean: Japan							61,60	11,20 - 165,00	[58]

Table 2 (Continued)

Tissue - Location	Cadmium		Lead		Mercury		Arsenic		References
	Mean ± SD	Min - Max	Mean ± SD	Min - Max	Mean ± SD	Min - Max	Mean ± SD	Min - Max	
Pacific Ocean: Japan							69,00 ± 52,00		[59]
Pacific Ocean: Australia*	0,08 ± 0,01	0,05 - 0,19	BDL		0,03 ± 0,004	BDL - 0,05	5,25 ± 1,47	0,32 - 14,01	[61]
Pacific Ocean: Australia	0,38 ± 0,05	0,24 - 0,90	BDL		0,14 ± 0,02	BDL - 0,24	25,00 ± 7,00	1,52 - 66,72	[61]
Blood									
Pacific Ocean: Australia	0,20 ± 0,05	0,005 - 0,61	0,10 ± 0,03	0,03 - 0,31	0,01 ± 0,003	0,001 - 0,04	22,24 ± 7,19	0,46 - 101,93	[61]
Fat									
Atlantic Ocean: Costa Rica	0,11 ± 0,02		0,06 ± 0,03						[24]
Pacific Ocean: Mexico	0,002	BDL - 1,47	0,03	BDL - 1,11					[42]
Pacific Ocean: Mexico					0,005	BDL - 0,01			[43]
Pacific Ocean: Japan	0,07		BDL		0,004				[51]
Pacific Ocean: China	BDL	(J)	0,09 ± 0,03	(J)	0,005 ± 0,003	(J)	1,27 ± 0,20	(J)	[56]
Stomach content									
Pacific Ocean: Mexico		2,22 - 11,11		BDL - 0,12					[44]
Pacific Ocean: Japan	0,37 ± 0,40	0,05 - 1,03	0,20 ± 0,05	0,13 - 0,28	0,03 ± 0,01	BDL - 0,05			[11]
Pacific Ocean: China	175,50 ± 248,00	(J)	0,60 ± 0,70	(J)	0,3 ± 0,06	(J)	4,06 ± 2,08	(J)	[56]
Pacific Ocean: Japan							6,10		[59]

Egg									
Pacific Ocean: Hawaiian Is.	Shell: 0,49						Shell: BDL		[38]
Pacific Ocean: Japan*	Egg: BDL		Egg: BDL		Egg: 0,001				[51]
Pacific Ocean: Japan	Yolk: BDL		Yolk: BDL		Yolk: 0,01				[51]
	Albumen: BDL		Albumen: BDL		Albumen: 0,002				[51]
	Shell: BDL		Shell: BDL		Shell: 0,002				[51]
Pacific Ocean: Hong Kong	Yolk: BDL		Yolk: 0,13 ± 0,03	0,08 - 0,37	Yolk: 0,005 ± 0,0003	0,003 - 0,005	Yolk: 6,68 ± 0,99	3,74 - 13,35	[57]
	Albumen: BDL		Albumen: 0,19 ± 0,04	BDL - 0,37	Albumen: 0,004 ± 0,001	0,004 - 0,07	Albumen: 6,30 ± 0,74	3,70 - 12,59	[57]
	Shell: 0,05 ± 0,02	BDL - 0,39	Shell: 0,27 ± 0,05	0,07 - 0,68	Shell: 0,002 ± 0,0005	0,0002 - 0,005	Shell: 0,54 ± 0,05	0,32 - 0,90	[57]
Pacific and Indian Ocean: Malaysia*	Egg:	BDL - 0,03	Egg:	BDL - 0,12	Egg: BDL		Egg:	BDL - 0,35	[60]
Liver									
Mediterranean Sea: Italy	102,34 ± 51,31	57,69 - 184,31	107,91 ± 41,75	60,22 - 168,38					[19]
Atlantic Ocean: Costa Rica	100,00 ± 11,00		82,50 ± 4,90		2482,00 ± 286,00		8,92 ± 0,93		[24]
Pacific Ocean: Hawaiian Is.	273,77 ± 192,69	4,06 - 590,63	95,55 ± 31,19	47,19 - 143,13	3661,01 ± 2410,57	290,00 - 7656,25	4,99 ± 2,03	0,47 - 8,72	[38]
Pacific Ocean: Mexico	60,04	6,79 - 133,00	62,91	1,32 - 166,00	14,35	BDL - 1765,00			[42]
Pacific Ocean: Mexico									[43]

Table 2 (Continued)

Tissue - Location	Cadmium		Lead		Mercury		Arsenic		References
	Mean ± SD	Min - Max	Mean ± SD	Min - Max	Mean ± SD	Min - Max	Mean ± SD	Min - Max	
Pacific Ocean: Mexico	76,52	6,79 - 128,00	90,95	41,81 - 109,00	350,00	BDL - 671,00	0,24	BDL - 5,31	[44]
Pacific Ocean: Australia**			124,06 ± 9,38	52,19 - 289,69					[48]
Pacific Ocean: Japan	156,88 ± 98,75	13,34 - 353,13	94,69 ± 22,28	54,69 - 147,19	1440,63 ± 812,50	102,19 - 3968,75	5,81 ± 2,63	2,19 - 16,91	[50]
Pacific Ocean: Japan	34,75		182,19		423,44		5,91		[51]
Pacific Ocean: Japan	139,00 ± 86,00	36,40 - 340,00	87,20 ± 30,60	42,60 - 166,00			4,74 ± 2,06	2,28 - 11,10	[11]
Pacific Ocean: Japan	155,94	57,50 - 406,25	104,06	69,38 - 176,88					[52]
Pacific Ocean: China	9,17 ± 2,89	(A)	211,60 ± 30,57	(A)			10,50 ± 4,03	(A)	[56]
	133,00 ± 148,60	(J)	128,90 ± 63,92	(J)			16,27 ± 13,81	(J)	[56]
Pacific Ocean: Australia	283,22 ± 36,38	119,44 - 479,38	113,44 ± 6,75	64,78 - 147,63					[61]
Kidney									
Mediterranean Sea: Italy	34,17 ± 17,50	20,00 - 59,59	109,97 ± 43,80	60,71 - 160,35					[19]
Atlantic Ocean: Costa Rica	8,34 ± 1,75		77,40 ± 1,90		300,00 ± 25,00		5,75 ± 0,28		[24]
Pacific Ocean: Hawaiian Is.	15,14 ± 11,41	4,58 - 43,75	92,89 ± 31,33	52,09 - 158,76	180,36 ± 228,22	36,67 - 745,89	4,00 ± 1,32	2,00 - 5,79	[38]
Pacific Ocean: Mexico	5,67	1,59 - 20,36	128,00	1,59 - 330,00	44,09	BDL - 516,00	0,31	BDL - 8,12	[42]
Pacific Ocean: Mexico	5,83	1,98 - 11,60	189,00	102,00 - 281,00	93,16	BDL - 547,00	1,51	BDL - 7,73	[44]

Pacific Ocean: Australia**			88,76 ± 2,92	64,17 - 132,51					[48]
Pacific Ocean: Japan	8,96 ± 3,58	3,96 - 16,42	123,34 ± 30,79	72,92 - 186,26	95,01 ± 60,84	47,50 - 246,69	5,04 ± 1,33	3,00 - 9,58	[50]
Pacific Ocean: Japan	6,33		141,68		57,92		5,50		[51]
Pacific Ocean: Japan	8,27 ± 4,06	2,41 - 18,90	169,00 ± 61,00	85,50 - 352,00			5,60 ± 1,38	2,78 - 8,36	[11]
Pacific Ocean: China	15,20 ± 7,22	(J)	143,10 ± 12,42	(J)			11,59 ± 4,79	(J)	[56]
Pacific Ocean: Australia	10,71 ± 1,00	5,83 - 18,96	119,80 ± 9,29	74,59 - 200,81					[61]
Muscle									
Pacific Ocean: Mexico	0,03	BDL - 13,76	38,26	10,44 - 134,00	20,99	BDL - 225,00	0,003	BDL - 7,75	[42]
Pacific Ocean: Japan	1,67 ± 1,95	0,38 - 12,62	41,86 ± 26,24	15,71 - 159,05	25,14 ± 21,91	4,62 - 140,96	0,52 ± 0,29	0,24 - 1,71	[50]
Pacific Ocean: Japan	1,24		46,05		54,24		1,29		[51]
Pacific Ocean: Japan	0,88 ± 0,42	0,54 - 2,06	47,70 ± 18,60	24,40 - 69,20			0,46 ± 0,10	0,32 - 0,61	[11]
Pacific Ocean: China	1,56 ± 0,11	(A)	238,70 ± 101,60	(A)			5,04 ± 3,14	(A)	[56]
	3,74 ± 1,64	(J)	147,70 ± 12,62	(J)			1,28 ± 0,52	(J)	[56]
Pacific Ocean: Australia	1,62 ± 0,33	0,71 - 5,05	58,86 ± 13,48	28,33 - 185,43					[61]
Blood									
Pacific Ocean: Australia	5,20 ± 0,50	2,01 - 8,26	40,39 ± 34,07	17,95 - 62,53					[61]
Fat									
Atlantic Ocean: Costa Rica	0,45 ± 0,09		62,10 ± 8,40		45,20 ± 6,10		0,83 ± 0,13		[24]

Table 2 (Continued)

Tissue - Location	Cadmium Mean ± SD	Cadmium Min - Max	Lead Mean ± SD	Lead Min - Max	Mercury Mean ± SD	Mercury Min - Max	Arsenic Mean ± SD	Arsenic Min - Max	References
Pacific Ocean: Japan	0,37		60,64		22,34		0,24		[51]
Pacific Ocean: Mexico	0,01	BDL - 9,48	49,82	19,51 - 163,00	2,63	BDL - 154,00	0,003	BDL - 0,79	[42]
Pacific Ocean: China	0,99 ± 0,38	(J)	105,00 ± 17,48	(J)			0,19 ± 0,06	(J)	[56]
Stomach content									
Pacific Ocean: Mexico		0,02 - 5,70		8,82 - 60,96		31,81 - 317,00		4,02 - 13,08	[44]
Pacific Ocean: Japan	9,41 ± 5,88	1,51 - 16,90	6,42 ± 2,37	3,79 - 9,53			7,63 ± 6,03	2,45 - 18,90	[11]
Pacific Ocean: China	10,42 ± 2,46	(J)	254,60 ± 121,40	(J)			28,82 ± 32,58	(J)	[56]
Egg									
Pacific Ocean: Hawaiian Is.	Shell: 34,89		Shell: 15,25		Shell: 34,40		Shell: 0,76		[38]
Pacific Ocean: Japan*	Egg: 0,78		Egg: 20,30		Egg: 10,90		Egg: 0,38		[51]
Pacific Ocean: Japan	Yolk: 1,68		Yolk: 126,02		Yolk: 65,15		Yolk: 1,52		[51]
	Albumen: 5,93		Albumen: 47,78		Albumen: 39,63		Albumen: BDL		[51]
	Shell: 11,57		Shell: 1,37		Shell: 4,83		Shell: 3,25		[51]
Pacific Ocean: Hong Kong	Yolk: 0,91 ± 0,11	0,45 - 2,06	Yollk: 120,15 ± 9,61	66,75 - 181,56	Yolk: 120,15 ± 7,74	77,43 - 176,22	Yolk: 0,72 ± 0,11	0,27 - 1,42	[57]

	Albumen: 2,22 ± 0,37	1,11 - 6,67	Albumen: 11,11 ± 2,22	1,48 - 27,78	Albumen: 140,75 ± 27,78	17,04 - 348,18	Albumen: 1,85 ± 0,37	0,74 - 5,19	[57]
	Shell: 3,17 ± 0,81	0,59 - 10,25	Shell: 2,93 ± 0,56	0,76 - 6,34	Shell: 5,61 ± 0,95	0,37 - 13,91	Shell: 0,66 ± 0,17	0,07 - 1,88	[57]
Pacific and Indian Ocean: Malaysia*	Egg:	0,06 - 1,07	Egg:	1,33 - 39,48					[60]

Liver							
Pacific Ocean: Hawaiian Is.		BDL	3,10 ± 2,85	0,43 - 10,59		BDL - 1,56	[38]
Pacific Ocean: Mexico	0,01	BDL - 7,40					[42]
Pacific Ocean: Mexico	0,00	BDL - 30,88					[44]
Pacific Ocean: Australia**			3,69 ± 0,50	0,22 - 8,38			[48]
Pacific Ocean: Japan		0,19 - 0,97					[50]
Pacific Ocean: Japan	0,22						[51]
Pacific Ocean: Japan			5,10 ± 2,30	2,00 - 9,90	2,20 ± 0,60	1,50 - 3,50	[11]
Pacific Ocean: Japan			10,31	2,72 - 23,44			[52]
Pacific Ocean: China	0,71 ± 0,03	(A)	12,43 ± 0,78	(A)	0,85 ± 0,21	(A) BDL - 1,11	[56]
	0,27 ± 0,25	(J)	25,65 ± 28,60	(J)		(J)	[56]
Pacific Ocean: Australia			12,34 ± 3,38	1,63 - 32,50			[61]
Kidney							
Pacific Ocean: Hawaiian Is.		BDL - 4,17	1,93 ± 1,78	0,66 - 6,58		BDL - 1,67	[38]
Pacific Ocean: Mexico	1,15	BDL - 26,43					[42]
Pacific Ocean: Mexico	3,19	1,19 - 25,13					[44]
Pacific Ocean: Australia**			2,46 ± 0,38	0,38 - 7,71			[48]
Pacific Ocean: Japan	2,58 ± 1,54	0,50 - 5,54					[50]
Pacific Ocean: Japan	2,13						[51]
Pacific Ocean: Japan			5,30 ± 2,40	2,10 - 11,00	2,20 ± 0,70	1,40 - 3,80	[11]

Table 2 (Continued)

Tissue - Location	Nickel Mean ± SD	Min - Max	Selenium Mean ± SD	Min - Max	Chromium Mean ± SD	Min - Max	*References*
Pacific Ocean: China	0,20 ± 0,13	(J)	5,89 ± 1,26	(J)	1,07 ± 0,40	(J)	[56]
Pacific Ocean: Australia			6,96 ± 2,04	1,21 - 21,29			[61]
Muscle							
Pacific Ocean: Mexico	0,03	BDL - 4,00					[42]
Pacific Ocean: Japan	BDL						[50]
Pacific Ocean: Japan		BDL - 0,29					[51]
Pacific Ocean: Japan			3,20 ± 1,00	1,00 - 4,80	1,40 ± 0,20	1,20 - 1,80	[11]
Pacific Ocean: China	1,07 ± 0,97	(A)	17,02 ± 20,17	(A)	2,18 ± 2,55	(A)	[56]
	0,22 ± 0,21	(J)	4,49 ± 1,09	(J)	2,71 ± 2,85	(J)	[56]
Pacific Ocean: Australia			5,57 ± 1,48	0,90 - 17,14			[61]
Blood							
Pacific Ocean: Australia			12,50 ± 3,21	0,35 - 46,34			[61]
Fat							
Pacific Ocean: Mexico	0,02	BDL - 13,42					[42]
Pacific Ocean: Japan		BDL - 0,07					[51]
Pacific Ocean: China	0,15 ± 0,05	(J)	0,72 ± 0,11	(J)	0,91 ± 0,24	(J)	[56]
Stomach content							
Pacific Ocean: Mexico		0,90 - 9,67					[44]
Pacific Ocean: Japan			1,00 ± 0,50	0,37 - 1,80	1,30 ± 0,80	0,53 - 2,80	[11]
						BDL - 1,19	
Pacific Ocean: China	0,39 ± 0,43	(J)	8,07 ± 1,78	(J)		(J)	[56]
Egg							
Pacific Ocean: Hawaiian Is.	Shell: BDL				Shell: 0,98		[38]
Pacific Ocean: Japan*	Egg: BDL						[51]
Pacific Ocean: Japan	Yolk: BDL						[51]

	Albumen: BDL						[51]
	Shell: BDL						[51]
Pacific Ocean: Hong Kong	Yolk: 0,51 ± 0,08	0,19 - 1,01	Yolk: 9,35 ± 1,60	3,47 - 20,29	Yolk: 2,51 ± 0,32	0,99 - 4,54	[57]
	Albumen: 0,74 ± 0,37	0,02 - 2,59	Albumen: 10,00 ± 2,22	2,96 - 27,04	Albumen: 1,85 ± 0,74	0,37 - 5,19	[57]
	Shell: 29,28 ± 3,66	6,34 - 51,24	Shell: 6,10 ± 2,03	1,10 - 26,84	Shell: 1,15 ± 0,24	BDL - 2,93	[57]
Pacific and Indian Ocean: Malaysia*			Egg:	0,05 - 0,84			[60]

BDL = bellow detection limit; * = $\mu g\ g^{-1}$ wet weight; ** = Mean ± SEM; J = juvenile; A = adult; Egg = whole egg.

In the case of mercury and cadmium, these levels followed the patterns of metal accumulation that were described for other marine vertebrates [65, 66]. As we observed in loggerhead sea turtles, mercury levels tended to be highest in liver tissue and cadmium levels tended to be highest in kidney tissue.

Several studies have emphasized that food is the main source of exposure to heavy metals as cadmium, and therefore carnivorous animals (as loggerhead sea turtle) often show metal levels comparatively higher than herbivorous (as green sea turtle) [17, 67]. However, in this review we observe similar ranges in cadmium levels in loggerhead and green sea turtles from Mediterranean Sea, Atlantic Ocean and Pacific Ocean (see Table 1 and 2). Both species showed the highest cadmium levels in specimens from Pacific Ocean, where a major cadmium enrichment can be present.

Concerning essential metals, a large variety of these have been measured in tissues of *Chelonia mydas*, although the most frequently analyzed are zinc, copper and manganese (in that order). The tissues mainly used on this matter are liver, kidney and muscle (see Table 2). The highest copper levels were found in liver [11, 19, 24, 38, 42, 44, 50, 51, 52, 56, 61], while bone and carapace contained considerable levels of zinc (176.50 and 319.50 $\mu g\ g^{-1}$ wet weight, respectively) [51]. The highest manganese levels were found in liver [24, 38, 50, 51, 56] and stomach content [11, 56]. If we observe the levels of these elements in whole eggs, we can see that zinc and iron exhibited the highest levels, while mercury, cadmium, lead, arsenic, copper, manganese, nickel and selenium showed the lowest ones [51, 60]. Metals in eggs were mainly accumulated in the yolk. Levels of lead, mercury, arsenic, copper, zinc, iron, manganese, nickel, selenium and chromium were higher in yolk than in albumen [51, 57], whereas eggshell contained the highest levels of copper, iron and nickel [38, 51, 57].

LEATHERBACK SEA TURTLE (*DERMOCHELYS CORIACEA*)

The available data about levels of metals and other inorganic elements (Cd, Pb, Hg, As, Cu, Zn, Fe, Mn, Ni, Se and Cr) in samples of leatherback sea turtle are shown in Table 3. Liver, kidney, muscle, blood, fat and eggs are the samples mainly analyzed. The elements most frequently analyzed in samples of this species are cadmium, copper and zinc, followed by lead, mercury and selenium. Referring to the eleven elements included in this chapter, there is no information about iron, manganese and chromium, which were not measured by any of the consulted studies.

Table 3. Heavy metal and essential element levels (µg g^{-1} dry weight) in tissues and other samples of *Dermochelys coriacea* from different locations

Tissue - Location	Cadmium		Lead		Mercury		Arsenic		References
	Mean ± SD	Min - Max	Mean ± SD	Min - Max	Mean ± SD	Min - Max	Mean ± SD	Min - Max	
Liver									
Atlantic Ocean: United Kingdom	0,22 ± 0,02		0,12 ± 0,02		0,39 ± 0,04		0,58 ± 0,11		[26]
Atlantic Ocean: France	21,38 ± 11,44	1,88 - 45,94							[10]
Kidney									
Atlantic Ocean: France	126,26 ± 117,09	35,29 - 258,35							[10]
Muscle									
Atlantic Ocean: United Kingdom	0,06 ± 0,01		0,31 ± 0,03		0,12 ± 0,06		0,21 ± 0,07		[26]
Atlantic Ocean: France	1,67 ± 0,95	0,76 - 4,76							[10]
Blood									
Atlantic Ocean: French Guiana	0,41 ± 0,15		0,92 ± 0,26		0,05 ± 0,02				[35]
Atlantic Ocean: US	0,36 ± 0,10	0,26 - 0,61	0,66	0,36 - 1,02	0,05 ± 0,00	0,05 - 0,10			[37]
Fat									
Atlantic Ocean: United Kingdom	BDL		0,04 ± 0,03		0,11 ± 0,02		1,28 ± 0,18		[26]

Table 3 (Continued)

Tissue - Location	Cadmium		Lead		Mercury		Arsenic		References
	Mean ± SD	Min - Max	Mean ± SD	Min - Max	Mean ± SD	Min - Max	Mean ± SD	Min - Max	
Egg									
Atlantic Ocean: French Guiana	Egg content:	0,1 ± 0,005	Egg content:	0,21 ± 0,005	Egg content:	0,05 ± 0,02			[35]
Pacific Ocean: Mexico	Shell:	0,90 ± 0,61	Shell:	11,60 ± 26,00					[39]
Liver									
Atlantic Ocean: United Kingdom	0,15 ± 0,04		2,62 ± 0,15		2,13 ± 0,16		1,41 ± 0,02		[26]
Atlantic Ocean: France	26,91 ± 13,75	3,28 - 61,56	91,25 ± 12,81	68,44 - 114,06					[10]
Kidney									
Atlantic Ocean: France	11,17 ± 1,38	9,83 - 12,58	107,09 ± 32,09	77,09 - 140,84					[10]
Muscle									
Atlantic Ocean: United Kingdom	0,26 ± 0,05		1,89 ± 0,10		1,62 ± 0,21		3,61 ± 0,48		[26]
Atlantic Ocean: France	4,52 ± 2,33	1,90 - 12,19	123,34 ± 28,10	87,14 - 177,62					[10]
Blood									
Atlantic Ocean: French Guiana	6,83 ± 1,43		56,61 ± 1,43				50,90 ± 0,26		[35]
Atlantic Ocean: US	5,05 ± 0,71	3,52 - 6,07					38,51 ± 14,43	20,15 - 68,49	[37]

Fat									
Atlantic Ocean: United Kingdom	0,06 ± 0,02		0,08 ± 0,01		0,07 ± 0,02		BDL		[26]
Egg									
Atlantic Ocean: French Guiana	Egg content:	3,28 ± 0,52	Egg content:	73,75 ± 11,61			Egg content:	7,50 ± 1,98	[35]
Pacific Ocean: Mexico	Shell:	8,90 ± 1,26	Shell:	11,90 ± 10,00	Shell:	7,90 ± 5,11			[39]

BDL = bellow detection limit.

The highest cadmium levels were found in kidney [10], whereas blood exhibited the highest lead concentrations [35, 37], although no data are available on lead concentrations in kidney. The majority of the mercury burden was found in liver [26].

In spite of the limited available data, cadmium and mercury followed the expected patterns of metal accumulation [17]: cadmium tends to accumulate mainly in kidney and mercury in liver. In the case of lead, a higher accumulation in calcified tissues would be expected, but no data on bones or carapaces of leatherback sea turtle are available.

The diet of leatherback sea turtles consists almost exclusively on jellyfish [67]. Caurant et al. [10] suggested that the ingestion of great quantities of jellyfish could imply a high exposure to cadmium for the leatherback sea turtles than for other fish-eating or herbivorous vertebrates [10, 17], although this prey has been very poorly studied. However, we observe that even the herbivorous green turtles can accumulate similar or higher cadmium levels than leatherback sea turtles in different areas of the world (range of cadmium levels in kidney of green turtles from Asia-Pacific area and Indian Ocean: 7.08 – 421.91 $\mu g\ g^{-1}$ dry weight, see Table 2). These changes in cadmium accumulation can be related with differences among the diet of sea turtles depending on different stages of development and different distribution areas. For these reasons, it is important to study sea turtles throughout their growth and in the different regions of the world where each species is distributed.

Concerning essential elements, the highest copper and selenium concentrations were found in liver and blood, respectively, while considerable zinc concentrations were found in muscle, kidney and liver [10]. Calcified tissues, as bone and carapace, could contain higher zinc concentrations [17], although currently no data on these samples have been found. On the other hand, fat tissue showed low zinc levels [26] in contrast with the higher ones detected in other species of sea turtles [18, 22, 24, 42, 51, 56].

Only two works have studied metal levels in leatherback sea turtle eggs: Vazquez et al. [39] analyzed eggshells from the Pacific coast of Mexico and Guirlet et al. [35] analyzed egg contents from the Atlantic coast of French Guiana. It is difficult to compare both studies due to they were carried out in different areas and dates. However, these results seem to reflect that lead and copper are mainly concentrated in eggshells, whereas zinc shows the highest levels in egg contents. In the case of zinc, this great presence in yolk and albumen may be in relation to necessary requirement for embryonic development [17]. Referring to lead, levels in eggshell [39] were two orders of magnitude higher than those found in other tissues and samples (see Table 3). This fact can reflect that eggshell is an important way for lead excretion in this species.

HAWKSBILL SEA TURTLE (*ERETMOCHELYS IMBRICATA*)

The available data about levels of inorganic elements (Cd, Pb, Hg, As, Cu, Zn, Fe, Mn, Ni, Se and Cr) in different tissues (liver, kidney, muscle, fat) and stomach content of hawksbill sea turtles are shown in Table 4. Nowadays, there are no data on levels of these elements in blood and egg of this species.

The tissues mainly analyzed are liver, kidney and muscle, whereas cadmium, lead, arsenic, copper and zinc are the metals mainly analyzed. Only one study measured iron and nickel levels [42] and another one measured chromium levels [11]. In spite of hawksbill sea turtles have a circumglobal distribution throughout waters of the Atlantic, Indian and Pacific Ocean [1], all the consulted studies were carried out in the Pacific coasts (Mexico, Japan and Australia).

The highest cadmium levels were detected in kidney (maximum level: 310.00 $\mu g\ g^{-1}$ dry weight) [11], whereas lead showed similar levels in all the studied samples (range: BDL – 1.12 $\mu g\ g^{-1}$ dry weight, see Table 4). The majority of the mercury burden was found in kidney and liver [11], and arsenic showed especially high levels in muscle [49, 59].

Cadmium levels which were found in the stomach contents of hawksbill sea turtles from Pacific coasts of Japan were greatly lower than those measured in kidney of these specimens (22.47 – 885.71 times lower, see Table 4). The interpretation of these results is not simple, due to the diet shifts that happen in sea turtles throughout their life cycle and among individuals [11, 50]. For example, although the concentration of cadmium in the diet of juvenile specimens could be higher than in adults, cadmium levels in the renal tissue of an adult specimen reflect intakes from the current and previous diet. This is due to the very long biological half-life that heavy metals present in the organisms. With respect to the trophic transfer of elements in this species, cadmium and mercury showed the highest trophic transfer coefficient (TTC) [11]. This fact suggests that bioaccumulation or biomagnification phenomenoms could be happening for these metals in hawksbill sea turtles. Besides, renal cadmium concentrations in some specimens were close to or above values that can cause renal damages [11].

Table 4. Heavy metal and essential element levels (μg g^{-1} dry weight) in tissues and other samples of *Eretmochelys imbricata* from different locations

Tissue - Location	Cadmium		Lead		Mercury		Arsenic		Refs.
	Mean ± SD	Min - Max	Mean ± SD	Min - Max	Mean ± SD	Min - Max	Mean ± SD	Min - Max	
Liver									
Pacific Ocean: Mexico	0,49		BDL						[42]
Pacific Ocean: Australia**		7,50 - 19,38				0,13 - 0,16		0,56 - 5,78	[48]
Pacific Ocean: Japan							15,30 ± 8,77	4,94 - 32,80	[49]
Pacific Ocean: Japan	7,05 ± 6,37	1,80 - 33,60	0,17 ± 0,13	0,02 - 0,53	0,87 ± 1,87	0,05 - 8,70			[11]
Pacific Ocean: Japan	3,84	1,97 - 5,69	0,09	0,03 - 0,19					[52]
Pacific Ocean: Japan							13,75 ± 7,81	2,06 - 23,44	[55]
Pacific Ocean: Japan							25,00 ± 32,00		[59]
Kidney									
Pacific Ocean: Mexico	4,2		BDL						[42]
Pacific Ocean: Australia**		15,00 - 52,92				0,13 - 0,17		0,54 - 3,88	[48]
Pacific Ocean: Japan							28,30 ± 9,82	8,62 - 36,60	[49]
Pacific Ocean: Japan	93,70 ± 76,30	19,10 - 310,00	0,27 ± 0,24	0,09 - 1,12	1,30 ± 1,20	0,08 - 5,00			[11]
Pacific Ocean: Japan							45,00 ± 38,00		[59]
Muscle									
Pacific Ocean: Mexico	1,02		0,38						[42]
Pacific Ocean: Japan							153,00 ± 65,10	23,10 - 205,00	[49]

Pacific Ocean: Japan	0,07 ± 0,04	0,02 - 0,15	0,04 ± 0,05	0,003 - 0,15	0,04 ± 0,03	BDL - 0,09			[11]
Pacific Ocean: Japan							210,00 ± 140,00		[59]
Fat									
Pacific Ocean: Mexico	0,43		BDL						[42]
Stomach content									
Pacific Ocean: Japan	0,59 ± 0,21	0,35 - 0,85	0,21 ± 0,09	0,06 - 0,31	0,04 ± 0,03	BDL - 0,09			[11]
Pacific Ocean: Japan							22,00		[59]
Liver									
Pacific Ocean: Mexico	2,47		25,89		71,88		74,00		[42]
Pacific Ocean: Australia**				55,31 - 94,69					[48]
Pacific Ocean: Japan	54,90 ± 116,00	8,28 - 570,00	109,00 ± 54,00	52,30 - 306,00			8,29 ± 4,26	3,76 - 17,90	[11]
Pacific Ocean: Japan	26,94	14,66 - 59,06	79,69	69,38 - 92,19					[52]
Kidney									
Pacific Ocean: Mexico	3,89		82,45		362,00		7,62		[42]
Pacific Ocean: Australia**				55,00 - 87,09					[48]
Pacific Ocean: Japan	7,04 ± 2,79	4,89 - 17,90	120,00 ± 32,00	81,20 - 200,00			13,20 ± 2,75	6,09 - 17,70	[11]
Muscle									
Pacific Ocean: Mexico	3,68		102,00		258,00		1,78		[42]
Pacific Ocean: Japan	0,96 ± 0,32	0,61 - 1,46	48,60 ± 26,10	25,80 - 111,00			0,70 ± 0,52	0,42 - 2,08	[11]

Table 4 (Continued)

Tissue - Location	Cadmium		Lead		Mercury		Arsenic		Refs.
	Mean ± SD	Min - Max	Mean ± SD	Min - Max	Mean ± SD	Min - Max	Mean ± SD	Min - Max	
Fat									
Pacific Ocean: Mexico	0,72		42,39		11,14		2,53		[42]
Stomach content									
Pacific Ocean: Japan	5,02 ± 1,60	3,22 - 7,10	24,20 ± 18,20	6,84 - 50,80			3,55 ± 2,17	0,94 - 6,30	[11]
Liver									
Pacific Ocean: Mexico	2,48								[42]
Pacific Ocean: Australia**					8,38 - 11,41				[48]
Pacific Ocean: Japan			49,00 ± 37,00		12,00 - 150,00	0,85 ± 0,68	0,21 - 2,10		[11]
Pacific Ocean: Japan			59,38		9,06 - 100,00				[52]
Kidney									
Pacific Ocean: Mexico	1,61								[42]
Pacific Ocean: Australia**					9,25 - 10,38				[48]
Pacific Ocean: Japan			28,00 ± 19,00		9,00 - 76,00	1,60 ± 0,80	0,30 - 2,90		[11]
Muscle									
Pacific Ocean: Mexico	BDL								[42]
Pacific Ocean: Japan			11,00 ± 10,00		2,60 - 33,00	1,10 ± 1,10	0,12 - 3,00		[11]
Fat									
Pacific Ocean: Mexico	BDL								[42]
Stomach content									
Pacific Ocean: Japan			14,00 ± 10,00		3,60 - 29,00	0,61 ± 0,19	0,40 - 0,89		[11]

BDL = bellow detection limit; ** = Mean ± SEM.

The essential elements copper and selenium showed the highest levels in liver [11, 42, 52], whereas zinc was mainly accumulate in kidney and muscle [11, 42].

The high copper levels that were found in liver of hawksbill sea turtles (maximum level: 570.00 μg g^{-1} dry weight) [11] exceeded in some cases the copper levels that are homeostatically controlled in other marine animals (10 – 100 μg g^{-1} dry weight in liver of marine mammals) [68]. These high levels could exert adverse effects in the biological systems of sea turtles, such as the formation of reactive oxygen species or the irreversibly and non-specifically association with thiol groups in proteins [69].

Besides the study of carcasses, the development of studies about metal levels in eggs and blood of this species in the future will be of great interest. These studies will contribute to the standardization of non-destructive methods for long-term monitoring of metals in hawksbill sea turtles.

Kemp's Ridley Sea Turtle (*Lepidochelys kempii*)

The available data about levels of serveral inorganic elements (Cd, Pb, Hg, Cu, Zn, Se and Cr) in different samples (liver, kidney, muscle and blood; no available data on fat, stomach content or eggs) of Kemp's ridley sea turtles are shown in Table 5. No data on arsenic, iron, manganese and nickel levels have been found.

All the published studies were carried out in North Atlantic [10, 30, 34, 36] since the distribution of this species is restricted to this area [1].

Blood is the most used sample for metal determination in the studies that included samples of Kemp's ridley sea turtles in their analysis. Cadmium, lead, mercury, copper and zinc are the most studied elements (see Table 5).

Regarding non-essential heavy metals, cadmium and mercury were mainly accumulated in liver and kidney (kidney > liver for cadmium and liver > kidney for mercury), whereas lead showed similar levels in all the analyzed soft tissues.

In the case of cadmium, levels in kidney (a target organ for cadmium accumulation) were considerably higher in other species of sea turtles (see Tables 1, 2, 3, 4 and 6) than cadmium levels measured in kidney of Kemp's ridley sea turtles (see Table 5), even for sub-adult and adult specimens (Kemp's ridley sea turtles SCL range: 33.90-64.00 cm) [34]. This carnivore species preys primarily upon crustaceans such as blue crabs (*Callinectes sapidus*). Blue crabs from American Atlantic coasts exhibited cadmium levels ranging from 0.001 to 1.76 μg

Table 5. Heavy metal and essential element levels (µg g^{-1} dry weight) in tissues and other samples of *Lepidochelys kempii* from different locations

Tissue - Location	Cadmium		Lead		Mercury		Copper		References
	Mean ± SD	Min - Max	Mean ± SD	Min - Max	Mean ± SD	Min - Max	Mean ± SD	Min - Max	
Liver									
Atlantic Ocean: US	2,13 ± 1,53	0,60 - 5,74	0,15 ± 0,15	0,03 - 0,48	1,23 ± 1,96	0,02 - 6,11	14,70 ± 7,65	2,42 - 26,70	[34]
Atlantic Ocean: US (J)					0,22 ± 0,22	0,06 - 0,66			[36]
Kidney									
Atlantic Ocean: US	2,41 ± 4,07	0,13 - 11,90	0,16 ± 0,15	0,03 - 0,47	0,74 ± 0,43	0,21 - 1,31	7,78 ± 3,35	3,19 - 13,00	[34]
Muscle									
Atlantic Ocean: France	0,43 ± 0,43	0,05 - 1,24					4,67 ± 2,38	2,10 - 8,81	[10]
Atlantic Ocean: US	0,05 ± 0,04	0,01 - 0,16	0,14 ± 0,18	0,01 - 0,59	0,43 ± 0,34	0,01 - 0,97	2,35 ± 1,54	0,33 - 5,89	[34]
Blood									
Atlantic Ocean: US			0,05	0,00 - 0,15	0,1	0,003 - 0,36	2,65	1,12 - 6,63	[30]
Atlantic Ocean: US	0,1	0,002 - 0,26	0,15	0,004 - 0,87	0,1	0,003 - 0,92	7,40	0,92 - 23,26	[34]
Atlantic Ocean: US (J)					0,10 ± 0,05	BDL - 0,20			[36]
Atlantic Ocean: US* plasma (J)	BDL						0,69 ± 0,68	0,33 - 3,18	[36]

BDL = bellow detection limit; * = µg g^{-1} wet weight; J = juvenile.

Table 5 (Continued)

Tissue - Location	Zinc		Selenium		Chromium		References
	Mean ± SD	Min - Max	Mean ± SD	Min - Max	Mean ± SD	Min - Max	
Liver							
Atlantic Ocean: US	69,00 ± 18,60	45,20 - 108,00			0,02 ± 0,01	0,006 - 0,05	[34]
Kidney							
Atlantic Ocean: US	134,00 ± 22,40	98,90 - 172,00			0,02 ± 0,01	0,01 - 0,04	[34]
Muscle							
Atlantic Ocean: France	78,10 ± 15,71	63,33 - 97,62					[10]
Atlantic Ocean: US	31,40 ± 13,80	10,30 - 60,80			0,05 ± 0,09	0,0005 - 0,30	[34]
Blood							
Atlantic Ocean: US	38,25	16,73 - 96,39					[30]
Atlantic Ocean: US	87,67	1,79 - 268,77			0,10	0,0005 - 1,07	[34]
Atlantic Ocean: US* plasma (J)	2,29 ± 3,55	0,51 - 13,07	0,49 ± 0,06	0,40 - 0,61			[36]

* = $\mu g\ g^{-1}$ wet weight; J = juvenile.

g^{-1} wet weight [34]. These levels are similar or lower than other ones measured in the stomach contents of other species of sea turtles (see Tables 2 and 4). The scarce and fragmented information available about cadmium levels in their preys are not enough to explain the important inter-specific differences that were observed in renal tissues of sea turtles. Deeper investigations are needed to understand if these differences are due to different cadmium levels in their preys or different pollution levels in the distribution areas. As well as metabolic and physiological differences among these species could be involved too [10].

Referring to lead, the burden of this metal in carapace samples of Kemp's ridley sea turtles from American Atlantic coasts (1.39 ± 2.61 μg g^{-1} dry weight) [34] was one order of magnitude higher than the lead burden measured in soft tissues of the same specimens (see Table 5). These results point out that calcified tissues such as carapace accumulate lead largely. It can be useful for long-term biomonitoring of this element, which is a quite reliable indicator for assess the presence of anthropogenic pollution [64].

Copper and zinc are the most studied essential elements and showed the higher levels in liver and kidney, respectively (see Table 5). Concerning copper, this pattern of major accumulation in liver is similar in all the species of sea turtles (see Tables 1, 2, 3, 4, 6), except for *Natator depressus* (no available data). Probably these high copper levels in liver are related to specific features that are present in sea turtles. These features could allow the accumulation of higher hepatic levels of copper in sea turtles than in other marine organisms [17].

Besides the results showed in Table 5, high zinc levels were detected in carapace samples of Kemp's ridely sea turtles from American Atlantic coasts (270 ± 170 μg g^{-1} dry weight) [34]. This fact indicates that, similar to lead, zinc presents high affinity for calcified tissues.

OLIVE RIDLEY SEA TURTLE (*LEPIDOCHELYS OLIVACEA*)

There is not abundant information about concentrations of heavy metals and essential elements in samples of olive ridley sea turtle. The available data (Cd, Pb, Hg, Cu, Zn, Fe, Mn, Ni and Cr) are present in Table 6.

The metals most frequently analyzed are cadmium, lead and mercury (in that order), and the tissues mainly analyzed are liver, kidney and muscle (see Table 6). As in other species of sea turtles, the highest cadmium and lead concentrations were found in kidney [41, 42, 48], and mercury tended to be higher in liver [43].

Several essential elements have been measured in samples of *Lepidochelys Olivacea*, although copper, zinc and nickel were the most studied (see Table 6). The highest copper levels were found in liver [41], while blood [45] and muscle [42] contained considerable levels of zinc. Manganese showed relatively high levels in kidney and fat tissue [42].

Concerning metal levels in eggs, some of them were mainly present in the albumen: levels of lead, copper and nickel were higher in albumen than in yolk [45, 46], whereas shell contained considerable levels of iron, nickel and lead [45, 46, 47].

FLATBACK SEA TURTLE (*NATATOR DEPRESSUS*)

The available information about levels of heavy metals and essential elements in this species is very scarce (see Table 7). Flatback sea turtle has been slightly studied in contrast with other species of sea turtles, since this species is present in a limited geographical area [1].

Several elements (Cd, Pb, Hg, As, Cu, Zn, Mn, Ni, Se and Cr) have been measured in samples of blood and egg [62]. Yolk and albumen contained considerable zinc levels, which could be related to requirements for embryonic development [17]. Cadmium, lead, mercury and nickel showed non-detectable levels (see Table 7).

CONCLUSION

Cadmium and copper showed especially high levels in renal and muscle tissues, respectively, and these levels could even cause adverse effects in some specimens of sea turtles. Zinc and lead tended to accumulate high concentrations in calcified tissues, as bone and carapace, although these tissues have not been enough studied in sea turtles. Further assessment of the presence of metals in these tissues is recommended, as well as the standardization of non-invasive procedures for sampling (use of carapace, non-viable eggs or blood), which will allow the development of long-term monitoring programmes for these protected species. Furthermore, a deeper knowledge of metal presence in some species of sea turtles is needed (especially *Dermochelys coriacea*, *Lepidochelys kempii* and *Natator depressus*), since the available information is really scarce.

Table 6. Heavy metal and essential element levels (µg g^{-1} dry weight) in tissues and other samples of *Lepidochelys olivacea* from different locations

Tissue - Location	Cadmium		Lead		Mercury		Copper		References
	Mean ± SD	Min - Max	Mean ± SD	Min - Max	Mean ± SD	Min - Max	Mean ± SD	Min - Max	
Liver									
Pacific Ocean: Mexico	13,12 ± 1,50		13,30 ± 2,10				33,50		[41]
Pacific Ocean: Mexico	17,89	4,98 - 148,00	BDL				36,73	16,99 - 100,00	[42]
Pacific Ocean: Mexico					0,21	0,06 - 0,80			[43]
Pacific Ocean: Australia**	20,00								[48]
Kidney									
Pacific Ocean: Mexico	15,84 ± 1,20		13,40 ± 1,90				17,16		[41]
Pacific Ocean: Mexico	60,03	0,81 - 274,00	0,03	BDL - 2,63			4,86	0,81 - 53,40	[42]
Pacific Ocean: Mexico					0,14	0,03 - 0,37			[43]
Pacific Ocean: Australia**	124,18								[48]
Muscle									
Pacific Ocean: Mexico	2,48 ± 0,40		8,90 ± 1,00				16,15		[41]
Pacific Ocean: Mexico	0,48	BDL - 8,85	BDL				1,28	0,70 - 4,37	[42]
Pacific Ocean: Mexico					0,05	0,02 - 0,14			[43]

Table 6 (Continued)

Tissue - Location	Cadmium		Lead		Mercury		Copper		References
	Mean ± SD	Min - Max	Mean ± SD	Min - Max	Mean ± SD	Min - Max	Mean ± SD	Min - Max	
Blood									
Pacific Ocean: Mexico			0,95 ± 0,18						[46]
Pacific Ocean: Mexico	0,45 ± 0,20						2,28 ± 0,40		[45]
Fat									
Pacific Ocean: Mexico	0,69	0,33 - 2,54	BDL				0,83	0,47 - 2,54	[42]
Pacific Ocean: Mexico					0,03	BDL - 0,16			[43] [43] [43]
Egg									
Pacific Ocean: Mexico			Yolk: 0,80 ± 0,10						[46]
			Albumen: 1,08 ± 0,20						[46]
			Shell: 1,05 ± 0,20						[46]
Pacific Ocean: Mexico	Yolk: 0,24 ± 0,10						Yolk: 2,20 ± 1,47		[45]
	Albumen: 0,22 ± 0,09						Albumen: 3,53 ± 2,87		[45]
	Shell: 0,47 ± 0,09						Shell: 7,48 ± 2,60		[45]
Indian Ocean: India	Shell:	2,44 - 4,88	Shell:	19,52 - 36,60			Shell:	14,64 - 19,52	[47]

Liver										
Pacific Ocean: Mexico	47,14	18,66 - 85,75	731,00	119,00 - 9201,00	0,10	BDL - 9,20	0,58	BDL - 3,88		[42]
Pacific Ocean: Australia**	46,25									[48]
Kidney										
Pacific Ocean: Mexico	6,68	0,43 - 114,00	193,00	40,37 - 667,00	5,31	3,93 - 7,52	0,02	BDL - 2,46		[42]
Pacific Ocean: Australia**	78,34									[48]
Muscle										
Pacific Ocean: Mexico	85,78	49,89 - 107,00	93,09	57,35 - 319,00	0,77	BDL - 4,34	0,01	BDL - 0,41		[42]
Blood										
Pacific Ocean: Mexico	58,40 ± 4,70						2,80 ± 1,30			[45]
Fat										
Pacific Ocean: Mexico	3,70	0,41 - 16,65	27,91	6,37 - 236,00	2,10	0,88 - 3,65	0,03	BDL - 0,51		[42]
Egg										
Pacific Ocean: Mexico	Yolk: 72,30 ± 10,90						Yolk: 3,30 ± 0,30			[45]
	Albumen: 33,60 ± 6,10						Albumen: 4,00 ± 3,90			[45]
	Shell: 12,40 ± 1,50						Shell: 48,50 ± 12,90			[45]
Indian Ocean: India	Shell:	14,64 - 36,60	Shell:	92,72 - 129,32	Shell:	4,88 - 14,64	Shell:	21,96 - 41,48	Shell: 7,32 - 26,84	[47]

BDL = bellow detection limit; ** = Mean ± SEM.

Table 7. Heavy metal and essential element levels (µg g^{-1} dry weight) in tissues and other samples of *Natator depressus* from different locations

Tissue - Location	Cadmium		Lead		Mercury		Arsenic		Copper		References
	Mean ± SD	Min - Max	Mean ± SD	Min - Max	Mean ± SD	Min - Max	Mean ± SD	Min - Max	Mean ± SD	Min - Max	
Blood Pacific Ocean: Australia	BDL		BDL		BDL				0,04 ± 0,0005		[62]
Egg Pacific Ocean: Australia	Egg content:	BDL	Egg content:	BDL	Egg content:	BDL	Egg content:	3,49 ± 0,10	Egg content:	2,14 ± 0,02	[62]
Blood Pacific Ocean: Australia	0,77 ± 0,005				BDL						[62]
Egg Pacific Ocean: Australia	Egg content:	54,16 ± 0,31	Egg content:	0,89 ± 0,005	Egg content:	BDL	Egg content: 1,56 ± 0,01		Egg content: 0,89 ± 0,02		[62]

BDL = bellow detection limit.

This study shows up that sea turtles are potential biological indicators for monitoring the inorganic marine pollution, since they are long-lived species that present several special ecological features. However, sea turtles will be more reliable indicators if the presence of metals is studied in specimens at different stages of development throughout their life cycle and they are studied in the different distribution areas.

REFERENCES

[1] International Union for Conservation of Nature. *IUCN Red List of Threatened Species* (version 2010.4) [Internet]. 2010 [cited 2010 Dec 15]. Available from: *http://www.iucnredlist.org*.

[2] IUCN Marine Turtles Specialist Group. Hazards [Internet]. 2010 [cited 2010 Dec 15]. Available from: *http://iucn-mtsg.org/about-turtles/hazards/*.

[3] Barba-Brioso C, Fernández-Caliani JC, Miras A, Cornejo J, Galán E. Multi-source water pollution in a highly anthropized wetland system associated with the estuary of Huelva (SW Spain). *Mar Pollut Bull* 2010; 60: 1259-1269.

[4] Yu R, Yuan X, Zhao Y, Hu G, Tu X. Heavy metal pollution in intertidal sediments from Quanzhou Bay, China. *J Environ Sci* 2008; 20: 664-669.

[5] Järup L. Hazards of heavy metal contamination. *Br Med Bull* 2003; 68: 167–182.

[6] Garrett RG. Natural source of metals of the environment. *Hum Ecol Risk Assess* 2000; 6: 945-963.

[7] Haynes D, Johnson JE. Organochlorine, heavy metal and polyaromatic hydrocarbon pollutant concentrations in the Great Barrier Reef (Australia): a review. *Mar Pollut Bull* 2000; 41: 267-278.

[8] O'Shea TJ, Geraci JR. Toxicology in marine mammals. In: Fowler ME, Miller RE, editors. *Zoo & Wild Animal Medicine: Current Therapy.* Philadelphia. WB Saunder Company; 1999.

[9] Cardellicchio N, Giandomenico S, Ragone P, Leo D. Tissue distribution of metals in striped dolphins (*Stenella coeruleoalba*) from the Apulian coasts, Southern Italy. *Mar Environ Res* 2000; 49: 55-66.

[10] Caurant F, Bustamante P, Bordes M, Miramand P. Bioaccumulation of cadmium, copper, and zinc in some tissue of three species of marine turtles stranded along the French Atlantic coasts. *Mar Pollut Bull* 1999; 38: 1085-1091.

[11] Anan Y, Kunito T, Watanabe I, Sakai H, Tanabe S. Trace element accumulation in hawsbill turtles (*Eretmochelys imbricata*) and green turtles (*Chelonia mydas*) from Yaeyama islands, Japan. *Environ Toxicol Chem* 2001; 20: 2802-2814.
[12] Mas A, Azcue JM. *Metales en Sistemas Biológicos*. Barcelona: PPU Eds.; 1993.
[13] Sakai H, Hichihashi H, Suganuma H, Tatsukawa R. Heavy metal monitoring in sea turtles using eggs. *Mar Pollut Bull* 1995; 30: 347-353.
[14] Storelli MM, Ceci E, Marcotrigiano GO. Distribution of heavy metal residues in some tissues of *Caretta caretta* (Linnaeus) specimens beached along the Adriatic Sea (Italy). *B Environ Contam Tox* 1998; 60: 546-552.
[15] Storelli MM, Ceci E, Marcotrigiano GO. Comparison of total mercury, methylmercury, and selenium in muscle tissues and in the liver of *Stenella coeruleoalba* (Meyen) and *Caretta caretta* (Linnaeus). *B Environ Contam Tox* 1998; 61, 541-547.
[16] Storelli MM, Marcotrigiano GO. Total organic and inorganic arsenic from marine turtles *Caretta caretta* beached along the Italian Coast (South Adriatic Sea). *B Environ Contam Tox* 2000; 65: 732-739.
[17] Storelli MM, Marcotrigiano GO. Heavy metal residues in tissues of marine turtles. *Mar Pollut Bull* 2003; 46: 397-400.
[18] Storelli MM, Storelli A, D'Addabbo R, Marano C, Bruno R, Marcotrigiano GO. Trace elements in loggerhead turtles (*Caretta caretta*) from the eastern Mediterranean Sea: overview and evaluation. *Environ Pollut* 2005; 135: 163-170.
[19] Storelli MM, Barone G, Storelli A, Marcotrigiano GO. Total and subcellular distribution of trace elements (Cd, Cu and Zn) in the liver and kidney of green turtles (*Chelonia mydas*) from Mediterranean Sea. *Chemosphere* 2008; 70: 908-913.
[20] Godley BJ, Thompson DR, Furness RW. Do heavy metal concentrations pose a threat to marine turtles from the Mediterranean Sea? *Mar Pollut Bull* 1999; 38: 497-502.
[21] Kaska Y, Furness RW. Heavy metals in marine turtle eggs and hatchlings in the Mediterranean. *Zool Middle East* 2001; 24: 127-132.
[22] Franzellitti S, Locatelli C, Gerosa G, Vallini C, Fabbri E. Heavy metals in tissues of loggerhead turtles (*Caretta caretta*) from the nothwestern Adriatic Sea. *Comp Biochem Physiol* 2004; 138: 187-194.
[23] Maffucci F, Caurant F, Bustamante P, Bentivegna F. Trace elements (Cd, Cu, Hg, Se, Zn) accumulation and tissue distribution in loggerhead turtles

(*Caretta caretta*) from Western Mediterranean Sea (southern Italy). *Chemosphere* 2005; 58: 535-542.

[24] Andreani G, Santoro M, Cottignoli S, Fabbri M, Carpene E, Isani G. Metal distribution and metallothionein in loggerhead (*Caretta caretta*) and green (*Chelonia mydas*) sea turtles. *Sci Total Environ* 2008; 390: 287-294.

[25] Jerez S, Motas M, Cánovas RA, Talavera J, Almela RM, Del Río AB. Accumulation and tissue distribution of heavy metals and essential elements in loggerhead turtles (*Caretta caretta*) from Spanish Mediterranean coastline of Murcia. *Chemosphere* 2010; 78: 256-264.

[26] Davenport J, Wrench J. Metal levels in a leatherback turtle. *Mar Pollut Bull* 1990; 21: 40-41.

[27] Torrent A, González-Díaz OM, Monagas P, Orós J. Tissue distribution of metals in loggerhead turtles (*Caretta caretta*) stranded in the Canary Islands, Spain. *Mar Pollut Bull* 2004; 49: 854-874.

[28] Stoneburner DL, Nicora MN, Blood ER. Heavy Metals in loggerhead sea turtle eggs (*Caretta caretta*): evidence to support the hypothesis that demes exist in the Western Atlantic population. *J Herpetol* 1980; 14: 171-175.

[29] Alam SK, Brim MS. Organochlorine, PCB, PAH, and metal concentrations in eggs of loggerhead sea turtles (*Caretta caretta*) from Northwest Florida, USA. *J Environ Sci Health* Part B 2000; 35: 705-724.

[30] Kenyon LO, Landry AM, Gill GA. Trace metal concentrations in blood of the Kemp's ridley sea turtle (*Lepidochelys kempii*). *Chelonian Conserv Biol* 2001; 4: 128-135.

[31] Day RD, Christopher SJ, Becker PR, Whitaker DW. Monitoring mercury in the loggerhead sea turtle, *Caretta caretta*. *Environ Sci Technol*. 2005; 39: 437-446.

[32] Day RD, Segars AL, Arendt MD, Lee AM, Peden-Adams MM. Relationship of blood mercury levels to health parameters in the loggerhead sea turtle (*Caretta caretta*). *Environ Health Perspect* 2007; 115: 1421-1428.

[33] Day RD, Keller JM, Harms CA, Segars AL, Cluse WM, Godfrey MH, Lee AM, Peden-Adams M, Thorvalson K, Dodd M, Norton T. Comparison of mercury burdens in chronically debilitated and healthy loggerhead sea turtles (*Caretta caretta*). *J Wildl Dis* 2010; 46: 111-117.

[34] Wang HC. Trace metal uptake and accumulation pathways in Kemp's ridley sea turtles (*Lepidochelys kempii*) [PhD thesis]. Texas, US: Texas A&M University; 2005.

[35] Guirlet E, Das K, Girondot M. Maternal transfer of trace elements in leatherback turtles (*Dermochelys coriacea*) of French Guiana. *Aquat Toxicol* 2008; 88: 267-276.
[36] Innis C, Tlusty M, Perkins C, Holladay S, Merigo C, Weber ES. Trace metal and organochlorine pesticide concentrations in cold-stunned juvenile Kemp's ridley turtles (*Lepidochelys kempii*) from Cape Cod, Massachusetts. *Chelonian Conserv Biol* 2008; 7: 230-239.
[37] Innis C, Merigo C, Dodge K, Tlusty M, Dodge M, Sharp B, Myers A, McIntosh A, Wunn D, Perkins C, Herdt TH, Norton T, Lutcavage M. Health evaluation of leatherback turtles (*Dermochelys coriacea*) in the Northwestern Atlantic during direct capture and fisheries gear disentanglement. *Chelonian Conserv Biol* 2010; 9: 205-222.
[38] Aguirre AA, Balazs GH, Zimmerman B, Galey FD. Organic contaminants and trace metals in the tissues of green turtles (*Chelonia mydas*) afflicted with fibropapillomas in the Hawaiian islands. *Mar Pollut Bull* 1994; 28: 109-114.
[39] Vazquez GF, Reyes MC, Fernandez G, Aguayo JEC, Sharma VK. Contaminations in marine turtle (*Dermochelys coriaca*) egg shells of Playon de Mexiquillo, Michoacan, Mexico. *Bull Environ Contam Toxicol* 1997; 58: 326-333.
[40] Presti S, Hidalgo AR, Sollod A, Seminoff J. Mercury concentration in the scutes of black sea turtles, *Chelonia mydas agassizii*, in the Gulf of California. *Chelonian Conserv Biol* 1999; 3: 531-533.
[41] Frías-Espericueta MG, Osuna-López JI, Ruiz-Telles A, Quintero-Alvarez JM, López-López G, Izaguirre-Fierro G, Voltolina D. Heavy metals in the tissues of the sea turtle *Lepidochelys olivacea* from a nesting site of the northwest coast of Mexico. *Bull Environ Contam Toxicol* 2006; 77: 179-185.
[42] Gardner SC, Fitzgerald SL, Vargas BA, Rodríguez LM. Heavy metal accumulation in four species of sea turtles from the Baja California peninsula, Mexico. *Biometals* 2006; 19: 91-99.
[43] Kampalath R, Gardner SC, Méndez-Rodríguez LC, Jay JA. Total and methylmercury in three species of sea turtles of Baja California Sur. *Mar Pollut Bull* 2006; 52:1816-1823.
[44] Talavera-Saenz A, Gardner SC, Riosmena Rodriquez R, Acosta Vargas B. Metal profiles used as environmental markers of green turtle (*Chelonia mydas*) foraging resources. *Sci Total Environ* 2007; 373: 94-102.

[45] Páez-Osuna F, Calderón-Campuzano MF, Soto-Jiménez MF, Ruelas-Inzunza JR. Trace metals (Cd, Cu, Ni, and Zn) in blood and eggs of the sea turtle *Lepidochelys olivacea* from a nesting colony of Oaxaca, Mexico. *Arch Environ Contam Toxicol* 2010; 59: 632-641.

[46] Páez-Osuna F, Calderón-Campuzano MF, Soto-Jiménez MF, Ruelas-Inzunza JR. Lead in blood and eggs of the sea turtle, *Lepidochelys olivacea*, from the Eastern Pacific: concentration, isotopic composition and maternal transfer. *Mar Pollut Bull* 2010; 60: 433-439.

[47] Sahoo G, Mohapatra BK, Sahoo RK, Mohanty-Hejmadi P. Ultrastructure and characteristics of eggshells of the olive ridley turtle (*Lepidochelys olivacea*) from Gahirmatha, India. *Acta Anat* 1996; 156: 261-267.

[48] Gordon AN, Pople AR, Ng J. Trace metal concentrations in livers and kidneys of sea turtles from south-eastern Queensland, Australia. *Mar Freshwat Res* 1998; 49: 409-414.

[49] Saeki K, Sakakibara H, Sakai H, Kunito T, Tanabe S. Arsenic accumulation in three species of sea turtles. *BioMetals* 2000; 13: 241-250.

[50] Sakai H, Saeki K, Ichihashi H, Kamezaki N, Tanabe S, Tatsukawa R. Growth-related changes in heavy metal accumulation in green turtle (*Chelonia mydas*) from Yaeyama Islands, Okinawa, Japan. *Arch Environ Contam Toxicol* 2000; 39: 378-385.

[51] Sakai H, Saeki K, Ichihashi H, Suganuma H, Tanabe S, Tatsukawa R. Species-specific distribution of heavy metals in tissues and organs of loggerhead turtle (*Caretta caretta*) and green turtle (*Chelonia mydas*) from Japanese coastal waters. *Mar Pollut Bull* 2000; 40: 701-709.

[52] Anan Y, Kunito T, Sakai H, Tanabe S. Subcellular distribution of trace elements in the liver of sea turtles. *Mar Pollut Bull* 2002; 45: 224-229.

[53] Kubota R, Kunito T, Tanabe S. Chemical speciation of arsenic in the livers of higher trophic marine animals. *Mar Pollut Bull* 2002; 45: 218-223.

[54] Kubota R, Kunito T, Tanabe S. Occurrence of several arsenic compounds in the liver of birds, cetaceans, pinnipeds and sea turtles. *Environ Toxicol Chem* 2003; 22: 1200-1207.

[55] Fujihara J, Kunito T, Kubota R, Tanabe S. Arsenic accumulation in livers of pinnipeds, seabirds and sea turtles: subcellular distribution and interaction between arsenobetaine and glycine betaine. *Comp Biochem Physiol C Toxicol Pharmacol* 2003; 136: 287-296.

[56] Lam J, Tanabe S, Chan S, Yuen E, Lam M, Lam P. Trace element residues in tissues of green turtles (*Chelonia mydas*) from South China Waters. *Mar Pollut Bull* 2004; 48: 164-192.

[57] Lam J, Tanabe S, Chan S, Lam M, Martin M, Lam P. Levels of trace elements in green turtle eggs collected from Hong Kong: Evidence of risks due to selenium and nickel. *Environ Pollut* 2006; 144: 790-801.

[58] Agusa T, Takagi K, Iwata H, Tanabe S. Arsenic species and their accumulation features in green turtles (*Chelonia mydas). Mar Pollut Bull* 2008; 57: 782-789.

[59] Agusa T, Takagi K, Kubota R, Anan Y, Iwata H, Tanabe S. Specific accumulation of arsenic compounds in green turtles (*Chelonia mydas*) and hawksbill turtles (*Eretmochelys imbricata*) from Ishigaki Island, Japan. *Environ Pollut* 2008; 153: 127-136.

[60] Van de Merwe JP, Hodge M, Olszowy HA, Whittier JM, Ibrahim K, Lee SY. Chemical contamination of green turtle (*Chelonia mydas*) eggs in peninsular Malaysia: implications for conservation and public health. *Environ Health Perspect* 2009; 117: 1397-1401.

[61] Van de Merwe JP, Hodge M, Olszowy HA, Whittier JM, Lee SY. Using blood samples to estimate persistent organic pollutants and metals in green sea turtles (*Chelonia mydas*). *Mar Pollut Bull* 2010; 60: 579-588.

[62] Ikonomopoulou MP, Olszowy H, Limpus C, Francis R, Whittier J. Trace element concentrations in nesting flatback turtles (*Natator depressus*) from Curtis Island, Queensland, Australia. *Mar Environ Res. Forthcoming* 2010 Sep 25.

[63] Witherington BE. Biological conservation of loggerheads: challenges and opportunities. In: Bolten AB, Witherington BE, editors. *Loggerhead Sea Turtles*. Washington DC. Smithsonian Institution; 2003.

[64] Bargagli R. Trace metals in Antarctica related to climate change and increasing human impact. *Rev Environ Contam Toxicol* 2000; 166:129-173.

[65] Vas P. Trace metal levels in sharks from British and Atlantic Waters. *Mar Pollut Bull* 1991; 22: 67-72.

[66] Law RJ, Fileman CF, Hopkins AD, Baker JR, Harwood J, Jackson DB, Kennedy S, Martin AR, Morris RJ. Concentrations of trace metals in the livers of marine mammals (seals, porpoises and dolphins) from waters around the British Isles. *Mar Pollut Bull* 1991; 28: 109-114.

[67] Eisenberg JF, Frazier J. A leatherback turtle (*Dermochelys coriacea*) feeding in the wild. *J Herpetol* 1983; 17: 82-86.

[68] Law RJ. Metals in marine mammals. In: Beyer WN, Heinz GH, Redmon-Norwood AW, editors. *Environmental Contaminants in Wildlife: Interpreting Tissue Concentrations*. Florida. CRC Boca Raton; 1996.

[69] Letelier ME, Sánchez-Jofré S, Peredo-Silva L, Cortés-Troncoso J, Aracena-Parks P. Mechanisms underlying iron and copper ions toxicity in biological systems: Pro-oxidant activity and protein-binding effects. *Chem Biol Interact* 2010; 188: 220-227.

In: Turtles: Anatomy, Ecology and Conservation ISBN: 978-1-61470-554-3
Eds: M. J. Cosgrove and S. A. Roe

Chapter 2

Anthropogenic Causes of Mortality of Sea Turtles in the Canary Islands: A Multidisciplinary Approach to the Conservation of Endangered Sea Turtles

Jorge Oros[1], Alberto Arencibia[1] and Patricia Monagas[2]

[1] Department of Morphology, Veterinary Faculty,
University of Las Palmas de Gran Canaria. Trasmontana s/n 35416
Arucas (Las Palmas), Spain

[2] Department of Environment, Canary Islands Government.
C/ Profesor Agustín Millares Carló, nº 18, 35071
Las Palmas de Gran Canaria, Spain

Abstract

Because all species of sea turtles are included on the Red List of the World Conservation Union, the efforts to conserve sea turtles, the advances in their medical management, and the studies on diseases and causes of mortality and/or stranding of sea turtles must to be increased. A multidisciplinary approach to this focus carried out by veterinarians, biologists, and scientists, is necessary in order to compile data that prove the main threats for these endangered reptiles, and to design adequate strategies of conservation.

This chapter lists the pathological findings and causes of mortality of 49 sea turtles (46 *Caretta caretta*, 2 *Chelonia mydas*, and 1 *Dermochelyscoriacea*) stranded on the coasts of the Canary Islands, Spain, between 2002 and 2009.Of these, 12 turtles (24.49%) had died of spontaneous diseases including different types of hepatitis, pneumonia, and septicemic processes. However, 37 turtles (75.51%) died from lesions associated with human activities: ingestion of hooks and monofilament lines (34.69%), entanglement in fishing nets (24.49%), and boat-strike injuries (16.33%).

In addition, polychlorinated biphenyls (PCBs 28, 31, 52, 101, 138, 153, 180, and 209) and DDT and its metabolites (*o,p'*-DDT, *o,p'*-DDE, and *o,p'*-DDD) were measured in liver and fat from 32 sea turtles. Tissues from these turtles contained higher levels of PCBs and *o,p'*-DDT than those reported in turtles from other geographical regions. Statistically, a positive correlation was detected between $\sum$PCBs concentration and cachexia, and between $\sum$DDT concentrations and cachexia. No histological lesions exclusively attributed to the acute effects of PCBs and DDTs were described. However, chronic effects of organochlorines can not be discarded.

INTRODUCTION

Two families and seven species of sea turtles are currently recognized (Pritchard 1997). The family Cheloniidae includes the green turtle (*Chelonia mydas*), loggerhead (*Caretta caretta*), hawksbill (*Eretmochelysimbricata*), Kemp's ridley (*Lepidochelyskempi*), olive ridley (*Lepidochelysolivacea*), and flatback turtle (*Natatordepressus*). The family Dermochelyidae includes only the leatherback (*Dermochelyscoriacea*). Only five species have been reported in the Canary Islands: loggerhead, green turtle, leatherback, hawksbill, and Kemp's ridley sea turtle (Mateo *et al.* 1997, Barbadillo*et al.* 1999). The most common species in the Canary Islands is the loggerhead sea turtle (Mateo *et al.* 1997). However, evidence of a decline in the population of loggerhead sea turtles in the Canary Islands has been reported (López-Jurado& González 1983, Blanco & González 1992).

All species of sea turtles are included on the Red List of the World Conservation Union (IUCN/SCC2010). The IUCN emphasized as a priority issue in its document entitled *A Global Strategy for the Conservation of Marine Turtles* (1995) the research on diseases of sea turtles.Thus in recent years, increased efforts have been devoted to the conservation of sea turtles, including medical management and pathological studies on stranded animals.

Sea turtles can be affected by several diseases. Some of these health problems are naturally occurring processes and are observed in both wild and captive turtles. Other conditions, such as certain nutritional deficiencies, are essentially a result of prolonged captivity in wildlife rehabilitation centers (George 1997).

There are previous reports of disease surveys of wild sea turtles in Australia (Glazebrook & Campbell 1990b, Raidal*et al.* 1998), Hawaii (Work *et al.* 2003, 2004), France (Duguy*et al.* 1998), Florida (Smith & Coates 1938), and the Canary Islands (Orós *et al.* 2004, 2005). However, usually these types of surveys do not include toxicological studies on stranded sea turtles. Regarding the previous studies on sea turtles stranded in the Canary Islands, it is remarkable that in addition to pathological, parasitological and microbiological studies, chemical analyses were carried out to determine the level of heavy metals in the tissues of 78 loggerhead sea turtles (Torrent *et al.* 2004).

Recent studies have reported baseline levels of persistent contaminants (polychlorinated biphenyls and organochlorine pesticides) on turtle populations of the Canary Islands (Monagas *et al.* 2008, Orós *et al.* 2009). Other studies on persistent contaminants have been focused on sea turtles from Long Island (Lake *et al.* 1994), Virginia (Rybitski*et al.* 1995), Scotland (Mckenzie*et al.* 1999), the Hawaiian Islands (Miao *et al.* 2001), the Baja California peninsula (Gardner *et al.* 2003), North Carolina (Keller *et al.* 2004, 2006), and the Mediterranean Sea (Corsolini*et al.* 2000, Storelli & Marcotrigiano 2000, Perugini *et al.* 2006, Storelli *et al.* 2007).

Polychlorinated biphenyls (PCBs) and DDT and its metabolites (OC-DDTs) have a particular significance because of their undesirable effects on environmental quality and animal health (Ahlborg*et al.* 1994). PCBs were manufactured from the 1930s to the 1970s for several industrial applications, such as liquid coolants for electrical transformers or as softeners in the production of plastics and as components of hydraulic fluids and lubricating oils. PCBs are able to bioaccumulate through the food chain and their effects have been reported on the immune, endocrine, and reproductive systems of different animal species (Fox 2001). OC-DDTs were used extensively all over the world as a domestic and agriculture pesticides (Ecobichon 1995). In most countries, DDT was banned in the 1970s due to its long residual life and its accumulation in food chains (Bolt &Degen 2002). In order to compile data that prove the main threats for sea turtles, and to design adequate strategies of conservation, it is necessary a multidisciplinary approach to the study of causes of mortality and/or stranding of sea turtles, carried out by veterinarians, biologists, and scientists. This chapter lists the pathological findings and causes of mortality of 49 sea turtles (46 *Caretta caretta*, 2 *Chelonia mydas*, and 1 *Dermochelyscoriacea*) stranded on the coasts of the Canary Islands, Spain, between 2002 and 2009. In addition, concentrations of polychlorinated biphenyls (PCBs 28, 31, 52, 101, 138, 153, 180, and 209) and DDT and its metabolites (*o,p'*-DDT, *o,p'*-DDE, and *o,p'*-DDD) in the liver and the adipose tissue of 32 sea turtles are presented.

TURTLES AND METHODOLOGY

Between January 2002 and December 2009, 49 sea turtles that stranded on the coasts of the Canary Islands [Gran Canaria (n=41; 83.67%), Tenerife (n=4; 8.16%), y El Hierro (n=2; 4.08%)] were submitted for necropsy to the Veterinary Faculty, University of Las Palmas de Gran Canaria (ULPGC). Some of them had been previously submitted to the Tafira Wildlife Rehabilitation Center (TWRC) for health evaluation, medical and surgical management, and possible rehabilitation.

Species identifications were made according to Frick (1996). Turtles belonging to three different species were examined: 46 loggerhead turtles (*Caretta caretta*) (93.87%), 2 green turtles (*Chelonia mydas*) (2.04%), and 1 leatherback turtle(*Dermochelyscoriacea*) (2.04%). Species, sex, age group, and biometrics data are shown in Table 1. Age group was determined on the basis of straight carapace length (SCL) (Bjorndal*et al.* 2001, Seminoff*et al.* 2004) and sexual maturity (estimated from the appearance of their gonads).

The toxicological study was limited to 32 sea turtles necropsied before 2005: 30 loggerhead turtles (93.70%), 1 green turtle (3.10%), and 1 leatherback turtle (3.10%).

Pathological Study

Necropsies were performed at the Veterinary Faculty (ULPGC). The gross postmortem examinations were carried out using the procedures previously described (Wolke& George 1981, Campbell 1996, Orós & Torrent 2001).

The condition of the turtles ranged from freshly dead to moderately autolyzed. Macroscopic lesions were recorded and tissue samples from all major organs were fixed in 10% neutral buffered formalin, embedded in paraffin, sectioned at 5 μm for light microscopy and stained with hematoxylin and eosin (HE). Special stains performed on selected cases included Gram-stain for bacteria, Ziehl Neelsen (ZN) for acid-fast organisms, periodic acid-Schiff (PAS) for protozoa and fungal hyphae, Grocott's methenamine silver nitrate (GMS) for fungi, and von Kossa stain for calcium (Bancroft & Stevens 1996).

Table 1. Species, sex, age group and biometric data (mean ± standard deviation) of the turtles analyzed

Species	Sex	Age group	Weight (kg)	SCL (cm)
Caretta caretta (n = 46)	Female (n = 39; 84.78 %) Males (n = 7;15.21 %)	Pelagic juvenile (n = 18; 39.13 %) Juvenile (n = 28; 60.86 %)	12.7 ± 6.6	43.2 ± 10.5
Chelonia mydas (n = 2)	Female	Juvenile	22 ± 1.1	51.5 ± 0.9
Dermochelys coriacea (n = 1)	Female	Adult	231.5	nm

SCL: straight carapace length.
n: number of turtles.
nm: not measured.

OrganochlorineCompounds Analysis

Polychlorinated biphenyls (IUPAC nos. 28, 31, 52, 101, 138, 153, 180, and 209) and OC-DDTs (*o,p'*-DDT, *o,p'*-DDE, and *o,p'*-DDD)were analyzed according to the method described by Tanabe *et al.* (1994). Tissues samples (liver and celomic fat) were collected during necropsy. After collection, the samples were wrapped in aluminium foil and stored at -20 °C until analysis.

Prior to analysis, tissue samples were homogenized using a commercial blender. The validity of analytical methods was confirmed with Standard Reference Materials (CARP-2: ground whole carp, *Cyprinuscarpio*) obtained from the National Research Council of Canada. Precision and accuracy are reported in Table 2.

Aliquots (4 -7 g) of the homogenized samples were ground with anhydrous sodium sulphate in a mortar, and extracted using Soxhlet apparatus for 6 h with 300 ml of diethyl ether:hexane (3:1) solvent mixture. Extracts were concentrated in volume to 10 ml in Kuderna-Danish and the aliquots (2 ml) were transferred to a glass column packed with 20 g of Florisil and dried by passing through nitrogen gas. Organochlorines adsorbed on Florisil were eluted with 150 ml of 20 % hexane-washed water in acetonitrile and transferred to a separatory funnel containing 600 ml of hexane-washed water and 100 ml of hexane. After partitioning, the hexane layer was concentrated, cleaned up with sulphuric acid, and passed through a 12 g Florisil packed glass column for separation.

Table 2. Precision and accuracy of analytical methods obtained using a certified ground whole carp Cyprinuscarpio (CARP-2)[a]

PCBs	CARP-2	
	Certified	Found[b]
PCB 28	34 ± 4.0	31 ± 3.1
PCB 52	138 ± 43	119.7 ± 13.9
PCB 101	145 ± 48	150.1 ± 24.1
PCB 138	103 ± 30	99.5 ± 18.7
PCB 153	105 ± 22	116 ± 16.6
PCB 180	53.3 ± 13.0	59.2 ± 10.7
PCB 209	4.6 ± 2.0	4.8 ± 3.6
o,p’-DDE	2.9 ± 0.5	2.4 ± 0.3
o,p’-DDD	21.8 ± 0.7	25.1 ± 1.1

[a]The concentrations are given in ng/g wet weight.

[b]Number de replicates is 4.

Final determination of organochlorine compounds PCBs was carried out using a Varian 3600 gas chromatograph fitted with an electron capture detector (GC-ECD). All analyses used a fused-silica capillary column Supelco (length = 30 m, inside diameter 0.53 mm and film thickness 0.50 μm). The column oven was programmed from 60 to 160 °C, held for 10 min, and then increased to 260 °C at a rate of 2 °C/min and held for 20 min. Injector and detector temperatures were set at 260 and 280 °C respectively. Nitrogen was used as carried gas with 63.3 ml/min. The patrons used as internal standards were the PCB-Mix 12, Iso-octane, and the *o,p'*-DDT and metabolites Mix 32500 (Lab. Dr. Ehrenstorfer) for PCBs and OCs respectively. Quantification interval used for organochlorine compounds was from 1 ng/g (instrumental detection limit) to 50000 ng/g (optimum linear limit). Concentrations of PCBs and OC-DDTs, means of four measurements, are presented as ng/g on a wet weight basis. The data were evaluated using the Statistics Software SPSS V. 13.0.

Mann-Whitney's U test was conducted to determine whether the difference in the levels of PCBs and OC-DDTs were related to the tissues. Spearman's rho correlation method was used to calculate the correlation between ∑PCB concentrations and ∑OC-DDT in both tissues and physical conditions such as cachexia and septicemia. According to the necropsy reports, values of 0, 1, and 2 were given to turtles with absence of cachexia, mild cachexia, and severe cachexia respectively. Values of 0, 1, and 2 were also given to turtles with absence of infections, infection in one organ, and severe septicemia.

PATHOLOGICALSTUDY

Table 3 lists the lesions observed in the surveyed sea turtles. Cachexia was diagnosed in 8 turtles (16.32%) based on muscle atrophy, sunken eyes and concave plastron.

Ulcerative skin lesions attributed to trauma were the most common gross lesions, occurring in 22.44% of turtles examined. The distribution of the skin lesions was multifocal, with multiple lesions (1 to 4 cm in diameter) especially on the dorsum of the neck, head, and flippers. Histologically skin lesions were characterized by the ulceration of the epidermis, necrosis of the epidermis and dermis with infiltrating degenerate granulocytes and bacteria, and an underlying layer of multinucleated giant cells. Bacterial colonies were detected in all cases of ulcerative skin lesions. In addition, in the 63.63% of the turtles suffering from ulcerative skin lesions, other lesions of bacterial origin were detected in other major organs. These traumatic injuries were the sources from which pathogenic bacteria entered the tissues and the bloodstream resulting in fatal septicemia. In a survey of the diseases of farmed sea turtles in northern

Australia (Glazebrook & Campbell 1990a), ulcerative dermatitis attributed to trauma was observed in 66/102 turtles (63.5%), being associated with opportunistic *Vibrio alginolyticus*, *Aeromonas hydrophila*, *Pseudomonas* sp., and *Chryseobacterium* sp. infection. *Vibrio alginolyticus* is regarded as a normal inhabitant of seawater (Roberts 1978), while *Aeromonas hydrophila* has long been recognized as an opportunistic pathogen of reptiles (Reichenbach-Klinke & Elkan 1965, Rosenthal &Mader 1996).

Multifocal granulomatous pneumonia was the most frequently observed respiratory lesion (10.20%), being associated with granulomatous pleuritisin several cases (6.12%) (Figure 1A). These lesions were characterized by the presence of well defined granulomas containing central areas of eosinophilic necrotic debris mixed with heterophils and macrophages. Multinucleated giant cells surrounding the necrotic debris were observed in some granulomas. All granulomas were negative for mycobacteria (ZN), and one case was positive for fungi (GMS, PAS)(Figure 1B). Exudative bronchopneumonia was characterized by the presence of numerous heterophils in the central lumen of bronchi and bacterial colonies. In the majority of the cases the pneumonic lesions were unilateral.

Table 3. Systemic lesions and diseases of sea turtles (n = 49) stranded in the Canary Islands

System	Lesions	n	%	System	Lesions	n	%
General status	Cachexia	8	16.3	Digestive			
	Anemia	7	14.3	Oral cavity	Ulcerative glossitis	1	2.04
Integument Skin				Esophagus	Ulcerative esophagitis	10	20.4
					Fibrinous esophagitis	2	4.1
	Traumatic ulcerative dermatitis	11	22.4				
	Subcutaneous edema	3	6.1		Traumatic esophageal perforation	5	10.2
Respiratory				Stomach	Catarrhal gastritis	2	4.1
Trachea	Tracheitis	1	2.04		Parasitic gastritis	4	8.2
Lungs	Granulomatous pneumonia	5	10.2		Traumatic gastric		

System	Lesions	n	%	System	Lesions	n	%
					perforation	2	4.1
	Interstitial pneumonia	1	2.04	Intestine	Catarrhal enteritis	3	6.1
	Exudative pneumonia	2	4.08		Necrotizing enteritis	5	10.2
	Pulmonary edema	3	6.1		Fibrinous enteritis	1	2.04
	Granulomatous pleuritis	3	6.1		Fibrinousserositis	2	4.1
	Fibrinouspleuritis	2	4.1		Intussusception	5	10.2
Cardiovasc. Heart				Liver	Fibrinousperihepatitis	3	6.1
	Hydropericardium	2	4.1		Granulomatous hepatitis	8	16.3
	Fibrinous pericarditis	2	4.1				
Blood vessels	Thrombosis	2	4.1		Hepatic lipidosis	3	6.1
Immune Spleen				Pancreas	Chronic interstitial pancreatitis	1	2.04
	Necrotizing splenitis	3	6.1	Celomicwall	Fibrinouscelomitis	4	8.2
	Granulomatous splenitis	2	4.1	Endocrine			
	Splenomegaly	2	4.1	Adrenal glands	Adenomegaly	1	2.04
Peyer's patches	Hyperplasia	5	10.2	Reproductive	Ovarian follicular cyst	1	2.04
Excretory				Muscle	Necrotizing myositis	4	8.2
Kidneys	Granulomatous nephritis	3	6.1		Cachetic myopathy	5	10.2
	Chronic interstitial nephritis	2	4.1	Bone	Fractures	4	8.2
	Tubulonephrosis	3	6.1		Traumatic amputation	6	12.2
Bladder	Catarrhal cystitis	1	2.04	Carapace/ Plastron	Lack of ossification	5	10.2
	Granulomatous serositis	2	4.1		Traumatic erosions	6	12.2
Salt glands	Purulent adenitis	5	10.2		Fractures	5	10.2

Table 3. (continued)

System	Lesions	n	%	System	Lesions	n	%
					Deformation (kyphosis/ lordosis)	2	4.1
				Nervous	Spinal cord compression	1	2.04
					Meningel hemorrhages	2	4.1
				Eye	Ulcerative keratitis	3	6.1
					Purulent kerato-conjunctivitis	4	8.2

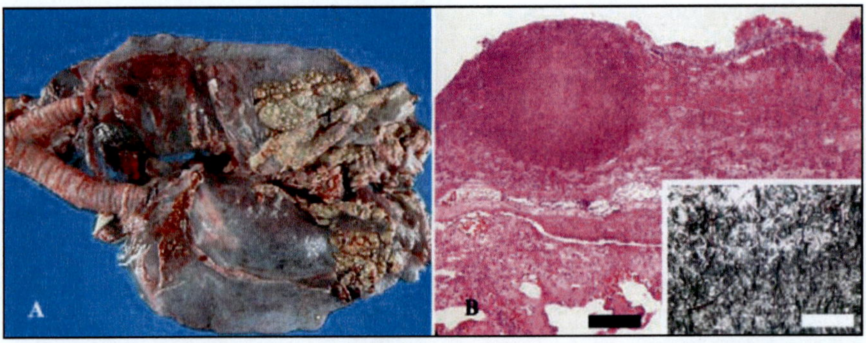

Figure 1. (A) Granulomatous pleuritis on the caudal area of both lungs of a loggerhead turtle. (B) Granulomatous inflammatory reaction in the pleura. HE. Bar = 200 µm. Inset: Micotic hyphae in the center of the granuloma. GMS. Bar = 60 µm.

In a previous study on sea turtles stranded in the Canary Islands, several bacteria including *A. hydrophila*, *Bacillus* sp., *Burkholderiacepacia*, *Citrobacter* sp., *Pasteurella* sp., *Proteus* sp., *Pseudomonas* sp., *Staphylococcus* sp., *Streptococcus* sp., and *V. alginolyticus* were associated with exudative bronchopneumonia and/or granulomatous pneumonia (Orós *et al.* 2005). Similar bacteria were isolated from bronchopneumonic lesions of turtles in Australia (Glazebrook & Campbell 1990a,b). *Vibrio alginolyticus*, *A. hydrophila*, and *Chryseobacterium* sp. have been repeatedly isolated from cases of ulcerative obstructive stomatitis and the rhinitis-pneumonia disease complex in hatchling and juvenile loggerhead and green turtles (Glazebrook *et al.* 1993). Granulomatous focal pneumonia has been frequently associated with *Paecilomyces* sp. infection in green turtles (Jacobson *et al.* 1979). The high prevalence of pulmonary lesions observed in sea turtles may indicate a predisposition to suffer from these pathologic processes. George (1997) considered the aspiration of seawater as one of the primary modes by which bacteria entered the respiratory system and caused clinical disease. In other cases the pulmonary lesions were secondary to traumatic injuries in the carapace, as a consequence of the position of the lungs, those being dorsally attached to the carapace and the vertebral column (Wyneken 2001). The majority of the pneumonic lesions observed in our survey were unilateral, and it seemed likely that the unilateral pneumonia in most cases was due to aspiration or full thickness shell trauma. Buoyancy problems observed in the turtles may indicate a pulmonary disease (Jacobson 1997). However, buoyancy disorders of sea turtles are not always associated with pulmonary lesions; they may also result from excessive gas in the gastrointestinal tract, such as occurs during an obstruction (Campbell 1996, Jacobson 1997).

Lesions in the cardiovascular system were not frequent. Fibrinous exudative pericarditis (4.08%) was associated with bacteria and fibrinous lesions in other major organs. Thus, these lesions were expression of a septicemic status, rather than a specific cardiovascular disease. The most common cardiovascular pathology described in sea turtles is spirorchid fluke infection (Glazebrook *et al.* 1989, George 1997, Gordon *et al.* 1998). High prevalences of sea turtle spirorchidiasis have been reported in other surveys around the world (Glazebrook *et al.* 1989, Glazebrook & Campbell 1990b, Gordon *et al.* 1998). Adult flukes have been found in the heart and major blood vessels, causing lesions such as mural endocarditis, arteritis, thrombosis or aneurysms. Dissemination of the fluke eggs throughout the vascular system results in vasculitis and microgranulomas wherever they lodge (Wolke*et al.* 1982, Glazebrook *et al.* 1989, Gordon *et al.* 1998). However, none of the sea turtles included in our study had intravascular adult flukes or trematode eggs in tissues. All loggerhead and green turtles included in our survey were juvenile and subadult specimens, whereas other studies in which spirorchidiasis was reported included adult turtles. Different alimentary habits of the juvenile and subadult turtles on the coasts of the Canary Islands can also explain the absence of spirorchid fluke infection. Sea turtles acquire the flukes by ingesting unknown cercaria-rich intermediate hosts. The life cycle of spirorchids in sea turtles is unknown but is probably similar to that of fresh water turtles (Holliman & Fisher 1968).

Multifocal necrotizing splenitis was the most frequent lesion that affected the spleen (6.12%).All cases of multifocal necrotizing splenitis were associated with septicemic processes of bacterial origin. Thus, the lesions involving the spleen were an expression of a septicemic status, rather than a specific splenic disease.

Several lesions were observed in the excretory system. Lesions included, granulomatous nephritis, chronic interstitial nephritis, and tubulonephrosis, affecting 6 turtles (12.24%). Chronic interstitial nephritis was characterized by the presence of aggregates of lymphocytes in the renal interstitium. Granulomatous nephritis was characterized by small, yellow-white, granulomatous foci scattered randomly throughout the kidney. Microscopically, granulomas were characterized by central foci of necrosis surrounded by macrophages and multinucleated giant cells. In two cases the renal lesions were secondary to traumatic injuries in the caudal aspect of the carapace, related to the anatomical location of the kidneys, retrocelomically and dorsally attached to the caudal carapace (Wyneken 2001).In two cases, the renal lesions were attributed to a septicemic status with multiorganic involvement.In a previous study on sea turtles stranded in the Canary Islands, granulomatous nephritis and renal abscesses were associated with *A. hydrophila*, *Citrobacter* sp., *E. coli*, *Proteus* sp., *V. alginolyticus*, and *Staphylococcus* sp. infections (Orós *et al.* 2005).

Exudative salt gland adenitis was observed in 9 loggerhead turtles (18.36%) (Figure 2).Lesions ranged from the presence of abundant eosinophilic material associated with bacterial colonies in the lumen of the central ducts of the glandular lobules to the destruction of the glandular tissue and presence of abundant eosinophilic material composed by heterophils and cell debris, lined by multinucleated giant cells. Multinucleated giant cells were detected around the inflammatory foci in all cases. Because the kidney of the sea turtles cannot produce hypertonic urine, excess salt entering the body down the concentration gradient, ingested in the food or by drinking seawater is largely excreted by paired salt glands (Reina *et al.* 2002). Salt glands are composed of specialized, secretory cells that concentrate sodium and chloride from the blood to the lumen of secretory tubules through an energy-dependent process (Abel & Ellis 1966). Sea turtle salt glands secrete a solution composed almost entirely of sodium chloride at approximately 1500-1800 mosmol/l (Marshall & Cooper 1988, Nicolson & Lutz 1989, Reina & Cooper 2000) in response to increasing plasma sodium concentration, and their activity is regulated by microcirculatory changes in or near the glands (Reina 2000).

Very few descriptions of salt gland diseases in wild sea turtles could be found in the literature. Common lesions include pale spots or calculi, which indicate malfunction in solute dissolution, which may occur in severely dehydrated marine turtles. This manifests as hard rugose deposits associated with necrosis of surrounding tissues (Flint *et al.* 2009). Granulomas associated with spirorchiid eggs, of variable severity, are also common in stromal tissue that surrounds central canals of lobules and may sometimes extend into and disrupt the glands themselves (Flint *et al.* 2009). Glazebrook & Campbell (1990a) reported *Pseudomonas* sp. infections in the salt glands of two farmed green sea turtles due to the removal of foreign material from the main excretory duct leading from the posterior orbit. The procedure involved the use of forceps which introduced bacteria and led to infection.

Ulcerative and fibrinous esophagitis and traumatic esophageal perforation were the most frequently observed lesions in the esophagus, being associated in all cases with ingestion of fishing hooks. Hooks were identified as belonging to the longline tuna fishery. Lesions were characterized by the ulceration of the esophageal mucosa, coating in eosinophilic fibrillar material consistent with fibrin, with infiltrating degenerate granulocytes and bacteria in the lamina propria and/or submucosa, and an underlying layer of multinucleated giant cells. The estimated total take of sea turtles by the longline tuna fishery in the international waters in the North of the Canary Islands for 1983 to 1991 was 3000 turtles (López-Jurado& González 1983, Blanco & González 1992). Although many of the turtles are liberated, it is estimated that approximately 15-50% of the turtles die due to the severe lesions caused by the fishing hooks (Lizana&Barbadillo 1997). Many of the turtles with traumatic lesions in the esophagus and stomach had also systemic lesions due to septicemia (Figure 3). The lesions induced by the fishing hooks in the esophagus and stomach were the primary routes by which pathogenic bacteria entered the tissues and the bloodstream resulting in fatal septicemia.

Larval nematodes from Anisakidae family caused gastritis in 4 turtles (8.16%)(Figure 4A). The migration of larvae from the lumen of the stomach to the celomic cavity was characterized by the presence of granulomas in the lamina propria, submucosa (Figure 4B), lamina muscularis and serosa of the stomach and mucosal ulceration. The diagnosis of a parasitic gastritis in sea turtles caused by the larval migration of nematodes belonging to the family Anisakidae has been described for the first time by Carr & Main (1973) and Rodgers & Burke (1982). Turtles act as paratenic hosts for third stage larvae unable to complete their life cycle when confined to the tissues of an ectothermic species (Glazebrook & Campbell 1990a). The migration of larvae from the lumen of the stomach to the celomic cavity resulted in ulceration of the gastric mucosa, and granulomatous inflammatory reactions. This pathologic condition was not considered a direct cause of mortality in any of the turtles examined in our survey.

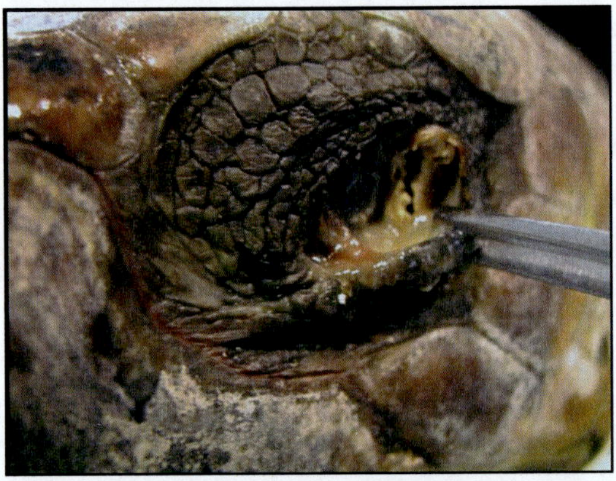

Figure 2. Purulent salt gland adenitis in a loggerhead turtle.

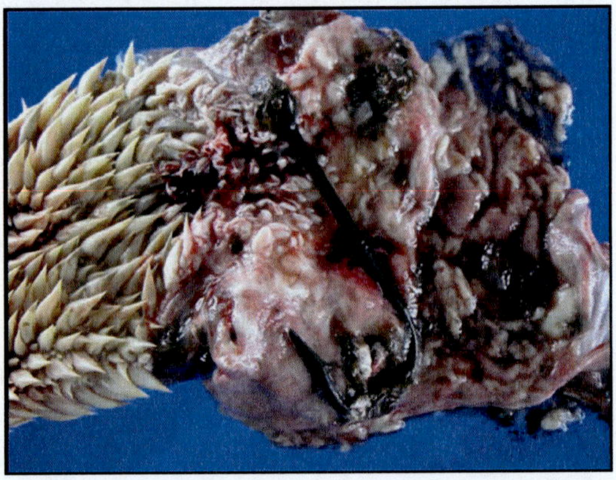

Figure 3. Lesion in the cardial area of the stomach of a loggerhead turtle caused by a fish hook.

Different histological types of enteritis included catarrhal enteritis, fibrinous enteritis, necrotizing enteritis, and fibrinous intestinal serositis, affecting 9 turtles (18.36%). In addition, all turtles with fibrinous intestinal serositis also demonstrated fibrinouscelomitis, being also associated with multisystemicsepticemic lesions. All cases of necrotizing enteritis were associated with intestinal intussusception caused by ingestion of monofilament fishing lines(Figure 5). Ingestion of monofilament fishing lines is a serious threat to sea turtles. Monofilament line can stop normal gut function and results in death when a strand that extends along a portion of the intestine becomes lodged at the anterior end.

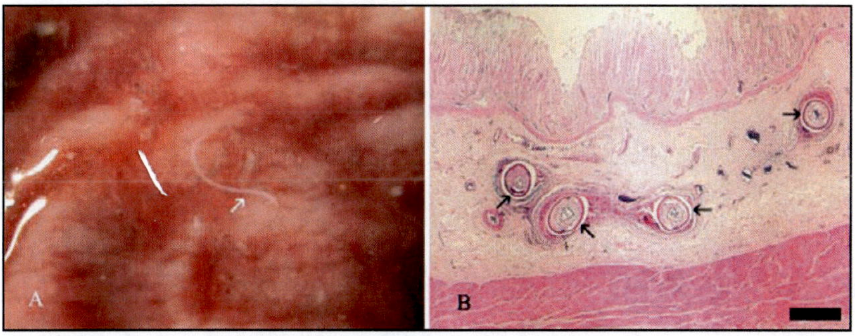

Figure 4. (A) Larval nematode (arrow) on the stomach of a loggerhead turtle. (B) Larval nematodes (arrows) in the gastric submucosa. Note the inflammatory reaction around the parasites. HE. Bar = 1200 μm.

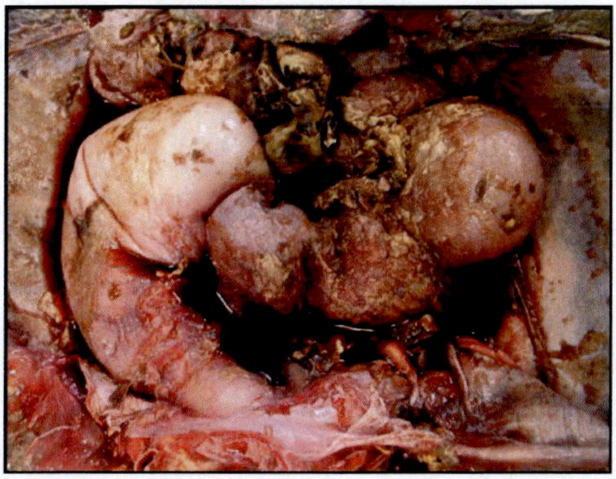

Figure 5. Intestinal intussusception and fibrinopurulentserositis associated to a monofilament fishing line in a loggerhead turtle.

Normal peristalsis of the intestine and movements of the digesta will result in the intestine becoming gathered along the length of the line (Bjorndal*et al.* 1994). Intestinal volvuli associated with different causes have been frequently reported in sea turtles (Schumacher *et al.* 1996, Hasbún*et a*l. 1998, Helmicket al. 2000). However, this lesion was not observed in the present survey.

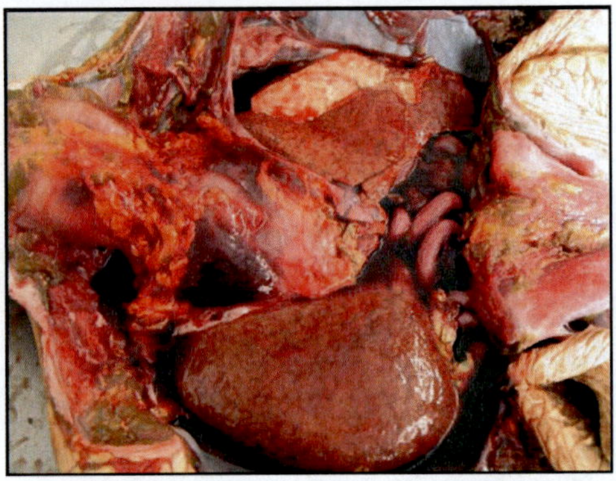

Figure 6. Fibrinopurulentperihepatitis (left hepatic lobe) in a loggerhead turtle.

Multifocal granulomatous hepatitis was the lesion most commonly observed in the liver (16.32%). Granulomatous hepatitis was characterized by small, yellowish-whitish, granulomas scattered randomly throughout the liver. Microscopically, these granulomas were characterized by necrotic centers surrounded by macrophages and multinucleated giant cells. These hepatic lesions were attributed to multisystemicsepticemic status (Figure 6).In only one case these lesions were associated with fungi. In a previous study on sea turtles stranded in the Canary Islands,necrotizing and/or granulomatous hepatitis were commonly observed (26/93 or 27.95%), being associated with *A. hydrophila*, *Citrobacter* sp., *E. coli*, *Proteus* sp., *Staphylococcus* sp., and *V. alginolyticus* infections (Orós *et al*. 2005).

Cachetic myopathy was characterized by generalized atrophy of skeletal muscles, affecting 5 turtles (10.20%).Lesions of focal necrotizing myositis (8.16%) were attributed to entanglement in fishing nets or boat strikes, and were characterized by necrosis of the myofibers with infiltrating degenerated heterophils and bacteria, and an underlying layer of multinucleated giant cells. These lesions were similar to those described by George (1997). Calabuig (1998) reported that 56% (45/72) of the turtles submitted to the TWRC in 1998 had similar lesions due to entanglement. These lesions have been previously described by others (Schroeder 1987, Teas 1994, George 1997).

Affections of the skeletal system included traumatic amputation of one or two flippers (12.24%), and multiple bone fractures (8.16%). These amputations were mainly related to entanglement in fishing nets.

Traumatic erosions and/or fractures of the carapace/plastron were frequently observed (22.44%)(Figure7). These lesions were mainly attributed to boat strikes.

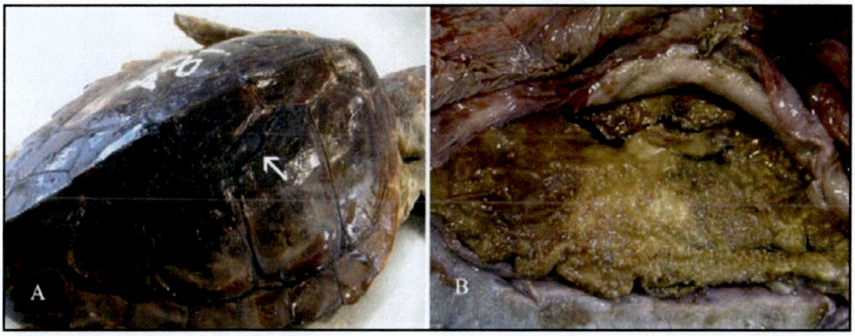

Figure 7. (A) Traumatism (arrow) in the carapace of a loggerhead turtle. (B) Purulent lesions in the right lung and thoracic cavity.

Diseases of the nervous system were uncommon and included compression of the spinal cord (associated with ciphosis) and meningeal hemorrhages associated with cranial trauma.

Eye lesions included heterophilic keratoconjunctivitis and ulcerative keratitis, affecting 6 turtles (12.24%). Heterophilic keratoconjunctivitis was characterized by the presence of large amounts of heterophils in the corneal stroma infiltrating the conjunctiva. Ulcerative keratitis was characterized by loss of epithelium of the cornea and the presence of large numbers of mixed inflammatory cells in the stroma. In all cases bacterial colonies were associated with the lesions. Histological examination of the ocular and periocular lesions did not reveal intranuclear inclusion bodies indicative of a herpesvirus infection (Jacobson *et al.* 1986).

The absence of fibropapillomatosis in the sea turtles stranded in the Canary Islands during this survey is remarkable because fibropapillomatosis of sea turtles is a disease of global distribution. However, fibropapillomatosis affects several populations in epidemic proportions throughout the world, and many observations have been reported in Florida, Hawaii, Puerto Rico, Barbados, Virgin Islands, Panama, Colombia, Venezuela, Belize, Australia and Indonesia (Balazs 1991, Jacobson 1991, Adnyana*et al.* 1997, Work *et al.* 2004). The prevalence started to increase in the 1980's at multiple sites including Florida and Hawaii. The prevalence of fibropapillomatosis in sea turtle populations in Hawaii has been observed as high as 92% (Balazs 1991). In the Indian River Lagoon System of east central Florida, 57% of green turtles collected between 1985 and 1986 hadfibropapillomatosis (Jacobson *et al.* 1989).

TOXICOLOGICAL STUDY

PCBs

PCB concentrations in both tissues (liver and fat) are shown in Table 4. The turtle tissues analyzed contained high levels of PCBs compared to concentrations reported in turtles collected in other locations around the world (McKim& Johnson 1983, Lake et al. 1994, Rybitski et al. 1995, Mckenzie et al. 1999, Corsolini et al. 2000, Storelli&Marcotrigiano 2000, Gardner et al. 2003, Keller et al. 2004, Perugini et al. 2006, Storelli et al. 2007).

Loggerheads showed the highest hepatic ∑PCB (IUPAC nos. 28, 52, 101, 138, 153, 180) mean concentration (1980 ± 5321 ng/g wet wt.), followed by the only specimen of leatherback turtle analyzed (445 ng/g wet wt.). ∑PCB concentration in the liver of the only specimen of green turtle analyzed was lower (116 ng/g wet wt.) than those detected in loggerheads and leatherback turtles. Loggerheads also showed the highest ∑PCB mean concentration in the adipose tissue (448 ± 1700 ng/g wet wt.).

The high degree of variation of concentrations of PCBs detected in our study may be attributed to the different exposure of individual turtles to pollutants and to the physical condition of the animals. Sea turtles visit different areas with different degree of contamination during migration, resulting in differences in exposure for each animal (Ziswiler 1986).According to recent studies, loggerhead turtles stranded on the coasts of the Canary Islands come from two different migration routes: from the Western Atlantic arriving to the Canary Islands by the Gulf Stream (Pérez-Jiménez 1997), and from Cape Verde (Monzón-Argüello 2009). In addition, turtles included in this study varied widely in physical condition, ranging from turtles that had been killed incidentally through human interaction, to turtles that showed severe emaciation and septicemia after prolonged illness.

Table 4.PCBs concentrations (range, mean ± standard deviation) (ng/g wet wt) in liver and fat of the three species of sea turtles analyzed

	Caretta caretta n = 30		*Chelonia mydas* n = 1		*Dermochelyscoriacea* n = 1	
	Liver	Fat	Liver	Fat	Liver	Fat
∑PCB	BDL – 34900 1980 ± 5320	BDL – 9828 448 ± 1700	116	144	445	77
PCB 28,31	BDL – 587 42 ± 131	BDL – 38 3.8 ± 9	BDL	5.7	71.5	9.7
PCB 52	BDL – 1012 98 ± 262	BDL – 56 5.5 ± 12.5	BDL	7.2	8.7	11.4
PCB 101	BDL – 671 32 ± 130	BDL – 84.5 6 ± 19	BDL	BDL	BDL	BDL
PCB 138	BDL – 13300 456 ± 2426	BDL – 2039 83± 370.5	BDL	BDL	BDL	BDL
PCB 153	BDL – 15440 915 ± 3178	BDL – 2604 133 ± 477	BDL	58	251	56
PCB 180	BDL – 3887 440 ± 1016	BDL – 5006 217 ± 907	116	73	114	BDL
PCB 209	BDL – 14877 1194 ± 3117	BDL – 218 15 ± 45.5	BDL	BDL	526	501

BDL: below detection limit.
∑PCB = sum of PCBs nos. 28, 52, 101, 138, 153, and 180.

Loggerheads showed the highest hepatic PCB 209 mean concentration (1194 ± 3117 ng/g wet wt.). Six loggerhead turtles had individual PCB concentrations higher than 2000 ng/g, and seven turtles showed individual PCB concentrations higher than 900 ng/g. All these turtles showed severe septicemia and/or severe cachexia. Statistically, a positive correlation was detected between ∑PCBs concentration and cachexia. Although sea turtles are long-lived high trophic organisms that would be expected to contain high concentrations of PCBs (Lake et al. 1994), the poor physical condition of these animals could be the reason of these high PCB levels. It has been described in cetaceans that many diseases affect metabolic centers and thus the capacity to metabolize or excrete pollutants may be affected, questioning whether PCB concentrations in the tissues of these specimens are representative of normal conditions (Aguilar et al. 1999).

As revealed by statistical analysis, turtles showed significant higher PCBs 28,31, PCB 153, PCB 209 and ∑PCBs concentrations in the liver than in the adipose tissue. However, no significant differences between concentrations in both tissues were detected for PCB 138 and PCB 180. According to other authors, the pattern of tissue distribution of PCBs reported in turtles and other marine species is the following: adipose tissue > liver > kidney > muscle (Tanabe et al. 1983, Martineau et al. 1987, Colborn& Clement 1992, Rybitski et al. 1995, Mckenzie et al. 1999). This is attributed to the highest total lipid content and triglycerides for the adipose tissues and the strong correlations between these and their burden of lipophilic pollutants (Cockroft et al. 1989). However, we detected in the liver ∑PCBs concentrations higher than in the adipose tissue. Thirteen turtles showed PCB levels in the liver higher than in the fat. All these turtles showed severe cachexia and/or septicemia. According to different studies, animals with abundant fat deposits can accumulate and tolerate higher concentrations of toxic chemicals because lipophilic pollutants are stored in the adipose tissue and are less available to target organs and receptors. In contrast, emaciated animals which have mobilized their lipid stores may be more susceptible to toxic effects as a result of remobilization of the pollutants, resulting in higher concentrations of PCBs in the remaining tissues like the liver (Bernhoft&Skaare 1994). Thus, the cachexia observed in our turtles may indicate a previous remobilization of the fat and the pollutants, and could explain the high concentration of PCBs in the liver. In addition, the septicemic condition reported in these turtles could be caused by a possible immunosupression as one of the most distinguished effects of organochlorine compounds.

The congener most frequently detected in the liver was PCB 180 (46.8%), followed by PCB 153 (31.2%), PCBs 28, 31 (25%), PCB 52 (25%), PCB 209 (15.6%), PCB 138 (15.6%), and PCB 101 (12.5%). The congeners most frequently detected in the adipose tissue were PCB 180 (50%) and PCB 153 (50%), followed by PCBs 28,31 (34.4%), PCB 52 (34.4%), PCB 209 (31.2%), PCB 138 (28.1%), and PCB 101 (18.7%).

PCB profiles in the loggerhead tissues were dominated by the higher chlorinated congeners. Hexachlorobiphenyls (PCBs 138 and 153) were predominant and accounted for 43.6% of the total PCBs, followed by decachlorobiphenyls (PCB 209) (33%) and heptachlorobiphenyls (PCB 180) (18%). Concerning individual congeners, the most abundant were PCB 209 (33%) followed by PCB 153 (29%), PCB 180 (18%), and PCB 138 (15%). The isomer pattern of PCBs in loggerhead turtles was different among liver and fat. In the liver the most abundant were PCB 209 (37.6%), followed by PCB 153 (29%), PCB 138 (14%) and PCB 180 (14%).In the fat the most abundant were PCB 180 (47%), followed by PCB 153 (28.6%), PCB 138

(18%) and PCB 209 (3%). Hexaclorobiphenyls and heptaclorobiphenyls are predominant in turtles and marine mammals (Muir et al. 1988, Rybitski et al. 1995, Colborn&Smolen 1996, Alam& Brim 2000, Corsolini et al. 2000, Storelli&Marcotrigiano 2000, Miao et al. 2001, Perugini et al. 2006, Storelli et al. 2007). However, Gardner et al. (2003) reported in Chelonia mydas a profile dominated by lower-chlorinated congeners. Differences in PCB patterns in turtles may be attributed to differences in the congener compositions of environmental media among regions, dietary differences or differences in the abilities of the various species and populations to metabolize PCBs (Gardner et al. 2003).

Regarding PCB profiles in the only specimen of green turtle analyzed, the most abundant was PCB 180 (73%) followed by PCB 153 (22%) and PCB52 (3%). Although PCB 180 was the most abundant in both tissues, this congener was the only PCB detected in the liver of this turtle.

Regarding PCB profiles in the only specimen of leatherback turtle analyzed, the most abundant was PCB 209 (66%) followed by PCB 153 (20%) and PCB 180 (7%). The isomer pattern of PCBs in this turtle was slightly different among liver and fat. In the liver the most abundant was PCB 209 (54%) followed by PCB 153 (26%) and PCB 180 (12%).In the fat the most abundant was PCB 209 (87%), followed by PCB 153 (10%), and PCB 52 (2%).

Hepatic ∑PCBs concentrations were positively correlated with cachexia ($r = 0.250$, $P = 0.04$). No significant correlation was observed between ∑PCBs concentrations and septicemia.

OC-DDTs

OC-DDTs concentrations measured in fat and liver and frequencies of detection of OC-DDTs in both tissues of loggerhead turtles are reported in Table 5. Due to the limited number of studies on tissue distribution of OC-DDTs in sea turtles, it is difficult to compare the results obtained in this study. The majority of surveys have been dedicated to *p,p'*-DDT compounds, and very little information has been published on *o,p'*-DDT compounds. Only Perugini*et al.* (2006) reported *o,p'*-DDT concentrations in the liver (BDL-21.7 ng/g) and fat (BDL-9.5 ng/g) of loggerhead turtles from the Adriatic Sea. In our study, *o,p'*-DDT concentrations in both tissues were higher than those detected by Peruginiet *al.* (2006). It may imply a relatively larger usage of technical DDT in our study area in recent years.

Table 5.*o,p'*-DDT concentrations (range, mean ± standard deviation) (ng/g wet wt) and frequencies of detection (percentage) in liver and fat of the three species of sea turtles analyzed

	Caretta caretta n = 30		*Chelonia mydas* n = 1		*Dermochelyscoriacea* n = 1	
	Liver	Fat	Liver	Fat	Liver	Fat
∑DDT	BDL – 810.8 78.8 ± 163.8 (43.47%)	BDL – 449 33.8 ± 84.3 (26.08%)	250.9	BDL	BDL	BDL
o,p'-DDT	BDL – 243.8 25.2 ± 60.7 (17.39%)	BDL – 327.2 26.7 ± 70.3 (21.73%)	250.9	BDL	BDL	BDL
o,p'-DDD	BDL – 798.8 49.8 ± 156.6 (30.43%)	BDL – 121.8 6.1 ± 24.6 (13.04%)	BDL	BDL	BDL	BDL
o,p'-DDE	BDL – 26.2 3.18 ± 6.75 (13.04%)	BDL	BDL	BDL	BDL	BDL

BDL: below detection limit (1 ng/g).
∑DDT = *o,p'*-DDT + *o,p'*-DDE + *o,p'*-DDD.

Although DDT was banned in the 1970s in North America and Europe, it has not been well documented when its use was definitely ended in the Canary Islands (Zumbado*et al.* 2005). It is well known that in underdeveloped African countries, OC-DDTs are still running, and it is necessary to have into account that the Canary Islands are located very close to Africa (Zumbado*et al.* 2005). In addition, studies of detection of organochlorine pesticides in pine needles (Villa *et al.* 2003) and serum samples from people from the Canary Islands (Zumbado*et al.* 2005) allow those authors hypothesize that DDT has been used recently in the Canary Islands, particularly in the islands devoted to intensive agriculture (Zumbado*et al.* 2005).

The high degree of variation of concentrations of OC-DDTs detected in our study may be attributed to the different exposure of individual turtles to pollutants and to the physical condition of the animals. Turtles included in this study varied widely in physical condition, ranging from turtles that had been killed incidentally through human interaction, to turtles that showed severe emaciation after prolonged illness.

As revealed by statistical analysis, OC-DDTs were differentially distributed in both tissues with the liver having the highest concentration. Regarding the pattern of tissue distribution of OC-DDTs in two grouped categories [emaciated turtles (n =8) and turtles with normal weight conditions (n= 24)], emaciated turtles showed higher average ∑DDT concentrations in the liver (98.56 ± 55.31ng/g) than in the fat (15.31 ± 45.56 ng/g), whereas turtles with normal body conditions showed higher average ∑DDT concentrations in the fat (39.76 ± 69.61ng/g) than in the liver (32.36 ± 57.13ng/g). The differences in the pattern of OC-DDTs among liver and fat are attributed to the hepatic metabolization.

The pattern of tissue distribution of OC-DDTs reported in sea turtles is the following: adipose tissue > liver > muscle (Rybitski*et al.* 1995, Mckenzie *et al.* 1999, Perugini*et al.* 2006). However, we detected in the liver ∑DDT concentrations higher than in the adipose tissue. Eleven turtles showed ∑DDT levels in the liver higher than in the fat. All these turtles showed severe cachexia. The cachexia observed in our turtles may indicate a previous remobilization of the fat and the pollutants, and could explain the high ∑DDT concentration in the liver.

Of the different organochlorine pesticides, *o,p'*-DDD was predominant and accounted for 53.15% of the total OC-DDTs, followed by *o,p'*-DDT (44.18%) and *o,p'*-DDE (3.76%). The pattern of OC-DDTs was different among liver and fat. In the liver the most abundant was *o,p'*-DDD(66.13%), followed by *o,p'*-DDT (30.09%), and *o,p'*-DDE (3.76%). In the fat the most abundant was *o,p'*-DDT (78.01%), followed by *o,p'*-DDD (22.08%). No detectable concentrations for *o,p'*-DDE were found in the fat samples.*o,p'*-DDD is a component of chlordane, a widely used pesticide. However, according to the bibliography, *p,p'*-DDE is, in general, the pesticide present in the greatest concentrations in sea turtles due to its highly persistent nature (Lake *et al.* 1994, Rybitsky *et al.* 1995, Mckenzie *et al.* 1999, Storelli & Marcotrigiano 2000, Keller *et al.* 2004, Perugini *et al.* 2006,Storelli *et al.* 2007). In our study, the predominance of *o,p'*-DDD may imply a species-specific metabolic capability and/or a specific exposure in reductive (anoxic) conditions. It is also remarkable that in the survey published by Gardner *et al.* (2003) *p,p'*-DDD contributed most to ∑DDT. The very low concentration for *o,p'*-DDE has been reported previously in human beings from the Canary Islands (Zumbado *et al.* 2005).

CONCLUSION

According to the stranding reports, clinical reports, gross and microscopic lesions, and toxicological results, we attempted to establish the causes of mortality of the turtles examined in this survey (Table 6). Only 12 turtles (24.49%) died from spontaneous diseases including different types of pneumonia, hepatitis, and septicemic processes. However, 37 turtles (75.51%) died from lesions associated with human activities such as ingestion of hooks and monofilament lines (34.69%), entanglement in derelict fishing nets (24.49%), and boat strike injuries (16.33%).

Table 6. Estimated causes of mortality in the stranded sea turtles in this study

Cause		n	%
Spontaneous diseases		12	24.49
Anthropogenic causes		37	75.51
	Hooks and monofilament lines	17	34.69
	Entanglement	12	24.49
	Traumatism	8	16.33

Net entanglement is the most important cause of sea turtle stranding in the Canary Islands and frequently the turtles land with a part of the net (Calabuig 1998). Entanglement in fishing nets resulted in severe ulcerative dermatitis, amputation of flippers, and septicemic processes. A comparison of our survey and the data provided by the sea turtle stranding registry of the TWRC indicates that turtles with lesions induced by entanglement usually have a better prognosis for rehabilitation that those with traumatic lesions in the carapace or plastron. Affection of vital organs such as lungs and kidneys, because the anatomical location, dorsally attached to the carapace, explain the generally poor prognosis for turtles with severe traumatic injuries in the carapace.

To assess the impact of organochhorine compounds in the turtles of our study is difficult because the most frequent causes of death were attributed to fishing activity. In addition, few toxicological studies have been published regarding organochlorines in reptiles and no safe concentrations have been established. It is remarkable that almost all turtles with severe septicemia showed very high levels of PCBs. Immunosupression as result of PCBs pollution has been previously described (Keller *et al.* 2004, 2006). However, the presence in our study of other factors different to PCBs concentrations, as incidental fishing, nutritional status, and exposition to different micro-organisms makes difficult to establish a clear association between PCB concentrations and causes of death. According to the histopathological study no lesions exclusively attributed to acute effects of PCBs were described. All the histological lesions were attributed to bacterial and/or fungal infections mainly associated with gross lesions caused by fishing activity. However, chronic effects of PCBs are much difficult to determine.

Histologically, we detected hepatic lesions in four loggerhead turtles with high $\sum$DDT concentrations. Lesions included diffuse vacuolar hepatic degeneration and hepatic lipidosis. These lesions were not exclusively attributed to acute effects of OC-DDTs because clinical history of these turtles also included severe septicaemia. However, chronic effects of OC-DDTs can not be discarded.

The detection of high levels of PCBs and OC-DDTs in these turtles makes necessary to continue to monitor the levels of these chemicals in turtles from the Canary Islands to contribute to the protection of these currently endangered species. In addition, studies of detection of other persistent organochlorines in loggerhead turtles stranded in the Canary Islands are necessary.

A multidisciplinary approach to this focus carried out by veterinarians, biologists, and scientists, is necessary in order to compile data that prove the main threats for these endangered reptiles, and to design adequate strategies of conservation.

ACKNOWLEDGMENT

The authors would like to thank members of the Tafira Wildlife Rehabilitation Center (Cabildo de Gran Canaria), specially P. Calabuig, for providing us the turtles. This study was partially supported by the Consejería de Medio Ambiente del Gobierno de Canarias. We are grateful to Elsevier for the permission to use the data of the toxicological studies previously published by us in the journals *Marine Pollution Bulletin* and *Chemosphere*.

REFERENCES

Abel, J. H. & Ellis, R. A. (1966).Histochemical and electron microscope observations on the salt secreting lachrymal glands of marine turtles.*American Journal of Anatomy, 118,* 337-358.

Adnyana, W., Ladds, P. W. & Blair, D. (1997). Observations of fibropapillomatosis in green turtles (*Chelonia mydas*) in Indonesia. *Australian Veterinary Journal, 75,* 737-742.

Aguilar, A., Borrel, A. & Pastor, T. (1999).Biological factors affecting variability of persistent pollutant levels in cetaceans. *Journal of Cetacean Research and Management, 1,* 83-116.

Ahlborg, U. G., Becking, G. C., Birnbaum, L. S., Brouwer, A., Derks, H. J. G. M., Feeley, M., Golor, G., Hanberg, A., Larsen, J. C., Liem, A. K. D., Safe, S. H., Schlatter, C., Waern, F., Younes, M. &Yrjänheikki, E. (1994). Toxic equivalency factors for dioxin-like PCBs: report on WHO-ECEH and IPCS consultation, December 1993. *Chemosphere, 28*, 1049-1067.

Alam, S. K. & Brim, M. S. (2000). Organochlorine, PCB, PAH, and metal concentrations in eggs of loggerhead sea turtles (*Caretta caretta*) from northwest Florida, USA. *Journal of Environmental Science and Health, 35,* 705-724.

Balazs, G. H. (1991) Current status of fibropapillomas in the Hawaiian green turtle, *Chelonia mydas*. In G. H. Balazs& S. G. Pooley (Eds.), *Research Plan for Marine Turtle Fibropapilloma*(pp. 47-57). Honolulu, HI: NOAA Technical Memorandum NMFS-SWFC-156, National Oceanic and Atmospheric Administration.

Bancroft, J. D. & Stevens, A. (1996). *Theory and Practice of Histological Techniques* (4th edition). New York, NY: Churchill Livingstone.

Barbadillo, L. J., Lacomba, J. I., Pérez-Mellado, V., Sancho, V. &López-Jurado, L. F. (1999). *Anfibios y Reptiles de la Península Ibérica, Baleares y Canarias*. Barcelona, Spain: Geoplaneta Editorial.

Bernhoft, A. &Skaare, J. U. (1994). Levels of selected individual polychlorinated biphenyls in different tissues of harbour seals (*Phocavitulina*) from the southern coast of Norway. *Environmental Pollution, 86,* 99-107.

Bjorndal, K. A., Bolten, A. B., Koike, B., Schroeder, B. A., Shaver, D. I., Teas, W. G. &Witzell, W. N. (2001). Somatic growth function for immature loggerhead sea turtles, *Caretta caretta*, in southeastern U. S. waters - Statistical Data included.*Fishery Bulletin, 99,* 240-246.

Bjorndal, K. A., Bolten, A. B. &Lagueux, C. J. (1994). Ingestion of marine debris by juvenile sea turtles in coastal Florida habitats. *Marine Pollution Bulletin, 28,* 154-158.

Blanco, J. C. & González, J. L. (1992). *LibroRojo de los Vertebrados Españoles*. Madrid, Spain: Ministerio de Agricultura, Pesca y Alimentación. ColecciónTécnica ICONA.

Bolt, H. M. &Degen, G. H. (2002). Comparative assessment of endocrine modulators with estrogenic activity: II. Persistent organochlorine pollutants. *Archives of Toxicology, 76,* 187-193.

Calabuig, P. (1998). *Recuperación de Tortugas Marinas Accidentadas en las Islas Canarias.Memoria de ActividadesRealizadas en el Centro de Rehabilitación de Fauna Silvestre de Tafira.* Las Palmas de Gran Canaria, Spain: Cabildo de Gran Canaria, Área de Medio Ambiente.

Campbell, T. W. (1996). Sea turtle rehabilitation. In D. R. Mader (ed.), *Reptile Medicine and Surgery* (pp. 427-436). Philadelphia, PA: W. B. Saunders Company.

Carr, A. F. & Main, A. R. (1973).*Turtle Farming Project in Northern Australia. Report on an Inquiry into Ecological Implications of a Turtle Farming Project.*Camberra, Australia: Union Offset Pty Ltd.

Cockcroft, V. G., De Kock, A. C., Lord, D. A. & Ross, G. J. B. (1989). Organochlorines in bottlenose dolphins (*Tursiopstruncatus*) from the east coast of South Africa. *South African Journal of Marine Science, 8,* 207-217.

Colborn, T. & Clement, C. (1992). *Chemically Induced Alterations in Sexual and Functional Development: The Wildlife/human Connection.* Princenton, NJ: Princeton Scientific Publishing.

Colborn, T. &Smolen, M. J. (1996).Epidemiological analysis of persistent organochlorine contaminants in cetaceans.*Reviews of Environmental Contamination and Toxicology, 146,* 91-172.

Corsolini, S., Aurigi, S. &Focardi, S. (2000). Presence of polychlorobiphenyls (PCBs) and coplanar congeners in the tissues of the Mediterranean loggerhead turtle *Caretta caretta.Marine Pollution Bulletin, 40,* 952-960.

Duguy, R., Morinière, P. & Le Milinaire, C (1998). Facteurs de mortalitéobservés chez les tortuesmarinesdans le golfe de Gascogne. *OceanologicaActa, 21,* 383-388.

Ecobichon, D. J. (1995). Toxic effects of pesticides: organochlorine insecticides. In C. D. Klaassen, M. O. Amdur& J. Doull(Eds.), *Casarett and Doull's Toxicology: The basic science of poisons* (pp. 649-655). New York, NY: McGraw-Hill.

Flint, M., Patterson-Kane, J. C., Limpus, C. J., Work, T. M., Blair, D. & Mills, P. C. (2009). Postmortem diagnostic investigation of disease in free-ranging marine turtle populations: a review of common pathologic findings and protocols. *Journal of Veterinary Diagnostic Investigation, 21,* 733-759.

Fox, G. A. (2001). Wildlife as sentinels of human health effects in the Great Lakes-St. Lawrence Basin. *Environmental Health Perspectives, 109,* 853-861.

Frick, M. G. (1996). *A Guide for the Identification of Stranded Dead Turtles: the Eastern United States and the Gulf of Mexico*. Savannah, GA: Savannah Science Museum Inc.

Gardner, S. C., Pier, M. D., Wesselman, R. &Juárez, J. A. (2003). Organochlorine contaminants in sea turtles from the Eastern Pacific.*Marine Pollution Bulletin, 46,* 1082-1089.

George, R. H. (1997). Health problems and diseases of sea turtles. In P. L. Lutz & J. A. Musick (Eds.), *The Biology of Sea Turtles* (pp. 363-386). Boca Raton, FL: CRC Press.

Glazebrook, J. S. & Campbell, R. S. F. (1990a).A survey of the diseases of marine turtles in northern Australia I. Farmed turtles.*Diseases of Aquatic Organisms, 9,* 83-95.

Glazebrook, J. S. & Campbell, R. S. F. (1990b).A survey of the diseases of marine turtles in northern Australia II.Oceanarium-reared and wild turtles.*Diseases of Aquatic Organisms, 9,* 97-104.

Glazebrook, J. S., Campbell, R. S. F. & Blair, D. (1989). Studies on cardiovascular fluke (Digenea: Spirorchiidae) infections in sea turtles from the great barrier reef, Queensland, Australia. *Journal of Comparative Pathology, 101,* 231-250.

Glazebrook, J. S., Campbell, R. S. F. & Thomas, A. T. (1993).Studies on an ulcerative stomatitis-obstructive rhinitis-pneumonia disease complex in hatchling and juvenile sea turtles *Chelonia mydas* and *Caretta caretta*.*Diseases of Aquatic Organisms, 16,* 133-147.

Gordon, A. N., Kelly, W. R. &Cribb, T. H. (1998). Lesions caused by cardiovascular flukes (Digenea: Spirorchidae) in stranded green turtles (*Chelonia mydas*). *Veterinary Pathology, 35,* 21-30.

Hasbún, C. R., Lawrence, A. J., Samour, J. H. & al-Ghais, S. (1998). Duodenal volvulus in free living green turtles from coastal United Emirates. *Journal of Wildlife Diseases, 34,* 797-800.

Helmick, K. E., Bennett, R. A., Ginn, P., Dimarco, N., Beaver, D. P. & Dennis, P. M. (2000). Intestinal volvulus and stricture associated with a leiomyoma in a green turtle (*Chelonia mydas*). *Journal of Zoo and Wildlife Medicine, 31,* 221-227.

Holliman, R. B. & Fisher, J. E. (1968). Life cycle and pathology of *Spirorchisscripta*Stunkard 1923 (Digenea: Spirorchiidae) in *Chrysemyspictapicta. Journal of Parasitology, 54,* 310-318.

IUCN/SSC (1995).*A Global Strategy for the Conservation of Marine Turtles.*Balmar, Arlington, VA: International Union for Conservation of Nature and Natural Resources.

IUCN/SSC (2010).2010 IUCN Red List of Threatened Species.www. iucnredlist.org. Accessed December 14, 2010.

Jacobson, E. R. (1991). An update on green turtle fibropapilloma. In G. H. Balazs& S. G. Pooley (Eds.), *Research Plan for Marine Turtle Fibropapilloma* (pp. 59-71). Honolulu, HI: NOAA-Tech Memo NMFS-SWFC-156.

Jacobson, E. R. (1997). Buoyancy problems in sea turtles: causes and diagnosis. In *Proceedings Annual Meeting of the American Association of Zoo Veterinarians* (p.10). Houston, TX: American Association of Zoo Veterinarians.

Jacobson, E. R., Gaskin, J. M., Roelke, M., Greiner, E. C. & Allen, J. (1986). Conjunctivitis, tracheitis, and pneumonia associated with herpesvirus infection in green sea turtles. *Journal of theAmerican Veterinary Medical Association, 189,* 1020-1023.

Jacobson, E. R., Gaskin, J. M., Shields, R. P. & White, R. H. (1979). Mycotic pneumonia in mariculture-reared green sea turtles.*Journal of the American Veterinary Medical Association, 175,* 929-932.

Jacobson, E. R., Mansell, J. L., Sundberg, J. P., Hajjar, L., Reichmann, M. E., Ehrhart, L. M., Walsh, M. &Murru, F. (1989). Cutaneous fibropapillomas of green turtles (*Chelonia mydas*).*Journal of Comparative Pathology, 101,* 39-52.

Keller, J. M., Kucklick, J. R., Harms, C. A. & McClellan-Green, P. D. (2004).Organochlorine contaminants in sea turtles: correlations between whole blood and fat. *Environmental Toxicology and Chemistry, 23,* 726-738.

Keller, J. M., McClellan-Green, P., Kucklick, J. R., Keil, D. E. &Peden-Adams, M. M. (2006). Effects of organochlorine contaminants on loggerhead sea turtle immunity: comparison of a correlative field study and in vitro exposure experiments. *Environmental Health Perspectives, 114,* 70-76.

Lake, J. L., Haebler, R., McKinney, R., Lake, C. A. &Sadove, S. S. (1994). PCBs and other chlorinated organic contaminants in tissues of juvenile Kemp's ridley turtles (*Lepidochelyskempi*). *Marine Environmental Research, 38,* 313-327.

Lizana, M. &Barbadillo, L. J. (1997). Legislación, protección y estado de conservación de los anfibios y reptilesespañoles. In J. M. Pleguezuelos (Ed.), *Distribución y Biogeografía de los Anfibios y Reptiles en España y Portugal* (pp. 477-516). Granada, Spain: Universidad de Granada Editorial.

López-Jurado, L. F. & González, S. (1983). Las tortugas en Canarias.*Aguayro, 147,* 29-31.

Marshall, A. T. & Cooper, P. D. (1988).Secretory capacity of the lachrymal salt gland of hatchling sea turtles, *Chelonia mydas.Journal of Comparative Physiology B, 157,* 821-827.

Martineau, D., Beland, P., Desjardins, C. &Legace, A. (1987). Levels of organochlorine chemicals in tissues of beluga whales (*Delphinapterusleucas*) from the St. Lawrence Estuary, Quebec, Canada.*Archives of Environmental Contamination and Toxicology, 16,* 137-147.

Mateo, J. A., Andreu, A. C. &López-Jurado, L. F. (1997). Las tortugas marinas de la Península Ibérica, Baleares, Azores, Madeira y Canarias: Introducción. In J. M. Pleguezuelos (Ed.), *Distribución y Biogeografía de los Anfibios y Reptiles en España y Portugal* (pp. 433-434). Granada, Spain: Universidad de Granada Editorial.

Mckenzie, C., Godley, B. J., Furness, R. W. & Wells, D. E. (1999). Concentrations and patterns of organochlorine contaminants in marine turtles from Mediterranean and Atlantic waters.*Marine Environmental Research, 47,*117-135.

McKim Jr., J. M. & Johnson, K. L. (1983). Polychlorinated biphenyls and *p,p'*-DDE in loggerhead and green postyearling Atlantic sea turtles. *Bulletin of Environmental Contamination and Toxicology, 31,* 53-60.

Miao, X., Balazs, G. H., Murakawa, S. K. K. & Li, Q. (2001).Congener-specific profile and toxicity assessment of PCBs in green turtles

(*Chelonia mydas*) from the Hawaiian Islands.*Science of the Total Environment, 281,* 247-253.

Monagas, P., Orós, J., Araña, J. & González-Díaz, O. M. (2008). Organochlorine pesticide levels in loggerhead turtles (*Caretta caretta*) stranded in the Canary Islands. *Marine Pollution Bulletin, 56,* 1949-1956.

Monzón-Argüello, C., Riuco, C., Carreras, C., Calabuig, P., Marco, A. &López-Jurado, L. F. (2009).Variation in spatial distribution of juvenile loggerhead turtles in the eastern Atlantic and western Mediterranean Sea.*Journal of Experimental Marine Biology and Ecology, 373,* 79-86.

Muir, D. C. G., Norstrom, R. J. & Simon, M. (1988).Organochlorine contaminants in Arctic marine food chain: accumulation of specific polychlorinated biphenyls and chlordane-related compounds. *Environmental Science and Technology, 22,* 1071-1079.

Nicolson, S. W. & Lutz, P. L. (1989). Salt gland function in the green sea turtle *Chelonia mydas.Journal of Experimental Biology, 144,* 171-184.

Orós, J., Calabuig, P. &Déniz, S. (2004). Digestive pathology of sea turtles stranded in the Canary Islands between 1993 and 2001. *Veterinary Record, 155,* 169-174.

Orós, J., González-Díaz, O. M. & Monagas, P. (2009). High levels of polychlorinated biphenyls in tissues of Atlantic turtles stranded in the Canary Islands, Spain.*Chemosphere, 74,* 473-478.

Orós, J. & Torrent, A. (2001). *Manual de Necropsia de Tortugas Marinas.* Las Palmas de Gran Canaria, Spain: Ediciones del Cabildo de Gran Canaria.

Orós, J., Torrent, A., Calabuig, P. &Déniz, S. (2005). Diseases and causes of mortality among sea turtles stranded in the Canary Islands, Spain (1998-2001).*Diseases of Aquatic Organisms, 63,* 13-24.

Pérez-Jiménez, A. (1997). *Caretta caretta* (Linnaeus, 1758).In J. M. Pleguezuelos (Ed.) *Distribución y Biogeografía de los Anfibios y Reptiles en España y Portugal* (pp. 435-437). Granada, Spain: Universidad de Granada Editorial.

Perugini, M., Giammarino, A., Olivieri, V., Guccione, S., Lai, O. R. &Amorena, M. (2006). Polychlorinated biphenyls and organochlorinepesticida levels in tissues of *Caretta caretta* from the Adriatic Sea.*Diseases of Aquatic Organisms, 71,* 155-161.

Pritchard, P. C. H. (1997). Evolution, phylogeny, and current status. In P. L. Lutz &Musick, J. A. (Eds.), *The Biology of Sea Turtles* (pp. 1-28). Boca Raton, FL: CRC Press.

Raidal, S. R., Ohara, M., Hobbs, R. P. & Prince, R. I. (1998).Gram-negative bacterial infections and cardiovascular parasitism in green sea turtles (*Chelonia mydas*).*Australian Veterinary Journal, 76,* 415-417.

Reichenbach-Klinke, H. & Elkan, E. (1965).*The Principal Diseases of Lower Vertebrates.Diseases of Reptiles.* London, UK: Academic Press.

Reina, R. (2000). Salt gland blood flow in the hatchling green turtle, *Chelonia mydas.Journal ofComparative Physiology B, 170,* 573-580.

Reina, R. D. & Cooper, P. D. (2000). Control of salt gland activity in the hatchling green sea turtle, *Chelonia mydas. Journal ofComparative Physiology B, 170,* 27-35.

Reina, R. D., Jones, T. T. &Spotila, J. R. (2002).Salt and water regulation by the leatherback sea turtle *Dermochelyscoriacea.Journal of Experimental Biology, 205,* 1853-1860.

Roberts, R. J. (1978). *Fish Pathology*. London, UK: Bailliere Tidall.

Rodgers, L. J. & Burke, J. B. (1982). Gastric ulceration associated with larval nematodes. Anisakis type I in pen-reared green turtles from the Torres Strait. *Journal of Wildlife Diseases, 18,* 41-46.

Rosenthal, K. L. &Mader, D. R. (1996). Microbiology. In D. R. Mader (Ed.), *Reptile Medicine and Surgery* (pp. 117-125). Philadelphia, PA: WB Saunders Company.

Rybitski, M. J., Hale, R. C. &Musick, J. A. (1995).Distribution of organo-chlorine pollutants in Atlantic sea turtles.*Copeia, 2,* 379-390.

Schroeder, B. A. (1987). *Annual Report of the Sea Turtle Stranding and Salvage Network Atlantic and Gulf coasts of the United States January-December 1986.* Miami, FL: NOAA-NMFS.

Schumacher, J., Papendick, R., Herbst, L. & Jacobson, E. R. (1996).Volvulus of the proximal colon in a hawksbill turtle (*Eretmochelysim-bricata*).*Journal of Zoo and Wildlife Medicine, 27,* 386-391.

Seminoff, J. A., Resendiz, A., Resendiz, B. & Nichols, W. J. (2004).Ocurrence of loggerhead sea turtles (*Caretta caretta*) in the Gulf of California, Mexico: Evidence of life- history variation in the Pacific Ocean. *Herpetological Review, 35,* 24-27.

Smith, G. M. & Coates, C. W. (1938).Fibro-epithelial growths of the skin in large marine turtles, *Chelonia mydas* (Linnaeus).*Zoology, 23,* 93-98.

Storelli, M. M., Barone, G. &Marcotrigiano, G. O. (2007). Polychlorinated biphenyls and other chlorinated organic contaminants in the tissues of Mediterranean loggerhead turtle *Caretta caretta. Science of the Total Environment, 373,* 456-463.

Storelli, M. M. &Marcotrigiano, G. O. (2000).Chlorobiphenyls, HCB, and organochlorine pesticides in some tissues of *Caretta caretta* (Linnaeus) specimens beached along the Adriatic Sea, Italy. *Bulletin of Environmental Contamination and Toxicology, 64,* 481-488.

Tanabe, S., Hidaka, H. &Tatsukawa, R. (1983).PCBs and chlorinated pesticides in Antarctic atmosphere and hydrosphere.*Chemosphere, 12,* 277-288.

Tanabe, S., Sung, J. K., Choi, D. Y., Baba, N., Kiyota, M. &Tatsukawa, R. (1994). Persistent organochlorine residues in northern fur seal from the Pacific coast of Japan since 1971. *Environmental Pollution, 85,* 305-314.

Teas, W. G. (1994).*Annual Report of the Sea Turtle Stranding and Salvage Network Atlantic and Gulf coasts of the United States January-December 1993.* Miami, FL: NOAA-NMFS.

Torrent, A., González-Díaz, O. M., Monagas, P. & Orós, J. (2004). Tissue distribution of metals in loggerhead turtles (*Caretta caretta*) stranded in the Canary Islands, Spain. *Marine Pollution Bulletin, 49,* 854-874.

Villa, S., Finizio, A., Díaz, R. &Vighi, M. (2003). Distribution of organochlorine pesticides in pine needles of an oceanic island: the case of Tenerife (Canary Islands, Spain). *Water Air Soil Pollution, 143,* 335-349.

Wolke, R. E., Brooks, D. R. & George, A. (1982).Spirorchidiasis in loggerhead sea turtles (*Caretta caretta*): pathology. *Journal of Wildlife Diseases, 18,* 175-185.

Wolke, R. E. & George, A. (1981).*Sea Turtle Necropsy Manual.* Kingston, RI: NOAA Technical Memorandum NMFS-SEFC-24, National Oceanic and Atmospheric Administration.

Work, T. M., Balazs, G. H., Rameyer, R. A. & Morris, R. A. (2004). Retrospective pathology survey of green turtles *Chelonia mydas* with fibropapillomatosis in the Hawaiian Islands, 1993-2003. *Diseases of Aquatic Organisms, 62,* 163-176.

Work, T. M., Balazs, G. H., Wolcott, M. & Morris, R. (2003).Bacteraemia in free-ranging Hawaiian green turtles *Chelonia mydas* with fibropapillomatosis. *Diseases of Aquatic Organisms, 53,* 41-46.

Wyneken, J. (2001). *The Anatomy of Sea Turtles.* Miami, FL: NOAA Technical Memorandum NMFS-SEFSC-470, National Oceanic and Atmospheric Administration.

Ziswiler, V. (1986).Clase Reptiles Reptilia. In V. Ziswiler (Ed.), *Zoología especial. Vertebrados. Tomo II: Amniotas*(pp. 23-72). Barcelona, Spain: Omega Ediciones.

Zumbado, M., Goethals, M., Álvarez-León, E. E., Luzardo, O. P., Cabrera, F., Serra-Majem, L. &Domínguez-Boada, L. (2005).Inadvertent exposure to organochlorine pesticides DDT and derivatives in people from the Canary Islands (Spain). *Science of the Total Environment, 339,* 49-62.

In: Turtles: Anatomy, Ecology and Conservation ISBN: 978-1-61470-554-3
Eds: M. J. Cosgrove and S. A. Roe

Chapter 3

THE PAINTED TURTLE AS A MODEL OF NATURAL ANOXIA TOLERANCE: ROLE OF THE NEUROTRANSMITTER GABA

***Leslie Thomas Buck*[*,1,2]*, David William Hogg*[1] *and Matthew Edward Pamenter*[1,3]**

[1]Department of Cell and Systems Biology, University of Toronto, Canada;
[2]Department of Ecology and Evolutionary Biology, University of Toronto, Canada
[3]Department of Paediatrics, University of California San Diego, CA, U. S.

ABSTRACT

Painted turtles range throughout the USA and Canada. The most northern member of this species is the western painted turtle *Chrysemys picta bellii;* which is also the most anoxia-tolerant vertebrate identified, surviving for up to 6 months buried in the mud of ice covered pond at 3°C. *C. picta* provides a working model of an anoxia-tolerant vertebrate, and of particular interest - a brain that functions without oxygen. We have recently been exploring the role of GABA, the primary vertebrate inhibitory neurotransmitter, in anoxia tolerance. Increases in endogenous GABAergic signalling appear to mediate neuronal electrical depression in

* Correspondence Address: Dr. L.T. Buck, University of Toronto, Dept. of Cell and Systems Biology and Ecology and Evolutionary Biology, 25 Harbord St., Toronto, ON M5S 3G5, Canada, Tel: (416) 978-3506, Fax: (416) 978-8532, E-mail: les.buck@utoronto.ca

C. picta and in a number of other anoxia-tolerant species. Unlike most other vertebrate species that undergo severe neural injury within minutes of an anoxic episode, *C. picta* neurons reduce action potential frequency and suppress metabolism rapidly with the onset of anoxia, and avoid injury. Anoxia-sensitive mammals lack a hypoxia/anoxia-mediated increase in GABA concentrations and suffer electrical hyper-excitability and excitotoxic cell death following the onset of an anoxic/ischemic episode. Experimental interventions aimed at increasing inhibitory GABAergic signalling prevent over-excitation and rescue mammalian neurons from anoxic/ischemic cell death. Drawing upon evidence from a number of studies on a variety of naturally anoxia-tolerant species we will explore in detail the role of GABA in neuroprotection. The anoxic response could be interpreted as a natural anaesthetic mechanism and is achieved by a reduction in excitatory neurotransmitter receptor activity (glutamate) and an increase in inhibitory receptor activity (GABA). An understanding of the complex regulation of GABAergic signaling in facultative anaerobes may provide insight into interventions that could prove efficacious against low oxygen situations in mammals, such as ischemic insult due to stroke.

Keywords: facultative anaerobe, shunting inhibition electrical depression, anoxia, spike arrest, snail, killifish

INTRODUCTION

Painted turtles (*Chrysemys picta*) range throughout the USA and into southern Canada from west to east and north of Lake Superior. The north-south distribution of painted turtles reveals an interesting relationship between latitudinal position and anoxia tolerance. The more northern species being much more anoxia-tolerant than southern species and this correlates with extended ice cover in northern waters. Indeed, the increased anoxia tolerance in the northern species is almost certainly the result of natural selection as they spread northward following the Wisconsinan glaciation about 18,000 to 20,000 years ago. The western painted turtle (*Chrysemys picta bellii*) occupies the most northern range and is the most anoxia-tolerant, surviving for up to 6 months submerged in nitrogen gassed water at 3°C (Ultsch and Jackson 1982). We and others have been studying the physiological and biochemical mechanisms by which this remarkable tolerance is achieved. Unlike most other vertebrate species that undergo severe neural injury within minutes of an anoxic episode, *C. picta* neurons reduce action potential frequency and suppress metabolism rapidly with the onset of anoxia and avoid injury. It is clear that the reversible suppression of metabolism by greater than 90% of normoxic resting levels is key; however, how this is achieved is less clear (Bickler and Buck 2007). Nevertheless, considerable progress has been made and in this review we will highlight recent advances and focus on a unique mechanism we have uncovered in the brain. We will also focus on the unique opportunity provided by an anoxia-tolerant brain to explore natural mechanisms by which mammal brain can be protected from the debilitating consequences of ischemic stroke.

Stroke is characterized by a sudden impairment of blood flow to a particular brain region, preventing delivery of oxygen and thus rendering the region functionally anoxic. This lack of oxygen availability presents severe challenges to the exquisitely oxygen-sensitive mammalian brain, which suffers ATP loss, membrane potential depolarization and irreversible cell damage within minutes of anoxia (Donnelly et al., 1992). During an anoxic episode, oxidative phosphorylation ceases and cells must rely entirely upon glycolysis to generate ATP, essentially sacrificing 90% of their ATP-producing capacity (Buck and Pamenter, 2006). Therefore, defense of ATP supplies is likely the most important aspect of surviving anoxic insults without detriment. In normoxic rat brain the primary consumers of ATP are excitatory electrical events, including action potential (AP) generation and propagation, and post-synaptic potentials (PSPs). Together these processes account for ~81% of neuronal energy expenditure (Attwell and Laughlin, 2001). Thus electrical depression of both APs and PSPs is critical to preventing ATP depletion and surviving transient low-oxygen insults in mammalian brain.

Some vertebrates, including freshwater turtles and some fish and amphibian species, display a remarkable tolerance to anoxic insults (Bickler and Buck, 2007). Anoxia-tolerance is achieved via multiple protective mechanisms including increased glycogen stores, enhanced buffering capacity, ion channel arrest, and perhaps most importantly: electrical depression (Bickler and Buck, 2007; Hochachka, 1986; Jackson, 2002; Perez-Pinzon et al., 1992a). One mechanism in particular that appears to be employed ubiquitously in facultative anaerobes is a large and transient elevation in extracellular GABA concentration $[GABA]_e$ during anoxic perfusion (see below). GABA is the predominant inhibitory neurotransmitter in the adult mammalian CNS and in anoxia-tolerant species thus far characterized, and mediates wide-spread electrical depression in most neurons (Lutz and Milton,

2004; Martyniuk et al., 2005; Turner and Whittle, 1983). GABA activates two specific receptors: $GABA_B$ receptors which act via an associated G protein, and $GABA_A$ receptors which contain a membrane-spanning chloride (Cl^-) channel that is also conducive to bicarbonate (HCO_3^-) (Kaila et al., 1993). Activation of either of these receptors is electrically inhibitory in adult mammalian neurons, and GABAergic interneurons make widespread connections throughout the nervous system and are therefore well positioned to regulate electrical activity. Furthermore, GABAergic interneurons may be relatively tolerant to hypoxia, potentially enabling them to regulate hypoxic-protective mechanisms without loss of function or cell death (Nitsch et al., 1989).

In this review we discuss GABA-based strategies of electrical depression utilized in the anoxia tolerance of several facultative anaerobes, and compare these strategies with the mechanisms of GABA-mediated electrical depression that have begun to emerge in mammalian models of stroke tolerance. Furthermore, we discuss the role of GABA inhibition in facultative anaerobes and immature mammalian neurons in which GABA is an excitatory neurotransmitter. Major findings from these studies are summarized in Tables 1 and 2 and are discussed below.

GABA AND ANOXIA-TOLERANCE: ENDOGENOUS DEFENSE MECHANISMS

Facultative anaerobes commonly experience prolonged hypoxia or anoxia in their natural environments. GABA-mediated inhibitory signaling is highly conserved in evolution and differential regulation of GABA receptors, GABA synthesis and uptake mechanisms, as well as cellular Cl^- gradients may underlie differences in oxygen sensitivity between facultative anaerobes and oxygen-sensitive adult mammals. Indeed mammalian neurons possess the same basic components of GABA-mediated signaling as facultative anaerobe neurons. Thus, mechanisms of neuroprotection that occur naturally in facultative anaerobes may be induced with experimental intervention in mammalian neurons, conferring neuroprotection (Figure 1). Anoxia-tolerant vertebrates exhibit common patterns of neurotransmitter release during anoxia. In general, excitatory amino acid (EAA) concentrations, in particular of the primary EAA glutamate, remain constant or decrease slightly with anoxia, while inhibitory amino acid (IAA) concentrations, in particular of the primary IAA GABA are greatly elevated during anoxia. These changes have been consistently observed in all identified facultative anaerobes in which GABA is inhibitory, including fish, reptile and amphibian species (see Table 1 and discussion below) and also persist during the developmental stage of at least one facultative anaerobe.

Table 1. GABA: Critical inhibitory neurotransmitter in endogenous mechanisms of anoxia tolerance and ischemic preconditioned neuroprotection

Anoxia Tolerant Species				
Facultative Anaerobe	**Change in GABA**	**GABA-mediated responses to anoxia**	**Neuroprotection**	**References**
Freshwater turtles (*C. Picta bellii, T. Scripta Elegans*)	↑ $[GABA]_e$ ↑ $GABA_A$ B_{Max}	↓ Electrical activity ↓ Glutamate release	Survives > 6 months of anoxia at low temperatures	(Lutz and Leone-Kabler, 1995; Nilsson et al., 1990; Thompson et al., 2007; Ultsch and Jackson, 1982)
Freshwater fish (*C. carassius, C. auratus*)	↑ $[GABA]_e$	↓ Electrical activity ↓ Glutamate release	Survives months of hypoxia at low temperatures	(Hylland and Nilsson, 1999; Johansson et al., 1997; Nilsson, 1990; Suzue et al., 1987)
Anoxic shore crab (*C. maenas)*	↑ $[GABA]_e$	↓ Electrical activity (putative)	Survives 12 hours in N_2-bubbled water at warm temperatures	(Nilsson and Winberg, 1993)
Killifish embryo (*A.* l*imnaeus*)	↑ $[GABA]_e$	↓ Electrical activity (putative)	Survives months in anoxic mud in a diapaused state	(Podrabsky et al., 2007)
Anoxia Intolerant Species				
Mammalian Ischemia Model	**Response to Ischemic Insult**	**GABA-targeted intervention**	**Effect due to ischemic insult**	**References**
Rat cortex (*R. norvegicus*)	↓ $GABA_A$ B_{Max} $[GABA]_e$ unchanged ↑ Glutamate release ↑ Cell death	↑ $[GABA]_e$ $GABA_A$ agonism	↓ Glutamate release	(Mielke and Wang, 2005; Ouyang et al., 2007)
		Ischemic Preconditioning*	↑ $[GABA]_e$ ↓ Glutamate release ↓ Cell death	(Johns et al., 2000)

Table 1 (Continued)

Anoxia Intolerant Species				
Mammalian Ischemia Model	**Response to Ischemic Insult**	**GABA-targeted intervention**	**Effect due to ischemic insult**	**References**
Rat hippocampus	$[GABA]_e$ unchanged ↑ Glutamate release ↑ Cell death	Ischemic Preconditioning	↑ $[GABA]_e$ ↓ Glutamate release ↓ Cell death	(Dave et al., 2005)
Gerbil hippocampus (*M. unguliculatus*)	$[GABA]_e$ unchanged ↓ $GABA_A$ mRNA ↓ $GABA_A$ B_{Max} ↑ Glutamate release ↑ AMPAR B_{Max} ↑ NMDAR B_{Max} ↑ Cell death	Ischemic Preconditioning*	↑ $GABA_A$ B_{Max} ↓ AMPAR B_{Max} ↓ NMDAR B_{Max} ↓ Cell death	(Li et al., 1993; Sommer et al., 2002; Sommer et al., 2003)
		Blockade of GABA uptake mechanisms	↑ $[GABA]_e$ ↓ Glutamate release ↓ Cell death	(Inglefield et al., 1995)
		↑ $[GABA]_e$ $GABA_A$ agonism	↑ $[GABA]_e$ ↓ Glutamate release ↓ Cell death	(Hall et al., 1997; Schwartz-Bloom et al., 2000; Sternau et al., 1989)

Comparison of endogenous GABA-mediated mechanisms of anoxic tolerance in facultative anaerobe brain with up-regulated GABA-mediated mechanisms of neuroprotection in ischemic mammalian brain. *GABAA receptor antagonists blocked neuroprotective effects attributable to ischemic preconditioning in these studies. Abbreviations: [GABA]e – extracellular GABA concentrations, BMax – receptor binding affinity.

Evidence for GABA-mediated mechanisms in facultative anaerobes will be discussed in the proceeding sections.

Anoxic GABA Changes in Fish: Roles in Developing and Adult Brain

Anoxia-tolerant fish have received considerable attention in the study of GABA-mediated mechanisms of anoxia tolerance. In their natural environment, Crucian carp (*Carassius carassius* L.) and their slightly less anoxia-tolerant cousins the common goldfish (*Carassius auratus*) survive in an active state for weeks of anoxia at low temperatures while over-wintering under frozen surfaces of lakes and ponds in northern Europe (Nilsson, 2001).

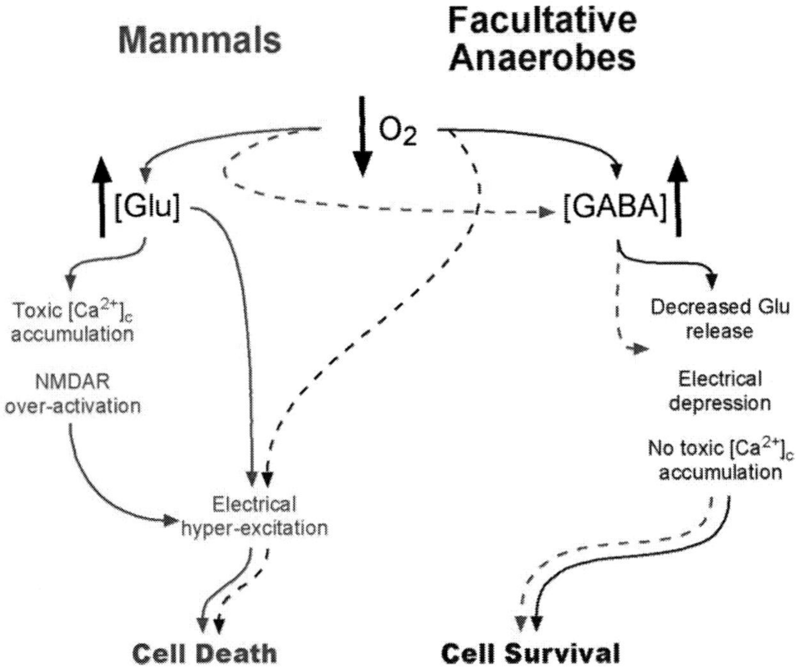

Figure 1. Schematic of inhibitory GABAergic mechanisms of neuroprotection in anoxia tolerant facultative anaerobe and preconditioned mammalian neurons. Ischemic preconditioning (IPC)- or GABA-induced neuroprotective mechanisms resemble endogenous mechanisms in facultative anaerobes: Facultative anaerobes (blue pathway) utilize endogenous GABA-mediated neuroprotective mechanisms (solid blue lines) when exposed to anoxia. Ischemic mammalian neurons (red pathway) undergo excitotoxic cell death (solid red lines) but can be rescued by experimental interventions including IPC, GABA perfusion or $GABA_A$ receptor agonism (dashed red lines). Blockade of $GABA_A$ receptors during anoxic or ischemic insult abolish neuroprotective mechanisms in facultative anaerobes (dashed blue lines) and in IPC- or GABA- pretreated mammalian neurons.

Anoxia tolerant fish rely entirely on glycolytic energy production and their brains undergo metabolic arrest in response to anoxia without significant depletion of ATP supplies (Johansson and Nilsson, 1995; Johansson et al., 1995). There is evidence supporting decreased electrical activity in the anoxic brain of these species from both visual and auditory inputs and this decreased excitability likely contributes to ATP conservation and may be mediated by enhanced GABAergic inhibitory signaling (Johansson et al., 1997; Suzue et al., 1987). Indeed, during anoxia $[GABA]_e$ in crucian carp brain doubles within 6 hours and increases a further 5-fold within two weeks of anoxic exposure (Hylland and Nilsson, 1999; Nilsson, 1990). These large increases in $[GABA]_e$ likely reduce electrical excitability and thus ATP demand to allow the crucian carp to maintain energy balance.

Interestingly, in both goldfish and crucian carp, expression of metabotropic glutamate (mGlu) receptors 2 and 3 are up-regulated in comparison to oxygen-sensitive fish species (i.e. trout) (Poli et al., 2003). These authors also report that in goldfish exposed to periods of anoxia sufficient to induce cell damage, the incidence of cell death was inversely proportional to the expression of mGlu2/3 receptors, and blockade of these receptors aggravated glutamate release and increased cell death due to anoxia. mGluRs have been linked to enhanced GABA release in preconditioned mammalian brain exposed to ischemic insult (Saransaari and Oja, 2004; Saransaari and Oja, 2005), and endogenously high expression of these receptors in the brain of facultative anaerobes may partially underlie rapid and prolonged elevations in GABA with anoxic exposure. Thus correlative evidence suggests a critical role for GABA in anoxia tolerance in fish, but the specific mechanisms of protection remain unexplored.

Facultative anaerobes may also utilize GABA-based mechanisms of neuroprotection during maturation and development. An example comes from the gulf killifish (*Fundulus grandis*), which has recently been identified as a facultative anaerobe (Virani and Rees, 2000). These fish are found in ephemeral ponds in the Maracaibo Basin of Venezuela and produce drought-resistant diapausing embryos during the dry season (Podrabsky et al., 2001). Diapaused embryos that are buried in mud experience severe hypoxia or even anoxia for months, and in response metabolic rate decreases 80-90% and protein synthesis is also depressed (Podrabsky and Hand, 1999; Podrabsky and Hand, 2000). Diapausing embryos also exhibit large increases in GABA that are maintained throughout the anoxic episode. This increase is accompanied by a decrease in glutamate, likely due to conversion to GABA (Podrabsky et al., 2007). These results indicate GABA may be an inhibitory neurotransmitter at all stages of development in facultative anaerobe brain. Thus even in early developmental stages, facultative anaerobes utilize endogenous GABA-based mechanisms of electrical inhibition to provide neuroprotection against low oxygen insults.

Anoxic GABA Changes in Freshwater Turtles

While the data supporting an increase in $[GABA]_e$ in anoxia-tolerant species is compelling, knowledge of the protective mechanisms regulated by GABA in these species is minimal and based primarily on mammalian studies of the inhibitory effects of GABA on nervous electrical activity. However, recent studies have begun to elucidate the mechanisms of GABA-mediated protection in a pair of facultative anaerobes: the painted and red-eared freshwater turtles (*Chrysemys picta belli* and *Trachemys scripta elegans*).

Freshwater turtles are the most anoxia tolerant vertebrate species identified. In nature, turtles over-winter in the anoxic mud beneath frozen ponds. In the laboratory, they can survive anoxic episodes lasting days at 25°C and up to six months at 3°C (Musacchia, 1959; Ultsch and Jackson, 1982). As with the facultative anaerobes discussed previously, anoxia causes rapid and prolonged elevations of $[GABA]_e$ in turtle brain (Hitzig et al., 1985; Nilsson et al., 1990; Nilsson and Lutz, 1991). Furthermore, $GABA_A$ receptor binding affinity (B_{Max}) is increased during anoxic exposure, likely due to increased receptor density on the plasma membrane (Lutz and Leone-Kabler, 1995; Sakurai et al., 1993).

Recently, studies have begun to examine the underlying mechanisms of GABA-mediated protection in the turtle brain. GABA appears to be involved in regulation of electrical depression in the anoxic cortex and also of glutamate release in early anoxia. Turtle brain exhibits a significant electrical depression with anoxic exposure, as characterized by reductions in: transmembrane potentials, Na^+ spike thresholds, AP frequency, field potentials, and EEG activity (Fernandes et al., 1997; Perez-Pinzon et al., 1992a; Perez-Pinzon et al., 1992b). Recent results from our lab indicate that GABA, acting through the $GABA_A$ receptor mediates electrical depression in the turtle cortex. Normoxic to anoxic transitions resulted in significant electrical depression including: increased whole-cell conductance, decreased AP frequency, increased AP firing threshold, increased inhibitory post-synaptic potential (IPSP) activity, and a mild membrane potential depolarization. GABA perfusion (30-750 μM) induced similar changes and both the GABA-mediated and the anoxia-mediated changes were blocked by the $GABA_A$ inhibitor gabazine. Furthermore, cells treated simultaneously with anoxia and gabazine underwent repeated seizure-like events and suffered severe membrane potential depolarizations and cell death characteristic of ischemic rat neurons (Pamenter et al., 2011).

These data suggest that elevated $[GABA]_e$ in the anoxic turtle brain activates $GABA_A$ receptors, which mediate an inhibitory Cl^- conductance (G_{Cl}). Their activation leads to increased whole-cell conductance and since the reversal potential for Cl^- is near neuronal resting membrane potential, increases in G_{Cl} oppose excitatory events and tend to clamp the membrane potential near the chloride ion's reversal potential (~ -80 mV in adult turtle brain). Thus $GABA_A$ receptor activation during anoxia opposes APs, as indicated by the decrease in AP frequency and increase in IPSP amplitude, by a process known as "shunting inhibition". Furthermore, increased cell death in anoxic neurons treated with gabazine highlights the critical role of GABA in anoxia tolerance in this facultative anaerobe and offers the first conclusive evidence that the elevation of GABA in an anoxia-tolerant species mediates specific protective mechanisms (Figure 1).

In addition to initiation of electrical depression, elevations in GABA may regulate glutamate release in the anoxic turtle brain. Glutamate release from turtle brain tissue decreases 40% in the first 1.5 hours of anoxia, and glutamate re-uptake mechanisms remain active (Milton et al., 2002a). During longer duration anoxic exposures (up to 5 hours), glutamate levels decrease further to 30% of control levels (Thompson et al., 2007). The early reduction is maintained by the coordinated action of ATP-sensitive K^+ channels and adenosine receptors; however, the later reduction is mediated by the coordinated action of $GABA_A$ receptors and adenosine (Milton et al., 2002b; Thompson et al., 2007). GABA-mediated changes in glutamate release have also been demonstrated in mammalian brain exposed to ischemic insult (see discussion below). Therefore, mechanisms regulating naturally occurring GABA-induced reductions in glutamatergic activity in the turtle may have homologous pathways in mammalian systems that can be switched on with appropriate interventions, and may explain some of the neuroprotection afforded by $GABA_A$ agonists in mammalian experiments (see below).

GABAergic Systems in Anoxia-sensitive Non-mammalian Vertebrates

Despite these examples, anoxia-tolerance is not a common trait in non-mammalian vertebrates. Interestingly, non-mammalian vertebrate species that are intolerant to anoxia exhibit changes in amino acids that closely resemble changes in anoxia-intolerant mammals, such that release of excitatory neurotransmitters is greater than release of inhibitory neurotransmitters. For example, in rainbow trout exposed to anoxic insult amino acid concentrations do not change until after the loss of ion homeostasis due to cell death as determined by K^+ efflux. Furthermore, glutamate levels increased prior to GABA release (Hylland et al., 1995). Another example comes from the frog *Rana pipiens*, which can survive hours of anoxia, but which does not exhibit elevations of amino acids until several hours into the anoxic period when all amino acids become elevated following total depletion of ATP (Lutz and Reiners, 1997). Thus while anoxic death of the frog is delayed, cell death does occur and in a fashion similar to mammalian neurons with regard to amino acid release patterns.

GABA and Ischemic Preconditioning: Models of Inducible Anoxia Tolerance

There is overwhelming correlative evidence suggesting a critical role for GABA in neuronal low oxygen survival in facultative anaerobes. GABA-mediated electrical inhibition is common in both ectothermic and mammalian vertebrate species and likely relies on similar mechanisms of GABA synthesis, uptake, and release (Martyniuk et al., 2007). Thus understanding of the protective mechanisms employed by facultative anaerobes may offer GABA-based strategies to combat ischemic damage in mammals. Recently, interest in GABA's role in ischemia-tolerance has increased and the results support the hypothesis that the mechanisms underlying anoxia tolerance in some species can be targeted and up-regulated to confer neuroprotection in mammals. (Table 1, Figure 1).

Ischemic preconditioning (IPC) is a phenomenon that occurs in anoxia-sensitive animals whereby mild anoxic insults induce mechanisms that protect cells against subsequent lethal anoxic insults (Murry et al., 1986). This phenomenon was originally described in canine heart but has proven to be transmittable to numerous models of neuroprotection (see below). Furthermore, IPC in the brain appears to be mediated by enhanced GABAergic inhibitory signaling. For example, rat cortical slices exposed to oxygen and glucose deprivation (OGD) exhibit glutamate levels that are elevated 5-fold while GABA levels are unchanged. However, in cells preconditioned with brief bouts of OGD, and subsequently exposed to continuous, otherwise lethal

OGD, GABA levels were considerably elevated and glutamate release and cell death were decreased (Johns et al., 2000).

A role for enhanced GABAergic activity in IPC-mediated neuroprotection is also supported by another interesting study that examined the binding affinity of glutamatergic and GABAergic receptors in ischemic hippocampal CA1 neurons. Five minutes of ischemia induced severe cell death; however, pretreatment with a 2.5-minute period of ischemia significantly reduced cell death induced by the subsequent 5-minute ischemic insult. In cells that survived ischemia the binding affinity (B_{Max}) of excitatory α-amino-hydroxy-5-methyl-4-isoxazole propionic acid (AMPA) and N-methyl-D-aspartate (NMDA) receptors decreased and that of the inhibitory $GABA_A$ receptor increased. In addition $GABA_A$ receptor B_{Max} increased progressively from 30 minutes to 48 hours after recirculation before decreasing (Sommer et al., 2002; Sommer et al., 2003). Furthermore, in rats preconditioned against OGD, elevations were observed in $[GABA]_e$ and in the activity of glutamate decarboxylase, the predominant source of GABA synthesis in the brain (Dave et al., 2005). Therefore, IPC renders mammalian neurons more anoxia-tolerant by mimicking endogenous changes observed in the anoxic turtle cortex (elevated $[GABA]_e$ due to increased synthesis and increased $GABA_A$ receptor B_{Max}).

Important from a clinical perspective, IPC can be induced pharmacologically by activation of GABA receptors. Pre-treatment with a variety of GABA-related pharmaceuticals is efficacious in preventing ischemic cell death in selectively vulnerable gerbil hippocampal neurons. GABA perfusion, specific activation of GABA receptors or prevention of GABA catabolism via inhibition of the GABA transaminase provide strong to moderate neuroprotection against bilateral cerebral ischemia (Grabb et al., 2002; Sternau et al., 1989). Furthermore, prevention of GABA reuptake reduces cell death in gerbil hippocampus (Inglefield et al., 1995).

There is particularly strong evidence for a central role for $GABA_A$ receptors in this mechanism. For example, activation of $GABA_A$ receptors with partial or complete agonists, such as diazepam, both prior to transient ischemia and immediately following the ischemic insult is neuroprotective against cell death in gerbil (Hall et al., 1997; Hall et al., 1998; Schwartz-Bloom et al., 1998; Schwartz-Bloom et al., 2000). Similarly, in rat hippocampal neurons exposed to 7-minutes of OGD, ATP concentrations decreased 70% and cytochrome c (an apoptotic factor released from the mitochondria prior to programmed cell death) concentrations in the cytosol were increased 400%. However, in cells treated with diazepam during reoxygenation, ATP levels were completely restored and cytochrome c release was prevented (Galeffi et al., 2000). In addition to studies demonstrating protection mediated by $GABA_A$ agonism, there is evidence that $GABA_A$ *antagonism* prevents the neuroprotective effects of IPC. Also $GABA_A$ receptor activation with GABA or specific agonists were all abolished by $GABA_A$ blockade (Grabb et al., 2002). Taken together these data indicate direct GABAergic manipulation is a good candidate for interventions against ischemic damage in mammalian neurons.

PRECONDITIONING IN ANOXIA-TOLERANT SPECIES?

Sensitivity to oxygen appears to follow a sliding scale that is closely connected to temperature effects. At one end are facultative anaerobes that survive for months without oxygen at low temperatures and days of anoxia at warmer temperatures. At the opposite end of the scale are relatively anoxia-intolerant mammals that are intolerant to anoxia and to low temperatures, but which can survive anoxic insults provided natural mechanisms of neuroprotection are up-regulated in advance or immediately following ischemic insult. Interestingly, the epaulette shark may represent a mid-point on this scale in that it is hypoxia-tolerant at high temperatures but it's tolerance may rely on naturally-occurring preconditioning protocols resembling experimental preconditioning interventions in mammals (Nilsson and Renshaw, 2004).

The hypoxic epaulette shark (*Hemiscyllium ocellatum*), is found in coral-rich offshore regions. During nocturnal low tide, these regions can become cut-off from the greater ocean and due to the high metabolic activity of the coral reef; the water can become extremely hypoxic overnight (Routley et al., 2002). Thus the epaulette shark experiences repeated bouts of severe hypoxia at regular intervals and does not exhibit apparent neuronal damage as a result of this intermittent hypoxia (Renshaw et al., 2002). Indeed intermittent hypoxia may serve to up-regulate defensive mechanisms against ischemic damage and render the epaulette shark more tolerant to subsequent bouts of anoxia. Epaulette shark brain pre-treated with short bouts of hypoxia exhibits increased $[GABA]_e$ and $GABA_A$ B_{Max}, decreased glutamate release and increased tolerance to anoxic insult (Nilsson and Renshaw, 2004; Renshaw et al., 2002). Thus even in hypoxia tolerant species preconditioning elevates neuroprotective pathways dependent on enhanced inhibitory GABAergic neurotransmission. This mechanism highlights the similarities between endogenous and induced anoxic protective mechanisms.

INHIBITORY OR EXCITATORY? CHLORIDE GRADIENTS AND $GABA_A$ RECEPTORS

Like most ions, chloride (Cl^-) is distributed unequally across the neuronal plasma membrane. In most mature neurons, extracellular Cl^- concentrations ($[Cl^-]_e$) are ~10-fold greater than intracellular concentrations and this gradient is maintained by the balance of extrusion of Cl^- though K^+-Cl^--cotransporters (KCC2), Cl^--bicarbonate exchangers, ATP-driven Cl^- pumps and voltage-sensitive Cl^- channels; with the accumulation of Cl^- through NKCC1 ($Na^+/K^+/2Cl^-$) (for review see (Delpire, 2000; Kaila, 1994). Primarily, it is the differences in expression of the opposing KCC2 and NKCC1 transporters that determine neuronal Cl^- gradients (E_{Cl}). Higher expression of KCC2 results in an E_{Cl} that is more negative than resting membrane potential. Thus increases in G_{Cl} results in inward Cl^- flux and membrane hyperpolarization. Therefore, $GABA_A$ receptor activation is inhibitory. Conversely, if NKCC1 expression is higher, E_{Cl} is reduced or even reversed and increases in G_{Cl} are depolarizing. Thus, $GABA_A$ receptor activation becomes excitatory.

In the CNS of adult mammals and most facultative anaerobes, the combination of expression of these transporters is such that the Cl^- reversal potential is generally negative with regard to resting membrane potential, and therefore activation of $GABA_A$ receptors hyperpolarizes the cells resting membrane potential and depresses cellular excitation (Table 1, Figure 2). Conversely, in some organisms at different stages in development, G_{Cl} is excitatory and GABA is a major excitatory neurotransmitter. In such systems, inhibition of GABAergic mechanisms is critical to preventing over-excitation during anoxic insults (Figure 2). Examples of this phenomenon are found in both the anoxia-tolerant and anoxia-intolerant species (Table 2).

Table 2. GABA is a critical excitatory neurotransmitter in pond snails and may regulate excitotoxicity in immature mammalian neurons exposed to ischemic insults

Species	Response to GABA during anoxia	Effect of GABAergic inhibition during anoxia	Neuroprotection	References
Anoxic pond snail (*L. stagnalis*)	Membrane potential depolarization ↑ Electrical activity	↓ $GABA_A$ activity ↓ NKCC1 activity ↓ Electrical activity	Survives days of anoxia at warm temperature	(Cheung et al., 2006; Rubakhin et al., 1996)
Immature rat (*R. norvegicus*) neohippocampus and neocortex	↑ $GABA_A$ activity ↑ Glutamate release ↑ Electrical excitation ↑ Cell Death ↑ $[Ca^{2+}]_c$	No information available	No information available	(Clayton et al., 1998a; Nunez et al., 2003; Zhao et al., 2005)

Comparison of endogenous inhibition of GABA-mediated electrical excitability in pond snail and of effects of GABA regulation on ischemic immature mammalian neurons. Interventions targetting depression of GABA mediated excitability have not been explored in mammalian brain but may prove to be effective against ischemic damage due to stroke. Abbreviations: $[GABA]_e$ – extracellular GABA concentrations, $[Ca^{2+}]_c$ - cytosolic calcium concentration.

Reduction in GABA-mediated Excitability in the Anoxia Tolerant Pond Snail

In the pond snail *Lymnaea stagnalis* GABA is an excitatory neurotransmitter. The reversal potential of Cl^- (and thus $GABA_A$ receptors) in these cells is ~ -20 mV and GABA perfusion causes membrane potential depolarization mediated by a $GABA_A$ receptor-dependent inward Cl^- current (Rubakhin et al., 1996). These snails are reported to survive up to 40-hours in N_2-bubbled water and when made hypoxic undergo a ~38% depression in action potential firing as well as a membrane potential hyper-polarization that is reversed by reoxygenation and mimicked by $GABA_A$ receptor blockade (Cheung et al., 2006). Interestingly, NKCC1 blockade also mimicked the hypoxic decrease in AP frequency, suggesting that in anoxia NKCC1 activity is reduced, elevating intracellular Cl^- ($[Cl^-]_i$) and thus decreasing the magnitude of GABA-mediated excitatory neurotransmission. Thus, just as anoxia-tolerant vertebrates that exhibit inhibitory responses to GABA possess endogenous mechanisms to upregulate inhibitory GABAergic systems, facultative anaerobes that exhibit excitatory GABAergic signaling possess endogenous mechanisms to *downregulate* excitatory GABA-mediated signaling. Furthermore, these mechanisms may be adaptable to immature mammalian brain to upregulate neuroprotection against excitatory GABAergic signaling during ischemia (see next section).

Cl^- Gradients and Excitotoxicity in Ischemic Developing Mammalian Neurons

While GABA is an inhibitory neurotransmitter in most mammalian neurons, there are a number of examples in which GABA becomes excitatory in the developing mammal and in regions of the adult mammalian CNS (for a review see (Ben-Ari, 2002). GABA is excitatory in some regions of the adult CNS where Cl^- gradients are rearranged. Depolarization due to GABA has been reported in neurons and interneurons from the hippocampus and dentate gyrus regions of the adult murine brain and can also be induced by repetitive stimulation or by prolonged exposure to GABA (Lambert and Grover, 1995; Michelson and Wong, 1991; Staley and Mody, 1992).

In the immature mammalian brain GABA is initially excitatory, primarily due to the low expression of KCC2, which results in high $[Cl^-]_i$ and a reversal potential for $GABA_A$ receptors that is depolarizing with regard to resting membrane potential (Figure 3). Thus, as with the pond snail, GABAergic signaling is excitatory and therefore undesirable during anoxia. This hypothesis is supported by data from embryonic murine cultures exposed to excitotoxicity, OGD or induced depolarization. Perfusion of extracellular GABA or direct agonism of GABAergic receptors increased neuronal cell death due to each of these treatments in immature neurons (Erdo et al., 1991; Lukasiuk and Pitkanen, 2000; Muir et al., 1996).

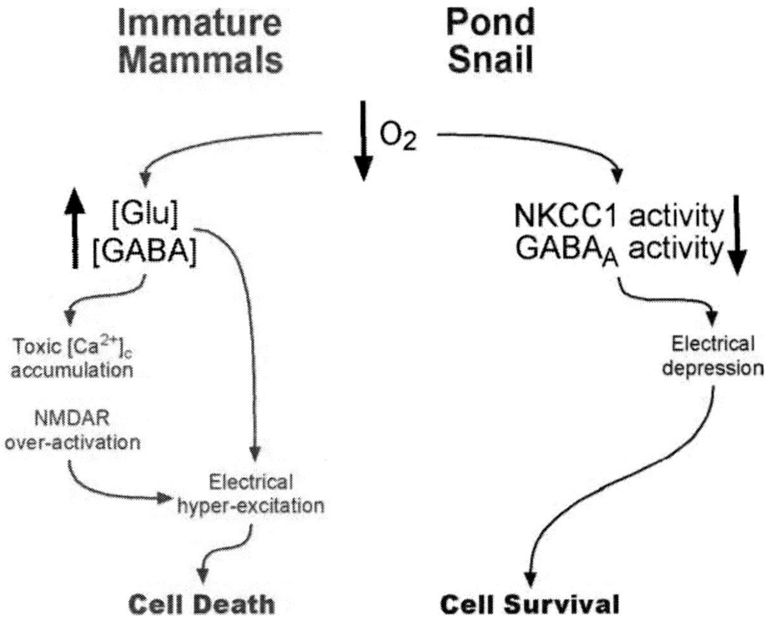

Figure 2. Schematic of excitatory GABAergic mechanisms in the anoxia- tolerant pond snail: potential neuroprotective roles in immature mammalian neurons? GABA is excitatory in neuronal circuits with small or reversed Cl^- gradients, such as in the pond snail and in immature mammalian neurons. In the anoxia tolerant snail, endogenous inhibition of GABAergic systems is neuroprotective (blue pathway). Conversely, ischemic immature mammalian neurons undergo GABA- and glutamate (Glu) mediated hyper-excitation and cell death (red pathway). The effect of GABA inhibition on survival in ischemic immature mammalian neurons has not been investigated.

In addition to inducing a hyper-excited state, inappropriate excitatory GABAergic signaling in immature rat cortex exposed to OGD leads to toxic accumulations of cytosolic Ca^{2+} ($[Ca^{2+}]_i$) that is prevented by NMDA receptors antagonists (Fukuda et al., 1998). Furthermore, glutamate induced $[Ca^{2+}]_i$ accumulation is increased in immature versus adult hippocampal neurons. The enhanced susceptibility to glutamate may be due to incomplete development of what will become inhibitory GABAergic inputs in dendritic regions (Azimi-Zonooz et al., 2006). Rather than underdeveloped inhibitor inputs an alternative explanation is that GABAergic neurons that will eventually become inhibitory are excitatory at this early stage of neuronal development.

As neurons mature, KCC2 expression increases and $[Cl^-]_i$ decreases, rendering GABA-mediated currents inhibitory (Figure 3) (Clayton et al., 1998a; Clayton et al., 1998b). Interestingly, expression of glutamic acid decarboxylase-67 (GAD67), an important enzyme in the production of GABA, is upregulated with brain development. Furthermore, ischemia results in a decrease in GAD67

Expression in immature brain but not in adult brain (Xu et al., 2006). These authors conclude that decreases in GAD67 expression represents a reduction in inhibitory signaling and thus reduced neuroprotection. However, it is more likely that the decrease in GAD67 expression in ischemic immature brain is a protective mechanism as GABA is excitatory at this stage of development. This interpretation is supported by the fact that perfusion of GABA (10-2000 mM) onto developing rat cortical neurons resulted in neuroprotection in mature neurons older than 20 days in culture, but enhanced cell death in less developed, immature neurons (Zhao et al., 2005). Therefore, just as inhibition of excitatory GABAergic signaling is neuroprotective in pond snail ganglia, GABAergic inhibition may also prove to be crucial to the protection of developing brain against hypoxic or ischemic insult, although they have received comparatively little study (Figure 2).

Cl^- Gradients and Hyper-excitability in Adult Mammalian Neurons Post-ischemia

While experiments with pond snails cannot be directly extrapolated to adult mammalian models, they offer insight into the rescue of critically damaged neurons that have already passed the window of conditioned protection or recovery based on GABA receptor manipulations. Mammalian cells that undergo ischemic insult experience periods of over-excitability during reoxygenation that may contribute to eventual cell death over the ensuing weeks following recovery, and appear linked to derangements in cellular GABA homeostasis and altered Cl^- gradients. Non-lethal administration of OGD or anoxic episodes induce transient elevations of $[Cl^-]_i$ and are not cytotoxic in rat neurons (Inglefield and Schwartz-Bloom, 1998; Inglefield and Schwartz-Bloom, 1999; Jiang et al., 1992).

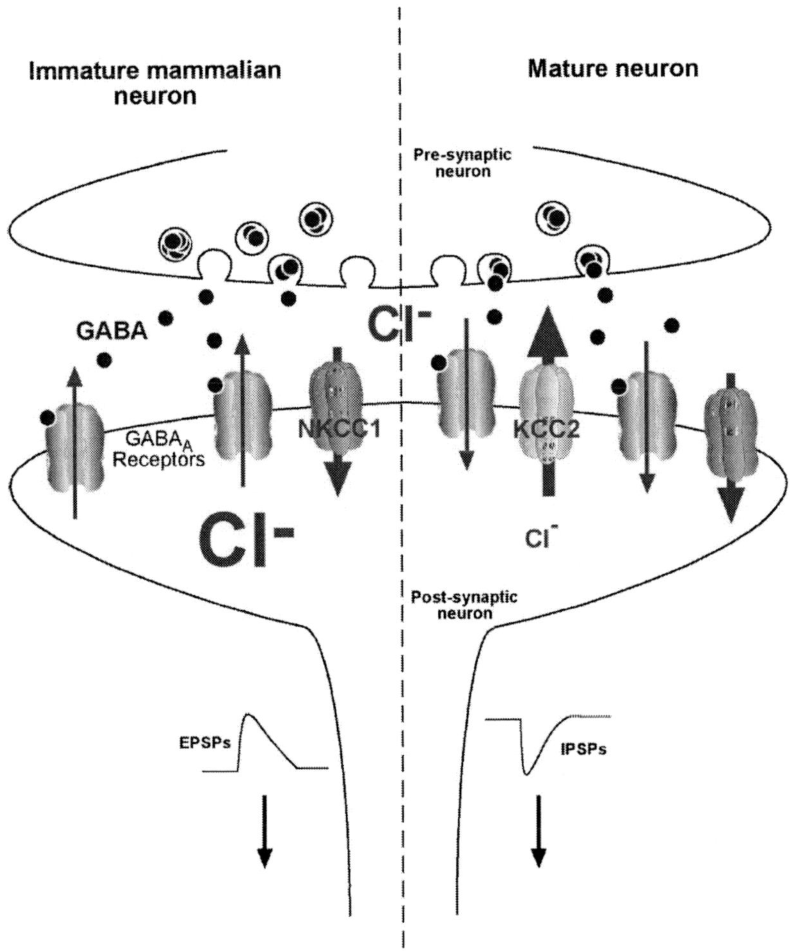

Figure 3. Schematic of the effect of chloride gradient reversal on the determination of GABAergic signaling in immature and mature mammalian neurons. In immature mammalian neurons (left side), $Na^+/K^+/Cl^-$ (NKCC1) is expressed but the K^+/Cl^- co-transporter (KCC2) is not and intracellular Cl^- (green) is high. Pre-synaptic GABA release (blue) leads to activation of post-synaptic $GABA_A$ receptors and passive extrusion of Cl^- ions out of the cell. This movement is depolarizing and results in the propagation of excitatory post-synaptic potentials (EPSPs) and increased neuronal excitation. In mature mammalian and facultative anaerobe neurons (right side), KCC2 expression is increased and intracellular Cl^- is low. Here, $GABA_A$ receptor activation results in the passive influx of Cl^- ions into the cell. This movement is hyperpolarizing and results in the propagation of inhibitory post-synaptic potentials (IPSPs) and decreased electrical excitability.

Conversely, longer durations of OGD or anoxic insult leads to loss of synaptic activity and a second, larger elevation in intracellular $[Cl^-]_i$ and cell death following reoxygenation (Galeffi et al., 2004; LoPachin et al., 2001; Zhu and Krnjevic, 1999). Pronounced elevations in $[Cl^-]_i$ can lead to reversal of $GABA_A$ receptors such that GABA becomes excitatory (Thompson and Gahwiler, 1989). Therefore, reversal of Cl^- gradients may contribute to hyper-excitability and cell death following ischemic insult.

These findings create a paradox in that excitatory GABA signaling enhances cell death attributable to ischemic insult, while cells treated with GABA agonists during or immediately post-ischemia experience neuroprotection. As with the pond snail, the answer may lie in the expression of the two opposing Cl^- transport mechanisms: KCC2 and NKCC1. In the reperfusion period following OGD, NKCC1 protein levels were unchanged but protein levels of the Cl^- extruder KCC2 were reduced by 30% at one hour and 70% by two hours post-insult. Interestingly, both the rise in intracellular Cl^- and the reduction in KCC2 protein levels were reversed by $GABA_A$ receptor agonism immediately following ischemic insult and cell death was avoided (Galeffi et al., 2004). These authors suggested that reduced KCC2 expression may be part of preprogrammed cell death and that $GABA_A$ receptor agonism may not directly affect KCC2 expression, but rather protects the cell and prevents cell death and the associated downstream mechanisms, including protein expression. Taken together, these data indicate that activation of $GABA_A$ receptors immediately after ischemic insults maintains KCC2 expression, which in turn is critical to maintaining neuronal Cl^- gradients. Failure to maintain this gradient likely not only removes the inhibitory mechanism mediated by GABAergic signaling, but may contribute to post-ischemic excitability by directly exciting neurons.

The loss of inhibitory signaling in post-ischemic neurons is likely further enhanced by decreased abundance of $GABA_A$ receptors and decreased expression of $GABA_A$ receptor subunit mRNA that is associated with extensive cell death in the recovering hippocampal area CA1 (Li et al., 1993). Mechanisms that limit GABA-mediated signaling in neurons that have deformed Cl^- gradients may help to reduce post-ischemic hyperexcitation and allow the cells to partially recover post anoxia. Thus a mechanism such as occurs in the pond snail, which reduces brain activity by limiting the uptake of Cl^- via NKCC1, may provide transient neuroprotection to neurons otherwise programmed to undergo apoptosis following ischemic insult.

CONCLUSION

The anoxia-tolerant western painted turtle provides an excellent model in which to explore natural neuroprotective mechanisms that may lead to improved stroke therapies; furthermore, our research indicates that GABA plays an import role in anoxia-tolerance and that GABA-based interventions against ischemic damage should be pursued. Previously, most treatments aimed at reducing ischemic cell death following stroke were targeted at excitatory NMDARs, which have been linked to neuronal death due to ischemia (Diemer et al., 1990). However, despite substantial evidence of the efficacy of NMDAR antagonists at preventing ischemic damage, no NMDAR antagonists have passed clinical trials due to a variety of psychotomimetic and toxic side effects, although recently mementine, a non-competetive partial antagonist of the receptor has shown promise (Lipton, 2004). Conversely, many of the compounds used to target GABA receptors in experimental models are already used clinically to treat other disorders including seizures (Caramia et al., 2000; Vincent et al., 2005), and may in fact regulate glutamate release following ischemic insults, thus circumventing the need to regulate NMDARs directly (Fukuda et al., 1998; Ouyang et al., 2007; Thompson et al., 2007).

Facultative anaerobes with high inwardly directed Cl^- gradients and hyperpolarizing responses to GABA exhibit a common response to anoxia involving moderate to extreme elevations of $[GABA]_e$, heightened $GABA_A$ B_{Max} and suppression of electrical activity including both AP and PSP frequencies. Mature mammalian neurons are sensitive to anoxia but experimental interventions that increase $GABA_A$ receptor activity prevent hyper-excitability and provide neuroprotection against ischemic insults. Thus, mimicry of the endogenous GABA-based defense mechanisms of anoxia-tolerant vertebrates results in increased anoxia-tolerance in adult mammalian neurons. A similar common mechanism may exist in the immature neurons of developing mammals in which GABA is an excitatory neurotransmitter. The pond snail possesses endogenous defense mechanisms that down-regulate the effect of excitatory GABA signaling, and mimicking these mechanisms may prove to be effective against ischemic insult in immature mammals.

It should be noted that in addition to elevated GABA, true facultative anaerobes employ a wide range of strategies to defend energy stores during anoxic insult (Bickler and Buck, 2007). Furthermore, these species are invariably ectothermic and are able to survive at low temperatures and therefore take advantage of considerable Q_{10} effects. However, GABA elevations are a hallmark of anoxia-tolerance in facultative anaerobes and likely underlie inducible neuroprotective mechanisms in hypoxic mammalian models. Thus, parallels between endogenous and induced GABAergic mechanisms highlight the potential of targeting GABA-based pathways in preventing cell death following ischemic insult due to stroke.

REFERENCES

Attwell, D. and Laughlin, S. B. (2001). An energy budget for signaling in the grey matter of the brain. *J Cereb Blood Flow Metab* 21, 1133-45.

Azimi-Zonooz, A., Shuttleworth, C. W. and Connor, J. A. (2006). GABAergic protection of hippocampal pyramidal neurons against glutamate insult: deficit in young animals compared to adults. *J Neurophysiol* 96, 299-308.

Ben-Ari, Y. (2002). Excitatory actions of gaba during development: the nature of the nurture. *Nat Rev Neurosci* 3, 728-39.

Bickler, P. E. and Buck, L. T. (2007). Hypoxia tolerance in reptiles, amphibians, and fishes: life with variable oxygen availability. *Annu Rev Physiol* 69, 145-70.

Buck, L. T. and Pamenter, M. E. (2006). Adaptive responses of vertebrate neurons to anoxia-Matching supply to demand. *Respir Physiol Neurobiol* 154, 226-240.

Caramia, M. D., Palmieri, M. G., Desiato, M. T., Iani, C., Scalise, A., Telera, S. and Bernardi, G. (2000). Pharmacologic reversal of cortical hyperexcitability in patients with ALS. *Neurology* 54, 58-64.

Cheung, U., Moghaddasi, M., Hall, H. L., Smith, J. J., Buck, L. T. and Woodin, M. A. (2006). Excitatory actions of GABA mediate severe-hypoxia-induced depression of neuronal activity in the pond snail (Lymnaea stagnalis). *J Exp Biol* 209, 4429-35.

Clayton, G. H., Owens, G. C., Wolff, J. S. and Smith, R. L. (1998a). Ontogeny of cation-Cl- cotransporter expression in rat neocortex. *Brain Res Dev Brain Res* 109, 281-92.

Clayton, G. H., Staley, K. J., Wilcox, C. L., Owens, G. C. and Smith, R. L. (1998b). Developmental expression of C1C-2 in the rat nervous system. *Brain Res Dev Brain Res* 108, 307-18.

Dave, K. R., Lange-Asschenfeldt, C., Raval, A. P., Prado, R., Busto, R., Saul, I. and Perez-Pinzon, M. A. (2005). Ischemic preconditioning ameliorates excitotoxicity by shifting glutamate/gamma-aminobutyric acid release and biosynthesis. *J Neurosci Res* 82, 665-73.

Delpire, E. (2000). Cation-Chloride Cotransporters in Neuronal Communication. *News Physiol Sci* 15, 309-312.

Diemer, N. H., Johansen, F. F. and Jorgensen, M. B. (1990). N-methyl-D-aspartate and non-N-methyl-D-aspartate antagonists in global cerebral ischemia. *Stroke* 21, III39-42.

Donnelly, D. F., Jiang, C. and Haddad, G. G. (1992). Comparative responses of brain stem and hippocampal neurons to O2 deprivation: in vitro intracellular studies. *Am J Physiol* 262, L549-54.

Erdo, S., Michler, A. and Wolff, J. R. (1991). GABA accelerates excitotoxic cell death in cortical cultures: protection by blockers of GABA-gated chloride channels. *Brain Res* 542, 254-8.

Fernandes, J. A., Lutz, P. L., Tannenbaum, A., Todorov, A. T., Liebovitch, L. and Vertes, R. (1997). Electroencephalogram activity in the anoxic turtle brain. *Am J Physiol* 273, R911-9.

Fukuda, A., Muramatsu, K., Okabe, A., Shimano, Y., Hida, H., Fujimoto, I. and Nishino, H. (1998). Changes in intracellular Ca2+ induced by GABAA receptor activation and reduction in Cl- gradient in neonatal rat neocortex. *J Neurophysiol* 79, 439-46.

Galeffi, F., Sah, R., Pond, B. B., George, A. and Schwartz-Bloom, R. D. (2004). Changes in intracellular chloride after oxygen-glucose deprivation of the adult hippocampal slice: effect of diazepam. *J Neurosci* 24, 4478-88.

Galeffi, F., Sinnar, S. and Schwartz-Bloom, R. D. (2000). Diazepam promotes ATP recovery and prevents cytochrome c release in hippocampal slices after in vitro ischemia. *J Neurochem* 75, 1242-9.

Grabb, M. C., Lobner, D., Turetsky, D. M. and Choi, D. W. (2002). Preconditioned resistance to oxygen-glucose deprivation-induced cortical neuronal death: alterations in vesicular GABA and glutamate release. *Neuroscience* 115, 173-83.

Hall, E. D., Andrus, P. K., Fleck, T. J., Oostveen, J. A., Carter, D. B. and Jacobsen, E. J. (1997). Neuroprotective properties of the benzodiazepine receptor, partial agonist PNU-101017 in the gerbil forebrain ischemia model. *J Cereb Blood Flow Metab* 17, 875-83.

Hall, E. D., Fleck, T. J. and Oostveen, J. A. (1998). Comparative neuroprotective properties of the benzodiazepine receptor full agonist diazepam and the partial agonist PNU-101017 in the gerbil forebrain ischemia model. *Brain Res* 798, 325-9.

Hitzig, B. M., Kneussl, M. P., Shih, V., Brandstetter, R. D. and Kazemi, H. (1985). Brain amino acid concentrations during diving and acid-base stress in turtles. *J Appl Physiol* 58, 1751-4.

Hochachka, P. W. (1986). Defense strategies against hypoxia and hypothermia. *Science* 231, 234-41.

Hylland, P. and Nilsson, G. E. (1999). Extracellular levels of amino acid neurotransmitters during anoxia and forced energy deficiency in crucian carp brain. *Brain Res* 823, 49-58.

Hylland, P., Nilsson, G. E. and Johansson, D. (1995). Anoxic brain failure in an ectothermic vertebrate: release of amino acids and K+ in rainbow trout thalamus. *Am J Physiol* 269, R1077-84.

Inglefield, J. R., Perry, J. M. and Schwartz, R. D. (1995). Postischemic inhibition of GABA reuptake by tiagabine slows neuronal death in the gerbil hippocampus. *Hippocampus* 5, 460-8.

Inglefield, J. R. and Schwartz-Bloom, R. D. (1998). Optical imaging of hippocampal neurons with a chloride-sensitive dye: early effects of in vitro ischemia. *J Neurochem* 70, 2500-9.

Inglefield, J. R. and Schwartz-Bloom, R. D. (1999). Fluorescence imaging of changes in intracellular chloride in living brain slices. *Methods* 18, 197-203.

Jackson, D. C. (2002). Hibernating without oxygen: physiological adaptations of the painted turtle. *J Physiol* 543, 731-7.

Jiang, C., Agulian, S. and Haddad, G. G. (1992). Cl- and Na+ homeostasis during anoxia in rat hypoglossal neurons: intracellular and extracellular in vitro studies. *J Physiol* 448, 697-708.

Johansson, D. and Nilsson, G. (1995). Roles of energy status, KATP channels and channel arrest in fish brain K+ gradient dissipation during anoxia. *J Exp Biol* 198, 2575-80.

Johansson, D., Nilsson, G., Ouml and Rnblom, E. (1995). Effects of anoxia on energy metabolism in crucian carp brain slices studied with microcalorimetry. *J Exp Biol* 198, 853-9.

Johansson, D., Nilsson, G. E. and Doving, K. B. (1997). Anoxic depression of light-evoked potentials in retina and optic tectum of crucian carp. *Neurosci Lett* 237, 73-6.

Johns, L., Sinclair, A. J. and Davies, J. A. (2000). Hypoxia/hypoglycemia-induced amino acid release is decreased in vitro by preconditioning. *Biochem Biophys Res Commun* 276, 134-6.

Kaila, K. (1994). Ionic basis of GABAA receptor channel function in the nervous system. *Prog Neurobiol* 42, 489-537.

Kaila, K., Voipio, J., Paalasmaa, P., Pasternack, M. and Deisz, R. A. (1993). The role of bicarbonate in GABAA receptor-mediated IPSPs of rat neocortical neurones. *J Physiol* 464, 273-89.

Lambert, N. and Grover, L. (1995). The mechanism of biphasic GABA responses. *Science* 269, 928-9.

Li, H., Siegel, R. E. and Schwartz, R. D. (1993). Rapid decline of GABAA receptor subunit mRNA expression in hippocampus following transient cerebral ischemia in the gerbil. *Hippocampus* 3, 527-37.

Lipton, S. A. (2004). Paradigm shift in NMDA receptor antagonist drug development: molecular mechanism of uncompetitive inhibition by memantine in the treatment of Alzheimer's disease and other neurologic disorders. *J Alzheimers Dis* 6, S61-74.

LoPachin, R. M., Gaughan, C. L., Lehning, E. J., Weber, M. L. and Taylor, C. P. (2001). Effects of ion channel blockade on the distribution of Na, K, Ca and other elements in oxygen-glucose deprived CA1 hippocampal neurons. *Neuroscience* 103, 971-83.

Lukasiuk, K. and Pitkanen, A. (2000). GABA(A)-mediated toxicity of hippocampal neurons in vitro. *J Neurochem* 74, 2445-54.

Lutz, P. L. and Leone-Kabler, S. L. (1995). Upregulation of the GABAA/benzodiazepine receptor during anoxia in the freshwater turtle brain. *Am J Physiol* 268, R1332-5.

Lutz, P. L. and Milton, S. L. (2004). Negotiating brain anoxia survival in the turtle. *J Exp Biol* 207, 3141-7.

Lutz, P. L. and Reiners, R. (1997). Survival of energy failure in the anoxic frog brain: delayed release of glutamate. *J Exp Biol* 200, 2913-7.

Martyniuk, C. J., Awad, R., Hurley, R., Finger, T. E. and Trudeau, V. L. (2007). Glutamic acid decarboxylase 65, 67, and GABA-transaminase mRNA expression and total enzyme activity in the goldfish (Carassius auratus) brain. *Brain Res* 1147, 154-66.

Martyniuk, C. J., Crawford, A. B., Hogan, N. S. and Trudeau, V. L. (2005). GABAergic modulation of the expression of genes involved in GABA synaptic transmission and stress in the hypothalamus and telencephalon of the female goldfish (Carassius auratus). *J Neuroendocrinol* 17, 269-75.

Michelson, H. B. and Wong, R. K. (1991). Excitatory synaptic responses mediated by GABAA receptors in the hippocampus. *Science* 253, 1420-3.

Mielke, J. G. and Wang, Y. T. (2005). Insulin exerts neuroprotection by counteracting the decrease in cell-surface GABA receptors following oxygen-glucose deprivation in cultured cortical neurons. *J Neurochem* 92, 103-13.

Milton, S. L., Thompson, J. W. and Lutz, P. L. (2002a). Mechanisms for maintaining extracellular glutamate levels in the anoxic turtle striatum. *Am J Physiol Regul Integr Comp Physiol* 282, R1317-23.

Milton, S. L., Thompson, J. W. and Lutz, P. W. (2002b). Mechanisms for maintaining extracellular glutamate levels in the anoxic turtle striatum. *American Journal of Physiology, (Regulatory Integrative Comparative Physiology)* 282, R1317-R1323.

Muir, J. K., Lobner, D., Monyer, H. and Choi, D. W. (1996). GABAA receptor activation attenuates excitotoxicity but exacerbates oxygen-glucose deprivation-induced neuronal injury in vitro. *J Cereb Blood Flow Metab* 16, 1211-8.

Murry, C. E., Jennings, R. B. and Reimer, K. A. (1986). Preconditioning with ischemia: a delay of lethal cell injury in ischemic myocardium. *Circulation* 74, 1124-36.

Musacchia, X. J. (1959). The viability of chrysemys picta submerged at various temperatures. *Physiol Zoo 1* 32, 57-50.

Nilsson, G. E. (1990). Long-term anoxia in crucian carp: changes in the levels of amino acid and monoamine neurotransmitters in the brain, catecholamines in chromaffin tissue, and liver glycogen. *J Exp Biol* 150, 295-320.

Nilsson, G. E. (2001). Surviving anoxia with the brain turned on. *News Physiol Sci* 16, 217-21.

Nilsson, G. E., Alfaro, A. A. and Lutz, P. L. (1990). Changes in turtle brain neurotransmitters and related substances during anoxia. *Am J Physiol* 259, R376-84.

Nilsson, G. E. and Lutz, P. L. (1991). Release of inhibitory neurotransmitters in response to anoxia in turtle brain. *Am J Physiol* 261, R32-7.

Nilsson, G. E. and Renshaw, G. M. (2004). Hypoxic survival strategies in two fishes: extreme anoxia tolerance in the North European crucian carp and natural hypoxic preconditioning in a coral-reef shark. *J Exp Biol* 207, 3131-9.

Nilsson, G. E. and Winberg, S. (1993). Changes in the brain levels of GABA and related amino acids in anoxic shore crab (Carcinus maenas). *Am J Physiol* 264, R733-7.

Nitsch, C., Goping, G. and Klatzo, I. (1989). Preservation of GABAergic perikarya and boutons after transient ischemia in the gerbil hippocampal CA1 field. *Brain Res* 495, 243-52.

Nunez, J. L., Alt, J. J. and McCarthy, M. M. (2003). A new model for prenatal brain damage. I. GABAA receptor activation induces cell death in developing rat hippocampus. *Exp Neurol* 181, 258-69.

Ouyang, C., Guo, L., Lu, Q., Xu, X. and Wang, H. (2007). Enhanced activity of GABA receptors inhibits glutamate release induced by focal cerebral ischemia in rat striatum. *Neurosci Lett.*

Pamenter, M.E., Ormond, J., Hogg, D.W., Woodin, M. and Buck, L.T. (2011) Endogenous $GABA_A$ and $GABA_B$ receptor-mediated electrical suppression is critical to neuronal anoxia tolerance. "in press" PNAS MS# 2011-02429R

Perez-Pinzon, M. A., Chan, C. Y., Rosenthal, M. and Sick, T. J. (1992a). Membrane and synaptic activity during anoxia in the isolated turtle cerebellum. *American Journal of Physiology, (Regulatory Integrative Comparative Physiology)* 263, R1057-R1063.

Perez-Pinzon, M. A., Rosenthal, M., Lutz, P. W. and Sick, T. J. (1992b). Anoxic Survival of the isolated cerebellum of the turtle pseudemis scripta elegans. *J. Comp. Physiol. B.* 162, 68-73.

Podrabsky, J. E., Carpenter, J. F. and Hand, S. C. (2001). Survival of water stress in annual fish embryos: dehydration avoidance and egg envelope amyloid fibers. *Am J Physiol Regul Integr Comp Physiol* 280, R123-31.

Podrabsky, J. E. and Hand, S. C. (1999). The bioenergetics of embryonic diapause in an annual killifish, austrofundulus limnaeus. *J Exp Biol* 202 (Pt 19), 2567-80.

Podrabsky, J. E. and Hand, S. C. (2000). Depression of protein synthesis during diapause in embryos of the annual killifish Austrofundulus limnaeus. *Physiol Biochem Zool* 73, 799-808.

Podrabsky, J. E., Lopez, J. P., Fan, T. W., Higashi, R. and Somero, G. N. (2007). Extreme anoxia tolerance in embryos of the annual killifish Austrofundulus limnaeus: insights from a metabolomics analysis. *J Exp Biol* 210, 2253-66.

Poli, A., Beraudi, A., Villani, L., Storto, M., Battaglia, G., Di Giorgi Gerevini, V., Cappuccio, I., Caricasole, A., D'Onofrio, M. and Nicoletti, F. (2003). Group II metabotropic glutamate receptors regulate the vulnerability to hypoxic brain damage. *J Neurosci* 23, 6023-9.

Renshaw, G. M., Kerrisk, C. B. and Nilsson, G. E. (2002). The role of adenosine in the anoxic survival of the epaulette shark, Hemiscyllium ocellatum. *Comp Biochem Physiol B Biochem Mol Biol* 131, 133-41.

Routley, M. H., Nilsson, G. E. and Renshaw, G. M. (2002). Exposure to hypoxia primes the respiratory and metabolic responses of the epaulette shark to progressive hypoxia. *Comp Biochem Physiol A Mol Integr Physiol* 131, 313-21.

Rubakhin, S. S., Szucs, A. and Rozsa, K. S. (1996). Characterization of the GABA response on identified dialysed Lymnaea neurons. *Gen Pharmacol* 27, 731-9.

Sakurai, S. Y., Lutz, P. L., Schulman, A. and Albin, R. L. (1993). Unchanged [3H]MK-801 binding and increased [3H]flunitrazepam binding in turtle forebrain during anoxia. *Brain Res* 625, 181-5.

Saransaari, P. and Oja, S. S. (2004). Metabotropic glutamate receptors modulate ischemia-induced GABA release in mouse hippocampal slices. *Neurochem Res* 29, 1511-8.

Saransaari, P. and Oja, S. S. (2005). Characteristics of GABA release in mouse brain stem slices under normal and ischemic conditions. *Neurochem Res* 30, 1549-56.

Schwartz-Bloom, R. D., McDonough, K. J., Chase, P. J., Chadwick, L. E., Inglefield, J. R. and Levin, E. D. (1998). Long-term neuroprotection by benzodiazepine full versus partial agonists after transient cerebral ischemia in the gerbil [corrected]. *J Cereb Blood Flow Metab* 18, 548-58.

Schwartz-Bloom, R. D., Miller, K. A., Evenson, D. A., Crain, B. J. and Nadler, J. V. (2000). Benzodiazepines protect hippocampal neurons from degeneration after transient cerebral ischemia: an ultrastructural study. *Neuroscience* 98, 471-84.

Sommer, C., Fahrner, A. and Kiessling, M. (2002). [3H]muscimol binding to gamma-aminobutyric acid(A) receptors is upregulated in CA1 neurons of the gerbil hippocampus in the ischemia-tolerant state. *Stroke* 33, 1698-705.

Sommer, C., Fahrner, A. and Kiessling, M. (2003). Postischemic neuroprotection in the ischemia-tolerant state gerbil hippocampus is associated with increased ligand binding to inhibitory GABA(A) receptors. *Acta Neuropathol* 105, 197-202.

Staley, K. J. and Mody, I. (1992). Shunting of excitatory input to dentate gyrus granule cells by a depolarizing GABAA receptor-mediated postsynaptic conductance. *J Neurophysiol* 68, 197-212.

Sternau, L. L., Lust, W. D., Ricci, A. J. and Ratcheson, R. (1989). Role for gamma-aminobutyric acid in selective vulnerability in gerbils. *Stroke* 20, 281-7.

Suzue, T., Wu, G. B. and Furukawa, T. (1987). High susceptibility to hypoxia of afferent synaptic transmission in the goldfish sacculus. *J Neurophysiol* 58, 1066-79.

Thompson, J. W., Prentice, H. M. and Lutz, P. L. (2007). Regulation of extracellular glutamate levels in the long-term anoxic turtle striatum: coordinated activity of glutamate transporters, adenosine, K (ATP) (+) channels and GABA. *J Biomed Sci* 14, 809-17.

Thompson, S. M. and Gahwiler, B. H. (1989). Activity-dependent disinhibition. II. Effects of extracellular potassium, furosemide, and membrane potential on ECl- in hippocampal CA3 neurons. *J Neurophysiol* 61, 512-23.

Turner, A. J. and Whittle, S. R. (1983). Biochemical dissection of the gamma-aminobutyrate synapse. *Biochem J* 209, 29-41.

Ultsch, G. R. and Jackson, D. C. (1982). Long-term submergence at 3C of the turtle *chrysemys picta bellii* in normoxic and severely hypoxic water I. rvival, gas exchange and acid-base status. *Journal of experimental biology* 96, 11-28.

Vincent, A. M., Backus, C., Taubman, A. A. and Feldman, E. L. (2005). Identification of candidate drugs for the treatment of ALS. *Amyotroph Lateral Scler Other Motor Neuron Disord* 6, 29-36.

Virani, N. A. and Rees, B. B. (2000). Oxygen consumption, blood lactate and inter-individual variation in the gulf killifish, Fundulus grandis, during hypoxia and recovery. *Comp Biochem Physiol A Mol Integr Physiol* 126, 397-405.

Xu, F. L., Zhu, C. L. and Wang, X. Y. (2006). Developmental changes of glutamate acid decarboxylase-67 in mouse brain after hypoxia ischemia. *Neurosci Bull* 22, 47-51.

Zhao, P., Qian, H. and Xia, Y. (2005). GABA and glycine are protective to mature but toxic to immature rat cortical neurons under hypoxia. *Eur J Neurosci* 22, 289-300.

Zhu, P. J. and Krnjevic, K. (1999). Persistent block of CA1 synaptic function by prolonged hypoxia. *Neuroscience* 90, 759-70.

In: Turtles: Anatomy, Ecology and Conservation ISBN: 978-1-61470-554-3
Eds: M. J. Cosgrove and S. A. Roe

Chapter 4

AN OVERVIEW OF THE THREATENED PHYLOGENETIC DIVERSITY OF LIVING TESTUDINES BASED ON A REVIEW OF THE COMPLEX EVOLUTIONARY HISTORY OF TURTLES

Josep Marmi[1] and Àngel H. Luján[2]
Institut Català de Paleontologia, C/ Escola Industrial 23, E-08201, Sabadell, Catalonia, Spain

ABSTRACT

The history of turtle lineage began around 225 Mya. Since then, this group of reptiles developed diverse ecological strategies and colonized a wide range of environments, from marine to fully terrestrial habitats. In spite of their great ecological diversity, the bauplan of turtles is peculiar and little variable, with a body encased in a rigid shell consisting of a dorsal carapace and a ventral plastron. This fact has made the morphological comparison of turtles with other vertebrates very complicate. Thus, the understanding of the origins and phylogenetic relationships of turtles within the Amniota and the evolution of turtle biology are stimulating challenges for researchers. Several attempts in

1 Correspondence address by e-mail: josep.marmi@icp.cat
2 Correspondence address by e-mail: angel.lujan@icp.cat

order to resolve these mysteries have been carried out from a multidisciplinary approach, but lacks consensus. The knowledge of the evolutionary history of turtle species and their inclusive groups is relevant for their conservation management, especially taking into account that many species are among moderate and high risk of extinction. In this chapter, we update and review the state of the art of the systematics of turtles from the point of view of several often conflicting disciplines, embryology, morphology, paleontology and molecular systematics. In addition, we sound out the amount of threatened phylogenetic diversity in the turtle tree of life based on fossil evidence, recent phylogenetic hypotheses and data from the red list of the International Union for the Conservation of Nature.

INTRODUCTION

The clade Testudines (= Chelonia *sensu* Gauthier et al., 1988) include about 289 species alive today split up into two primary groups: Pleurodira and Cryptodira (Joyce et al., 2004). Nowadays, cryptodires are more diverse than pleurodires (Figure 1), the latter being more restricted geographically, living only in the southern hemisphere (Gaffney et al., 2006). Cryptodires are also adapted to a wider range of habitats, from open marine to arid environments, whereas pleurodires are restricted to freshwater (Gaffney et al., 2006).

Turtles are mysterious animals for the biologists, especially as far as their obscure evolutionary history is concerned. The origins and phylogenetic relationships of turtles with the remaining amniotes have been explored from different approaches such as paleontology, comparative anatomy, developmental biology, and molecular phylogenetics. However, there is no consensus among different source of data making the position of turtles in the amniote tree of live still rather confusing (see below and Werneburg & Sánchez-Villagra, 2009 and references therein). Uncertain phylogenetic relationships remain unresolved among fossil and extant turtles and even among several living turtle families and higher taxa (Shaffer et al., 1997; Gaffney et al., 2006; Barley et al., 2010). In addition, the evolution of adaptation of turtles to terrestrial and aquatic environments is poorly known because of uncommon preservation of appendicular skeletal elements in extinct turtles and their incomplete fossil record (Scheyer & Sander, 2007). As an example, there has been an intense debate about the terrestrial (e.g. Joyce & Gauthier, 2004;

Scheyer & Sander, 2007) versus semiaquatic (e.g. Gaffney, 1990) origins of turtles.

The knowledge of the evolutionary history of organisms is crucial to characterizing the biological diversity. Phylogeny is a record of how biodiversity has come about and the understanding of the origins of a group of organisms can assist in their conservation by contrasting current versus historical patterns (Harvey et al., 1996; Purvis et al., 2005). Moreover, phylogenies can inform about the risk of lineage extinction and assess the amount of evolutionary history lost if extinction occurs (Purvis et al., 2000). Thus, the extinction of species-poor or monotypic highly divergent lineages containing endangered species would entail the loss of a significant part of the evolutionary history of the inclusive clade. It is also possible to extrapolate from the past and present to the future, in order to predict the state of biodiversity under different scenarios (Rosenzweig, 2001). In addition, well resolved phylogenies are quite important in wildlife management because provide legal basis for conservation programs (O'Brien, 1994).

According to IUCN (2010), near a half of turtle species are threatened. Among them, 59 species are classified as vulnerable, 40 species as endangered and 30 species are critically endangered, representing the 20%, 14% and 10% of the total number of turtle species respectively. Thus, a better understanding of the evolutionary history of turtles is urgently needed in order to estimate the amount of turtle diversity that is at risk of extinction. In this chapter, the evolutionary significance of turtles is discussed by means of a review of the most recent advances in the knowledge of turtle origins, systematics, and the evolution of ecological adaptation. In addition, available data on endangered species is analyzed from a evolutionary point of view in order to obtain a general view of the amount of evolutionary history of turtles that may be lost if no efficient conservation measures are taken.

The Mysterious Phylogenetic Relationships of Turtles within Amniota

The origins of turtles and their body plan are one of the great riddles of reptile evolution. Turtles are characterized by a horny beak rather than teeth and a shell composed of an upper carapace and a lower plastron, joined together by a bony bridge. Their shell is like that of no other animal and it is a composite structure derived from ribs, parts of the shoulder girdle and specialized dermal bones (Reisz & Head, 2008). The peculiar body plan of turtles and their highly derived anatomical features make their phylogenetic position within the amniotes obscure. In fact, there has been an intense debate if turtles evolved from a lineage within Parareptilia —a well supported clade of stem amniotes including pareiasaurs and procolophonoids among others— or belong to Eureptilia which include diapsid reptiles (Tsuji & Müller, 2009).

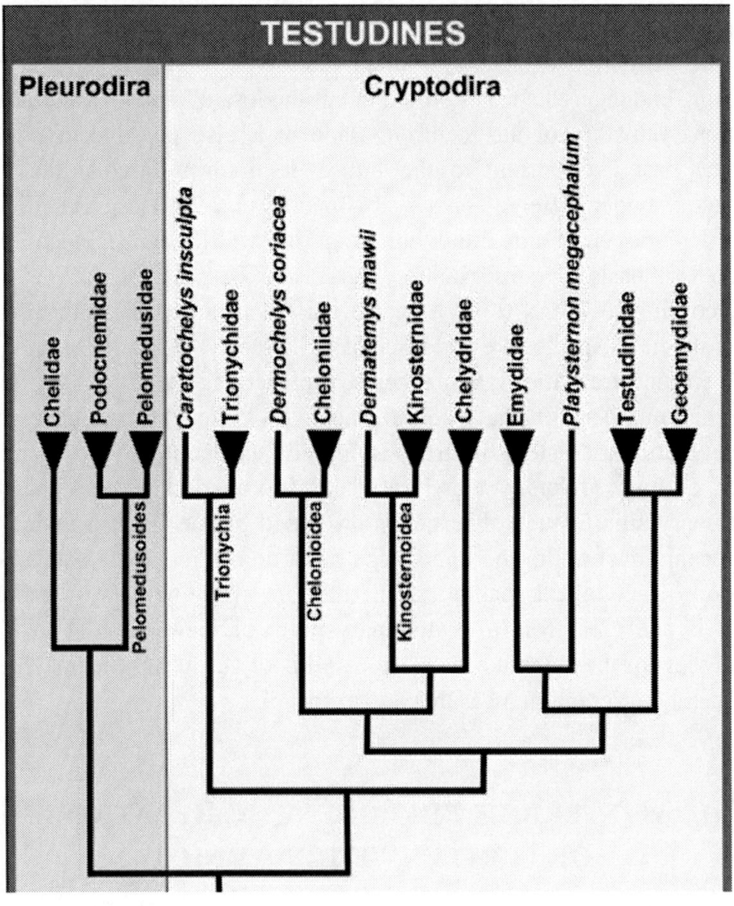

Figure 1. Phylogenetic tree of the main lineages of living turtles based on Joyce et al. (2004) and Barley et al. (2010).

Until today, up to ten phylogenetic hypotheses have been proposed, but none of them has a general acceptance (Figure 2). For instance, it has been suggested that turtles are sister of Thecodontia (including Mammalia and Archosauria) (Gardiner, 1993), Sauria (crown diapsids) (Caspers et al., 1996), Sphenodontia (Fushitani et al., 1996), Archosauria (Kumazawa & Nishida, 1999), Crocodylia (Hedges & Poling, 1999), Lepidosauria (Zardoya & Meyer, 2000), and Aves (Pollock et al., 2000). However, most of these phylogenetic hypotheses are constructed ignoring extinct taxa. Extensive data including both extinct and extant amniotes suggest that turtles are part of a clade of basal terrestrial "anapsid" reptiles (Gauthier, 1994), are related to procolophonoids (Laurin & Reisz, 1995), or pareiasaurs (Lee, 1997) or they are sister taxa of sauropterygians within Eureptilia (de Braga & Rieppel, 1997). Disagreements among all these hypotheses probably are due to both morphological and molecular data are sensitive to taxon sampling, homology issues, rate heterogeneity and missing data (Lee et al., 2008).

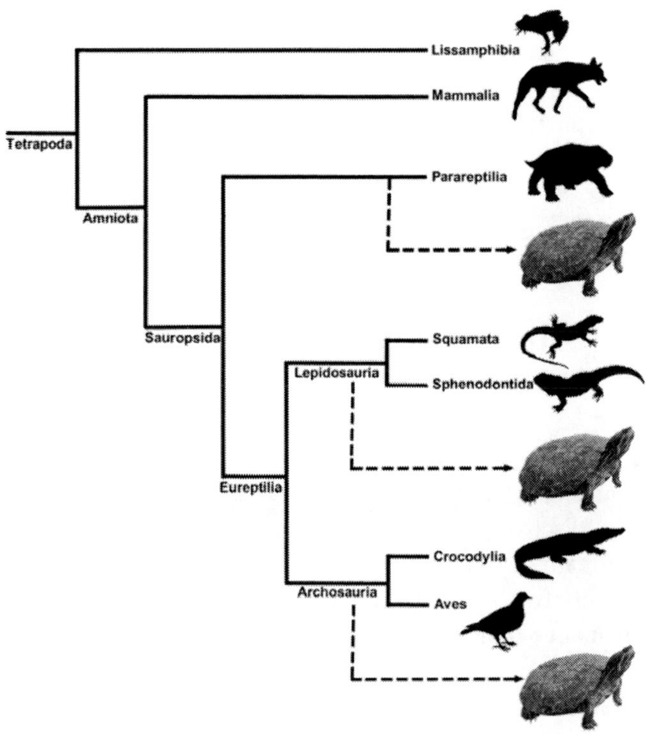

Figure 2. Three (out of ten) main hypotheses for the phylogenetic relationships of turtles with remaining amniotes.

Recent studies have not shed much light to the phylogenetic relationships of turtles within Amniota. New molecular analyses strongly support a turtle + archosaur relationship within Eureptilia (e.g. Hugall et al., 2007; Shedlock et al., 2007). In fact, since the beginnings of 2000s there has been decreasing support to the hypothesis of a turtle-parareptile relationship which is no longer considered valid among the majority of evolutionary biologists (see references in Tsuji & Müller, 2009). However, new embryological data (Werneburg & Sánchez-Villagra, 2009), combined molecular and morphological data (Lee et al., 2008) and fossil evidence (Lyson et al., 2010) place again turtles outside Diapsida. Specifically, in this latter study, turtles nest within parareptiles, forming a well supported clade with the genus *Eunotosaurus*, as the sister group to Diapsida (Figure 3). According to Lyson et al. (2010), basal-most stem turtles *Odontochelys semitestacea* and *Proganochelys quenstedti* have all six unequivocal synapomorphies diagnostic of Parareptilia which were listed by Tsuji & Müller (2009): 1) absence of a lacrimal-nasal contact, 2) absence of a caniniform region, 3) shortened postorbital region, 4) single median embayment of the posterior margin of the skull roof, 5) absence of a supraglenoid foramen and 6) absence of a subtemporal process of the jugal. Thus, the debate about the phylogenetic affinities of turtles with other amniotes seems still open and further analyses are needed to confirm the turtle-parareptile hypothesis as suggest these last studies.

Oldest Known Turtles and the Evolution of Shell

In spite of major controversies on their phylogenetic relationships, the evolutionary history of turtles began in the early Late Triassic, around 220 Mya. Since its discovery in 1884, *Proganochelys quenstedti* from the Late Triassic of Germany was considered the oldest turtle known (Gaffney, 1990). Nevertheless, only two years ago, *Proganochelys* was dethroned as the basal-most Testudinata due to new evidence from China. *Odontochelys semitestacea* is a relatively small turtle, about 40 cm long, approximately five million of years older than *Proganochelys* that was discovered near Guanling, Guizhou Province (southwestern China) (Li et al., 2008). According to these authors, *Odontochelys* is evolutively more primitive than *Proganochelys* in having teeth on premaxilla, maxilla and dentary; absence of fully formed carapace;

absence of acromial process on scapula and presence of long tail, among other characters. *Odontochelys* also shares primitive features with *Proganochelys* (see Li et al., 2008 for details).

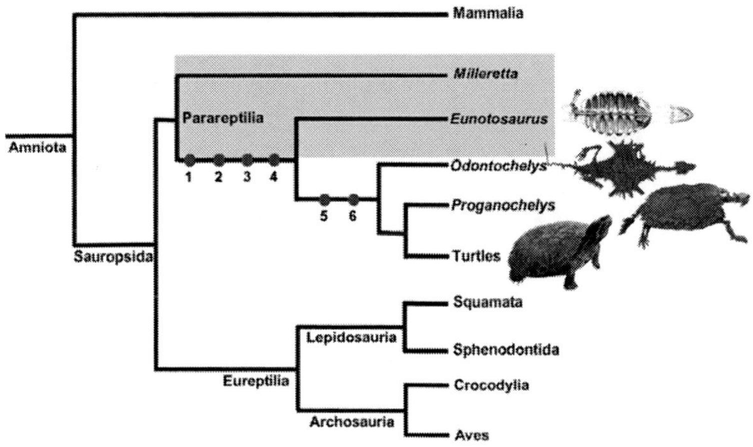

Figure 3. According to new phylogenetic analyses carried out by Lyson et al. (2010) turtles are closely related to the parareptile *Eunotosaurus* based on the following shared characters: 1, broad T-shaped ribs; 2, ten elongate trunk vertebrae; 3, cranial tubercles; 4, wide trunk. Two additional features are shared by *Odontochelys*, *Proganochelys* and extant turtles: 5, presence of plastron and 6, presence of neurals.

Turtle shell represents an evolutionary novelty and some authors propose that turtle evolution might be radical or saltatory instead of a gradual change (Gilbert et al., 2001; Rieppel, 2001). The comparison of rib cage of the parareptile *Eunotosaurus* and the shell of *Odontochelys* and *Proganochelys* provides valuable information about the evolution of turtle shell (Lyson et al., 2010). The dorsal part of turtle shell, or carapace, is derived from the ribs. The scapula is found under the carapace, contrary to the pattern of remaining amniotes in which the scapula is outside the rib cage (Figure 4 and Nagashima et al., 2009). In the Late Permian, *Eunotosaurus* had broadened ribs with apparent metaplastic ossification of the dermis and ribs joined by a second ossification. In the next stage, 44 Myr after, *Odontochelys* developed neural plates and a fully ossified plastron, containing portions of the shoulder girdle and gastralia. The carapace of *Odontochelys* resembles embryonic stages of extant turtles in that there is only some broadening and consequent flattening of the dorsal ribs, which do not expand into costal plates (Li et al., 2008;

Nagashima et al., 2009). Anterior ribs grew posteriorly in *Odontochelys* and the scapula remained anterior to the ribs (Figure 4). The dorsal ribs of *Odontochelys* seem arrested axially and they do not bend ventrally, suggesting that a caparacial-like ridge might be acquired in the flank of their embryos but not persisted and encircled the carapace margin in later development, as in modern turtles (Nagashima et al., 2009). The completion of the shell may have occurred rapidly, approximately in five million of years, which is the timespan that separate *Odontochelys* and *Proganochelys* (Lyson et al., 2010).

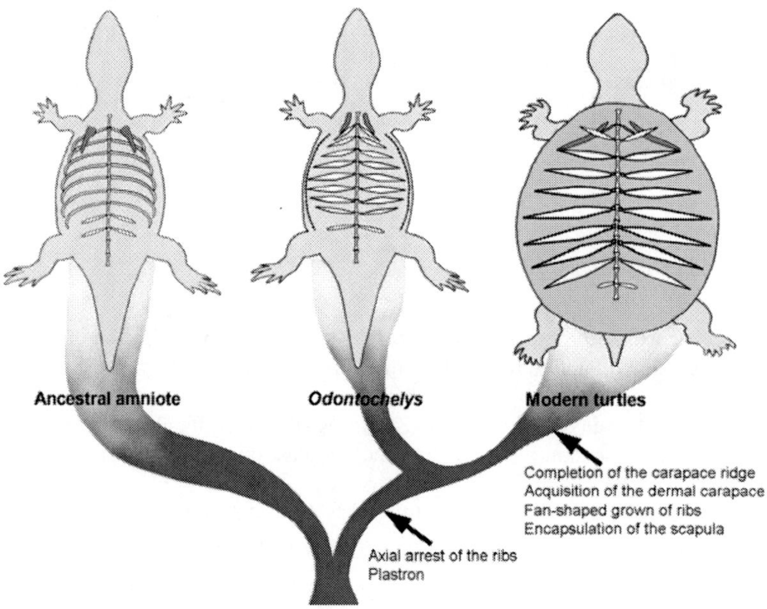

Figure 4. Evolution of turtle body plan. In the general pattern of amniotes (excluding turtles), scapulas (in dark grey) are outside the rib cage. In *Odontochelys*, scapulas are anterior to the ribs whereas in modern turtles they are encapsulated under the carapace. Dorsal ribs are arrested axially in *Odontochelys* and in modern turtles ribs are placed in a fan-like arrangement. The broken black line in *Odontochelys* indicates a carapacial-like ridge that it might be present in their embryos. Modified from Nagashima et al. (2009).

The genus *Proganochelys* is well known from several skeletons. These ancient turtles are characterized by a massive shell and spiked armor on the neck and tail (Reisz & Head, 2008). The shell is fully developed with carapace and plastron enclosing shoulder and pelvic girdles (Gaffney, 1990). The carapace consists of fused ribs, neural bones with fused thoracic vertebrae and marginal bones. The plastron is formed from interclavicles, clavicle and five paired bones fused together. Other Late Triassic stem turtles are *Proterochersis robusta* from several localities of Germany (Lapparent de Broin, 2001) and *Palaeochersis talampayensis* from northwestern Argentina (Rougier et al., 1995).

TERRESTRIAL VERSUS AQUATIC ORIGINS OF TURTLES

Extant turtles inhabit a wide diversity of terrestrial and fresh- and sea-water environments such as land, ponds, lakes, streams, large rivers, estuaries and ocean. An aquatic origin of turtles has been suggested when they are considered to be closely related to extinct marine sauropterygian reptiles (de Braga & Rieppel, 1997). However, other authors indicate that oldest turtles were clearly terrestrial (Joyce & Gauthier, 2004; Scheyer & Sander, 2007).

The lifestyle of extinct turtles may be inferred from different source of evidence. The gross morphology of shell, shoulder girdle and limbs reveals adaptations to terrestrial or aquatic environments (Depecker et al., 2006). Terrestrial turtles have domed shell and long scapular prong and short coracoid associated with a mode of locomotion in which walking is predominant (Figure 5a). By contrast, in both highly aquatic freshwater and marine turtles the scapular prong is short and the coracoid is long and they are associated with flat shells and swimming locomotion (Figure 5b). However, Joyce & Gauthier (2004) pointed out that some exceptions exist to this general rule. For instance, the aquatic Asian box turtle (*Cuora amboinensis*) has highly domed shell and the terrestrial African pancake tortoise (*Malacochersus tornieri*) has a greatly flattened shell (Ernst & Barbour, 1989). Other commonly used indicator is the depositional environment. Although it is easy to discard marine habitat for turtles discovered in terrestrial sediments, turtles discovered in fluvial or marine sediments may be either aquatic or terrestrial because rivers can bury remains of terrestrial animals or transport them to marine environments (Joyce & Gauthier, 2004). In these cases, detailed taphonomic analysis is needed in order to infer the degree of transport undergone by turtle remains before burying. For instance, isolated plates with evident abrasion marks indicate long distance transport (Figure 6a), whereas articulated shells and lack of abrasion, scavenging and tramping marks suggest short biostratinomic history and little transport (Figure 6b, c) (Brand et al.,

2000; Bertini et al., 2006). In this latter example, the habitat of the fossil turtle can be inferred from the depositional setting (e.g. Marmi et al., 2009).

Figure 5. Adaptations to terrestrial and aquatic lifestyles: a, terrestrial *Testudo hermanni* with a moderate domed shaped carapace; b, semiaquatic *Trachemys scripta* with a flat shell; c, thin section of left xiphiplastron of *Testudo hermanni* showing wide and low vascularized external (ECO) and internal (ICO) cortices; d, thin section of a right xiphiplastron of *Trachemys scripta* with thin and highly vascularized external and internal cortices. The cancellous bone (CB) consists of short trabeculae and small to medium sized vascular spaces in both species. The pictures (a) and (b) have been taken at the Centre de Recuperació d'Amfibis i Rèptils de Catalunya (CRARC).

In living turtles, forelimbs usually reflect their lifestyle (Ernst & Barbour, 1989). Morphometric analysis of forelimbs revealed a close relationship between relative hand length and habitat preference (Joyce & Gauthier, 2004). Terrestrial turtles are short-handed to facilitate digital rollover during walking and highly aquatic turtles such as chelonioids are long-handed for swimming (Figure 7). However, after death, skull and limb bones separate from the body early and these elements are generally not preserved in fossil turtles (Brand et al., 2000; 2003). In these cases, histology provides an alternative way for testing the ecology of extinct turtles since the bone structure of the shell of living aquatic and terrestrial turtles reveal histological differences (Scheyer & Sander, 2007). Shell bones of terrestrial turtles exhibit a diploe structure with well developed external and internal cortices, weak vascularization of the compact bone layers and dense interior cancellous bone with overall short traveculae (Figure 5c). On the contrary, aquatic turtles increase overall vascularization of the bone tissue reducing cortical bone layers and creating medium to large sized vascular spaces in the cancellous bone delimited by long and slender traveculae (Figure 5d). This pattern reported in aquatic turtles may be interpreted as an adaptation to aquatic environments in order to increase buoyancy. However, several exceptions of the general pattern exist, with terrestrial taxa showing moderate levels of vascularization (e.g. *Kinixys homeana* and *Chelonoidis carbonaria*) and aquatic taxa showing terrestrial-like bone microstructures (e.g. *Mauremys mutica* and *Pangshura tentoria*).

Figure 6. Different examples of preservation of turtle remains from the Late Cretaceous of Pyrenees: a, isolated plate with evident signs of abrasion in the external (top) and articulation (bottom) surfaces suggesting long distance transport (Barranc de Torrebilles site, Isona, southeastern Pyrenees); b, cast of a *Solemys* sp. carapace with remains of peripheral plates (arrowed) from Mina Esquirol site (Vallcebre, southeastern Pyrenees); c, partial shell of a bothremydid with carapace and plastron articulated from Barranc de Torrebilles site. Short transport or autochthony have been inferred for (b) and (c) and the depositional setting suggests that these extinct turtles inhabited lagoonal and fluvial environments, respectively (Marmi et al., 2009 and Marmi et al. under study). The pencil in (b) measures 14.6 cm.

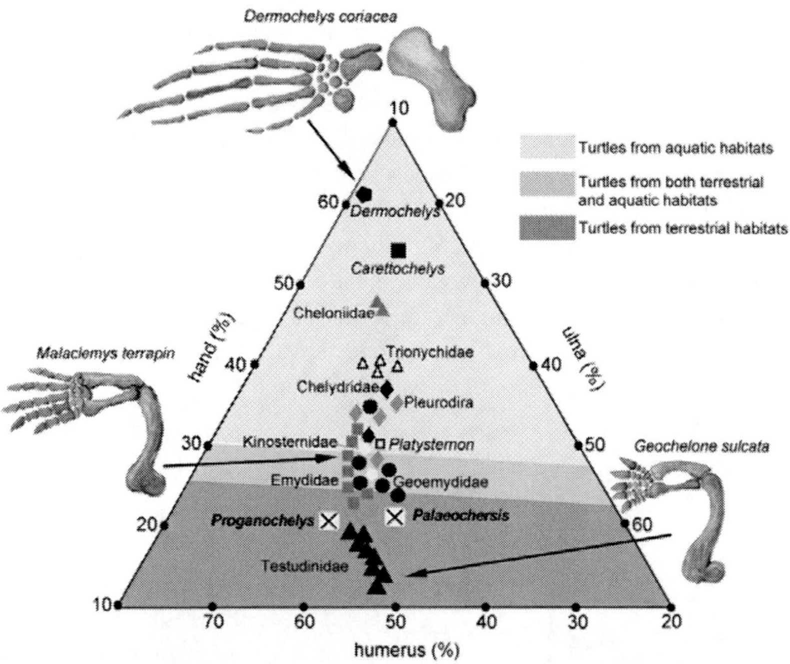

Figure 7. Distribution of taxa and habitat preferences on a ternary plot using forelimb measurements. There is large difference in hand sizes between specialized taxa such as marine *Dermochelys coriacea* and terrestrial *Geochelone sulcata*. Late Triassic turtles *Proganochelys* and *Palaeochersis* are included within the group of terrestrial taxa. Modified from Joyce & Gauthier (2004).

Gaffney (1990) suggested that *Proganochelys* inhabited freshwater as a bottom walker, being not exclusively aquatic or terrestrial, based on interpretations of the depositional environment and features of limb morphology. The analysis of forelimb proportion and depositional environment also indicate aquatic habits for *Odontochelys* and suggest a possible aquatic origin of turtles (Li et al., 2008). According to a recent phylogenetic analysis, stem turtles are closely related to the parareptile *Eunotosaurus* which lacks obvious aquatic adaptations and is only known from terrestrial sediments (Gow 1997; Lyson et al., 2010). In addition, forelimb proportions, shape of shell and histological data provide strong evidence for a terrestrial lifestyle in *Proganochelys* and *Palaeochersis* (Joyce & Gauthier, 2004; Scheyer & Sander, 2007). Thus, the putative marine ecology of *Odontochelys* might be interpreted as independently derived and not ancestral to the subsequent radiation of turtles, supporting an origin of stem turtles in terrestrial environments (Lyson et al., 2010).

The Scarce Jurassic Fossil Record and the Origins of Crown Turtles

The Jurassic is an outstanding period for understanding the evolution of turtles but their fossil record is fragmentary and sparse from the Late Triassic to the Late Jurassic, with only few known species spread around the world (Sterli, 2008 and references therein). However, as a whole, turtle remains have been collected from almost all continents —southern Africa, southern South America, North America, Europe, India and Central Asia (Sterli, 2008)— suggesting that chelonians were already distributed throughout the world. The Jurassic is also a key period to decipher the timing of the origin of crown groups of turtles (i.e. the clades delimited by living representatives). In this sense, two main hypotheses have been proposed (Figure 8).

The first hypothesis (Figure 8a) suggests that all turtles with the exception of *Proganochelys*, *Palaeochersis* and *Australochelys africanus*, from the Early Jurassic of South Africa, belong to one of the main groups of living turtles, Pleurodira or Cryptodira (Gaffney et al., 2007 and references therein). Based on this hypothesis, the crown turtles originated in the Late Triassic and *Proterochersis* would be interpreted as a stem pleurodire. On the contrary, other authors such as Joyce (2007) and Sterli (2008) interpret Late Triassic to Middle Jurassic turtles as stem groups suggesting that the origins of crown turtles were more recent, during the Middle to Late Jurassic (Figure 8b). According to Sterli & de la Fuente (2010) and references therein, Gaffney's hypothesis was supported by misleading interpretation of some key characters considered as synapomorphies of Pleurodira and Cryptodira. For instance, Gaffney (1975) includes the presence of a processus trochlearis oticum, a vertical flange on the transverse process of the pterygoid, and the presence of an epipterygoid among the synapomorphies that define Cryptodira. However, according to Joyce (2007) and Sterli & Joyce (2007) the former character may be a synapomophy of a more inclusive clade than Cryptodira and the two latter characters should be considered symplesiomorphies of turtles.

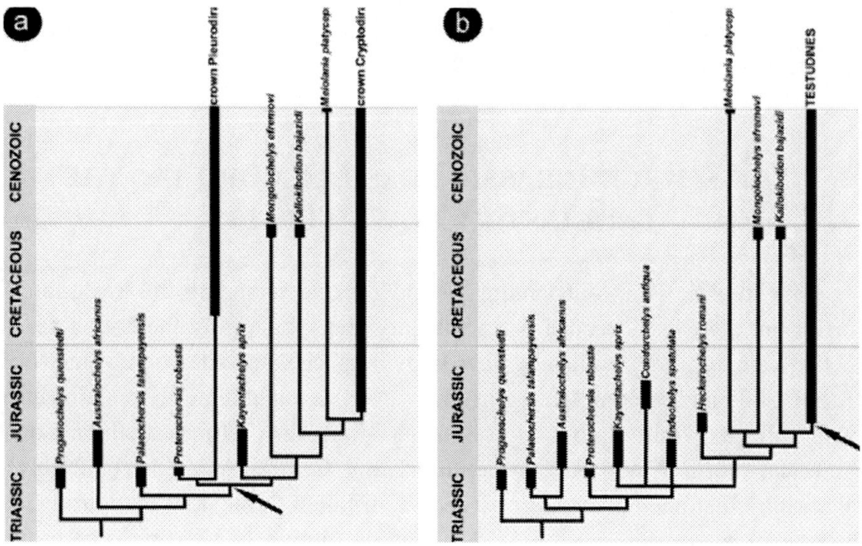

Figure 8. Two phylogenetic hypotheses about the origins of crown turtle taxa: a, lineages leading living turtles originated in the Late Triassic according to Gaffney et al. (2007); b, the origin of living turtle lineages occurred between the Middle and Late Jurassic, according to Joyce (2007) and Sterli (2008). Arrows indicate the crown group. Modified from Sterli & de la Fuente (2010).

Origins and Diversification of Pleurodira

Extant pleurodires or "side necked turtles" are informally characterized by bending the neck in a horizontal plane (Gaffney et al., 2006). Nowadays, the lineage contains two major crown clades: Chelidae and Pelomedusoides that includes Pelomedusidae and Podocnemidae (Joyce et al., 2004). However, the extensive fossil record has revealed that pleurodires were significantly more complex and diverse in the past (Gaffney et al., 2006). Following Joyce et al. (2004) taxonomic proposal for turtles, only three Late Jurassic fossil species may be situated unambiguously along the phylogenetic stem of pleurodires: *Platychelys oberndoferi* from Europe, *Caribemys oxfordiensis* from Cuba, and *Notoemys laticentralis* from Argentina (de la Fuente & Iturralde-Vinent, 2001; Joyce et al., 2004). The large geographic distances among these three stem taxa indicate that the evolutionary history of the pleurodire lineage started in the Late Jurassic or even before.

The oldest representatives of living clades are reported since the Cretaceous. The clade Chelidae includes around 52 species of freshwater pleurodire turtles living in South America and Australasia that diversified in South Gondwana (Broin, 1988). Classically, the fossil record of this group was considered extremely poor and restricted to Tertiary (Joyce et al., 2004 and references therein). However, findings carried out in South America at the beginnings of 2000s have firmly established the presence of this crown clade in the Late Cretaceous. Thus, the oldest representatives of Chelidae are of Santonian (*Lomalatachelys*), Turonian-Campanian (*Bonapartemys* and *Prochelidella*) and Campanian-Maastrichtian (*Palaeophrynops* and *Yaminuechelys*) ages (De la Fuente et al., 2001; Lapparent de Broin & de la Fuente, 2001). However, Gaffney et al. (2006) recognizes this family since the Early Cretaceous.

Nowadays, Pelomedusidae are less diverse than Chelidae and contains eighteen living species (Joyce et al., 2004). The fossil record attributed to this group is scarce and rather confusing and specimens exhibit few diagnostic features (Joyce et al., 2004; Gaffney et al., 2006). *Teneremys lapparenti*, from the Aptian of Niger, is the only well recognized stem taxa of Pelomedusidae (Lapparent de Broin, 2000a). The crown clade is known at least since the Miocene (Lapparent de Broin, 2000b). According to Gaffney et al. (2006) the Podocnemidinura is sister of the extinct family Bothremydidae. Within the former, Podocnemidae are closely related to *Brasilemys* from the Albian of Brazil and *Hamadachelys* from the Cenomanian of Morocco. The crown taxa of Podocnemidae are recognized since the Late Cretaceous (Joyce et al., 2004). At present, Podocnemidae include eight living species.

ORIGINS AND DIVERSIFICATION OF CRYPTODIRA

Cryptodira is a morphologically diverse clade of turtles that differs from Pleurodira in the manner that retracts their head into the shell, bending their neck in a vertical plane to withdraw the head backwards. Thus, the majority of Cryptodira can put the head straight back into the shell, with some exceptions such as sea turtles or *Platysternon megacephalum*. Nowadays, the Cryptodira is the most extensive and rich group of Testudines that informally includes freshwater turtles, bigheaded turtles, Central American river turtles, pig noise turtles, snapping turtles, mud and musk turtles, box and wood turtles, tortoises, soft shell turtles, and sea turtles. The oldest taxa primitively inhabited freshwater habitats, but later invaded marine environments at least once and at least four independent lineages diversified in fully terrestrial habitats (Joyce & Gauthier, 2004). Nowadays, the lineage contains eleven clades that can be grouped into four or five major crown clades: Chelonioidea, Trionychoidea, Chelydridae, Kinosternoidea and Cryptoderinea (Joyce et al., 2004; Barley et al., 2010). Although, the fossil record is extensive and fairly complete, the debate about the phylogenetic affinities of extinct Cryptodira with the major crown clades remains still open and further analyses are needed (Gaffney et al., 1987; Gaffney & Meylan, 1988; Joyce et al., 2004; Danilov & Parham, 2006; Anquetin et al., 2009; Joyce et al., 2009).

According to recent hypotheses, the phylogenetic stem of Cryptodira contains a series of common fossil taxa, such as Baenidae, Meiolanidae, Pleurosternidae, Plesiochelydiae, Sinemydidae, and Macrobaenidae (Gaffney, 1996; Gaffney et al., 1998; Parham & Hutchison, 2003; Joyce et al., 2004). However, other analyses suggest that most of these groups diverged before the diversification of Testudines (Sterli, 2010). Following Gaffney et al. (1987), only *Kayentachelys aprix* from the Early Jurassic of North America, may be situated unambiguously along the phylogenetic stem of Cryptodira, taking into account the following characters: presence of the processus trochlearis oticum and processus pterygoideus externus projecting posteriorly with a flat, vertical plate. Nevertheless, other authors propose Sinochelydae, from the Early Cretaceous of China and Mongolia, as the oldest reported Cryptodira based on characteristics of the shell (Hirayama et al., 2000).

Nowadays, the clade Chelonioidea includes seven or eight species of sea turtles belonging to Dermochelyidae and Cheloniidae. All species of this group are characterized by hard-shelled bodies, except for *Dermochelys coriacea* that is covered with horny scutes. The extensive fossil record has revealed that Chelonioidea was more abundant and diverse in the past, including several extinct families, such as Toxochelyidae, Osteopygidae, Thalassemyidae and Protostegidae (e.g. Gaffney et al., 2006). This last family of sea turtles contains the probable oldest representative of the living clade, *Santanachelys gaffneyi* from the Early Cretaceous of Brazil (Hirayama, 1998). This species had a primitive flipper that still possessed movable digits and a specialized skull with large interorbital foramina (Hirayama, 1998). As no other turtles are known from the phylogenetic stem, *Santanachelys gaffneyi* may also represent the oldest Chelonioidea (Joyce et al., 2004). However, it is important to note that new phylogenetic analyses place *Santanacheys* out of the Testudine clade (Sterli, 2010). The oldest remains of dermochelyids are reported from the Campanian of North America and Japan (Joyce et al., 2004 and references therein) and *Toxochelys latiremys*, from the Late Cretaceous of North America, may be among the oldest stem cheloniids (Hirayama, 1998).

Trionychia is the sister taxa of different basal forms from the Early Cretaceous such as Adocidae and Nanhsiungchelyidae (Danilov & Parham, 2006). According to these authors, *Yehguia tatsunensis*, from the Late Jurassic of China, is placed on the stem of Trionychia, near the clades Adocidae and Nanhsiungchelyidae. Among the list of potential trionychians, *Sandownia harrisi* from the Early Cretaceous (Aptian) of Europe may be the oldest (Meylan et al., 2000). The crown clade Trionychia comprises the families Trionychidae and Carettochelyidae, informally called softshell and pig nosed turtles. The fossil record of the Carettochelyidae was well represented in the Tertiary period but actually only persists a single species, *Carettochelys insculptata*. The oldest representative of Carettochelyidae is *Kizylkumemys schultzi* from the middle Cretaceous (Albian and Cenomanian) of Central Asia (Nessov, 1977; Meylan, 1988). Trionychidae contains 23 especies of living softh-shelled turtles but has a poor fossil record. Despite this, the oldest hypothesized representatives of the trionychid clade are *Trionyx riabinin* and *"Trionyx" kansaiensis*, both reported from the Late Cretaceous (Santonian-early Campanian) of Asia and Kazakhstan (Vitek & Danilov, 2010), and *Aspideretoides allani, A. foveatus, A. splendidus* and *Apalone latus* from the Late Cretaceous (middle Campanian) of North America (Gardner et al., 1995).

Phylogenetic relationships among Kinosternoidea and other clades of cryptodires are unclear. Joyce et al. (2004) places Kinosternoidea within Trionychoidea, following Gaffney (1975). On the contrary, new molecular data supports a deep sister group relationship between Kinosternoidea and Chelydridae (Barley et al., 2010). The sister group relationship between *Dermatemys mawii* and Kinosternidae was only proposed within the last two decades, but currently it is strongly supported by morphological and molecular data (e.g., Gaffney & Meylan, 1988; Hutchison, 1991; Shaffer et al., 1997; Sterli, 2010). The oldest representatives along the stem lineage of *Dermatemys mawii* are *Haplochelys clark* and *Agomphus pectoralis* both from the Late Cretaceous (Maastrichtian) of North America (Knauss et al., 2010). This phylogenetic placement breaks with the tradition started by Hutchison & Bramble (1981), which suggest that *Agomphus* spp. is a stem-kinosternoid and *Hoplochelys* spp. is a representative of the kinosternid stem (Hutchison, 1991; Meylan & Gaffney, 1989; Joyce, 2007). The genus *Baptemys*, from the Eocene of North America, has been allied to *Dermatemys mawii* within

Dermatemydidae (Knauss, 2006). On the other hand, the oldest putative members of stem kinosternidae may be some undescribed remains from the Campanian of North America (Hutchison et al., 1998).

Taxonomic composition of Chelydridae and its relationships with remaining clades within Cryptodira are also still under debate. This family of turtles has a long fossil history from North America, Asia and Europe, far outside its present range. According to Eaton et al. (1999a, 1999b), the oldest crown chelydrid is *Protochelydra zangerli* from the Eocene of North Dakota and the oldest stem chelydrid is *Emarginachelys cretacea* from the Late Cretaceous (Turonian) of Montana. Following to Knauss et al., (2010), *Baltemys staurogastros*, *Xenochelys formosa* and *X. lostcabinensis,* from the Early Eocene of Wyoming, also belong to the crown clade Chelydridae. However, previous works placed *Baltemys* and *Xenochelys* within crown Kinosternidae (Joyce et al., 2004).

The crown clade Cryptoderinea contains *Platysternon megacephalum* and the crown taxa Testudinoidea (Joyce et al., 2004). *Platysternon megacephalum* is an extant freshwater turtle from Asia that forms a monotypic lineage sister to the Testudinoidea (Joyce et al., 2004; Krenz et al., 2005; Sterli, 2010). The fossil record of the *Platysternon* lineage is very fragmentary, with few and partial specimens coming from the Paleocene and Oligocene of Kazakhstan (Joyce et al., 2004 and references therein). Testudinoidea contains three major clades, Testudinidae, Emydidae and Geoemydidae called informally land tortoises and generally pond turtles. In addition to the last principal crown groups, Testudinoidea includes different basal families from the Cretaceous of Asia such as Haichemydidae, Sinochelyidae and Lindholmemydidae because of the development of an ossified bridge connecting the plastron with the carapace (Sukhanov, 2000; Danilov & Sukhanov, 2001).

The crow Testudinidae is represented by 43 extant species characterized by exclusive terrestrial habitats. This clade has an excellent fossil record but the majority of the basal forms, such as *Hadrianus, Stylemys, Manouria, Cheirogaster, Ergilemys* and others, lack skulls making phylogenetic analyses difficult. The oldest representatives are *Hadrianus majuscula* from the Early Eocene (Wasatchian) of New Mexico and *Achilemys cassouleti* from the Early Eocene (Ypresian) of France.

The crow Geoemydidae contains 61 species of freshwater, swamp, lagoon and humid environment turtles living in all the continents except to Australia and Antarctica.

Table 1. Conservation status of living turtles according to the International Union for Conservation of Nature (IUCN) database (IUCN, 2010). Numbers in parenthesis indicate the number of species within each turtle family. Remaining numbers are the number of species per genus classified within each threat category: CE, critically endangered; E, endangered; V, vulnerable; LR/NT, lower risk/near threatened; LR/LC, lower risk/least concern; DD, data deficient; NL, not listed in the IUCN Red List. The taxonomy of genera follows Joyce et al. (2004) appendixes

Major taxa		Genus	IUCN threat category						
			CE	E	V	LR/NT	LR/LC	DD	NL
Pleurodira	Chelidae (52)	*Acanthochelys*			1	3			
		Batrachemys	1						5
		Chelodina	1	1	1	3	2		5
		Chelus							1
		Elseya		1	1		1	2	3
		Elusor		1					
		Emydura					1		4
		Hydromedusa			1				1
		Mesoclemmys			1				1
		*Phynops**		1		2			4
		Platemys							1
		Pseudemydura	1						
		Rheodytes			1				
	Pelomedusidae (17)	*Pelomedusa*							1
		Pelusios			1		3	1	11
	Podocnemidae (8)	*Erymnochelys*	1						
		Podocnemis		1	3	1			1
		Peltocephalus			1				

Table 1 (Continued)

Major taxa		Genus	IUCN threat category						
			CE	E	V	LR/NT	LR/LC	DD	NL
Cryptodira	Dermatemydidae (1)	*Dermatemys*	1						
	Kinosternidae (25)	*Claudius*				1			
		Kinosternon			4	2	6	3	3
		Staurotypus				2			
		Sternotherus			1				3
	Trionychidae (24)	*Amyda*			1				
		Apalone					2		1
		Aspideretes			3				
		Chitra	1	1					
		Cyclanorbis				2			
		Cycloderma				2			
		Dogania					1		
		Lissemys					1	1	
		Nilssonia		1					
		Palea		1					
		Pelochelys		1	1				
		Pelodiscus			1				
		Rafetus	1	1					
		Trionyx							1
	Carettochelyidae (1)	*Carettochelys*			1				
	Chelydridae (3)	*Chelydra*			1		1		
		Macroclemys			1				
	Dermochelyidae (1)	*Dermochelys*	1						
	Cheloniidae (6)	*Caretta*		1					

Major taxa	Genus	IUCN threat category						
		CE	E	V	LR/NT	LR/LC	DD	NL
	Chelonia		1					
	Eretmochelys	1						
	Lepidochelys	1		1				
	Natator						1	
Platysternidae (1)	*Platysternon*		1					
Testudinidae (43)	*Chersina*							1
	Dipsochelys							1
	*Geochelone***	3		6		1		1
	Gopherus			3		1		
	Homopus			1	1			3
	Indotestudo		2	1				
	Kinixys			1	1		1	2
	Malacochersus			1				
	Manouria		1	1				
	Psammobates		1					2
	Pyxis	2						
	Testudo	1		2	1	1		
Geoemydidae (61)	*Batagur*	2						
	Cuora	8	2	1				
	Cyclemys				1			3
	Geoclemys			1				
	*Geoemyda****		3					
	Hardella			1				
	Heosemys	2	1	1				
	Hieremys		1					
	Kachuga	1	2					

Table 1 (Continued)

Major taxa	Genus	IUCN threat category						
		CE	E	V	LR/NT	LR/LC	DD	NL
	Leucocephalon	1						
	Malayemys			1				
	Mauremys	1	3		1			2
	Melanochelys			1	1			
	Morenia			2				
	Notochelys			1				
	Orlitia		1					
	Pangshura		1		1	2		
	Rhinoclemmys				5			4
	Sacalia		2					
	Siebenrockiella			1				
Emydidae (48)	*Actinemys*			1				
	Chrysemys							1
	Clemmys			1				
	Deirochelys							1
	Emyoidea				1			
	Emys				1		1	
	Glyptemys		1	1				
	Graptemys		2	1	5			4
	Malaclemys				1			
	Pseudemys		1		2			4
	Terrapene		1		2		1	
	Trachemys		2	5	3			5

*, Includes genus *Rhinemys*

**, Includes genera *Astrochelys, Centrochelys* and *Chelonoidis.*

***, Includes genus *Vijayachelys.*

This group of turtles, informally called Eurasian pond, river turtles and Neotropical wood turtles, comprises a significant portion of the diversity of extant turtles and the major number of freshwater species. Nowadays, the oldest representative of geoemydids probably is *Echmatemys* ssp. from the Eocene of North America and Asia (Hirayama, 1985). Other undefined stem geoemydids are known from the Paleocene to Oligocene of various North American, Asian and probably European sites. Moreover, some basal taxa such as *Grayemy-Hokouchelys*, *Clemmydopsis*, *Epiemys* and *Elkemys* have been attributed to Geoemydidae but phylogenetic relationships among these taxa and extant forms are unclear (Mlynarsky, 1976).

The crow Emydidae contains 48 species of extant turtles most of which living in the New World, with the exception of *Emys orbicularis* from Europe, north Africa and western Asia. This group of turtles, informally called pond turtles, developed diverse ecological strategies and colonized a wide range of environments, from aquatic to fully terrestrial habitats. Nowadays, the oldest fossil emydid is *Gyremys sectabilis* from the Late Cretaceous of Judith River Formation of Montana, North America (Ernst & Barbour, 1989). Traditionally, a number of fossils have been associated (Mlynarsky, 1976) with this group but none has been integrated into a cladistic analysis and consequently cannot be referred with any confidence to Emydidae (Joyce et al., 2004). Recently, phylogenetic analyses have been carried out to clarify the phylogeny of some primitive forms, such as *Hummelemys, Palaeochelys, Juvemys, Merovemys* and others (Hervet, 2004). However, results do not elucidate their relationships with the Geoemydidae and Emydidae stem.

An Overview of the Turtle Threatened Phylogenetic Diversity

Posadas et al. (2001) highlighted the fact that conservation of biodiversity requires the knowledge of its history. Phylogenies provide the ways to measure biodiversity, to quantify the evolutionary history of a set of species and to assess conservation priorities (Mace et al., 2003). Despite the large number of the recently published papers, key questions about the origins and evolutionary history of turtles remain unanswered, as it has been explained in the above paragraphs. Phylogenetic analyses have not resolved satisfactorily if turtles are sister to parareptiles (e.g. Lyson et al., 2010) or diapsids (e.g. Shedlock et al.,

2007). The Jurassic turtle fossil record is fragmentary and sparse making the understanding of the origins of extant lineages difficult (see Sterli, 2008). The phylogenetic relationships of extant Testudines are also unclear, with controversial results about the placement of Pleurodires (see Sterli, 2010 for an example) and other major lineages such as Chelonioidea, Kinosternoidea and Chelydridae (Barley et al., 2010 and references therein) in the tree of life of turtles. Some factors can explain disagreement in phylogenetic analyses carried out with different source of data (e.g. fossil evidence, morphology, DNA sequences). Firstly, morphological variation of turtles is still poorly understood from an evolutionary point of view. This is enhanced by large gaps in their fossil record, especially in key periods such as Jurassic, Cretaceous and Paleocene. Moreover, fossil turtles usually lack parts of the skeleton with characters of great systematic value (e.g. skulls). On the other hand, molecular trees show short internal branches, often weakly supported, and long terminal branches suggesting a rapid radiation of the living groups of turtles since the Late Jurassic (Sterli, 2010, but see Barley et al., 2010). This complicates the determination of interrelationships and the position of the root of major lineages of extant turtles.

Despite this bleak picture, we have estimated the amount of threatened phylogenetic diversity of turtles based, in part, on the results reported by Sterli (2010) and data from IUCN (2010) red list (Table 1, Figure 9). However, due the large disagreement among current studies, relationships among largest clades of living turtles are represented as a polytomy, especially within Cryptodira. Most of living major lineages of turtles are recognized in the fossil record since the Middle-Late Cretaceous (Figure 9). Only three of fourteen living lineages can be considered slightly endangered, two within Pleurodira (Chelidae and Pelomedusidae) and one within Cryptodira (Kinosternidae), containing 25% or less threatened species (Figure 9). At the opposite end, Podocnemidae, Carettochelyidae, Dermochelyidae, Cheloniidae, Dermatemydidae and Platysternidae are at high risk of extinction with 75% or more species threatened. The status is critical for Dermochelyidae, Dermatemydidae and Platysternidae because they are represented by a single species classified as endangered or critically endangered (Table 1).

This overview of the conservation status of the turtle tree of life clearly shows that most of the phylogenetic diversity generated during 90 million of years or more is in a real risk of extinction and the situation is extreme for an important number of lineages. Turtles agree with the prediction of Purvis et al. (2000) in that species on long branches (i.e. without close relatives) are particularly likely to be at risk.

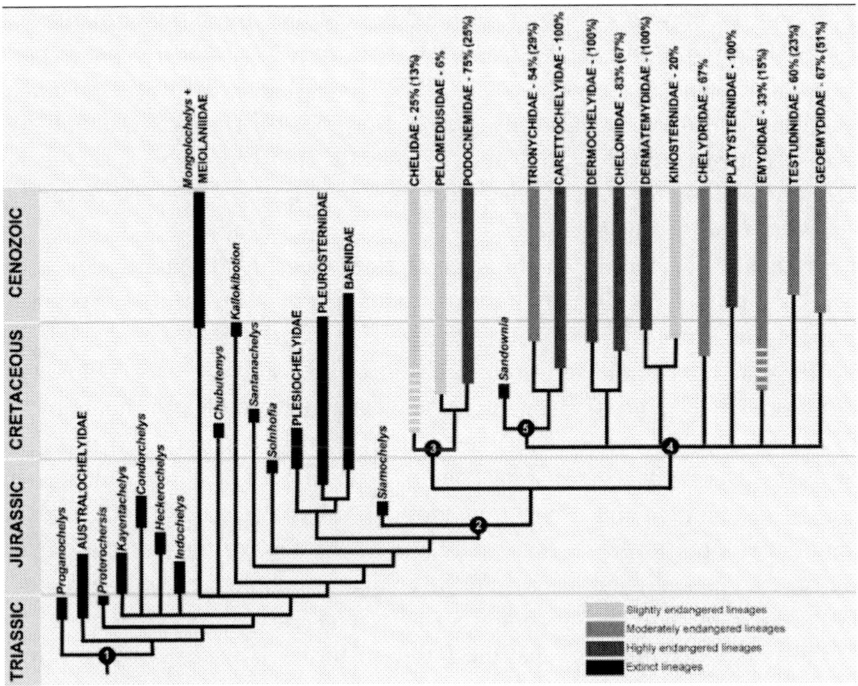

Figure 9. Evolutionary tree of turtles with phylogenetic relationships of stem taxa based on Sterli (2010). Phylogenetic relationships within Testudines (node 2) are not clarified because the lack of consensus among different authors (see text for details). The percentage of threatened species (vulnerable, endangered and critically endangered—see Table 1) is shown for each lineage. Percentages of endangered plus critically endangered species are within parenthesis. A quantification of the risk of lineage extinction is indicated based on Table 1: slightly endangered lineages (25% or less of species threatened), moderately endangered lineages (between 25% and 75% of species threatened), highly endangered lineages (75% or more of species threatened). Numbers in nodes mean: 1, Testudinata; 2, Testudines; 3, Pleurodira; 4, Cryptodira; 5, Trionychia.

Moreover, according to these authors, fossil record demonstrates that some clades (e.g. Pleurodira, Chelonioidea) were more diverse in the past and probably have suffered considerable extinction through geological time due to unknown phylogenetic factors promoting a greater risk of extinction for these lineages.

Nee & May (1997) noted that the importance assigned to a species is not necessarily proportional to the amount of evolutionary history it represents. For instance, conservation concern is often focused on their behavior, potential usefulness or their role in an ecosystem rather than simply on overall measures of evolutionary history. Moreover, from a systematic perspective, there has been debate on what may be a greater loss: the demise of the sole survivor of an ancient lineage (for instance, the Central American river turtle, *Dermatemys mawii*) or the demise of a member of a rich species flock (for instance, the Roti Island snake-necked turtle, *Chelodina mccordi*) (Nee & May, 1997). Distinct taxa and 'living fossils' contribute disproportionately to overall biodiversity because the large amount of evolutionary history that may represent (Bowen, 1999). However, other authors have argued that these 'sole survivors' are a dead-end and conservation efforts should be focused on the species flock because its future evolutionary potential to restock biodiversity if a mass extinction episode occurs (Erwin, 1991). According to Bowen (1999), the leatherback sea turtle (*Dermochelys coriacea*) may illustrate the complexity of the decision making in conservation biology and that several perspectives (systematic, ecological and evolutionary) should be taken into account. This critically endangered species is the last survivor of a lineage that traces back to the Late Cretaceous or before. Thus it has a high relevance from the systematic point of view. In addition, it has a remarkable ecological and evolutionary significance because it is one of the few vertebrates that feeds on jellyfish (Scyphozoa) and has a suite of unique morphological and physiological adaptations allowing it to forage in freezing waters (Bowen, 1999).

This study demonstrates that the future of turtles is uncertain if no effective conservation measures are carried out. Turtles represent an exclusive group of reptiles and amniotes because their peculiar body plane and long evolutionary history, traced back since at least 225 million years ago, that justifies conservation efforts. Moreover, most endangered species of turtles inhabit aquatic environments. Their conservation may guarantee the conservation of fragile habitats such as rivers and wetlands that frequently suffer the impact of human activities.

ACKNOWLEDGMENT

Authors agree colleages of the Mesozoic Research Team (Institut Català de Paleontologia) and the Centre de Recuperació d'Amfibis i Rèptils de Catalunya (CRARC) for their support in research activities. This study has been funded by the project CGL2008-06533-C03-01/BTE (Ministerio de Ciencia e Innovación, Spain). J.M. has a grant of the Juan de la Cierva Program (JCI-2007-187-1593).

REFERENCES

Anquetin, J., Barrett, P.M., Jones, M.E.H., Moore-Fay, S. & Evans, S.E. (2009). A new stem turtle from the Middle Jurassic of Scotland: new insights into the evolution and palaeoecology of basal turtles. *Proceedings of the Royal Society B*, 276, 879-886.

Barley, A.J., Spinks, P.Q., Thomson, R.C. & Shaffer, H.B. (2010). Fourteen nuclear genes provide phylogenetic resolution for difficult nodes in the turtle tree of life. *Molecular Phylogenetics and Evolution*, 55, 1189-1194.

Bertini, R.J., Miloni-Santucci, R., Vieira-Toledo, C.E. & Costa-Menegazzo, M. (2006). Taphonomy and depositional history of an Upper Cretaceous turtle-bearing outcrop from the Adamantina Formation, southwestern Scyphozoa São Paulo state. *Revista Brasileira de Paleontologia*, 9, 181-186.

Bowen, B.W. (1999). Preserving genes, species or ecosystems? Healing the fractured foundations of conservation policy. *Molecular Ecology*, 8, 5-10.

Brand, L.R., Goodwin, H.T., Ambrose, P.D. & Buchheim, H.P. (2000). Taphonomy of turtles in the Middle Eocene Bridger Formation, SW Wyoming. *Palaeogeography, Palaeoclimatology, Palaeoecology*, 162, 171-189.

Brand, L.R., Hussey, M. & Taylor, J. (2003). Taphonomy of freshwater turtles: decay and disarticulation in controlled experiments. *Journal of Taphonomy*, 1, 233-245.

Broin, F. de (1988). Les tortues et le Gondwana. Examen des rapports entre le fractionnement du Gondwana et la dispersion géographique des tortues pleurodires à partir du Crétacé. *Studia Palaeocheloniologica*, II, 103-142.

Caspers, G.J., Reinders, G.J., Leunissen, J.A.M, Wattel, J. & de Jong, W.W. (1996)._Protein sequences indicate that turtles branched off from the Amniote tree after mammals. *Journal of Molecular Evolution*, 42, 580-586.

Danilov, I.G. & Sukhanov, V.B. (2001). New data on lindholmemydid turtle *Lindholmemys* from the Late Cretaceous of Mongolia. *Acta Palaeontologica Polonica*, 46, 125-131.

Danilov, I.G. & Parham, J.F. (2006). A redescription of '*Plesiochelys*' *tatsuensis* from the Late Jurassic of China, with comments on the antiquity of the crown clade Cryptodira. *Journal of Vertebrate Paleontology*, 26, 573-580.

De Braga, M. & Rieppel, O. (1997). Reptile phylogeny and the interrelationships of turtles. *Zoological Journal of the Linnean Society*, 120, 281-354.

De la Fuente, M.S. & Iturralde-Vinent, M. (2001). A new pleurodiran turtle from the Jagua Formation (Oxfordian) of western Cuba. *Journal of Paleontology*, 75, 860-869.

De la Fuente, M.S., Lapparent de Broin, F. de & Manera de Bianco, T. (2001). The oldest and first nearly complete skeleton of a chelid, of the *Hydromedusa* sub-group (Chelidae, Pleurodira), from the Upper Cretaceous of Patagonia. *Bulletin de la Société Géologique de France*, 172, 105-112.

Depecker, M., Berge, C., Penin, X. & Renous, S. (2006). Geometric morphometrics of the shoulder girdle in extant turtles (Chelonii). *Journal of Anatomy*, 208, 35-45.

Eaton, J.G., Hutchison, J.H., Holroyd, P.A., Korth, W.W. & Goldstrand, P.M. (1999a). Cretaceous vertebrate faunas from Kaiparowits Plateau, south-central Utah. In D.D. Gillette (Ed.), *Vertebrate Paleontology in Utah Miscellaneous Publication*. Pp. 345-353. Salt Lake City: Utah Geological Survey.

Eaton, J.G., Hutchison, J.H., Holroyd, P.A., Korth, W.W. & Goldstrand, P.M. (1999b). Vertebrates of the Turtle Basin local fauna, middle Eocene, Sevier Plateau, south-central Utah. In D.D. Gillette (Ed.), *Vertebrate Paleontology in Utah Miscellaneous Publication*. Pp. 463-468. Salt Lake City: Utah Geological Survey.

Ernst, C.H. & Barbour, R.W. (1989). *Turtles of the world*. Washington, DC: Smithsonian Institution Press.

Erwin, T.L. (1991). An evolutionary basis for conservation strategies. *Science*, 253, 750-752.

Fushitani, K., Higashiyama, K., Moriyama, E.N., Iami, K. & Hosokawa, K. (1996). The amino acid sequences of two alpha chains of hemoglobins from Komodo dragon *Varanus komodoensis* and phylogenetic relationships of amniotes. *Molecular Biology and Evolution*, 13, 1039-1043.

Gaffney, E.S. (1975). A phylogeny and classification of the higher categories of turtles. *Bulletin of the American Museum of Natural History*, 155, 389-436.

Gaffney, E.S. (1990). The comparative osteology of the Triassic turtle *Proganochelys*. *Bulletin of the American Museum of Natural History*, 194, 1-263.

Gaffney, E.S. (1996). The postcranial morphology of *Meiolania platyceps* and a review of the Meiolaniidae. *Bulletin of the American Museum of Natural History*, 229, 1-166.

Gaffney, E.S. & Meylan, P.A. (1988). A phylogeny of turtles. In M. Benton (Ed.), the phylogeny and classification of tetrapods, Vol. 1, *Amphibians, reptiles, birds*. Pp. 157-219. Oxford: Clarendon Press.

Gaffney, E.S, Hutchison, J.H., Jenkins, F.A. & Meeker, L.J. (1987). Modern turtle origins: the oldest known cryptodire. *Science*, 237, 289-291.

Gaffney, E.S., Kool, L., Brinkman, D.B., Rich, T.H. & Vickers-Rich, P. (1998). *Otwayemys*, a new cryptodiran turtle from the Early Cretaceous of Australia. *American Museum Novitates*, 3233, 1-28.

Gaffney, E.S., Tong, H. & Meylan, P.A. (2006). Evolution of the side-necked turtles: the families Bothremydidae, Euraxemydidae, and Araripemydidae. *Bulletin of the American Museum of Natural History*, 300, 1-700.

Gaffney, E.S., Rich, T.H., Vickers-Rich, P., Constantine, A., Vacca, R. & Kool, L. (2007). *Chubutemys*, a new eucryptodiran turtle from the Early Cretaceous of Argentina, and the relationships of Meiolaniidae. *American Museum Novitates*, 3599, 1-35.

Gardiner, B.G. (1993). Haematothermia: warm-blooded amniotes. *Cladistics*, 9, 369-395.

Gardner, J.D., Russell, A.P. & Brinkman, D.B. (1995). Systematics and taxonomy of soft-shelled turtles (family Trionychidae) from the Judith River Group (mid- Campanian) of North America. *Canadian Journal of Earth Sciences*, 32, 631-643.

Gauthier, J.A. (1994). The diversification of the amniotes. In D.R. Prothero & R.M. Schoch (Eds.), *Major features of vertebrate evolution*. Pp. 129-159. Knoxville: Paleontological Society.

Gauthier, J., Estes, R. & de Queiroz, K. (1988). A phylogenetic analysis of *Lepidosauromorpha*. In R. Estes & G. Pregill (Eds.), *Phylogenetic relationships of the lizard families*. Pp. 15-98. Standford: Standford University Press.

Gilbert, S.F., Loredo, G.A., Brukman, A. & Burke, A.C. (2001). Morphogenesis of the turtle shell: the development of a novel structure in tetrapod evolution. *Evolution and Development*, 3, 47-58.

Gow, C.E. (1997). A reassessment of *Eunotosaurus africanus* Seeley (Amniota: Parareptilia). *Palaeontologia Africana*, 34, 33-42.

Harvey, P.H., Leigh-Brown, A.J., Maynard-Smith, J. & Nee, S. (1996). *New uses for new phylogenies*. Oxford: Oxford University Press.

Hedges, S.B. & Poling, L.L. (1999). A molecular phylogeny of reptiles. *Science*, 283, 998-1001.

Hervet, S. (2004). Systématique du groupe « *Palaeochelys* sensu lato – *Mauremys* » (Chelonii, Testudinoidea) du Tertiaire d'Europe occidentale: principaux résultats. *Annales de Paléontologie*, 90, 13-78.

Hirayama, R. (1985). Cladistic analysis of batagurine turtles. *Studia Palaeocheloniologica*, 1, 140-157.

Hirayama, R. (1998). Oldest known sea turtle. *Nature*, 392, 705-708.

Hirayama, R., Brinkman, D.B. & Danilov, I.G. (2000). Distribution and biogeography of non-marine Cretaceous turtles. *Russian Journal of Herpetology*, 7, 181-198.

Hugall, A.F., Foster, R. & Lee, M.S.Y. (2007). Calibration choice, rate smoothing and the pattern of tetrapod diversification according to the long nuclear gene RAG-1. *Systematic Biology*, 56, 543-563.

Hutchison, J.H. (1991). Early Kinosterninae (Reptilia: Testudines) and their phylogenetic significance. *Journal of Vertebrate Paleontology*, 11, 145-167.

Hutchison, J.H. & Bramble, D.M. (1981). Homology of the plastral scales of the kinosternidae and related turtles. *Herpetologica*, 37, 73-85.

Hutchison, J.H., Eaton, J.G., Holroyd, P.A. & Goodwin, M.B. (1998). Larger vertebrates of the Kaiparowits Formation (Campanian) in the Grand Staircase-Escalante National Monument and adjacent areas. In L.M. Hill & J.J. Koselak (Eds.), *Learning from the Land. Grand Staircase-Escalante National Monument Science Symposium Proceedings*. Pp. 391-398. Washington D.C.: U.S. Department of the Interior, Bureau of Land Management.

IUCN (2010). IUCN Red List of Threatened Species. Version 2010.4. *www.iucnredlist.org*. Downloaded on 01 December 2010.

Joyce, W.G. (2007). Phylogenetic relationships of Mesozoic turtles. *Bulletin of the Yale Peabody Museum*, 48, 3-102.

Joyce, W.G. & Gauthier, J.A. (2004). Palaeoecology of Triassic stem turtles sheds new light on the turtle origins. *Proceedings of the Royal Society of London* B, 271, 1-5.

Joyce, W.G., Parham, J.F. & Gauthier, J.A. (2004). Developing a protocol for the conversion of rank-based taxon names to phylogenetically clade names, as exemplified by turtles. *Journal of Paleontology*, 78, 989-1013.

Joyce, W.G., Lucas, S.P., Scheyer, T.M., Heckert, A.B. & Hunt, A. P. (2009). A thin-shelled reptile from the Late Triassic of North America and the origin of the turtle shell. *Proceedings of the Royal Society of London B*, 276, 507-513.

Knauss, G.E. (2006) Morphological description of *Baptemys wyomingensis* and an analysis of its phylogenetic relationship within kinosternoidea (testudines). *Geological Society of America Abstracts with Programs*, Vol. 38, 4, 16.

Knauss, G.E., Joyce, W.C., Lyson, T.R. & Pearson, D. (2010). A new kinosternoid from the Late Cretaceous Hell Creek Formation of North Dakota and Montana and the origin of the *Dermatemys mawii* lineage. *Paläontologische Zeitschrift*, 84, DOI: 10.1007/s12542-010-0081-x.

Krenz, J.G., Naylor, G.J.P., Shaffer, H.B. & Janzen, F.J. (2005). Molecular phylogenetics and evolution of turtles. *Molecular Phylogenetics and Evolution*, 37, 178-191.

Kumazawa, Y. & Nishida, M. (1999). Complete mitochondrial DNA sequences of the green turtle and blue-tailed mole skink: statistical evidence for archosaurian affinity of turtles. *Molecular Biology and Evolution*, 16, 784-792.

Lapparent de Broin, F. de (2000a). The oldest pre-Podocnemidid turtle (Chelonii, Pleurodira), from the early Cretaceous, Ceará state, Brasil, and its environment. *Treballs del Museu de Geologia de Barcelona*, 9, 43-95.

Lapparent de Broin, F. de (2000b). African chelonians from the Jurassic to the present: phases of development and preliminary catalogue of the fossil record. *Palaeontologia Africana*, 36, 43-82.

Lapparent de Broin, F. de (2001). The European turtle fauna from the Triassic to the Present. *Dumerilia*, 4, 155-217.

Lapparent de Broin, F. de & de la Fuente, M.S. (2001). Oldest world Chelidae (Chelonii, Pleurodira), from the Cretaceous of Patagonia, Argentina. *Compte Rendu des Académie des Sciences de la Terre et des Planètes*, 333, 463-470.

Laurin, M. & Reisz, R.R. (1995). A reevaluation of early amniote phylogeny. *Zoological Journal of the Linnean Society*, 113, 165-223.

Lee, M.S.Y. (1997). Pareiasaur phylogeny and the origin of turtles. *Zoological Journal of the Linnean Society*, 120, 197-280.

Lee, M.S.Y., Reeder, T.W., Slowinski, J.B. & Lawson, R. (2008). Resolving reptile relationships: molecular and morphological markers. In J. Cracraft & M.J. Donoghue (Eds.), *Assembling the tree of life*. Pp. 451-467. New York: Oxford University Press.

Li, C., Wu, X.C., Rieppel, O., Wang, L.T. & Zhao, L.J. (2008). An ancestral turtle from the Late Triassic of southwestern China. Nature, 456, 497-501.

Lyson, T.R., Bever, G.S., Bhullar, B.A.S., Joyce, W.G. & Gauthier, J.A. (2010). Transitional fossils and the origin of turtles. Biology *Letters*, 6, 830-833.

Mace, G.M., Gittleman, J.L. & Purvis, A. (2003). Preserving the tree of life. *Science*, 300, 1707-1709.

Marmi, J., Vila, B. & Galobart, À. (2009). *Solemys* (Chelonii, Solemydidae) remains from the Maastrichtian of Pyrenees: evidence for a semi-aquatic lifestyle. *Cretaceous Research*, 30, 1307-1312.

Meylan, P.A. (1988). *Peltochelys* Dollo and the relationships among the genera of the Carettochelyidae (Testudines: Reptilia). *Herpetologica*, 44, 440-450.

Meylan, P.A. & Gaffney, E.S. (1989). The skeletal morphology of the Cretaceous cryptodiran turtle, *Adocus*, and the relationships of the Trionychoidea. *American Museum Novitates*, 2941, 1-60.

Meylan, P.A., Moody, T.J., Walker, C.A. & Chapman, S.D. (2000). *Sandownia harrisi*, a highly derived trionychoid turtle (Testudines: Cryptodira) from the Early Cretaceous of the Isle of Wight. *Journal of Vertebrate Paleontology*, 20, 522-532.

Mlynarski, M. (1976). Testudines. *Handbuch der Paläoherpetologie*, Number 7. 130 Pp.

Nagashima, H., Sugahara, F., Takechi, M., Ericsson, R., Kawashima-Ohya, Y., Narita, Y. & Kuratani, S. (2009). Evolution of the turtle body plan by the folding and creation of new muscle connections. *Science*, 325, 193-196.

Nee, S. & May, R.M. (1997). Extinction and the loss of evolutionary history. *Science*, 278, 692-694.

Nessov, L.A. (1977). A new genus of pitted-shelled turtle from the Upper Cretaceous of Karakalpakia. *Paleontological Journal*, 11-96-107.

O'Brien, S. (1994). Genetic and phylogenetic analyses of endangered species. *Annual Review of Genetics*, 28, 467-489.

Parham, J.F. & Hutchison, J.H. (2003). A new eucryptodiran turtle from the Late Cretaceous of North America (Dinosaur Provincial Park, Alberta, Canada). *Journal of Vertebrate Paleontology*, 23, 783-798.

Posadas, P., Miranda-Esquivel, D.R. & Crisci, J.V. (2001) Using phylogenetic diversity measures to set priorities in conservation: an example from southern South America. *Conservation Biology*, 15, 1325-1334.

Pollock, D.D., Eisen, J.A., Doggett, N.A. & Cummings, M.P. (2000). A case for evolutionary genomics and the comprehensive examination of sequence biodiversity. *Molecular Biology and Evolution*, 17, 1776-1788.

Purvis, A., Agapow, P.M., Gittleman, J.L. & Mace, G.M. (2000). Nonrandom extinction and the loss of evolutionary history. *Science*, 288, 328-330.

Purvis, A., Gittleman, J.L. & Brooks, T. (2005). *Phylogeny and conservation.* Cambridge: Cambridge University Press.

Reisz, R.R. & Head, J.J. (2008). Turtle origins out of sea. *Nature*, 456, 450-451.

Rieppel, O. (2001). Turtles as hopeful monsters. *Bioessays*, 23, 987-991.

Rosenzweig, M.L. (2001). Loss of speciation rate will impoverish future diversity. *Proceedings of the National Academy of Sciences of the USA*, 98, 5404-5410.

Rougier, G.W., de la Fuente, M.S. & Arcucci, A.B. (1995). Late Triassic turtles from South America. *Science*, 268, 855-858.

Scheyer, T.M. & Sander, P.M. (2007). Shell bone histology indicates terrestrial palaeoecology of basal turtles. *Proceedings of the Royal Society of London B*, 274, 1885-1893.

Shaffer, H.B., Meylan, P. & McKnight, M.L. (1997). Tests of turtle phylogeny: molecular, morphological and paleontological approaches. *Systematic Biology*, 46, 235-268.

Shedlock, A.M., Botka, C.W., Zhao, S., Shetty, J., Zhang, T., Liu, J.S., Deschavanne, P.J. & Edwards, S.V. (2007). Phylogenomics of nonavian reptiles and the structure of the ancestral amniote genome. *Proceedings of the National Academy of Sciences USA*, 104, 2767-2772.

Sterli, J. (2008). A new, nearly complete stem turtle from the Jurassic of South America with implications for turtle evolution. *Biology Letters,* 4, 286-289.

Sterli, J. (2010). Phylogenetic relationships among extinct and extant turtles: the position of Pleurodira and the effects of the fossils on rooting crown-group turtles. *Contributions to Zoology*, 79, 93-106.

Sterli, J. & Joyce, W.G. (2007). The cranial anatomy of the lower Jurassic turtle *Kayentachelys aprix*. *Acta Palaeontologica Polonica*, 52, 675-694.

Sterli, J. & de la Fuente, M.S. (2010). Anatomy of *Condorchelys antiqua* Sterli, 2008, and the origin of the modern jaw closure mechanism in turtles. *Journal of Vertebrate Paleontology*, 30, 351-366.

Sukhanov, V.B. (2000). Mesozoic turtles of the middle and central Asia. In M.J. Benton, M.A. Shishkin, D.M. Unwin & E.N. Kurochkin (Eds.), *The age of dinosaurs in Russia and Mongolia*. Pp. 309-367. Cambridge: Cambridge University Press.

Tsuji, L.A. & Müller, J. (2009). Assembling the history of the Parareptilia: phylogeny, diversification, and a new definition of the clade. *Fossil Record*, 12, 71-81.

Vitek, N.S. & Danilov, I.G. (2010). New material and a reassessment of soft-shelled turtles (Trionychidae) from the Late Cretaceous of Middle Asia and Kazakhstan. *Journal of Vertebrate Paleontology*, 30, 383-393.

Werneburg, I. & Sánchez-Villagra, M.R. (2009). Timing of organogénesis support basal position of turtles in the amniote tree of life. *BMC Evolutionary Biology*, 9, 82.

Zardoya, R. & Meyer, A. (2000). Mitochondrial evidence on the phylogenetic position of caecilians (Amphibia: Gymnophiona). *Genetics*, 155, 765-775.

In: Turtles: Anatomy, Ecology and Conservation ISBN: 978-1-61470-554-3
Eds: M. J. Cosgrove and S. A. Roe

Chapter 5

HEMATOLOGY OF THE LOGGERHEAD TURTLE, *CARETTA CARETTA*

Alessandra Pica[1], Filomena Basile[1] and Chester A. Glomski[2]

[1]Department of Biological Sciences, University of Naples, Federico II
[2]Department of Pathology and Anatomical Sciences, State University of New York at Buffalo, U. S.

ABSTRACT

The study of the hematological characteristics of loggerhead turtles can be very helpful in the rescue and rehabilitation of this endangered species. Normal hemocellular and biochemical values are the basis for the evaluation of a subject's health status. The identification of the blood cells through cytochemical or immunocytochemical methods is the first step to obtain a correct classification of the cells. The following blood cell types were found: lymphocytes, monocytes, heterophils (the counterparts of mammalian neutrophils), eosinophils, basophils (rarely observed), erythrocytes along with some of their precursors, and thrombocytes. Lymphocytes and monocytes are easily distinguishable, as they are similar to the mammalian counterparts, whereas heterophils (which are the equivalent of mammalian neutrophils) and eosinophils both contain eosinophilic granules in their cytoplasm, making their identification difficult. The thrombocytes are nucleated cells, whose shape is variable in dry film smears of circulating blood, either round or oval; they take part in the hemostatic process, as suggested by their tendency to form aggregates. The erythrocytes are also nucleated cells, showing an ellipsoidal and flattened configuration. In all specimens of this marine reptile, both healthy and unhealthy, the mature erythrocytes contain one characteristic inclusion in their cytoplasm, which was recognized as a Heinz body, formed by the precipitation of hemoglobin. The loggerhead's hemoglobin was studied in depth and revealed some specific characteristics, one of which is that under physiologic conditions it precipitates due to its instability.

While no data are available about the erythropoiesis in the primary organs, circulating erythropoiesis was revealed to occur in the loggerhead's blood, just as it also occurs at this site in other non-mammalian vertebrates.

Loggerhead blood infestation by spirorchids was reported by Wolke *et al.* (1982), while Di Santi *et al.* (2010) found two specimens of loggerheads hemoparasitized by *Theileria*.

MAIN BODY

The loggerhead *Caretta caretta* is an endangered species, according to the International Union for Conservation of Nature and Natural Resources, due to human activities disturbing or destroying marine environment and turtle nesting beaches throughout the world. In the last decades, sea turtle populations have been drastically reduced by hunting for meat, eggs and hide, incidental capture in fishing nets or by hooks, especially those used in longline fishing practices, uncontrolled coastal development, vehicular traffic on beaches and sea pollution.

The first step in the evaluation of the health of any animal, besides the veterinary examination, consists of blood tests which are aimed at establishing its hematological and biochemical profile. As blood is a special tissue, which comes in contact and interacts with all other tissues, the study of its components can lead to vital information regarding the subject's health status. In recent times, hematological and biochemical data have been collected to define the normal, healthy values of marine turtles for scientific reasons *per se* as well as for evaluating the data of turtles that have been more and more frequently injured due to human activities. Internationally, many centers are currently active in the research and conservation of this and other sea turtle species as well as caring for sick and injured specimens. Since all blood parameters may vary along with the environmental temperature, geographical distribution of the population, age, sex and physiological condition of the specimen, as well as the capture technique and the treatment of the blood sample, complete baseline ranges of blood parameters of healthy *Caretta caretta* from all geographical areas are still lacking. An assessment of reference intervals for the blood parameters of this sea turtle is clearly needed.

As in other poikilotherms, loggerhead hemogram values are influenced by the environmental temperature, so that the season and the geographical region consistently affect the rate of hemopoiesis. As might be anticipated, hemopoiesis is enhanced in warm seasons as opposed to the cold ones (Glomski and Pica, 2006*)*. Erythropoiesis, specifically, is subject to seasonal fluctuations: environmental temperature affects red cell production in all poikilotherms, so that the most active erythroid proliferation and maturation occur in warm seasons. In poikilotherms, the erythropoietic processes can be expected to be dormant in winter. In spring, with the awakening of subdued physiological functions, various percentages of erythroblasts appear in the circulating blood. Thus the relative level of erythroblasts in the circulating blood count can be used as an indicator of concurrent activity of erythropoietic bone marrow. Consistent with these considerations, the temperature of ocean waters is lower than that of inland seas and consequently the major erythrogramic values (RBC count, hematocrit and hemoglobin concentration in the blood of loggerhead turtles) in these *loci* have been identified as lower than their counterparts residing in the warmer inland seas, such as the Mediterranean Sea. Since sea turtles cannot regulate their body temperature, if the water temperature drops too quickly, they can become *cold-stunned*. This mainly happens to juvenile specimens, which should move to warmer waters before the temperature drops. Thus frequently, in cold seasons sea turtles are found *stranded* on the beaches or just offshore. They are lethargic and their immune system is depressed, consequently, infections can spread throughout their bodies. A case report about the recovery from a widely distributed and deep skin necrosis in a *cold-stunned* loggerhead specimen has been presented by Occhiello *et al.* (2009).

Biochemical blood and hemocellular values also vary according to the age of the specimens. Bradley *et al.* (1998) conducted monthly morphologic evaluations of the blood cells and hemograms of 10 Cape Romaine, South Carolina-collected loggerheads for the first year of these turtles' life. Lithium heparin was used as the anticoagulant for the hemogram. Smears obtained from un-anticoagulated blood were used to observe and describe the circulating blood cells: heterophils, eosinophils, basophils, lymphocytes, monocytes, immature red blood cells, hematogones (extruded nuclei) and thrombocytes. Monitoring of hemogram values revealed an increase of the white cell counts, mean hemoglobin concentration (MCH) and mean corpuscular hemoglobin concentration (MCHC). Kakizoe *et al.* (2007) evaluated the changes in hematological characteristics and plasma chemistry values during the first three years of life of five juvenile loggerhead turtles, taken from the same clutch and grown in an indoor artificial nesting beach. The RBC count was stable during the whole period, while the WBC count showed variability in the first months, reaching the maximum value in the 7th month of life, which was approximately double of that obtained in adult females. In regard to WBC differential counts, the percentage of heterophils decreased and the percentage of lymphocytes increased until to the 10th month of life and remained stable for the rest of the observation period. Monocytes were observed after the 3rd month of life and their percentage fluctuated thereafter. Basophils were rarely observed. The authors did not observe or describe eosinophils. All biochemical values analyzed by the authors showed a wide variability throughout the three-year investigation. While many clinical findings and blood parameters showed significant differences between foraging and nesting turtles, as observed by Deem *et al.* (2009) in a comparative study over 83 loggerheads collected along the coast of Georgia (USA), parameters such as hemoglobin concentration, PCV, MCHC and plasma Fe^{2+} were stable during nesting season in 29 healthy loggerheads from Arabian Sea in the nesting season (Alkindi & Mahmoud, 2002).

Stamper *et al.* (2005) used body condition and hematological parameters to compare the health status of 42 migratory and 15 resident juvenile loggerhead sea turtles in North Carolina, and to also assess whether a relationship between the carapace barnacle load and the health status of the turtle could be established. A total of 18 migratory and 15 resident turtles were examined for barnacle epibiota load. Significant differences were found between hematological and biochemical values in migratory versus residential specimens, probably due to physiological activities correlatable to migration. An association between turtle weight and hematocrit was described and rationalized on the basis that longer dives were performed by larger turtles, thereby necessitating a higher hematocrit for a greater supply of oxygen. Unexpectedly, there was no correlation between carapace barnacle load and hematological values. The authors hypothesized that this result (i.e. the lack of an identifiable difference) may have been due to the limited number of specimens or to the limitation of the analysis of barnacle load.

The capture techniques may affect blood values as seen by Harms *et al.* (2003). They evaluated the effects of two capture techniques for 16 specimens of sea turtles, i.e., by trawl and by pound nets, in North Carolina, South Carolina and Georgia. In the first case, turtles were forcedly submerged for ≤ 30 minutes, while in the second one they were free to surface and to feed at will. Both capture systems caused alterations of many parameters, such as blood gas, acid-base and lactate status. However, more significant perturbations were induced in trawl-captured turtles.

In regard to blood withdrawal, in reptile plasma is preferred over serum because of the variability of the time for clot formation and, consequently, the significant changes in the chemical composition of the sample that may occur in the interval (Bolten *et al.*, 1992). In fact, the comparison between analysis of plasma and serum samples from 23 dually-analyzed specimens led to different results. Plasma samples anticoagulated with lithium heparin or sodium heparin showed no differences in concentration for each analyte. Furthermore, a remarkable source of variability is due to the different biochemical autoanalyzers, as the values obtained by using two different autoanalyzers varied more between themselves than among the results derived from samplings processed following three different post-phlebotomy methods of handling the aspirate (lithium or sodium heparin, or no anticoagulant). Sodium or lithium heparin are preferred as anticoagulants, as reptilian blood may be hemolyzed by EDTA (Jacobson, 1987; Grumbles *et al.*, 1990). A variability in the results for hematocrit and concentration or enzymatic activities of many biochemical analytes was similarly observed by Wolf *et al.* (2008) while testing five different analyzers with blood collected from 22 captive juvenile sea turtles, 18 of which were loggerheads. Moreover, for biochemical analysis, it is important to separate plasma from heparinized whole blood as soon as possible or at least within 24 hours post-collection, otherwise biochemical analyte values may be affected (Eisenhawer *et al.*, 2008).

Casal and Orós (2007) provided a morphological description of circulating blood cells of *Caretta caretta* (erythrocytes, heterophils, eosinophils, basophils, lymphocytes, monocytes and thrombocytes) and also studied their cytochemical characteristics in 35 juvenile loggerhead sea turtles. Some of these workers' cytochemical observations, with the exception of the basophil, differed from those reported for other sea turtle species. Moreover, Casal *et al.* (2007) evaluated the ultrastructural characteristics of the circulating blood cells (except basophils) in 15 juvenile specimens of *Caretta caretta*. A peculiarity in the red blood cells was observed by Eiras *et al.* (2000), who found intracytoplasmic, mostly single inclusions in the mature erythrocytes of 20 juvenile loggerheads from Madeira, Portugal. The identity of the above inclusions was not definitively established, but the authors theorized they could be viral or rickettsial. More recently, Basile *et al.* (2011, a) identified these intraerythrocytic inclusions as Heinz bodies formed through the precipitation of loggerhead's unstable hemoglobin, which had been previously studied and sequenced by Petruzzelli *et al.* (1996).

In addition to the aforementioned investigations of the blood picture of the loggerhead turtle, many other studies have been conducted during the last decade to establish its hematological and biochemical parameters in specimens collected on a worldwide basis. In some instances however, the investigations embraced a small number of subjects. Plasma protein fractions were analyzed electrophoretically by Gicking *et al.* (2004), in a group of 41 healthy, wild loggerheads (juvenile and adult, male and female) from Florida Bay, USA. In 29 of the 41 turtles, four protein fractions were determined: albumin, alpha globulin, beta globulin and gamma globulin; 11 of the 41 specimens showed a beta-gamma bridging and just one turtle of 41 presented an additional prealbumin fraction Pires *et al.* (2006) determined hemogram and plasma protein values in eight captive loggerheads from Bahia, Brasil. Jacobson *et al.* (2007) collected data from specimens of *Caretta caretta* and *Chelonia mydas* (green turtle) entering the Port St Lucie Power Plant canal system in Florida, US, in order to monitor the wild populations of these species and to determine the normal hematocrit values, several plasma biochemical parameters and protein electrophoretogram fractions. Some hematological values of loggerheads from the Atlantic ocean were determined (Pires *et al.*, 2006; Jacobson *et al.*, 2007; Casal *et al.*, 2009; Casal & Orós, 2009; Deem *et al.*, 2009; Osborne *et al.*, 2010), some from the Pacific Ocean specimens (Flint *et al.*, 2010) and some from the Mediterranean sea population (Gelli *et al.*, 2009; Fazio *et al.*, 2010; Basile *et al.*, 2011, b). Hemocellular and plasma protein electrophoretogram fraction values collected by the aforementioned authors are reported in Table 1, biochemical values are summarized in Table 2. Values in tables 1 and 2 are mean values, medians or reference intervals, according to the original data presented by the authors. In case of studies including two or more groups of turtles or subsets, the lowest and highest values were reported.

Table 1. Hemogram values and electrophoretogram fractions in the loggerhead turtles (See References)

Hematological Parameters	Bolten et al 1992	Alkindi and Mahmoud 2002	Gicking et al 2004	Stamper et al 2005	Pires et al 2006	Jacobson et al 2007	Kakizoe et al 2007	Eisenhawer et al 2008
RBC (10^3/μL)				380-880	275		325-429	
Hb (g/dL)		8.87			8.65			
Hct (%)		31.78		28-32	33.12		16.0-26.2	
MCV (fL)					1214			
MCH (pg)					317			
MCHC (%)		27.94			26			
WBC (10^3/μL)				14.3-15.8	3.656		4.7-10.2	
Het (%)				27-60	59.4		35.8-73.3	
Eos(%)				2-6	10.4			
Bas (%)				0	0.13		0.00-0.17	
Lym (%)				36-65	29.3		24.6-58.6	
Mon (%)					0.9		0.0-2.2	
Azurophils (%)				3-4				
Het (10^3/μL)				3.48-6.33	2.16		2.508-5.120	
Eos (10^3/μL)				0.23-0.88	0.367			
Bas (10^3/μL)				0.00	0.003			
Lym (10^3/μL)				4.29-9.92	1.111		1.321-5.340	
Mon (10^3/μL)					0.019			
Azurophils (10^3/μL)				0.75-1.52				
TBC (10^3/μL)					10.968			
Total protein (g/L)	36-41		43	36-40	65	26.0-40.8	15.6-33.3	0.047-0.048
Pre-albumin (g/L)						0.0-0.4		
Albumin (g/L)	6-11		10	11-23		5,7-11.2	5.6-13.7	0.015
Globulins (g/L)				22-28				0.032-0.033
Alpha (g/L)			4.8			2.0-6.8		
Alpha-1 (g/L)								
Alpha-2 (g/L)								
Beta (g/L)			8.0			6.9-36.1		
Gamma (g/L)			19.4			9.6-17.6		
A:G ratio			0.33			0.24-0.51		

Table 1. (continued)

Hematological Parameters	Wolf et al 2008	Casal et al 2009	Casal and Orós 2009	Deem et al 2009	Gelli et al 2009	Fazio et al 2010	Flint et al 2010	Osborne et al 2010	Basile et al 2011
RBC (10^3/μL)		94-187	189	520					410-570
Hb (g/dL)									8.0-14.2
Hct (%)	24.5-28.0	28-40	29	32			13.9-47.3		23-34
MCV (fL)									487-723
MCH (pg)									170-261
MCHC (%)									34-42
WBC (10^3/μL)		1.6-5.9	5.8	9.005			2.63-50.91		17.5-23.8
Het (%)			78.9						41.0-62.4
Eos(%)			3.2						1.9-2.6
Bas (%)			0.0						0
Lym (%)			16.9						32.4-56.6
Mon (%)			1.3						0.2-2.6
Azurophils (%)									
Het (10^3/μL)		1.1-4-6		3.677			0.18-48.95		
Eos (10^3/μL)		0.2-0.3		1.152			0.09-3.45		
Bas (10^3/μL)		0.000001		0					
Lym (10^3/μL)		0.3-1.0		2.725			0.55-33.94		
Mon (10^3/μL)		0.01-0.07		0.960			0.05-3.22		
Azurophils (10^3/μL)				0					
TBC (10^3/μL)		42.6-44.3	43.8				0.07-17.49		2.0-4.9
Total protein (g/L)	31-35	24-41	28	37	42.8	41.9	29.0-72.2	33	
Pre-albumin (g/L)				0.0					
Albumin (g/L)	5-15	11-17	11	7.9	10.7	7.5	7.5-15.9	9.9	
Globulins (g/L)		13-24	17	29			21.2-59.0		
Alpha (g/L)								4.6	
Alpha-1 (g/L)				1.4					
Alpha-2 (g/L)				1.2					
Beta (g/L)				9.9				0.84	
Gamma (g/L)				15.7				0.97	
A:G ratio								0.43	

Table 2. Biochemical values in the loggerhead turtle (See References)

Hematochemical Parameters	Bolten et al 1992	Alkindi and Mahmoud 2002	Stamper et al 2005	Jacobson et al 2007	Kakizoe et al 2007	Eisenhawer et al 2008	Wolf et al 2008
Glucose (mmol/L)	4.42-4.45		4.39 -5.53	4.06-6.17	6.77-9.16	8.72-9.33	5.89-6.86
Cholesterol (mmol/L)	2.50-2.74			1.37-5.90	2.68-5.00	3.62-3.75	
Triglycerides (mmol/L)	0.91-0.98				0.27-1.68		
Urea (mmol/L)							
BUN (mmol/L)	17.89-18.73		22.5-28.57	10.22-34.32	0.015-0.051	52.94-54.77	37.97-43.99
Uric Acid (mmol/L)	0.041-0.065		0.02-0.04	0.030-0.083	0.008-0.057	0.048-0.059	0.024-0.036
Creatinine (μmol/L)	17.68-26.52				38.90-100.19		
Total Bilirubin (μmol/L)	0.34-5.13				2.05-8.21		
AST (U/L)	170-187		129-206	149-345	75.0-150.6	342-352	229-264
ALT (U/L)	2.0-12.2				1.2-6.6		
GGT (U/L)				8-31	1.2-6.0	16-24	
Amylase (U/L)					226.2-1036.2		
Lipase (U/L)							
CPK (U/L)				319-1742	590-2,927	1,022-1,419	299.5-972.0
LDH (U/L)	90-128		120-227		58.2-176.4		65-206
ALP (U/L)	10.5-14.0		9.0-23.0	7-22	833-5,041	36-38	20-45
Sodium (mmol/L)	155.9-162.0		151.0-158.0	148-168	143.20-154.40	155-161	148-156
Potassium (mmol/L)	4.0-4.2		3.30-4.50	3.5-7.9	3.20-4.48	4.0-4.3	3.8-4.4
Chloride (mmol/L)	116.5-119.3		112-118	103-128	103.00-120.60	119-126	118-123
Calcium (mmol/L)	1.57-1.72		1.55-2.20	1.07-1.82	1.49-2.12	1.72-1.85	1,35-1,77
Phosphorus (mmol/L)	2.56-2.62		2.03-2.26	1.92-3.04	1.98-5.42	2.85-2.94	2.18-2.59
Magnesium (mmol/L)			2.17-2.26	1.28-2.55		1.8-2.2	
Iron (μmol/L)	5.73	13,620					
Carbon dioxide (mmol/L)	21.0-23.7						
Anion gap			10.8-17.5				

Table 2. (continued)

Hematochemical Parameters	Casal et al 2009	Casal and Orós 2009	Deem et al 2009	Gelli et al 2009	Fazio et al 2010	Flint et al 2010	Basile et al 2011
Glucose (mmol/L)	3.3-7.2	7.10	5.88	6.06		4.3-8.9	5.38-9.10
Cholesterol (mmol/L)	3.6-8.7	3.69	1.94	1.99			1.92-3.74
Triglycerides (mmol/L)	1.3-7.4	0.85	0.62	0.60			
Urea (mmol/L)	7.2-36.3			6,87	10.34	21.7-48.4	
BUN (mmol/L)		36.1	29.63				12.78-73.02
Uric Acid (mmol/L)	0.06 - 0.10	0.06	0.041	346,000		0.017-0.157	0.059-0.159
Creatinine (μmol/L)	31.8-39.7	26.5	26.52	3,540,000	44.20	14.5-41.2	
Total Bilirubin (μmol/L)	3.4-17.2	3.42		850,000		0.9-2.9	3.42-6.84
AST (U/L)	123-194	202.2	165	468	37.15	78.7-274.6	44-184
ALT (U/L)	11-24	14.1	16	13.32	98.75		6
GGT (U/L)			9.0	1.11			
Amylase (U/L)			263				
Lipase (U/L)			1.0				
CPK (U/L)			534	3,704		157.4-2,211.1	
LDH (U/L)	<100-310	101.2	572	461.32		29.4-391.8	
ALP (U/L)	67-103	64.7		59.53		10.5-132.9	
Sodium (mmol/L)			156			141-158	
Potassium (mmol/L)			5.1			3.2-5.7	
Chloride (mmol/L)			130			107-125	
Calcium (mmol/L)	2.0-3.1	2.43	1.85	1.58		1.2-2.0	1.67-2.17
Phosphorus (mmol/L)			2.07	2.59		1.6-3.2	
Magnesium (mmol/L)						1.7-4.5	1.48-2.22
Iron (μmol/L)							
Carbon dioxide (mmol/L)			15				
Anion gap							

Morphology and Cytochemical Characteristics of Blood Cells

The identification of loggerhead's leukocytes is still a matter of debate due to the occurrence of rather similar eosinophilic cytoplasmic granulation in both heterophils and eosinophils, which can lead to misinterpretation of the differential count. Controversial results have been reported by various authors depending on the morphological or cytochemical methods employed to classify white blood cells in the differential count, so they often cannot be compared. Unlike the granulocytes, lymphocytes and monocytes are easily distinguishable, as they are similar to their mammalian counterparts. Casal & Orós (2007) have provided morphological and cytochemical assessment of the loggerhead's white blood cells. The predominant lymphocytes are small and round with a similarly shaped nucleus that occupies the majority of the cytoplasm (nucleus: cytoplasm ratio 0.6) and which is also characterized by its dense, solid aggregation of chromatin (Fig 1b). The cytoplasm is basophilic, moderately granular and yields a positive NB-esterase reaction (in the absence of sodium fluoride). The monocytes are round cells, 33% larger than the lymphocytes, and maintain an eccentrically located, kidney-shaped nucleus whose chromatin pattern is less dense than that of the lymphocytes (Fig 1c). The cytoplasm is slightly basophilic, sometimes vacuolated, and it is weakly stainable with NBE (without fluoride). The nucleus:cytoplasm ratio is smaller than that of lymphocytes (0.5). Casal and Orós (2007) found rare basophils displaying the typical feature of mammalian counterparts such as intensely basophilic granules overlying the nucleus; they were stainable only with Toluidine Blue. The loggerhead's heterophils are large, round cells that present a round and central or oval and eccentric nucleus, containing clumped chromatin and a prominent amount of slightly eosinophilic cytoplasm full of weakly eosinophilic granulation (Fig 1b,c). They are stainable by acid phospatase (ACP) with and without tartrate, by peroxidase (PER), by chloroacetate esterase (CAE), Sudan Black B (SBB) and slightly stained by the periodic acid Schiff reaction (PAS). Eosinophils (Fig 1b) display a round configuration with a generally eccentrically placed round or oval nucleus containing clumped chromatin. The abundant cytoplasm is weakly basophilic (a characteristic feature of the eosinophils of at least some other species) and contains a few to a modest number of round eosinophilic granules stainable by ACP (with and without tartrate), NBE (with sodium fluoride), CAE and PAS and moderately by SBB.

The thrombocytes (Fig 1b), which take part in the hemostatic process, are complete, intact cells, morphologically different but functionally analogous to the mammalian platelets. This is revealed by their aggregation into small clumps observed in smears made from fresh, unanticoagulated blood. They were described as round or oval-shaped cells (Casal & Orós, 2007). In the oval cells the nucleus was ovoid and the transparent cytoplasm was accumulated at the two poles of the cell. In the case of the round forms, a round nucleus was surrounded by a thin rim of cytoplasm.

The erythrocyte of loggerhead turtles has the same fundamental structure that is universally maintained by the other poikilothermic vertebrates. It is a flattened, biconvex, permanently nucleated ovoid cell (ranging from 17-19 μm x 9-10 μm in size), whose configuration is recognized to facilitate passive transport in flowing plasma and offers a large surface area that enhances oxygen exchange (Fig 1a). This turtle's red cell, however, does present some species-related attributes comparable to those observed in the red cells of constituents of other *taxa*. These include an overall shape, dimensions, numerical quantity per volume unit of blood and amount of hemoglobin in a given cell which in sum characterize the erythrocytes of this marine species and other members of this general group. Thus marine turtles have comparatively large erythrocytes, approximately 400-700 fL in volume with a wide oval shape as opposed to terrestrial reptiles such as *Iguana iguana*, the common green iguana, which maintains smaller (~300 fL) narrower red cells which, in addition, have a greater cellular length to width ratio. This latter aspect confirms that the loggerhead's erythrocytes are not only larger but also proportionally wider than those of the iguana. It is considered that the voluminous red cell of the turtle coupled with its large amount of hemoglobin is less efficient in rapid exchange of oxygen than a smaller erythrocyte with a proportionately equivalent of hemoglobin because of the (mathematically documentable) relatively greater cell surface area per hemoglobin content of the lesser-sized cell. The paradigm presented by the turtle's erythrocyte yields a slower, prolonged release of oxygen that would be of benefit for an air-breathing species that engages in long dives.

The ultrastructural characteristics of the circulating blood cells of loggerhead turtles (n=15) were investigated and detailed by Casal *et al.* (2007). Heterophils, eosinophils, lymphocytes, monocytes and thrombocytes were clearly identified in this study, while the basophils were so rarely observed that their number was insufficient for analysis. These workers reported that all the leukocytes are generally equivalent to their counterparts from other species of sea turtles. An exception was observed in the eosinophils which tended to be uniform in size, unlike the eosinophils of the green turtle, *Chelonia mydas* (Work *et al.*, 1998), and also do not present any crystalline structure in their granules. The crystalloid center is a hallmark of the eosinophil granules in other species. The ultrastructure of eosinophil reveals that this cell is spherical and maintains a round or oval-shaped with variable amounts of heterochromatim. Some eosinophils have multiple small sized or one large electron-lucent cytoplasmic vacuole; as would be anticipated mitochondria, endoplasmic reticulum and Golgi complexes are routinely identified.

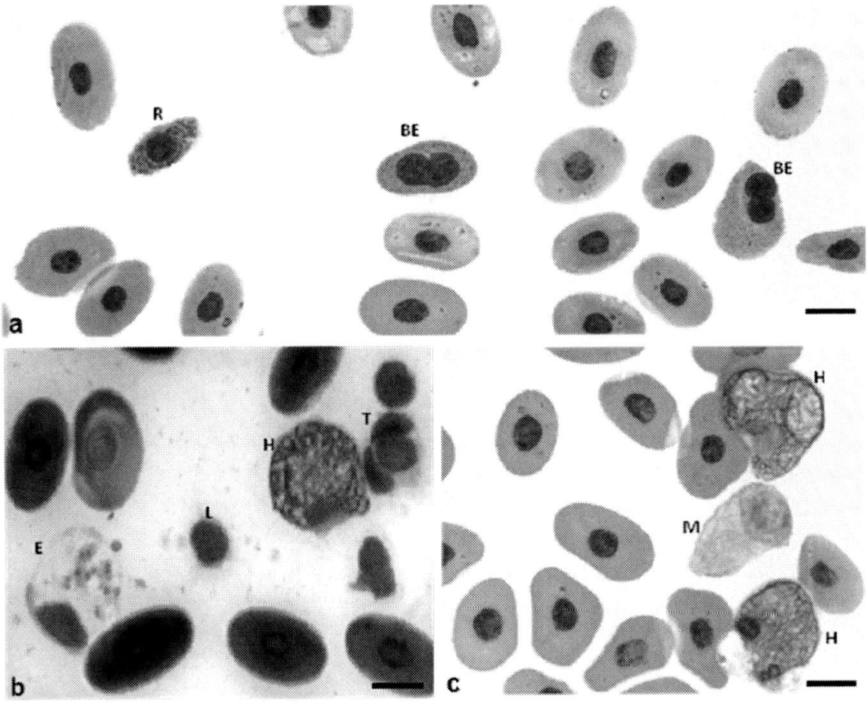

Figure 1. a. Circulating erythropoiesis in the loggerhead's blood: a reticulocyte (R) and two basophilic erythroblasts (BE) among mature erythrocytes (scale bar = 8 μm). b. Eosinophilic granulocyte (E), lymphocyte (L), heterophilic granulocyte (H), thrombocytes (T) (scale bar = 8 μm). c. Monocyte (M) and two heterophils (H) (scale bar = 7 μm).

Heterophils are round in shape and have an eccentrically placed spherical, heterochromatin-rich nucleus. Numerous electron-dense, round or elongated granules are dispersed throughout its abundant cytoplasm. A limited population of variably electron-dense, pleomorphic granules is also present. Organelles typically presented in the cytoplasm of an animal's cell (e.g., mitochondria) are identifiable. On the basis of the foregoing characterization the heterophils of the loggerhead resemble those of the fellow marine turtle *Chelonia mydas*, the green turtle. They are, however, unlike the heterophils of some other reptiles which concurrently present two apparently distinct populations of heterophils (different levels of maturity?) in their circulation. One version of this granulocyte has homogenous electron-dense granules in its cytoplasm, while the other has granules some of which are electron dense while the others do not display this morphologic feature. The lymphocytes morphologically correspond to those of the green turtle and other reptiles with their irregularly round shape and spherical, often-indented nucleus containing a prominent mass of dense heterochromatin that fills most of the nucleoplasmic space. The nucleus is encircled by a halo of cytoplasm containing some small electron-dense granules accompanied by a few polyribosomes, mitochondria and sparse endoplasmic reticulum. The monocytes of the loggerhead turtle and the rarely described monocytes of the green turtle have similar ultrastructure. Loggerhead's monocytes are round or fusiform and their spherical nucleus contains a modest amount of heterochromatin. Some small electron-dense granules, a prominent Golgi complex and representatives of typical organelles such as mitochondria are present in the cytoplasm. Thrombocytes are usually oval shaped and sometimes round. The nuclei of oval shaped thrombocytes are correspondingly ovoid but the heterochromatin of these cells is not centrally located in the nucleus. Small granules of variable electron-density are observable in the cytoplasm together with canalicular structures (perhaps plasmalemmal infoldings as obtained in mammalian platelets) which are a common feature of reptilian thrombocytes.

Inclusion Bodies of Erythrocytes

The erythrocytes, which are permanently nucleated cells, have an oval nucleus with dense heterochromatin. A certain electron-dense cytoplasmic inclusion is identifiable in some erythrocytes. Its origin has been a matter of debate and hypotheses have proposed that it could be viral or rickettsial in nature or a degenerating organelle. It has finally been concluded that it is a Heinz body derived from the precipitation of the loggerhead's unstable hemoglobin. In a series of studies of healthy and unhealthy, captive and wild free-living, juvenile and mature *Caretta caretta* conducted on circulating blood of specimens housed at the Sea Turtle Rescue and Rehabilitation Center of the Zoological Station Anton Dohrn (Naples, Italy) this specific inclusion was identified in a variable number of circulating erythrocytes in each of the assessed turtles (Basile *et al.*, 2011). It was characterized in MGG-stained dry film blood smears as spherical in configuration, approximately 1-3 μm in diameter and somewhat dark blue in color. The inclusions were only seen in mature erythrocytes and were generally located along the major axis of the cell between the nucleus and plasmalemma. One inclusion body was identified in a given cell. In one evaluation of six juvenile and nine adult subjects, 3-82% of the erythrocytes presented this cytoplasmic incorporation. Heinz bodies are well documented in inframammalian and mammalian vertebrates. Tinctorial attributes of the Heinz bodies in the loggerhead include positive staining by brilliant cresyl blue, toluidine blue and the periodic acid-Schiff reaction. These loggerhead's Heinz bodies are Feulgen reaction-negative thereby indicating that they are not DNA. The etiology of the development of Heinz bodies in the loggerhead remains undetermined. Since they have been seen in healthy turtles, including a few captive healthy representatives that have been maintained in ideal normal environments for 2-4 years and also in sick turtles, it is apparent that they are not life threatening or disabling. The hemoglobin of the loggerhead has been theorized to be less stable than that of other species' hemoglobin. Its degradation under normal circumstances in a presumed long-lived cell or after possible exposure to pollution may consequently lead to the development of Heinz bodies. The removal of the Heinz body from the turtle's erythrocyte without destroying the latter has been thought to occur through a process known as "pitting", that is the elimination of only the inclusion body from the cytoplasm of the erythrocyte, so that the cell temporarily takes the shape of a teardrop (dacryocyte) and then reverts back to its original shape (Basile *et al.*, 2011, a).

ERYTHROPOIESIS

Circulating erythropoiesis or at least maturation of very immature erythroblasts is identifiable in the blood of the loggerhead. The youngest recognizable immature precursor of the erythrocyte in this *locus* is the basophilic erythroblast. It is a spherical cell with a centrally located round nucleus surrounded by a relatively narrow band of very basophilic cytoplasm.

With continued maturation the cell evolves into an intermediate form termed the acidophilic erythroblast, which presents a flattened ellipsoidal configuration. The cytoplasm is acidophilic due to the accumulation of hemoglobin, while the nucleus is oval and the chromatin is not as compact as that of a mature red cell. The reticulocyte is a further developed cell and is the immediate antecedent of the mature erythrocyte. It is identified by the detection of a small amount of residual cytoplasmic RNA by supravital staining with brilliant cresyl blue.

HEMOPARASITIZATION IN LOGGERHEAD TURTLES

Reptilian erythrocytes, like those of the pisces and amphibians, are susceptible to intracellular parasitization. The inhabitance of a parasite in a red cell is frequently an unexpected or incidental finding during the examination of a blood film stained with MGG or comparable Romanowsky type dye. This apparently holds true for the erythrocytes of *Caretta caretta*. A benign infestation by the intraerythrocytic protozoan *Theileria sp.* has been recently detected for the first time in two specimens of Mediterranean loggerhead turtles. The vector of the parasite was thought to be the leech *Ozobranchus margoi*. The hemogramic data of these two turtles suggested that the infestation had an undetectable or minimal impact on the erythrocytic and leukocytic parameters or apparent health of the hosts (Di Santi *et al.*, 2010). Various other members of the Phylum Apicomplexa have been detected in loggerhead blood and are capable of invading leukocytes and erythrocytes. The previously noted *Theileria* is a protozoan that invades both types of cells. When erythrocytes are hosts for parasites, the affected red cells along with their inhabitants may be subjected to erythrophagocytosis (Jaensh & Raidal, 2006). In most infestations by apicomplexans, the parasitizations are well tolerated and the host vertebrates do not present clinical signs of disease. Diagnosis of infection is based on the recognition of the trophozoite, multinucleate schizont, or gametocyte inhabiting red cells (Telford, 1984; Barnard & Upton, 1994).

Since loggerheads are ectotherms their body temperature is more subjected to fluctuations than that of homeotherms, such as mammals and birds, thus making it easier for pathogens to grow and proliferate (Alfaro *et al.*, 2006). In sea turtle diseases, bacteria act both as primary pathogens and as secondary invaders, when the host's immune system is weak. The bacteria found in turtles are commonly isolated in localized infections but also play a key role in epizootics which lead to the spreading of bacteria in the host's blood (bacteraemia) and consequent septicaemia (Cooper, 1983). Infestation by three species of spirorchids (*Carettacola*, *Hapalotrema*, and *Neospirorchis*) was detected in the heart and blood vessels of mainly sub-adult loggerhead sea turtles. This infestation causes severe debilitation and mortality (Wolke *et al.*, 1982). As the current information about infestations and infections in sea turtles is incomplete, more research about turtle parasites and hemoparasites are still needed to fully understand the interactions between turtles and pathogens.

Although many studies have been carried out to date and others are in progress, the hematology of this sea turtle still remains a fascinating subject and one of which only a few aspects are definitively comprehended and clarified. At least four major areas require further investigation.

Baseline and complete values of the hematologic parameters have to be established for the world wide, diversely distributed populations of the loggerhead turtle. This would be very helpful in sea turtle rescue and conduction of rehabilitation programs.

In regard to leukocytes, no data are available about the physiological roles of lymphocytes in the loggerhead immune system, so it would be very interesting to know whether B and T-lymphocytes are detectable in sea turtle blood and which duties they carry out.

A precise, accurate technique to routinely morphologically distinguish between the eosinophils and heterophils is needed for the performance of leukocyte differential counts.

Finally, data regarding loggerhead hemopoiesis are unavailable. This species is classified as endangered and thereby needs intensive evaluation. Any investigation of its hemopoiesis requires studies of the tissues in which this occurs. The maturational process and kinetics of the leukocytes and the thrombocytes cannot be studied in the peripheral blood since the progenitors of these cells are not normally present in circulation. In the case of the erythrocyte, although reticulocytes and some erythroblasts are identifiable in the circulation, a complete understanding of erythropoiesis requires examination of erythroid proliferation in the hemopoietic tissues.

REFERENCES

Alfaro, A; Køie, M; Buchmann, K. Synopsis of infections in sea turtles caused by virus, bacteria and parasites: an ecological review [online]. 2008 Available from: *http://www.latinamericanseaturtles.org*

Alkindi, AYA; Mahmoud, IY. Hematological survey in two species of sea turtles in the Arabian Sea during nesting season. *Pakistan Journal of Biological Sciences* 2002 5(3): 359-361.

Barnard, SM; Upton, SJ. *A veterinary guide to the parasites of reptiles.* Protozoa Krieger Publishing Co. Malabar, FL. 1994 Vol. 1 pp35-63.

Basile, F; Di Santi, A; Caldora, M; Ferretti, L; Bentivegna, F; Pica, A. [a] Inclusion bodies in loggerhead erythrocytes are associated with unstable hemoglobin and resemble human Heinz bodies. *Journal Exp Zool A* 2011 DOI: 10.1002/jez.687, May 2, 2011.

Basile, F; Di Santi, A; Ferretti, L; Bentivegna, F; Pica, A. [b] Hematology of the Mediterranean population of sea turtle (*Caretta caretta*): comparison of blood values in wild and captive, juvenile and adult animals. *Comparative Clinical Pathology* 2011. *In press.*

Bolten, AB; Jacobson, ER; Bjorndal, KA. Effects of anticoagulant and autoanalyzer on blood biochemical values of loggerhead sea turtles (*Caretta caretta*). *Am. J. Vet. Res.* 1992 Vol. 53, No. 12.

Bradley, TA; Norton, TM; Latimer, KS. Hemogram values, morphological characteristics of blood cells and morphometric study of loggerhead sea turtles, *Caretta caretta*, in the first year of life. *ARAV* 1998 Volume 8, No. 3.

Casal, AB; Orós, J. Morphologic and cytochemical characteristics of blood cells of juvenile loggerhead turtles (*Caretta caretta*). *Research in Veterinary Science* 2007 82, 158-165.

Casal, AB; Freire, F; Bautista-Harris, G; Arencibia, A; Orós, J. Ultrastructural characteristics of blood cells of juvenile loggerhead turtles (*Caretta caretta*). *Anat. Histol. Embryol.* 2007 36, 332-335.

Casal, AB; Camacho, M; López-Jurado, LF; Juste, C; Orós, J. Comparative study of hematologic and plasma biochemical variables in Eastern Atlantic juvenile and adult nesting loggerhead sea turtles (*Caretta caretta*). *Veterinary Clinical Pathology* 2009 38/2; 213-218.

Casal, AB; Orós, J. Plasma biochemistry and haematology values in juvenile loggerhead sea turtles undergoing rehabilitation. *Veterinary Record* 2009 164, 663-665.

Cooper, JE. Diseases of the Reptilia. Volume 1 and 2. Edited by John E. Copper and Oliphant F. Jackson. Academic Press 1983.

Deem, SL; Norton, TM; Mitchell, M; Segars, A; Alleman, AR; Cray, C; Poppenga, RH; Dodd, M; Karesh, WB. Comparison of blood values in foraging, nesting and stranded loggerhead turtles (*Caretta caretta*) along the coast of Georgia, USA. *Journal of Wildlife Diseases* 2009 45(1), 41–56.

Di Santi A; Basile F; Ferretti L; Bentivegna F; Pica A. Hemoparasitization by *Theileria* in the loggerheads *Caretta caretta* of the Mediterranean Sea. Comp Clinical Pathology 2010 DOI 10.1007/s00580-010-1065-7

Eiras, JC; Dellinger, T; Davies, AJ; Costa, G; Alves de Matos, AP. Intraerythrocytic inclusion bodies in the loggerhead sea turtle *Caretta caretta* from Madeira. *Journal of the Marine Biological Association of the UK* 2000 80:5, 957-958.

Eisenhawer , E; Courtney, CH; Raskin, RE; Jacobson, E. Relationship between separation time of plasma from heparinised whole blood on plasma biochemical analytes of loggerhead sea turtles (*Caretta caretta*). *Journal of Zoo and Wildlife Medicine* 2008 39 (2): 208-215.

Fazio, E; Liotta, A; Medica, P; Giacoppo, E; Ferlazzo, A. Effects of different health status on blood haematochemical values of loggerhead sea turtles (*Caretta caretta*). *Comp Clin Pathol* 2010 DOI: 10.1007/s00580-010-1070-x.

Flint, M; Morton, JM; Limpus, CJ; Patterson-Kane, JC; Mills, PC. Reference intervals for plasma biochemical and hematologic measures in loggerhead sea turtles (*Caretta caretta*) from Moreton Bay, Australia. *J Wildl Dis* 2010 46 (3): 731-741.

Gelli, D; Ferrari, V; Zanella, A; Arena, P; Pozzi, L; Nannarelli, S; Vaccaro, C; Bernardini, D; Romagnoli, S. Establishing physiological blood parameters in the loggerhead sea turtle (*Caretta caretta*). *Eur J Wild Res* 2009 55(1): 59-63.

Gicking, JC; Foley, AM; Harr, KE; Raskin, RE; Jacobson, E. Plasma protein electrophoresis of the Atlantic loggerhead sea turtle, *Caretta caretta*. *Journal of Herpetological Medicine and Surgery* 2004 14 (3): 13-18.

Glomski, CA; Pica, A. Erythrocytes of the Poikilotherms: a Phylogenetic Odyssey. Foxwell & Davies (London, UK) Ltd- Scientific Publisher 2006.

Grumbles, J; Rostal, D; Alvarado, J; Owens, D. Hematology study on the black turtle, *Chelonia agassizi*, at Playa Colola, Michoacan, Mexico. In: Richardson T. H., Richardson J. I., Donnelly M., compilers, in *Proceedings. 10th Annu Workshop Sea Turtle Biology Conservation.* Miami, Fla: NOAA Tech Mem NMFS-SEFC-278, 1990; 235-237.

Harms, CA; Mallo, KM; Ross, PM; Segars, A. Venous blood gases and lactates of wild loggerhead sea turtles *(Caretta caretta*) following two capture techniques. *Journal of Wildlife Diseases* 2003 39(2): 366-374.

Jacobson, ER. Reptiles. in Harness J., ed. Vet Clin North Am Small Anim Pract. Philadelphia: WB Saunders Co, 1987; 1203-1225.

Jacobson, E; Bjorndal, K; Bolten, A; Herren, R; Harman, G; Wood, L. Establishing plasma biochemical and hematocrit reference intervals for sea turtles in Florida. 2007. Available at: http://accstr.ufl.edu/blood_chem.htm.

Jaensh, SM; Raidal, SR. Peripheral erythrophagocytosis in two reptiles. *Comp Clin Pathol* 2006 15:113–116.

Kakizoe, Y; Sakaoka, K; Kakizoe, F; Yoshii, M; Nakamura, H; Kanou, Y; Uchida, I. Successive changes of hematologic characteristics and plasma chemistry values of juvenile loggerhead turtles (*Caretta caretta*). *Journal of Zoo and Wildlife Medicine* 2007 38(1): 77–84.

Occhiello, A; Bentivegna, F; Borrelli, A; Schiattarella, A; Mancini, A; Pica. A. Skin necrosis in sea turtle cold stunning: regeneration following rMnSOD topic treatment in a specimen of C*aretta caretta*. *Comp Clinical Pathology* 2009 pp.365- 369 Vol.18,

Osborne, AG; Jacobson, ER; Bresette, MJ; Singewald, DA; Scarpino, RA; Bolten, AB. Reference intervals and relationships between health status, carapace length, body mass, and water temperature and concentrations of plasma total protein and protein electrophoretogram fractions in Atlantic loggerhead sea turtles and green turtles. *J Am Vet Med Assoc.* 2010 237(5):561-567.

Petruzzelli, R; Aureli, G; Lania, A; Galtieri, A; Desideri, A; Giardina, B. Diving behaviour and haemoglobin function: the primary structure of the α- and β-chains of the sea turtle (*Caretta caretta*) and its functional implications. *Biochem J* 1996 15:959–965.

Pires, TT; Rostan, G; Guimarães, JE. Hematologic examination and total protein values of sea turtles of species *Caretta caretta* (Linnaeus, 1758), in captivity, from Praia do Forte, Mata de São João – Bahia. Braz. *J. vet. Res. anim. Sci.* 2006 São Paulo, v. 43, n. 3, p. 348-353.

Stamper, MA; Harms, C; Epperly, SP; Braun-McNeill, J; Avens, L; Stoskopf, MK. Relationship between barnacle epibiotic load and hematologic parameters in loggerhead sea turtles (*Caretta caretta*), a comparison between migratory and residential animals in Pamlico Sound, North Carolina. *Journal of Zoo and Wildlife Medicine* 2005 36(4), 635-641.

Telford, SR Jr. Haemoparasites of reptiles. In: *Diseases of Amphibians and Reptiles*. Hoff G. L., Frye F.L. and Jacobson E. R. (eds). Plenum press. NY. 1984 pp385-517

Wolf, KN; Harms, CA; Beasley, JF. Evaluation of five clinical chemistry analyzers for use in health assessment in sea turtles. *Journal of the American Veterinary Medical Association* 2008 233 (3), 470-475.

Wolke, RE; Brooks, R; George, A. Spirorchidiasis in loggerhead sea turtles (*Caretta caretta*): pathology. *Journal of Wildlife Diseases* 1982 Vol. 18. No. 2, April., 175-185.

Work, TM; Raskin, RE; Balazs, GH; Whittaker, SD. Morphologic and cytochemical characteristics of blood cells from Hawaiian green turtles. *AJVR* 1998 Vol 59, No10, October.

In: Turtles: Anatomy, Ecology and Conservation ISBN: 978-1-61470-554-3
Eds: M. J. Cosgrove and S. A. Roe

Chapter 6

RESPONSES OF FRESHWATER TURTLES TO DROUGHT: THE PAST, PRESENT AND IMPLICATIONS FOR FUTURE CLIMATE CHANGE IN AUSTRALIA*

***John Roe*[1] *and Arthur Georges*[2]**
[1] Indiana-Purdue University, IN, U. S.
[2] University of Canberra, Australia

ABSTRACT

Australia was not always arid, and the freshwater turtles that have survived to this day are either relicts residing in river systems that have themselves had a history or continuity through the drying of the continent, or they are species that have adapted in some way to meet the challenges of drought. In this chapter, we report on how Australian freshwater turtles cope with periodic loss of habitat through drought at a range of spatial and temporal scales, and consider how they might fare under changing climates predicted to occur through global warming. Global warming is not occurring in isolation of other environmental changes at a landscape scale, and we look at what interactions there might

* A version of this chapter was also published in Meltdown: Climate Change, Natural Disasters and other Catastrophes - Fears and Concerns for the Future, edited by Kathryn Gow, published by Nova Science Publishers, Inc. It was submitted for appropriate modifications in an effort to encourage wider dissemination of research.

be between climate change and the increasing demands of agriculture, industry and our cities for water, in presenting challenges for our unique freshwater turtle fauna.

Keywords: turtles, drought, climate change, habitat loss, freshwater fauna

INTRODUCTION

Australia has the lowest precipitation of any vegetated continent – only Greenland and Antarctica have less – but few people realize how dry Australia actually is. To put this in perspective, the average annual discharge from all of Australia's rivers is 237,582 gigaliters (Australian Bureau of Rural Sciences)[1] which is less than half that of the Mississippi River in North America. Consequently, much of the Australian continent is arid or semi-arid, and wetland habitats are small and poorly developed compared to other continents. As air and sea current patterns changed when Australia broke off from Antarctica, major episodes of progressive aridification occurred 30-35 Mya[2], 15 Mya and 2 Mya. The climate became not only drier, but also less predictable, and the biota that persisted had to contend with drought or periodic loss of free-standing water across much of the continent.

Drought is most simply defined as a departure below 'normal' precipitation (Dracup, Lee and Paulson, 1980; Wilhite and Glantz, 1985), although droughts can fall anywhere along a spectrum of climactic variability from long-term aridity at one extreme to annual cycles of wet and dry at the other. Aridity occurs where environmental water inputs are persistently insufficient to balance loss, whereas seasonally variable regions accrue a relatively temporary water deficit. In a sense, aridity can be considered as a long-term drought spanning millennia, whereas seasonally wet-dry climates cycle in and out of drought on a shorter term basis. The distinction between these scenarios is only a temporal one. Drought in the conventional sense falls somewhere in between, defined by water deficits spanning several years or decades compared to expectations derived over a period of historical record keeping, typically spanning decades or centuries.

---

[1] http://www.daff.gov.au/brs/social-sciences.

[2] Million years ago.

From the perspective of wildlife responses to drought, this temporal distinction is important to grasp. At either end of the drought continuum, water deficits (whether short or long term) can be a relatively predictable feature of the regional climate. That is not to say that annual fluctuations in the occurrence, amount, timing, and duration of wet and dry periods do not occur – but over the balance of several years the timing and duration of water shortage occurs in a predictable sequence that has persisted for several thousands or millions of years. Such protracted periods of relatively stable climactic patterns can shape biological communities through evolutionary responses so that morphology, physiology and behaviour are suitably matched to the conditions (Ligon and Peterson, 2002; Withers, 1993). However, an abrupt departure from the normal precipitation pattern that spans several years or decades (i.e., drought in the conventional sense) may cause the existing biological communities to struggle to keep pace with the changing climate, even if environmental changes are brought about by natural climactic cycles.

The impact of drought on wildlife should not be discussed without also considering the human dimensions. Human activities may result in the imposition or exacerbation of drought conditions, and our infrastructure in developed landscapes can alter the natural responses of wildlife to drought. For instance, anthropogenic global climate change threatens to increase aridity in several regions of Australia, as well as to increase stochasticity[3] in climate, which can influence the frequency, severity, and duration of extreme weather events such as droughts (Bates et al., 2008; IPCC, 2007). Human water resource development alters water flow and hydrological regimes, in a sense simulating and hastening the onset of drought conditions and attenuating seasonal cycles of wet and dry. Roads, feral predators, and other dimensions of human society may further hamper the ability of wildlife to respond to drought.

In this chapter, we introduce the Australian freshwater turtle fauna and explore how they cope (or in some cases, do not cope) with periodic loss of aquatic habitat through drought at a range of spatial and temporal scales, and consider how they might fare under changing climates predicted to occur through global warming. Global climate change is not occurring in isolation to other environmental alterations at a landscape level, and thus in assessing some of the challenges for Australia's unique freshwater turtle fauna, we examine what interactions there might be between climate change and the

[3] Stochasticity implies irregular and unpredictable variability.

increasing demands of agriculture, industry and cities for water and other natural resources.

THE FRESHWATER TURTLES OF CONTEMPORARY AUSTRALIA

There are no strictly terrestrial turtles or tortoises that inhabit Australia. Instead, the Australian turtle fauna is restricted to freshwaters and the surrounding ocean. In this chapter, we focus only on those that occupy the inland freshwater systems, which are dominated by species belonging to the family Chelidae. This family of side-necked turtles is restricted to the Australasian region and South America - not even fossil forms are found outside this range (Gaffney, 1991; Williams, 1953a, b) - so it is a group with clear Gondwanan origins. Their fossil record extends back to the late Cretaceous (de Broin, 1987) in South America and the Miocene in Australia (Gaffney, Archer and White, 1989). Among extant forms in Australia and New Guinea are the river turtles in the genera *Emydura* (5 species) and *Elseya* (7 species) of northern and eastern Australia and New Guinea, and *Rheodytes* (1 species) and *Elusor* (1 species) from east coastal rivers of Queensland. The critically endangered *Psuedemydura umbrina* resides in seasonally dry wetlands of south Western Australia. Australia is also renowned for its spectacular snake-necked turtles (12 species) that range in habitat from permanent waters (*Chelodina expansa*) to highly ephemeral[4] streams and ponds (*Chelodina steindachneri*). The pig-nosed turtle (*Carettochelys insculpta)* is the only species of Australian freshwater turtle outside of the family Chelidae, belonging to the family Carettochelydidae (the only extant species in this family) that lives in the major rivers and associated wetlands of northern Australia and southern New Guinea (Georges and Rose, 1993).

Not surprisingly, given this diversity, Australia's freshwater turtles occupy a wide range of habitats - from the flooded forests and extensive swamps of the Fly River and Kikori Delta in Papua New Guinea (Georges et al., 2006a, 2008) to the dryland rivers of Australia's arid interior (Georges et al., 2006b); from the tropical north to the snow country around Cooma; from dilute waters of the sandstone plateaus of Arnhem Land and the Kimberleys (Thomson et al., 2000), to the coastal floodplains of northern Australia and the saline or

---

[4] Ephemeral bodies of water tend to dry during drought and flood again following precipitation.

brackish coastal wetlands of the Styx River in Queensland and Lake Alexandria at the mouth of the Murray River in South Australia. However, Australia was not always arid (Quilty, 1994), and the freshwater turtles that have survived to this day are either relicts residing in watersheds that have themselves had a history of continuity through the drying of the continent, or they are species that have adapted in some way to meet the challenges of drought (Georges and Thomson, 2006). Below, we explore several ways that Australian turtles respond naturally to the drying of their aquatic habitat.

Natural Responses to Habitat Drying

Along with temperature, precipitation patterns that determine flooding and drying of surface waters are likely to be the most powerful drivers of freshwater turtle distributions and community structure[5] in Australia. Several of Australia's wetlands are under the influence of natural flood-drought cycles, and in these contexts, many turtles have adaptations to drought that allow them to successfully cope with habitat drying by moving to more permanent waters (escape in space) or by remaining in or near the dry waterbody in an inactive state to await the return of flooding (escape in time). However, several species do not regularly contend with the drying of their waterholes and may lack any evolved responses to cope successfully with drying.

Aestivation

In regions where water bodies seasonally or periodically dry, many Australian turtles have evolved specializations to survive extended periods of drought by entering a state of inactivity called aestivation (Burbidge 1981, Kennett and Christian 1994, Roe and Georges, 2007, 2008, Roe et al., 2008). The Western Swamp Turtle, *P. umbrina*, inhabits temporary swamps that flood during winter rains, but by late spring or early summer the swamps have dried and may remain so for 6-9 months (Burbidge, 1981). Several species of turtles in the genus *Chelodina* (*C. canni, C. longicollis, C. rugosa, C. steindachneri*) also inhabit wetlands that dry for several months. In some

[5] Community structure reflects the number and types of species in an area and their relative abundances.

extreme circumstances, *C. longicollis* can survive aestivation for 13–16 months (Stott 1987; Roe and Georges, 2007), while others (*C. steindachneri*) may occasionally need to survive for longer time periods out of water (Cann, 1998), although in this latter case no empirical work has yet confirmed this.

Aestivation is a strategy to survive the challenges of drought by reducing the body's energy and water demands. This may involve a suite of morphological, physiological, and behavioural traits. When drought occurs, turtles may respond by leaving the wetland and seeking refuge in naturally occurring holes in the ground, or by burrowing several centimeters under soil, leaf litter, woody debris, or other structures (Burbidge 1981; Chessman, 1983; Rees 2008; Roe and Georges, 2007, 2008). Alternatively, some burrow in wetland sediment where they remain entombed until rains return (Cann, 1998; Kennett and Christian 1994). Morphological adaptations for terrestrial aestivation are perhaps best demonstrated in *P. umbrina*, which has an extensively roofed skull and an expanded shell that allows for the complete withdrawal of their head and limbs to reduce water loss. Other turtles employ physiological specializations to further conserve energy and water. For instance, *C. rugosa* reduces metabolism by 72% during dry season aestivation (Kennett and Christian, 1994), while *C. longicollis* is slow to lose water to evaporation and can further conserve water by altering the composition of excreta (Chessman, 1984; Rogers, 1966; Roe et al., 2008). While these behavioral, morphological and physiological specializations highlight adaptations that can greatly extend survival during drought, turtles are nevertheless vulnerable at this time. Predation rates can be high (Fordham et al., 2006b), and if waters do not re-flood before energy stores are depleted, turtles will eventually die of starvation or dehydration (Roe and Georges, 2008, Roe et al., 2008). Thus aestivation is only a temporary means of increasing the probability of surviving drought, and the duration that a waterhole remains dry is a critical factor in whether turtles can persist in the area.

Movement to other Waterholes

As an alternative to aestivation, turtles may move to more permanent waterholes when drought strikes. While it is not uncommon to see turtles of several species emigrating from wetlands in Australia (i.e., when they cross roads, golf courses, lawns, or paddocks), this is one of the least well understood aspects of turtle behaviour (Roe and Georges, 2007). We have studied overland movements in *C. longicollis* at Booderee National Park and in the suburbs of Canberra (Graham et al., 1996; Kennett et al. 1990; Rees, 2008; Roe and Georges, 2007). At both sites, drought cues mass emigrations from drying wetlands into nearby permanent waterbodies. These permanent waterbodies offer temporary refuge during drought, where turtles remain (sometimes for years) before staging a return to temporary wetlands following floods. Turtles regularly travel up to three kilometers overland to reach permanent water, but occasional movements of more than five kilometers overland have also been reported in *C. longicollis* (Parmenter, 1976; Roe, 2007). However, just as the temporal duration of dry conditions is a limiting factor for survival through aestivation, an ability to escape to freshwater drought refuges depends upon the availability of waterbodies in the immediate vicinity that are somehow more resilient to drought. Most freshwater turtles could presumably traverse short distances (i.e., a few hundred meters) overland, but when farther distances must be traversed or when the intervening terrain is especially harsh, few would be capable of making such travels. Even the most basic capacities for (and limitations to) overland travel remain largely unknown for the majority of Australia's freshwater turtles.

Life History Specializations to Drying

The onset and duration of habitat drying and the return of floods can vary predictably (i.e., seasonally) or unpredictably (i.e., stochastically). It is in those species, that have to contend with unpredictability in flood-drought cycles, that we see some of the most interesting life history and behavioural "solutions" for survival. *Chelodina rugosa* deals with unpredictability in the recession of wet season flooding by laying its eggs underwater where development is suspended until the ground dries (Kennett et al., 1993a, b). Development is then adjusted so that embryos can compensate for differing durations of inundation to hatch at the optimal time prior to the onset of the ensuing rainy season (Fordham et al., 2006a). *Carettochelys insculpta* overcomes unpredictability in the onset of the wet season for its hatchlings by late-term embryonic aestivation. Eggs develop rapidly in anticipation of the wet-season floods then enter a state of torpor within the eggshell (Webb et al., 1986). Oxygen deprivation brought about by torrential rain that floods the nest cavity brings the hatchlings out of their slumber and they hatch and make their frenzied way into the rising floodwaters (Doody et al., 2001; Georges, 1992). Hatchlings of other species, like *C. longicollis*, may emerge from eggs but remain within the nest chamber until heavy rain softens the soil and allows their escape. *Chelodina longicollis* adults have a range of adaptations that make them well-suited for both overland travel to refugial waterbodies and for extended bouts of terrestrial aestivation. This flexibility allows them to cope with drought and other unpredictable occurrences of aquatic habitat loss in a variety of ways depending on the context (Roe and Georges, 2008).

Permanent Water Specialists

Permanent water specialists[6] represent the vast majority of turtle genera in Australia, and it is these that are least likely to tolerate the drying of their aquatic habitats for any length of time. Permanent waters include lakes, ponds, rivers, and streams that typically remain inundated or maintain water flow without drying. Permanent water specialists include *Emydura*, *Elseya*, *Rheodytes*, *Elusor*, and *Carettochelys*, but *C. expansa* and *C. oblonga* also rely on habitats that do not typically dry out.

Turtles that are rarely exposed to drying conditions typically lack adaptations that would allow them to successfully contend with habitat drying by evacuating or aestivating (Chessman, 1984; Christiansen and Bickham, 1989; Gibbons et al., 1983). However, permanent water specialists may occasionally enter areas that periodically dry, such as when river flows force them out of channels and into shallow backwaters or downstream waterholes that can quickly become isolated when flows subside. In these scenarios, turtles can become unexpectedly stranded by drying floodwaters and many perish as a result (see Figure 1).

Nevertheless, some of Australia's permanent water specialists successfully inhabit arid regions, with some of the most widely fluctuating water flows on earth, and where most water can vanish for long periods. The Cooper Creek turtle, *Emydura macquarii emmottii*, has a most interesting life history, as it manages to persist in areas where other turtles cannot. This species occupies the permanent waterholes of the Cooper Creek floodplain in central Australia. Precipitation over much of the Cooper catchment is exceptionally low

---

[6] Habitat specialists tend to occupy a narrow range of environmental conditions, whereas habitat generalists can occupy a wide range of conditions.

(approximately 330 mm per year, Australian Bureau of Meteorology,[7] but monsoonal rainfall in its headwaters results in intermittent and widespread flooding of this dryland river.

Photo: Queensland EPA Roma. Photo reproduced with permission.

Figure 1. When freshwater refuges with a long history of water permanence dry, permanent water specialists typically perish, such as in Lake Numalla, Queensland, where some 10,000 Murray turtles were estimated to have died.

An exceptionally flat landscape results in an extensive network of distributary channels, peppered with isolated and deeper waterholes. It is in these waterholes that this species resides between episodes of floods, often in exceptionally high population densities (Georges et al., 2006b). Should waterholes dry (which happens during the worst of our droughts) the turtles perish, but when the system floods once again others are able to reinvade from refugial waterholes that did not dry. This dynamic allows this river turtle to persist in Australia's arid centre. The scale of this landscape is immense – waterholes are separated by hundreds of kilometers of arid and semi-arid land, and the temporal scale of flooding is measured in years instead of months, attributes presumably unsuited to the drought survival strategies of any other Australian turtle.

---

7 http://www.adl.brs.gov.au/water2010/pdf/monthly_reports/awap_3_report.pdf.

DROUGHT IMPACTS AND THE HUMAN DIMENSION

The human dimensions of drought relevant to freshwater turtles fall into two broad categories. The first is creation of conditions that either cause drought or produce drought-like conditions, and the second includes activities that limit the natural ability of turtles to respond to drought. To assess the impact of drought on turtles, we must understand how these anthropogenic dimensions interact to directly and indirectly influence their populations.

The most visible consequence of drought in freshwater systems occurs when a water body dries completely, and consequences are most severe when water bodies that typically serve as aquatic refugia also dry out. Surface waters can recede and eventually dry from decreased runoff from rainfall, inflow from rivers, or supply by groundwater. Though rainfall ultimately drives all three processes, our demands on water for irrigation, industry and municipal activities hasten the pace of drying and extends the temporal and spatial impact of drought into areas that were once more extensively flooded or served as permanent water refuges (Kingsford and Thomas, 1995; Kingsford, 2000). Reports of mass turtle mortality in drying wetlands of Australia are increasing (see Figures 1 and 2), though such reports have not yet made their way into the scientific literature, as they have in other countries (Christiansen and Bickham, 1989, Bodie and Semlitsch, 2000). Many turtles die of starvation or dehydration during or following wetland drying, but drought can indirectly cause mortality as well. Turtles emigrating from a drying waterbody are vulnerable to predators including the introduced European Fox (Spencer, 2002), or they may encounter vehicular traffic in developed areas resulting in mass mortality on roads in Australia (Chessman, 1988; Rees, 2008), see Figure 3) and elsewhere (Aresco, 2005; Ashley and Robinson, 1996). Exclusion fences can also entrap turtles on land where they eventually die of exposure to predators, high temperatures, or dehydration (Anonymous 1941; JHR and AG pers. obs.).

Feral pigs can even target and depredate turtles aestivating in the mud of dry wetlands, threatening the persistence of turtle populations in northern Australia (Fordham et al., 2006b, 2007). All of the above are examples of how human alterations to natural systems interact with drought to exacerbate impacts upon turtles.

Photo: John Roe.

Figure 2. Receding waters can concentrate turtles, resulting in high densities. Disease and competition for food increase under such conditions, killing many individuals even before the water body dries completely such as in Lake McKenzie, Jervis Bay Territory.

Photo: Damien Michael. Photo reproduced with permission.

Figure 3. Turtles forced to emigrate from drying waterbodies frequently cross roads in developed areas, often leading to mass mortality from vehicular collisions. This *Chelodina expansa* was killed on the Riverina Highway adjacent to the Wonga Wetlands and the Murray River.

Drought can also result in important changes to habitat heterogeneity with subsequent impacts on turtles. For instance, large river systems naturally include not only the deeper and permanent channel, but during floods the aquatic habitat expands to encompass adjacent floodplain wetlands (e.g., oxbows, billabongs, marshes, and sloughs). Such flooding events resupply these highly productive and important wetlands with water and nutrients and provide a conduit for turtles to travel into them (Doody et al., 2002; Georges et al., 2006). Low rainfall and high use of upstream water during drought reduces both the extent and duration of flooding in these adjacent wetlands such that, in Australia's most developed rivers, these habitats may no longer flood at all (Kingsford, 2000; Roshier et al., 2001).

Turtle communities in such systems are often spatially organized according to species habitat preferences (Chessman, 1988), and in Australia several species of turtles including *C. insculpta* (Doody et al., 2002). *C. longicollis* (Chessman, 1988), and *C. expansa* (D. Bower, University of Canberra, unpubl. data) make extensive use of these flooded backwaters. Not only would turtle community diversity decline should river floodplains fail to flood, but our work in Booderee National Park in the Jervis Bay Territory indicates that such drought-induced changes in hydrology can also occur in wetlands isolated from river flows, with subsequent effects on turtle growth, reproduction and regional abundance (Kennett and Georges, 1990, Roe and Georges, 2008).

The retention of water in reservoirs is a common way to meet societal water demands in times of drought, but ironically, this too can have negative effects on some turtles. For example, the Fitzroy turtle (*Rheodytes leukops*) inhabits shallow riffle zones (Tucker et al., 2001), a habitat affinity linked to its gill-like cloacal respiration and diet of aquatic invertebrate prey (Gordos, 1999; Priest, 1997). Dams that transform free flowing rivers to lake-like conditions or that otherwise alter natural flow regimes and increase sedimentation rates have degraded much riffle habitat in the Fitzroy River catchment, which has certainly contributed to the listing of *R. leukops* as a Vulnerable species (Georges, 1993). A change in turtle community diversity (typically a reduction in the number of species) often follows the construction of dams, with river (lentic) turtles being replaced by pond (lotic) turtles or species with more general habitat preferences (DonnerWright et al. 1999; Tucker et al., 2001). Our changes to natural hydrology patterns in aquatic systems are not limited to rivers, but also include modifications to still-water wetlands for increased water retention during drought (i.e., farm dams, Brock et al., 1999). In this case, humans' influence on some species may be positive, as farm dams, irrigation channels and reservoirs for water retention in agricultural and pastoral regions now offer drought refuges. The proliferation of these artificial aquatic habitats has perhaps augmented populations of *C. longicollis* in dry regions (Rees, 2008) and may even have facilitated its expansion into areas that were previously too dry (Beck, 1991). However, the persistence of these "unnatural" wetlands and the benefits they offer aquatic turtles into the future is uncertain under changing climate scenarios.

Photo: Sally Grundy. Photo reproduced with permission.

Figure 4. Increased salinization of freshwater habitats has led to the encrustation of turtles' shells by estuarine tube worms. In severe cases encrustation can impede turtle movements and limit the ability to retract their head and limbs, such as in this *Emydura* in Lake Alexandrina, South Australia.

The rising salinity of freshwaters is another major issue with which turtles must contend in Australia. Rising soil salinities brought about by land clearing, intensive irrigation and reduced rates of evapotranspiration[8] have allowed water tables to rise and contaminate freshwater systems (Lemly et al., 2000), but the impact on Australian turtles in these systems has not yet been examined. We know that in some cases increasing salinities have brought

8 Evapotranspiration is the movement of water from the ground to the atmosphere via evaporation and plant transpiration.

saline-tolerant taxa[9] into unwelcome contact with turtles, as in the case of turtle fouling by encrusting estuarine worms in Lake Alexandria, South Australia (see Figure 4). Many nominally freshwater turtles may successfully utilize saline environments (Dunson et al. 1986, Georges and Rose, 1993, Moll and Moll, 2004); however, the physiological, behavioral, and population level responses of Australian turtles to rapidly increasing salinities are in immediate need of assessment.

THE FUTURE OF AUSTRALIA'S TURTLES IN A CHANGING CLIMATE

Predictions for changes to Australia's climate include a reduction in precipitation across much of the continent, particularly in the southern temperate regions. This will result in expansion of arid and semi-arid zones that are less hospitable to turtles, and an increase in the frequency, duration and intensity of drought. The recent drought in south-eastern Australia, whether or not it is part of a natural cycle, serves as a harbinger of impending climate change and focuses our attention on what might be in store for the future. Turtles in the Murrumbidgee River and elsewhere have perished in waterholes not known to dry since European colonization. The Murray-Darling River now seldom flows into the sea, and the combination reduced water flows with marine incursion and increased salinization is presenting major challenges to the freshwater turtle communities of Australia's largest river system. In the south-west, *P. umbrina*, one of the rarest turtles in the world is already on the brink of extinction, being restricted to a few small ephemeral swamps. Any further reduction in precipitation will require continual human intervention to maintain this species in the wild (Kuchling and Dejose, 1988, Kuchling et al., 1991).

If there is any reason to be optimistic, it is that this unfolding crisis has been experienced to some degree by the Australian biota before, and those species that are with us today are there because they have survived the changes. The challenge will be to determine what humans have done to reduce the capacity of the biota to survive climate change as it has done in the past. In

---

9 Saline-tolerant taxa are those animals that can successfully inhabit waters with higher concentrations of dissolved salts than in normal freshwater. Freshwater typically has dissolved salt concentrations below 0.5 parts per thousand.

the case of *P. umbrina*, the answer is clear. We have dramatically reduced the extent of available habitat through reclamation of swampland for urbanization and agriculture, rendering this species extremely vulnerable to climate change. In the Murray-Darling drainage and other watersheds that are currently experiencing greatly reduced flow, we need to know where the refugial areas were during past periods of drying so that we can protect their integrity during this drying event.

These refugia are areas in which the turtles and other aquatic biota persisted during extended periods of low precipitation, only to expand out to their current distributions when precipitation increased. On a shorter timescale, responses to drought may involve a dynamic of localized extinction and repopulation. What have we done - in terms of fragmenting our freshwater systems through habitat degradation, construction of weirs, dams and roads, and alteration of flow intensity, duration and timing - to disrupt this dynamic? These considerations apply at many geographic scales, from the Cooper Creek watershed in the arid centre to individual wetlands and wetland networks, such as at Booderee National Park in the temperate south-east.

CONCLUSION

Clearly, failure to address several issues, including the maintenance of adequate water during periods of reducing precipitation, the isolation of off-channel waterbodies in the lower Murray to address human demands for water quantity and quality, the construction of more dams and storage facilities without exploring alternatives, and our increasing take from aquifers with little regulation, are all responsible for environmental changes that are likely to have major (and potentially adverse) impacts on freshwater communities, of which turtles are an integral part.

Regardless of the ultimate cause of recent droughts in Australia and other parts of the world, the challenge before us is to improve water resource management in a way that allows for more efficient use of water to meet societal demands, while at the same time ensuring that the natural capacity for biota to respond to drought is maintained.

REFERENCES

Anonymous. (1941). Death on the plains-where instinct fails. *Wild life* (Melbourne), 3, 13-14.

Aresco, M.J. (2005). Mitigation measures to reduce highway mortality of turtles and other herpetofauna at a north Florida lake. *Journal of Wildlife Management,* 69, 549-560.

Ashley, E.P., and J.T. Robinson. (1996). Road mortality of amphibians, reptiles, and other wildlife on the long point causeway, Lake Erie, Ontario. *Canadian Field-Naturalist,* 110, 403-412.

Bates, B.C., Z.W. Kundzewicz, S. Wu, and J.P. Palutikof (Eds.). (2008). Climate Change and Water. Technical Paper of the Intergovernmental Panel on Climate Change. IPCC Secretariat, Geneva.

Beck, R.G. (1991). The Common Longnecked Tortoise *Chelodina longicollis* (Shaw 1802) (Testudinata: Chelidae): A comparative study of the morphology and field behaviour of disjunct populations. *South Australian Naturalist,* 66, 4-22.

Bodie, J.R. and Semlitsch, R.D. (2000). Size-specific mortality and natural selection in freshwater turtles. *Copeia 2000,* 732-739.

Brock, M. A., Smith, R.G.B. and Jarman, P.J. (1999). Dam it, drain it: alteration of water regime in shallow wetlands on the New England Tableland of New South Wales, Australia. *Wetlands Ecology and Management*, 7, 37-46.

Burbidge, A.A. (1981). The ecology of the western swamp tortoise *Pseudemydura umbrina* (Testudines: Chelidae). *Australian Wildlife Research,* 8, 203-223.

Cann, J. (1998). *Australian Freshwater Turtles.* Beaumont Publishing, Singapore.

Chessman, B.C. (1983). A note on aestivation in the snake-necked turtle, *Chelodina longicollis* (Shaw) (Testudines: Chelidae). *Herpetofauna,* 14, 96.

Chessman, B.C. (1984). Evaporative water loss from three south-eastern Australian species of freshwater turtle. *Australian Journal of Zoology*, 32, 649-655.

Chessman, B.C. (1988). Habitat preferences of freshwater turtles in the Murray valley, Victoria New South Wales. *Australian Wildlife Research,* 15, 485-491.

Christiansen, J.L., and Bickham, J.W. (1989). Possible historic effect of pond drying and winterkill on the behavior of *Kinosternon flavescens* and *Chrysemys picta*. *Journal of Herpetology,* 23, 91-94.

de Broin, F. (1987). The late Cretaceous fauna of Los Alamitos, Patagonia, Argentina. Part IV. Chelonia. Revista del Museo Argentino de Ciencias Naturales, Bernardino Rivadavia. *Paleontologia*, 3, 131–139.

DonnerWright, D.M., Bozek, M.A., Probst, J.R. and Anderson. E.M. (1999). Responses of turtle assemblages to environmental gradients in the St. Croix River in Minnesota and Wisconsin, U.S.A. *Canadian Journal of Zoology,* 77, 989-1000.

Doody, J.S., Georges, A., Young, J.E., Pauza, M.D., Pepper, A.L., Alderman, R.L and Welsh. M.A. (2001). Embryonic aestivation and emergence behaviour in the pig-nosed turtle, *Carettochelys insculpta. Canadian Journal of Zoology,* 79, 1062-1072.

Doody, J. S., Young, J. E. and Georges. A. (2002). Sex differences in activity and movements in the pig-nosed turtle, *Carettochelys insculpta*, in the wet-dry tropics of Australia. *Copeia 2002,* 93-103.

Dunson, W.A. (1986). Estuarine populations of the snapping turtle (*Chelydra*) as a model for the evolution of marine adaptations in reptiles. *Copeia 1986,* 741-756.

Dracup, J. A., Lee, K. S. and Paulson, E.G. (1980). On the definition of droughts. *Water Resources Research,* 16, 297-302.

Fordham, D. A., Georges, A. and Corey. B. (2006a). Compensation for inundation-induced embryonic diapause in a freshwater turtle: achieving predictability in the face of environmental stochasticity. *Functional Ecology*, 20, 670-677.

Fordham, D. A., Georges, A. and Corey. B. and Brook, B.W. (2006b). Feral pig predation threatens the indigenous harvest and local persistence of snake-necked turtles in northern Australia. *Biological Conservation,* 133, 379-388.

Fordham, D., Georges, A. and Brook. B.W. (2007). Indigenous harvest, exotic pig predation and local persistence of a long-lived vertebrate: managing a tropical freshwater turtle for sustainability and conservation. *Journal of Applied Ecology,* 45, 52-62.

Gaffney, E.S. (1991). The fossil turtles of Australia. In P.Vickers-Rich, J.M. Monaghan, R.F., Baird, and T.H. Rich (Eds.). *Vertebrate Paleontology of Australasia* (pp. 704-720). Pioneer Design Studio and Monash University Publications Committee, Melbourne.

Gaffney, E.S., Archer, M. and White, A. (1989). Chelid turtles from the Miocene freshwater limestones of Riversleigh Station, Northwestern Queensland, Australia. *American Museum Novelties,* 2959, 1-10.

Georges, A. (1992). Thermal characteristics and sex determination in field nests of the pig-nosed turtle *Carettochelys insculpta* (Chelonia: Carettochelydidae) from northern Australia. *Australian Journal of Zoology,* 40, 511-521.

Georges, A. (1993). Settling conservation priorities for Australian freshwater turtles. In D. Lunney and D. Ayers (Eds.), *Herpetology in Australia, a Diverse Discipline* (pp. 453-476). Royal Zoological Society of New South Wales.

Georges, A., Alacs, E., Pauza, M., Kinginapi, F., Ona, A., and Eisemberg, C. (2008). Freshwater turtles of the Kikori Drainage, Papua New Guinea, with special reference to the pig-nosed turtle, *Carettochelys insculpta. Wildlife Research,* 35,700-711.

Georges, A., Guarino, F., and Bito, B. (2006a). Freshwater turtles of the TransFly region of Papua New Guinea - notes on diversity, distribution, reproduction, harvest and trade. *Wildlife Research,* 33, 373-384.

Georges, A., Guarino, F., and White, M. (2006b). Sex-ratio bias across populations of a freshwater turtle (Testudines: Chelidae) with genotypic sex determination. *Wildlife Research,* 33, 475-480.

Georges, A. and Rose, M. (1993). Conservation biology of the Pig-nosed Turtle, Carretochelys insculpta. *Chelonian Conservation and Biology,* 1, 3-12.

Georges, A. and Thomson, S. (2006). Evolution and zoogeography of Australian freshwater turtles. In *Evolution and Biogeography of Australian Vertebrates* (pp. 281-300). J.R. Merrick, M. Archer, G.M. Hickey, and M.S.Y. Lee (Eds.). Oatlands, NSW: Australian Scientific Publishing.

Gibbons, J.W., J.L. Greene, and Congdon, J.D. (1983). Drought-related responses of aquatic turtle populations. *Journal of Herpetology,* 17, 242-246.

Gordos, M.A. (1999). Physiological ecology of the Fitzroy Turtle, *Rheodytes leukops*. Honours Thesis, University of Queensland.

Graham, T.E., Georges, A., and McElhinney, N. (1996). Terrestrial orientation by the eastern long-necked turtle, *Chelodina longicollis*, from Australia. *Journal of Herpetology*, 30, 467-477.

Intergovernmental Panel on Climate Change. (2007). Synthesis Report. R.K. Pachauri, and A. Reisenger (Eds.). IPCC, Geneva, Switzerland. Available from, http://www.ipcc.ch/ ipccreports/ar4-syr.htm (accessed March 2009).

Kennett, R. and Christian, K. (1994). Metabolic depression in estivating long-neck turtles. *Physiological Zoology,* 67,1087-1102.

Kennett, R.M. and A. Georges. (1990). Habitat utilization and its relationship to growth and reproduction of the eastern long-necked turtle, *Chelodina longicollis* (Testudinata: Chelidae), from Australia. *Herpetologica* 46, 22-33.

Kennett, R., Christian, K., and Pritchard, D. (1993a). Underwater nesting by the freshwater turtle *Chelodina rugosa* from tropical northern Australia. *Australian Journal of Zoology,* 41, 47-52.

Kennett, R., Georges, A., and Palmer-Allen, M. (1993b). Early developmental arrest during immersion of eggs of a tropical freshwater turtle, *Chelodina rugosa* (Testudinata: Chelidae) from northern Australia. *Australian Journal of Zoology*, 41, 37-45.

Kingsford, R. T. (2000). Ecological impacts of dams, water diversions and river management on floodplain wetlands in Australia. *Austral Ecology*, 25, 109-127.

Kingsford, R.T., and Thomas, R.F. (1995). The Macquarie Marshes in arid Australia and their waterbirds: A 50 year history of decline. *Environmental Management,* 19, 867-878.

Kuchling, G. and Dejose, J.P. (1988). A captive breeding operation to rescue the critically endangered Western Swamp Tortoise *Pseudemydura umbrina* from extinction. *International Zoo Yearbook*, 31, 103-109.

Kuchling, G, Dejose, J.P. Burbidge, A.A, and Bradshaw, S.D. (1991). Beyond captive breeding: the Western Swamp Tortoise *Pseudemydura umbrina* recovery programme. *International Zoo Yearbook*, 31, 37-41.

Lemly, A. D., Kingsford, R.T., and Thompson, J.R. (2000). Irrigated agriculture and wildlife conservation: conflict on a global scale. *Environmental Management*, 25, 485-512.

Ligon, D.B. and Peterson, C.C. (2002). Physiological and behavioral variation in estivation in mud turtles (*Kinosternon* spp.). *Physiological and Biochemical Zoology*, 75, 283-293.

Moll, D. and Moll, E.O. (2004). *The Ecology, Exploitation, and Conservation of River Turtles.* Oxford University Press, New York.

Parmenter, C.J. (1976). The natural history of the Australian freshwater turtle *Chelodina longicollis* Shaw (Testudinata, Chelidae). PhD Thesis, University of New England.

Priest, T. (1997). Bimodal respiration and dive behaviour of the Fitzroy River Turtle, *Rheodytes leukops*. Honours Thesis, University of Queensland.

Quilty, P.G. (1994.) The background: 144 million years of paleoclimate and paleogeography. In *History of the Australian Vegetation, Cretaceous to Present* (pp. 14-44). R. Hill (Ed.). Cambridge: Cambridge University Press.

Rees, M. (2008). Impacts of urbanization on the eastern long-necked turtle *Chelodina longicollis* (Testudines:Chelidae). Honours thesis, University of Canberra.

Roe, J.H. (2007). The terrestrial ecology of a freshwater turtle, *Chelodina longicollis*, in Booderee National Park, Australia. PhD Thesis, University of Canberra.

Roe, J.H., and Georges, A. (2007). Heterogeneous wetland complexes, buffer zones, and travel corridors: landscape management for freshwater reptiles. *Biological Conservation,* 135, 67-76.

Roe, J.H., and Georges, A. (2008). Maintenance of variable responses for coping with wetland drying in freshwater turtles. *Ecology,* 89, 485-494.

Roe, J.H., Georges, A., and Green, B. (2008). Energy and water flux during terrestrial aestivation and overland movement in a freshwater turtle. *Physiological and Biochemical Zoology*, 81, 570-583.

Rogers, L.J. (1966). The nitrogen excretion of *Chelodina longicollis* under conditions of hydration and dehydration. *Comparative Biochemistry and Physiology,* 18, 249-260.

Roshier, D. A., Whetton, P.H., Allan, R.J., and Robertson, A.I. (2001). Distribution and persistence of temporary wetland habitats in arid Australia in relation to climate. *Austral Ecology,* 26, 371-384.

Spencer, R.J. (2002). Experimentally testing nest site selection: fitness tradeoffs and predation risk in turtles. *Ecology,* 83, 2136-2144.

Stott, P. (1987). Terrestrial movements of the freshwater tortoise *Chelodina longicollis* Shaw as monitored with a spool tracking device. *Australian Wildlife Research,* 14, 59-567.

Thomson, S., Kennett, R., and Georges, A. (2000). A new species of long-necked turtle (Testudines: Chelidae) from the Arnhem Land plateau, Northern Territory, Australia. *Chelonian Conservation and Biology,* 3, 675-685.

Tucker, A. D., Limpus, C. J., Priest, T.E., Cay, J., Glen, C., and. Guarino, E. (2001). Home ranges of Fitzroy River turtles (*Rheodytes leukops*) overlap riffle zones: potential concerns related to river regulation. *Biological Conservation,* 102, 171-181.

Webb, G., D., Choquenot, D., and Whitehead, P. (1986). Nest, eggs and embryonic development of *Carettochelys insculpta* (Chelonia: Carettochelidae) from Northern Australia. *Journal of Zoology, London (B) 1*, 521-550.

Wilhite, D. A. and Glantz, M. H. (1985). Understanding the drought phenomenon: the role of definitions. *Water International,* 10, 111-120.

Williams, E.E. (1953a). Fossils and the distribution of Chelyid turtles (2). Breviora *(Museum of Comparative Zoology),* 13, 1-8.

Williams, E.E. (1953b). Fossils and the distribution of Chelyid turtles (2). Breviora *(Museum of Comparative Zoology)*, 32, 1-6.

Withers, P.C. (1993). Metabolic depression during estivation in the Australian frogs Neobatrachus and Cyclorana. *Australian Journal of Zoology,* 41, 467-473.

INDEX

A

B

C

D

E

F

G

H

I

J

K

L

M

N

O

P

T

U

V

W

Y

Z